Pollination Biology, Vol. 1

Dharam P. Abrol

Pollination Biology, Vol. 1

Pests and Pollinators of Fruit Crops

Springer

Dharam P. Abrol
Division of Entomology
Faculty of Agriculture
Sher-e-Kashmir University of Agricultural
Sciences and Technology
Jammu, India

ISBN 978-3-319-21084-1 ISBN 978-3-319-21085-8 (eBook)
DOI 10.1007/978-3-319-21085-8

Library of Congress Control Number: 2015944076

Springer Cham Heidelberg New York Dordrecht London

Printed on acid-free paper

Springer International Publishing AG Switzerland is part of Springer Science+Business Media (www.springer.com)

To my family
Asha
Vitasta
Rajat

Foreword

M.S. Swaminathan

Our population is growing, but the per capita availability of land and water resources are going down. It is obvious that we will have to produce more and more from less and less land and water. This can be achieved only through the evergreen revolution pathway which can help us to increase productivity in perpetuity without associated ecological harm.

The present book by Dr. D.P. Abrol is a timely one since it deals with the fundamentals of reproductive biology, namely, pollination. There is widespread concern about malnutrition in our country, and it is now clear that for every nutritional malady, there is a horticultural remedy. We should ensure that food crops not only yield more but also are free of pesticide residues. For example, DDT has a long residual toxicity, and this was captured in the famous book *Silent Spring* by Rachel Carson (1962). Fortunately, this book covers pest management strategies which will protect the safety of pollinators. The book gives an overview of the pest problems and the pollinator needs. Many pests and pollinators are important in sub-tropical and tropical food crops. In order to enhance productivity in an environmentally sustainable manner, this book provides integrated information on different management strategies. This is the basic requirement for successful horticulture.

We owe a deep sense of gratitude to Dr. D.P. Abrol for his labour of love for the continued growth in the productivity and quality of temperate, tropical and subtropical fruits. I hope the book will be widely read and utilized by policymakers, scientists, research scholars and farmers.

Third Cross Street
Taramani Institutional Area
Chennai, 600 113, India

M.S. Swaminathan Research Foundation
E-mail: founder@mssrf.res.in

Foreword

Pardeep K. Sharma

The major challenge before agricultural scientists in the twenty-first century is to sustain higher productivity in agriculture to meet the target of 300 million tonnes from perhaps 100 million hectares of land without compromising the future in terms of resource degradation and depletion. The world is facing food deficit coupled with instability of climatic cycles. The growing population pressure has hastened the environmental degradation, ultimately posing a threat to natural resources and fast approaching famine. In the next 50 years, the global population is expected to reach nine billion, doubling the food, feed, and crop demand. Concomitantly, this situation has further aggravated because of pollinator decline worldwide resulting in pollination crisis. These problems have been further aided and abetted by a lack of prophylactic progress in the conservation of biodiversity and increased agricultural production.

Insect pollinators play an important role in producing crops in global agriculture. Pollinator-dependent crops contribute to maintaining a healthy variety in the human diet and often have a high market value, beneficial for local or regional economies. Pollinators play a crucial role in flowering plant reproduction and in the production of most fruits and vegetables. Most of the fruit crops are highly cross pollinated and depend upon insects or benefit from insect pollination for fruit set. Insect pests on the other hand cause major economic damage on fruit crops in tropics, subtropics, and temperate. Evidently, pest management in fruit crops on one hand and safety to the pollinators on the other are a challenging task in the context of increasing horticultural productivity without upsetting the ecological balance. The book *Pollination Biology: Pests and Pollinators of Fruit Crops* Vol. 1 by Dr D P Abrol is therefore a timely contribution. The publication of the famous book *Silent Spring* by Rachel Carson in 1962 was a turning point in awakening the world about the deleterious effects of agrochemicals. On reading this book, Albert Schweitzer remarked "Man has lost the capacity to foresee and to forestall. He will end by destroying the earth," What is now needed is anticipatory research in relation to the impact of pests on crops in an era of climate change, as well as participatory research with farm families on methods of integrating the ecological prudence of farmers with the technological advances of today. This book aims to integrate and develop pest control strategies in a way to minimize their impact on beneficial insect species such as natural enemies and pollinators to enhance fruit production and quality.

The book covers interplay between pest management strategies and safety of pollinators. It gives an overview of pest problems and pollinator needs of the crops followed by information on pollination scenario and mechanism of pollination in fruit crops. This book integrates two diverse yet mutually linked organisms for the benefit of quality and quantity production of horticultural crops. Detailed information is provided on pests and pollinators of temperate, subtropical, and tropical fruit crops integrating production inhibiting and production promoting components for sustainable and enhanced productivity. The main objective of this book is to provide the synthesis and critical analysis of different management strategies having bearing on agriculture, sustainability and environmental protection, and pollination needs of different crops which result in boosting quality and productivity to encourage beekeeping interventions, helping people to strengthen livelihood and ensure maintenance of habitat and biodiversity. This is, indeed, an awesome task, and we must congratulate Professor Abrol for bringing out this book for the benefit of the global community.

Sher-e-Kashmir University of Agricultural Sciences and Technology
Jammu 180009, India

Pardeep K. Sharma

Preface

Fruits play a significant role in economic development, nutritional security, employment generation and overall economic growth of a country. They provide a dynamic tool for enhancing economic returns, creating employment avenues and ensuring ecological sustainability. Production constraints include two diverse but mutually interlinked approaches such as attack of pests on the one hand and inadequate pollination on the other. The widespread use of pesticides in modern agriculture throughout the world has become necessary for the protection of the plants against insect pests and diseases to obtain higher yields to meet the food requirement of an increasing population, but the injudicious use of pesticides has resulted in contamination of agro-ecosystem and agriculture produce including nectar and pollen and caused heavy losses to the pollinators. Such contaminated nectar and pollen when brought to hive may cause damage to brood besides the contamination of the stored honey.

The world human population disproportionately growing at an alarming rate is a major challenge facing agriculture in the twenty-first century. The human population has already crossed the seven billion mark and is expected to rise to nine billion or more during the middle of this century requiring raising of food productivity by some 70–100 % through horizontal expansion. Production in the developing countries would have to be doubled as compared to the present. The cropping systems and pollinator demographics are changing throughout the world, and any deviation in pollinator sufficiency will prove increasingly inadequate for meeting the demands for an abundant, high-quality food supply in the twenty-first century. Evidence for decline of pollinators throughout the world is a matter of serious concern for agricultural productivity. It is manifestly clear that bee pollinators are a valuable and limited natural resource that should be conserved and encouraged at all costs. This awareness owes its origin to an apparent decline of domesticated honeybees throughout the globe due to several causes such as attack of parasitic mites such as *Varroa destructor*, which has a devastating effect. The parasite occurs now in every continent on which *A. mellifera* is kept, except Australia, and it is considered the most serious health threat to apiculture. Indiscriminate use of pesticides for controlling crop pests has resulted in large-scale killing of pollinating bees and other beneficial insects including natural enemies. Evidently, an updated book on applied bee management and conservation for crop pollination was considered necessary.

Most of the fruit crops are highly cross-pollinated and depend upon insects or benefit from insect pollination for fruit set. Insect pests, on the other hand, cause major economic damage on fruit crops in tropics, subtropics and temperate. Evidently, pest management in fruit crops on the one hand and providing safety to the pollinators on the other is a challenging task in the context of increasing horticultural productivity without upsetting the ecological balance. The book *Pollination Biology* in two volumes (*Pests and Pollinators of Fruit Crops* Vol. 1 and *Pests and Pollinators of Vegetable and Oilseed Crops* Vol. 2) aims to integrate and develop pest control strategies in a way to minimize their impact on beneficial insect species such as natural enemies and pollinators to enhance fruit production and quality.

The first volume of the book *Pollination Biology: Pests and Pollinators of Fruit Crops* covers interplay between pest management strategies and safety of pollinators in fruit crops. The introductory chapter gives an overview of pest problems and pollinator needs of the crops followed by information on pollination scenario and mechanism of pollination in fruit crops. Detailed information is provided on pests and pollinators of temperate, subtropical and tropical fruit crops.

The aim of this book is to fill the gap by providing the synthesis and critical analysis of different management strategies having a bearing on agriculture, sustainability and environmental protection. The compilation of this book is unique in the sense that it does not deal with the conventional way of discussing pest management for different crops but also takes into consideration the role of pollinators and their profitable utilization in the larger context of ecologically based pest management for management of pests on the one hand and safety of pollinators on the other. Though number of books on pests as well as pollinators on fruits are available, they have a limited approach covering one aspect or the other. The proposed book provides complete information not only on tropical but subtropical and temperate fruits as well. This book fulfils the gap in literature, putting all the information about fruits in one book.

The goal of this book is to synthesize the latest scientific literature into principles and practices that are relevant to workers in crop pollination in tropical and temperate areas. The book aims to promote a large, diverse, sustainable and dependable bee pollinator workforce that can meet the challenge for optimizing food production well into the twenty-first century.

A vast spectrum of people has helped in one way or the other in the writing of this book, which would have remained a distant dream without their active help and support. This book is the outcome of my personal experiences and the contributions of several workers which have been incorporated. I express my humble and profound thanks to all of them whose hard work has enabled me to compile the suitable information in such a manner that it would be useful to those interested in basic and applied pollination and pest management. The illustrations and figures are either original or redrawn from other sources, which have been cited individually in the figure legends. All the authors whose work has been used/referenced deserve special appreciation and heartiest acknowledgements. I am particularly thankful to Professor Dr. Raghavendra Gadagkar, Centre for Ecological Sciences, Indian Institute of Science, Bangalore, who has always been a source of inspiration, needed help, guidance and encouragement. I thank my university authorities for the excellent working atmosphere and needed encouragement for compiling such a voluminous book. I also thank Dr. Uma Shankar and Dr. Debjyoti Chatterjee for their help and support. I am also extremely thankful to **Zuzana Bernhart,** Senior Publishing Editor, and **Mariska van der Stigchel,** Publishing Editor/Assistant of Springer, who took great pains and keen interest in the publication of this book in a very impressive way. I am also highly thankful to Prof. Dr. M. S. Swaminathan, the father of green revolution in India, for writing the foreword of this book.

Last but not the least, my sincere thanks are due to my wife Professor Dr. Asha Abrol, daughter Er. Vitasta and Dr. Pardeep KL.Sharma, Hon'ble Vice-Chancellor of Sher-e-Kashmir University of Agricultural Sciences & Technology of Jammu and son Er. Rajat for their endurance and help while writing this book.

Jammu, India D.P. Abrol

Contents

About the Book

The book covers pest management strategies and safety of pollinators in fruit crops. Detailed information is provided on pests and pollinators of temperate, subtropical and tropical fruit crops. The information provided in this book will be useful to pollination biologists; honeybee biologists in entomology departments; students; teachers; scientists of agriculture, animal behaviour, botany, conservation, biology, ecology, entomology, environmental biology, forestry, genetics, plant breeding, horticulture, toxicology and zoology; seed growers and seed agencies. The book provides critical synthesis and analysis of different management strategies having a bearing on agriculture, sustainability and environmental protection taking into consideration the role of pollinators and their profitable utilization in the larger context of ecologically based pest management for management of pests on the one hand and safety of pollinators on the other for optimizing food production well into the twenty-first century.

About the Author

Dr. D.P. Abrol

Professor and Head, Division of Entomology, Sher-e-Kashmir University of Agricultural Sciences and Technology of Jammu is a known apiculture scientist and entomologist of national and international repute. Dr. Abrol has authored 14 books and 8 manuals and published over 200 original research papers, 10 reviews, 20 book chapters and over 64 popular articles on honeybee diseases, pollination biology, toxicology and integrated pest management. He has chaired several national and international symposia/conferences and delivered lead/guest lectures. He is a member of several scientific societies in India and abroad. He is referee of the various national and international journals and expert in various selection/screening and evaluation committees of scientific bodies/institutions of the country. He has completed several externally funded research projects and has collaborative research projects with international organizations in Poland and Switzerland. He has visited South Korea, Malaysia and several other countries as special invitee to these countries. Besides receiving letters of appreciation from different organizations, he is recipient of Young Scientist Award 1992 conferred by the Jammu and Kashmir State Council for Science and Technology – a prestigious State Award – for his outstanding contributions in the field of Agricultural Sciences. Besides, he is also recipient of Pran Vohra Award 1993 – a prestigious Young Scientist Award conferred by Indian Science Congress Association, Calcutta, for his outstanding and innovative research in the field of agricultural sciences. He was also conferred the Prof. T. N. Ananthakrishnan Award 1997–1998 – a prestigious National Award for his outstanding contributions in the field of Entomology by T N Ananthakrisnan Foundation, G .S Gill Research Institute, Chennai. He won the Dr. Rajinder Prasad Puruskar 1999–2000 – a prestigious National Award from Indian Council of Agricultural Research, New Delhi – for his Hindi book on beekeeping entitled *Madhmakhi Palan-Sidhant Evam Vidhian* and the 11th Apicultural Association Award 2010 for outstanding contributions in Apiculture.

Chapter 1
Introduction

The fruiting plants and human societies came in contact much earlier long before the discovery of agriculture around 10,000 years ago (8,000–10,000 BC). The nomadic man had fruits collected from wild as the main source of food which provided minerals, vitamins, fibres and compounds of medicinal value. Cereals were probably the first plants cultivated by humans practicising hunterer gather style. The domestication of fruiting plants may have resulted from seeds thrown in waste. With the advent of time selection of fruiting plants and their exchanges took place in different parts of the world (Samson 1986).

The world population is increasing at a very fast speed and has already crossed seven billion mark and is expected to exceed nine billion by 2050 which will necessitate production of more food without increasing acreage from less land and water and labour. Rice and wheat are the two cereals which provide staple food for the world but diversity of fruits, vegetables and beans is essential to quality food rich in minerals, vitamins and compounds of medicinal importance. Both cereals and fruits need equal emphasis for food and nutritional security. About half of world population suffering from malnution lives in Asia.

Efforts made in sixties led to green revolution in production of food grains which mainly resulted from high yielding varieties but horticultural crops received little attention. Sustainable horticultural crop production needs to be considered in association with agricultural production on long term basis to end poverty and improve nutritional food security.

India has a wide variety of climate and soils on which a large number of fruits and vegetables can be grown. After achieving green revolution in food grains in sixties, it became clear that horticulture for which agro-climatic conditions of India and topography are well suited could be an ideal method of achieving sustainability of small holdings, poverty alleviation and malnutrition. India is the second largest producer of fruits and vegetables after China contributing 10 % and 13.38 % of the total world's fruits and vegetables, respectively. Total fruit production in India was 68.47 million tonnes from an area of 6.1 million hectares (Indian Horticulture Data Base 2009). India is first in the production of mango, citrus, sapota and banana.

D.P. Abrol, *Pollination Biology, Vol. 1*, DOI 10.1007/978-3-319-21085-8_1

About 40 types of fruits are commercially produced in India, however, the major share comes from tropical and subtropical fruits like mango (28.17 %), banana (24.93 %), citrus (15.56 %), guava (3.79 %) and grapes (2.76 %). Among temperate fruits, apple occupies major share (2.98 %) (Jain 2004).

Fruit crops are mostly cross-pollinated or benefit from bee/insect pollination for significant fruit production. At the same time, insect pests are a major threat for quality and quantity of fruit production and account for 40 % of reduction in crop yields (SP-ISM 2010). Major damage occurs either due to direct feeding or indirectly by vectoring different diseases. The major weapon to minimize these losses is to rely on insecticides which not only kill beneficial insects including honeybees and other natural enemies but also result in pesticides residues rendering fruits unfit for human consumption and resulting in health hazards.

Understanding pollinator needs of different fruit crops, their associated pollinators and pest problems including basic biology of pests shall be useful to development management strategies for safety of pollinators on one hand and management of pests on the other.

1.1 Fruit Crops and Food Security

The distribution of world economic fruit crops is given in Table 1.1. Fruits play a significant role in economic development, nutritional security, employment generation and overall economic growth of a country. Horticulture provides a dynamic tool for enhancing economic returns, attaining nutritional security, creating employment avenues and ensuring ecological sustainability. India is the largest producer of fruits (45.2 million tonnes) and vegetables (84.82 million tonnes) next to China. India' share in world production is 9.9 % fruits, 9.9 % vegetables, 7.8 % potatoes and 11.5 % in onions. In world production India ranks 3rd in potatoes (923.16 million tonnes), 2nd in onion (4.2 million tonnes), brinjal (8.0 million tonnes), cabbage (5.39 million tonnes), cauliflower (4.4 million tonnes) and pea (1.71 million tonnes). In fruits, India ranks 1st in banana (13.30 million tonnes), grapes (7.39 million tonnes), mandarin orange (18.53 million tonnes) and mango (12.73 million tonnes), 10th in apples (1.35 million tonnes), papaya (0.15 million tonnes) and 4th in pine apple (1.17 million tonnes). Inspite of largest producer of fruits, the per capita availability of fruits in India is 62 g per day while FAO recommends a consumption of 85 g per day. The targeted fruit production in India is 200 million tonnes by 2020.

The average productivity of temperate fruits in the world is 7.4 t/ha in comparison to 5.8 t/ha in India. Among temperate fruits apple accounts for highest production. Its average productivity is 26 t/ha in USA is much higher than India (5.8 t/ha). Even in other fruit crops the productivity in USA is much higher than India. In almonds, walnut, peaches and nectarines the productivity is 3.8, 3.4 and 19.3 t/ha, respectively in comparison to 0.4, 1.0 and 8.1 t/ha. In apples, the highest productivity of 10.3 t/ha was obtained in Jammu and Kashmir followed by 6.3 t/ha in Himachal Pradesh and 1.1 t/ha in Uttarakhand. In walnut, the highest productivity

Table 1.1 Distribution of world economic fruit crops (production in tonnes) (Source: Jackson et al. 2007)

Crop	Africa	Asia	Europe	North and central America	Oceania	South America	World
Almonds	174,020	328,370	403,955	409,113	8,500	3,960	1,327,918
Apples	1,476,620	27,515,097	16,967,550	6,239,690	780,647	3,200,705	56,180,309
Apricots	307,730	1,057,183	776,889	123,590	38,000	55,040	2,358,432
Avocados	193,000	240,745	59,710	1,375,119	21,636	408,200	2,298,410
Bananas	7,158,915	24,364,113	371,400	8,225,294	705,977	17,469,251	58,294,950
Chestnuts	260	345,073	149,320			43,858	538,511
Citrus fruit NES	2,665,745	1,335,718	21,400	32,325	9,655	86,217	4,151,060
Currants		200	595,655	0	2,465		598,320
Dates	1,634,563	3,165,699	8,000	21,132		145	4,829,539
Grapefruit and pummel	360,167	1,045,047	47,300	3,148,081	23,320	429,471	5,053,386
Grapes	28,94,400	13,165,638	28,401,029	5,955,508	1,019,723	4,734,381	56,170,679
Hazelnuts (filberts)	140	600,672	129,296	14,970			745,078
Lemons and limes	616,810	3,051,536	1,383,480	2,183,880	44,116	1,934,271	9,214,093
Mangoes	1,925,592	18,183,745		2,026,577	38,262	890,082	23,064,258
Oranges	4,724,031	12,144,337	5,569,281	17,799,726	383,940	25,711,064	66,332,379
Papayas	774,155	1,259,524		603,336	19,858	2,167,843	4,824,716
Peaches and nectarines	438,250	4,367,632	3,903,337	1,496,000	109,000	773,444	11,087,663
Pears	350,240	8,481,698	3,543,492	884,980	197,022	911,264	14,368,696
Pineapples	2,006,033	6,660,046	2,000	1,305,074	147,248	2,363,939	12,484,340
Pistachios	1525	256,300	8,000	88,450			354,275
Plums	166,320	3,710,135	2,956,364	899,016	34,500	232,115	7,998,450
Raspberries	140	600	250,927	56,960	896		309,523
Strawberries	51,000	508,322	1,176,868	867,332	15,435	65,036	2,683,993
Tangerines, mandarins and others	1,031,295	11,978,972	2,419,399	845,776	96,060	1,493,641	17,865,143
Walnuts	6,600	517,426	301,673	218,200	70	25,150	1,069,119

of 1.36 t/ha was recorded in Jammu and Kashmir in comparison to 0.5 t/ha in Uttarakhand and 0.5 t/ha in Himachal Pradesh (Anon 2005).

1.2 Temperate Fruits

Apples pears strawberry, plums, peaches, cherries and raspberries belonging to family Rosaceae are important fruits grown in temperate areas. Apples and pears are referred to as pome fruits because the edible fleshy part is a combination of outer ovary wall and basal part of the flower.

Plums, peaches, and cherries come from different species of the genus *Prunus*. They share a fruit type known as a drupe, consisting of a fleshy mesocarp and a single seed inside of a hard endocarp. While there are native species of Prunus in the New World, the domesticated species are native to Eurasia. The modern cultivated strawberry is a hybrid that actually an aggregation of fruits, or aggregate fruit. Each tiny seed is itself a fruit. The large, succulent, mass is the swollen top of the stem on which the flower was borne. Raspberries are also aggregated fruits, but each globular segment of the raspberry is itself a fruit, called a drupelet. The caps of drupelets pull free of the stem tips when the berry is picked. Grapes are the second most widely cultivated fleshy fruit (on a tonnage-produced basis). However, the majority of grapes are not eaten as fruit but are turned into other foods, such as vinegar, liqueurs, raisins, and wine. The most widely cultivated species of grape is *Vitis vinifera* (Vitaceae), a woody perennial vine native to middle Asia. There are hundreds of varieties of grapes that vary in the color of the skin, flesh, flavor, and sweetness of the berries. Nuts are dry fruits, each of which contains a single seed that is free inside the ovary wall, except for an attachment at one end called the funiculus. The pericarp (the walls of the ovary) is hard and fibrous. Commercially grown nuts include filberts, pecans, walnuts, and macadamia nuts, sold both for eating and for cooking.

1.3 Tropical Fruits

Tropical fruits are grown in great diversity of climates ranging from perennial to herbaceous species (Alexander and Possingham 1984; Akora 1998). The herbaceous group contains important crop plants such as banana, pineapple, and papaya while the woody or perennial plants include tree species, shrubs, and vines (Verheij and Coronel 1992)

Day length in the tropics varies too little to noticeably affect the annual course of radiation or temperature. Thus, most climatic cues of plant development are not as firmly tied to calendar months as is the case at higher latitudes. The phenological cycle in the tropics may shift a few months from 1 year to the next. This complicates phenological studies, but should make it easier to detect which environmental factor triggers a certain growth and development response. Because of their diversity, not all tropical fruit follow the same strategy for growth and development. Strategies are based on differences in tree form and branching habits. For instance, single-stemmed plants such as pineapple, papaya, and banana are contrasted with freely branching fruit trees (Verheij and Coronel 1992). Most tropical fruit are considered as perennial plants which persist for several years without abrupt, major changes other than seasonal leaf formation, flowering, and fruit development (Bennett et al. 1976). However, pineapple, papaya, and passion fruit are grown for shorter periods of time and arthropod management on these crops is influenced by their duration in the field.

Tropical fruits has traditionally been produced in great quality and quantity in areas near their centre of their origin. For instance best quality and quantity of mango is produced in India as compared to other countries such as Mexico, Guatemala and Israel. Similarly, Litchis are traditionally grown in southern China, but now they are grown in the USA, Australia, and elsewhere. A surprisingly wide range of tropical fruits including bananas, papayas, and pineapples are grown in high altitude areas (2000–2500 m) near the equator in countries such as Kenya, Colombia, Ecuador, Peru, Indonesia, and India. In nearly all cases, these fruits flourish in unusual environmental niches which are frost free and receive a high level of solar radiation.

According to FAO (2001), world production of tropical fruit (avocado, banana, citrus, mango, papaya, and pineapple) increased by 36–54 % from 1975 to 2000 (Table 1.2), and production of minor crops (*Annona*, guava, passion fruit, etc.) increased by 24–29 % during this period (Table 1.2). However, Verheij and Coronel

Table 1.2 Major insect and mite pests affecting fruit crops (Source: Rieger 2006)

Insect	Crops affected	Damage
Codling moth (*Cydia pomonella*)	Apple, pear, plum, walnut	Fruit feeding, fruit drop
Oriental fruit moth (*Grapholita molesta*)	Peach, plum, apricot, almond, apple	Shoot dieback, fruit feeding, fruit drop
Plum curculio (*Conotrachelus nenupar*)	Peach, plum, apple, cherry, blueberry	Surface scarring, catfacing, fruit feeding, fruit drop
Leaf rollers (*Platynoda, Argyrotaenia spp*)	Apple, pear, peach, plum, grape, citrus, strawberry	Leaf and bud feeding, damage; webbing on fruit; fruit damage and subsequent rot
Scales (*Quadraspidiotus*)	Most fruit crops	Fruit scarring, cosmetic damage; leaf feeding; limb and twig dieback, tree decline; honeydew secretion and sooty mold development
Stink bugs and Plant bugs (*Leptoglossus, Lygus*)	Most fruit crops	Fruit catfacing, spotting
Aphids (*Aphis*)	Most fruit crops	Leaf and shoot feeding, distortion; honeydew secretion and sooty mold development; virus transmission
Leafminers (*Lithocolletis*)	Citrus, apple	Leaf feeding by tunneling
Mites (*Tetranychus, Panonychus*)	Most fruit crops	Leaf feeding, stippling, distortion; webbing at shoot tips

(1992) stated that these figures could be misleading, because most fruit comes from trees scattered in home gardens, making it difficult to compile reliable statistical data.

India is a major tropical fruit producing country in the world followed by Brazil, China, Mexico and Philippines (FAO 2001). The other leading in tropical fruit production are Ecuador, Colombia, Cost Rica and Cuba.

Tropical fruit production in Africa is apparently concentrated in Nigeria, Burundi, South Africa, Congo, Cameroon, Egypt, and Angola. Few tropical fruits are grown in Europe and only relatively small amounts in the USA, Australia, and Israel. Many equatorial tropical countries including those in Africa, the Americas, India, and South-East Asia are beginning to strengthen research programmes on propagation, genetic improvement, control of flowering, control of pathogens and arthropods, irrigation and nutrition, postharvest handling, and storage.

1.4 Integrated Pest Management of Fruit Crops

1.4.1 Pest Problems

Insect pests play havoc on crop productivity and cause considerable damage. Post harvest losses are estimated to range from 25 % to 30 % of fruit production. Pest management needs to be environmentally safe ecologically sound and economically sustainable for horticultural crop production. There are hundreds of thousands of species caising damage to fruit crops but only few are of serious concern. Rieger (2006) has listed several species of insects and mites causing damage to fruit crops.

Mites injure plant tissues by feeding on the surface, the leaves loose photosynthetic activity and fall prematurely. The attacked fruits develop blemishes rendering fruit unfit for consumption and marketing. Besides, insects and mites, nematodes are yet another group of serious pests of fruit crops causing significant damage. The most important nematode species include the root knot (Meloidogyne), ring (*Criconemella*), dagger (*Xiphinema*), root lesion (*Pratylenchus*), and cyst (*Heterodera*). They feed on roots causing stunting, wilting and chlorosis. Their presence can be identified in the form of root galls, knots or lesions. The losses caused by nematodes can be resuced using several control methods in an integrated manner.

The W. Ebeling's book, *Subtropical Entomology* written in 1950 still continues to be cited as an essential reference for students and researchers of tropical and subtropical fruit entomology. However, tropical crop production entomology has undergone major changes since the book was published. Several other texts dealing with pest management systems for tropical crops (Lamb 1974; Hill 1975; Swaine et al. 1985; Braga et al. 1998; Tandon 1998; Mariu 1999) have since been written. These publications are in response to substantial growth of tropical fruit production and increased concern for food safety. Many books dealing with integrated pest management have concentrated on field crop IPM with the occasional exception of temperate fruits, such as pear and apple (Boethel and Eikenbary 1979), citrus (Anonymous 1991) or nuts and almonds (Anonymous 1985).

Many techniques are employed in an integrated manner to keep pests below a certain threshold. The W. Ebeling's book, *Subtropical Entomology* written in 1950 still continues to be cited as an essential reference for students and researchers of tropical and subtropical fruit entomology. However, tropical crop production entomology has undergone major changes since the book was published. Several other texts dealing with pest management systems for tropical crops (Lamb 1974; Hill 1975; Swaine et al. 1985; Braga et al. 1998; Tandon 1998; Mariu 1999) have since been written. These publications are in response to substantial growth of tropical fruit production and increased concern for food safety. Many books dealing with integrated pest management have concentrated on field crop IPM with the occasional exception of temperate fruits, such as pear and apple (Boethel and Eikenbary 1979), citrus (Anonymous 1991) or nuts and almonds (Anonymous 1985).

Integrated control of a pest or pest complex in a tropical crop requires a full investigation of relationships and interactions within the whole ecosystem (Entwistle 1972). Because of the specifically complex society of plants and arthropods inhabiting the area where tropical fruits originate, tropical fruits may have a greater diversity of tropically associated arthropods than any temperate crop. For instance, Entwistle (1972) maintains that the 1400 arthropod species listed for cocoa can only be the skeleton crew of the total cocoa fauna. Thus, information on tropical fruit entomology is regionally uneven both in quantity and quality; this being the result of national policies or the absence of them, and the dominant nature of certain pest problems. Moreover, few authors have yet taken an integrated attitude to arthropods of tropical fruit crops. Globalization of tropical fruit and rapid expansion of export markets has resulted in exotic plants grown outside of their centres of origin. For instance, citrus originated in Asia and is grown in the Neotropics while papaya and passion fruit originated in the Neotropics and now are grown in Asia, Australia, etc.

Thus, upon arrival in their new area of domestication, exotic fruit plants are populated by an indigenous regional arthropod fauna, and by exotic arthropods that have also migrated from the crop's centre of origin. Thus, arthropod systems (pollinators, pests, and natural enemies) for introduced plant species will be somewhat different from those observed in the crop's area of origin. For instance, the European honeybee *Apis mellifera* did not co-evolve with avocado but it is used as a pollinator of avocado in Israel and Australia. Original pollinators from the area of origin of avocado in Mexico and Guatemala could be other more efficient hymenopteran species. In the Neotropics several insects have adapted to introduced plant species, such as citrus. Citrus is one of the preferred hosts for *Diaprepes abbreviates* and *D. abbreviates*. The later species is a key pest of citrus in Florida and Caribbean region, however still there is no report of its occurrence in China and Asia.

Tropical fruit crops provide a relatively long-term and stable environment, offering continuing habitats for both pests and natural enemies. This environment provides excellent opportunities for biological control and alternative pest management programmes (Bennett et al. 1976). However, the extensive use of broad spectrum pesticides disrupts this stable environment, and leads to instability of arthropod densities (Hoyt and Burts 1974). According to Hoyt and Burts (1974) any attempt to develop integrated control programmes in fruit crops must take into account the following: (i) knowledge of native or resident arthropod fauna; (ii) arthropod fauna affecting the tree crop in its area of origin or domestication; and (iii) the presence of natural enemies. The basis for integrated pest management includes the pest's biology and ecology, sampling and monitoring techniques, economic thresholds, and the application of management tactics, i.e. chemical, biological, autocidal, plant resistance, etc.

In order to apply IPM concepts to fruit protection and production, the biology and ecology of these arthropods particularly in their relationship with fruits must be known. As climatic and geographical variation within the range of tropical fruit cultivation is known to affect the composition of insects on each tropical crop. For instance, *Bephratelloides* spp. (Hymenoptera: Eurytomidae) are the most common insects affecting *Annona* crops in the Neotropics. Within this group, *Bephratelloides cubensis*, a key pest species present in Florida, has a female-biased sex ratio, while a similar species, *B. pomorum*, dominant in *Annona*growing regions of Brazil, has an almost 50:50 sex ratio. In the same way, the moth *Cerconota annonella* is found affecting *Annona* in the warmer regions of South America and the Caribbean but is absent from subtropical Florida. Insects can be classified as autochthonous species with a more local distribution, or as inter-regional species which may affect crops in different areas either by exceptional powers of dispersal coupled with the host and environmental tolerance, or as homophylous, i.e. tending to distribution by man. These classifications are not necessarily mutually exclusive and it may be difficult, as for example with such mealybugs as *Maconellicoccus hirsutus* and the aphid *Toxoptera aurantii*, to distinguish the main agents of dispersal (Entwistle 1972).

Most arthropods affecting tropical fruit are polyphagous (mites, fruit flies, scales, whiteflies, and thrips) and studies on their biology have been published extensively (White and Elson-Harris 1994; Jeppson et al. 1975; Helle and Sabelis 1985a, b; Rosen 1990a, b; Parker et al. 1995).

Management practices for insect pests of different tropical fruits vary from one fruit crop to another in terms of spray timings and volumes of insecticides used. Pantoja et al. report that most papaya producers apply insecticides on a calendar basis for controlling fruit flies and leafhoppers in papaya. Such sprays on regular basis can cause can cause heavy outbreaks of pests. Some of the pesticides recommended for control of citrus pests, when applied to other crops without proper knowledge may exacerbate the crop losses due to phytotoxicity, induced surge of population densities of other pests and reduction of natural enemy populations etc. Unfortunately indiscriminate use of non-selective pesticides continues but there is a strong urge to evaluate the new molecules of new generation pesticides, their impact on pests and predators, proper timing of spray and selection of spray applications. There has been little information on the deleterious effects of pesticides on pollinators. In a very interesting study, Aguiar-Menezes et al. (2002) suggested that in case of passion fruits, pollinator poisoning can be minimized by adopting spray timings according to the opening time of different cultivars. For instance, purple passion cultivars open during early morning if sprayed during late afternoon pollinator losses can be escaped. Similarly yellow cultivars whose flowers open in the late afternoon should be sprayed early in the morning.

Non-selective traps such as black light and bait traps are commonly used in tropical fruit because they can provide useful information for monitoring purposes. Furthermore, development of selective traps (baited with attractants or pheromones) often to the pest species level, has been a tremendous boost to IPM of temperate fruit (Beers et al. 1994) and it may provide the same benefit available for tropical fruit pests. Use of feeding and sexual attractants (secondary pheromones) in tropical fruit has been largely limited to fruit flies. Few species-specific attractants are available for other insect groups such as weevils (Curculionidae) (Gold et al. 2002) and Lepidoptera (Bailey et al. 1988).

Biocontrol programmes need initiation of activities much earlier before the pests assume epidemic proportions. For instance, efforts for biocontrol of mealy bug, *Maconellicoccus hirsutus*, and the carambola fruit fly, *Bactrocera carambolae*, in the Caribbean started only when the pests had already assumed major threats to valuable crops. If the pest involved is stenophagous such as avocado weevils, papaya fruit fly, avocado thrips, and *Annona* fruit moths affecting single commodity, the efforts to develop biocontrol strategies are rather poorly undertaken.

Generally, if a stenophagous insect (i.e. avocado weevils, papaya fruit fly, avocado thrips, and *Annona* fruit moths) is the major constraint to production of a single commodity, then efforts to develop biological control are pursued halfheartedly. The exceptions to this rule are the current strong efforts towards biocontrol of banana weevil, and the search for natural enemies of avocado thrips. Efforts also differ among countries. For instance, in Australia and Israel biological control is an important component of *Annona* pest management, while other *Annona*-producing countries, i.e. Brazil, Ecuador, Venezuela, Colombia, and Mexico, rely on chemical pest control.

It is likely that with time the use of biological control agents for tropical fruit protection will increase extensively. The main constraints on this trend are the lack of commitment and money. Currently, very few people in tropical countries are researching these areas, and very few specialists receive adequate financial support. While major emphasis has been given to staple crops (i.e. rice and cassava), with the exception of banana, little or no emphasis has been given to fruit crops.

Host plant resistance offers considerable promise as a tactic in pest management. Even though it appears that most efforts are concentrated in developing genetic resistance to plant pathogens, use of this tactic merits attention for numerous crops against numerous arthropod pest species. Evaluation of tropical fruit germplasm collections for resistance to arthropod pests should become a high priority. Most efforts are directed at insect vectors (*citrus, papaya*, and pineapple), and pests of mango, avocado, and guava.

Cultural control concerns the employment of cultural management methods to minimize insect damage. The use of trap crops, host plant removal, removal of pests, and the reduction of pests' habitats are considered beneficial by some and controversial by others. For instance, Hansen and Amstrong (1990) reported that field sanitation did not reduce infestation by the mango nut weevil, *Sternochetus mangiferae* (Coleoptera: Curculionidae) in Hawaii, while in the Neotropics, removal of *Annona* spp. Fruit infested with the seed borer *Bephratelloides cubensis* is considered to be a regular practice to avoid future infestations. Untreated backyard trees and neglected plantings are considered to be major sources of pests such as fruit flies. Thus, in Australia, hygiene and attention to alternative host plants that can increase pest pressure on the *Annona* (custard apple) orchard, are important. Mature fruit infested with yellow peach moth or with the Queensland fruit fly should be collected and destroyed. Preferred fruit fly hosts like guava and loquat should not be planted as trap crops in or near the *Annona* orchard. According to Hoyt and Burts (1974), cultural practices generally do not offer a direct means for controlling pests but, when used properly, they can enhance natural enemy activity or retard pest population growth to a degree that is important in integrated control programmes. With polyphagous insects, the wild host plants may be so numerous that special eradication would be impracticable, yet removal of related wild host plants may be very beneficial where insect pests have restricted host ranges. Gold et al. described crop sanitation as an important factor for maintaining low banana weevil densities in banana plantings. However, crop sanitation has not been firmly established for controlling important pests of avocado (i.e. weevils) *Annona* (annona seed borer) and mango (mango seed weevil)

1.5 Pollination

Pollination is the transfer of pollen from the anther of a flower to the stigma of the same or another flower. This process begins seed production in flowering plants. In some regions the need for pollination has long been understood: the practices of artificial pollination and the encouragement of natural pollination originated in the Middle East. Since ancient times humans ensured that female date palms are productive by hand-pollinating them with pollen from male inflorescences. It is a long established procedure for owners of fig groves to introduce capri figs, which although inedible, produce pollen and harbour the tiny wasps that pollinate other fig flowers. This knowledge of pollination even extended into religious life. An Assyrian Palace relief from 2750 years ago depicts an eagle-headed god who appears to be pollinating a sacred tree with pollen from a fir cone. The bag he is carrying possibly contains more fir cones and pollen. So our subject is not only important to humans, it is a concern of the gods.

The floral morphology of a fruit crop is important in determining the mode of pollination (i.e., wind or insect) and type of fruit that will arise when the ovary matures. A complete flower is one possessing all four fundamental appendages: sepals, petals, stamens and pistils. An incomplete flower lacks one or more of these features. The position of the base of the pistil, or ovary, with respect to the other three appendages is important in identification, and also partially determines what the fruit type will be. The two most common positions are inferior (epigynous) and superior (hypogynous). A third possibility is "half-inferior" (perigynous), as found in the stone fruits (*Prunus* spp.). A perfect flower possesses both male (stamens) and female (pistil) parts, whereas an imperfect flower may be either staminate (functionally male, having only stamens) or pistillate (functionally female, having only a pistil or pistils). The style is very short or lacking in some pistillate flowers, like in pecan, but the stigma and ovary are always present if the pistil is functional. If staminate and pistillate flowers are borne on the same plant but in different locations, the species is termed monoecious. If staminate and pistillate flowers occur only on different plants, the species is termed dioecious. One can see the ramifications for pollination and orchard design: a dioecious species such as pistachio or kiwifruit must be planted in an orchard with male plants near females for pollination. Aside from pollination, the male plants are useless since they do not possess ovaries that will ripen into fruit. An inflorescence is a cluster of flowers, and there are several terms for specific inflorescences. Generally, inflorescences fall into two categories, determinate and indeterminate. In a determinate inflorescence, the top-most flower is the most mature, and generally opens first, whereas the top-most flower in an indeterminate inflorescence is the least mature and last to appear. The most common inflorescence types in fruit crops are indeterminate (spikes, racemes, panicles, umbels, corymbs), with the cyme being the most common determinate inflorescence.

1.6 Fruits Crops and Pollination

Fruit crops benefits greatly from insect pollination. McGregor (1976) reported that the bees are essential for fruit set in majority of the cultivated fruit crops. For instance, some varieties of strawberry are highly self sterile and depend highly on insect pollination for any significant fruit production (Houbaert et al. 1992; Hofmann and Haufe 1995; Ahn et al. 1989; Singh 1979; Gundia et al. 1995). Houbaert et al. (1992) found that percentage of misshapen fruits decreased from 44 % to 2 %, when a colony of bees was introduced in strawberry fields. They further found that yield also increased from 0.2 to 0.3 kgs per plant and financial returns were made than double per plant, when pollinated by bees. In a similar study, Chagnon et al. (1989) found that honey bee pollination is essential for commercial fruit production in strawberry. Similar is the situation with almost all the fruit crops.

The different chemicals used for controlling crop pests have a negative effect on the beneficial insects including honeybees which severely affects the pollination process and consequently fruit production. With the magnified increase in production of insecticides, the accidental killing of bees has become a major problem for beekeepers worldwide (Johansen 1977; Mayer 2003; Masoodi 2003). Poisoning of insecticides to honey bees is generally more pronounced because of their long hours of working on the crop flowers for pollen and nectar collection and long flights with pollen loads. Bee poisoning is more chronic in areas which lack of sufficient wild pollen and nectar plants to sustain the number of colonies required for crop pollination increasing efforts towards crop improvement and obviating the use of pesticides (Day 1987), agriculture remains heavily depended on these chemicals (Rana and Goyal 1991). So, there is an urgent need of the integration of the two, killing pests on one hand and ensuring safety of pollinators on the other for sustainable crop production (Mayer and Johansen 1983). Conservation of honeybees for crop pollination is vital to agricultural production (Smirle 1990; Gupta and Sharma 1998; Kremen et al. 2002). Despite

In countries with highly mechanized agriculture, the use of bees for pollination has increased greatly during the twentieth century and it is now an integral part of crop production. In recent times far less attention has been paid to the pollination of tropical crops. This is partly because other factors, such as inadequate soil, water, and poor soil fertility are also limiting crop production. In these cases pollination as does occur is sufficient to provide for the plant's bearing capacity. With the use of improved cultivars and irrigation, pollination may easily become the limiting factor, leading to increased need for pollinating insects.

The effect of pollination on the yield of most crops grown in temperate climates is well-established. There is relatively little information about the pollination of tropical crops. More data is urgently needed, especially as the results of pollination research can be of direct benefit to growers, who can take immediate action without the need for complex equipment, or large-scale investment, and reap an almost immediate reward. Some flowers self-pollinate without the aid of insect visitors, although their structure indicates that in the evolutionary past they were insect-pollinated. For example, placing bags over flowers (so that insects cannot reach them) of paw, sweet pepper, hot pepper and egg plant usually does not decrease fruit set. So growers need not be too concerned about the pollination of these crops: improvement in yield would seem to depend solely upon increasing the bearing capacity of the plant by favourable nutrition and by plant breeding.

Some apparently insect-pollinated flowers are parthenocarpic. Examples are various cultivars of banana and Citrus. In contrast, plants traditionally supposed to be wind pollinated may benefit from insect pollination. Thus we now know that the oil palm is primarily insect-pollinated and **not** wind-pollinated as previously supposed. There are indications that the following tropical crops benefit from insect pollination: allspice, avocado, cashew, coconut, coffee, cotton, lychee, mango, melon, passion fruit, safflower, sunflower, and many varieties of Citrus and Cucurbits. It is likely that some (for example safflower, mango, okra, and opium poppy) give only moderately increased yield when pollinated by insects while others (for example cashew and guava) give greatly increased yields. However, more research is needed. The pollination requirements of other crops including tropical legumes, which are beginning to be used in large-scale production, are mostly unknown. Because of the lack of time and facilities, too many of the studies on tropical crop pollination have been of a preliminary nature only. Careful studies must be instigated to determine: (a) whether fruit or seed set is usually adequate; (b) whether the crops benefit from pollination; and (c) which insects pollinate the flowers.

1.7 Pollination Requirements of Crops

A standard method to determine whether a crop benefits from insect pollination is to compare the yields of plants that are grown under three different conditions: (a) covered by nylon screen cages containing honeybee colonies; (b) covered by cages to exclude insects; (c) not caged. If cages are not available useful information may be obtained by bagging individual flowers or flower heads to exclude pollinating insects. Some of the bagged flowers must be hand-pollinated to find the maximum set possible under these conditions. Experiments must also determine whether the crops benefit from self- or cross-pollination. Cross-pollination is obviously needed when the sexes are segregated on different plants (for example melon) or by different periods of flowering of the same plant (for example avocado). Cross-pollination may also increase yields of plants that can be self-pollinated. Cross-pollination can only occur when the insect moves from one plant to another, and this usually happens on only a small proportion of flower visits. Production of hybrid seed crops on a commercial scale is creating a special need for cross-pollination by insects, and a high population of pollinating insects is needed to carry pollen from rows of male plants to rows of female plants. Many varieties of fruit tree need cross-pollination. When an insect moves from a tree of a polliniser variety to a tree of a main variety, it pollinates only the first few flowers it visits, so parts of trees adjacent to the polliniser tree get most fruit set. To obtain an even fruit set, orchards should be planned so that the main variety trees are surrounded by polliniser trees.

Pollination is a critical phase of the production cycle for most of fruit and vegetable crops. Bees are the most important delivery vehicle for pollen, and their activity ensures that the flower stigma receives sufficient pollen for fertilization to occur. Well pollinated crops ripen earlier, produce larger and more even fruit, and improve grower profit. Most of the commercially grown fruits require insect pollination. Instead of seed production the goal here is to obtain the edible fruit. Among the important nut crops only almonds depend on insects for pollination, most of the others rely on wind. Most fruit crops require pollination and seed formation for their fruit to develop, in case of oranges, where development is wholly parthenocarpic. On the other hand in case of Rosaceous group where pollination requires special attention. Included here are the almonds, berries, pome fruits and stone fruits. Therefore, a commercial variety is basically one genotype, with the rare exception in which plants were grown from mutant buds. The Rosaceae are rarely auto-self pollinating and even if they are self-fruitful, they require insects to transfer the pollen from the anthers to the stigma. The phrase "self-fruitful" is more accurate here than self-fertile because the fruit or seed that results from pollination does not have to be fertile. Some varieties are completely self-fruitful, some are only partially so and many are mostly self-unfruitful. If there is any reduction in productivity when a variety has been self-fertilized it is probably in the self-unfruitful category. From a horticultural viewpoint self-fruitful varieties have three advantages over self-unfruitful varieties. They do not have to be interplanted with other varieties, they require fewer pollinating insects, and the entire blooming cycle is effective for pollination as the flowers are perfect and their blooming period is in synchrony.

Self-unfruitful varieties present several problems. Different varieties must be interplanted in order to provide a foreign pollen source. The flowering of the pollinizer variety has to be in synchrony with that of the pollinated. The pollinizer must be genetically compatible with the pollinated, and ideally the pollinizer should have fruit of marketable value. The latter also requires that the pollinizer be readily pollinated by the primary commercial variety (= reciprocal pollination).

With interplanting of varieties they must be close enough so that insect pollinators will include both varieties on a single foraging flight. When compared with self-fruit varieties, pollinating insects need to be present or provided for in greater numbers because only flower visits subsequent to visits on other varieties are effective. Placing pollen in beehives can alleviate some of the problems (Fig. 1.1).

Fruitfulness among members of the Rosaceae varies with the species and variety. Apples are mostly self-unfruitful, while pears are partially self-unfruitful. Plums can be either partially or wholly self-unfruitful. Peaches and nectarines are mostly self-fruitful, but some have either scarce or poor pollen. In almonds most varieties are self-unfruitful. Apricots are mostly self-fruitful. Sour cherries are at least partially self-fruitful but yields are increased with crossing. Sweet cherries are mostly self-unfruitful. Raspberries are partially self-unfruitful, while blackberries, loganberries and dewberries are mostly self-fruitful. Strawberries can be either self-fruitful or self-unfruitful.

Decisions on how to interplant in cases of self-unfruitful varieties naturally involve both horticulture and entomology. The economic value of the pollinizer variety must be considered. There are problems with cultivation and harvesting two or more interplanted varieties. The species and number of pollinating insects and

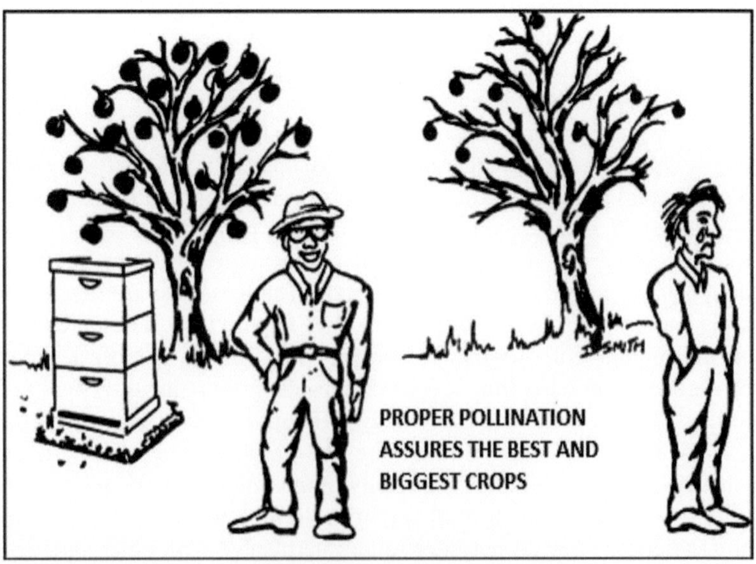

Fig. 1.1 Model exhibiting impact of pollination on production of fruit crops

how much territory will individuals cover while foraging are of primary concern. These considerations vary for different crops and in different localities.

In order to determine how many pollinating insects are required requires many considerations. Are the varieties self-fruitful or unfruitful? What is the type of interplanting in the case of self-unfruitful varieties? During the hours of foraging what kind of weather might be expected. What is the number of flowers that need to be pollinated because small fruits need to set a greater percentage of fruit? What kind of competition is expected with other blossoms in the area during the critical hours for effective pollination? What are the number and variety of pollinating insects already present in the area and do they persist year after year. What is the efficiency of the insects provided, which would usually involve honeybees? What is the distance of the crop from apiaries? Usually hives are placed in groups within a planting. In general practice as many pollinating insects as possible are provided. Some plantings within or adjacent to wild country already have an abundance of wild bees, but their presence should be known to prevail over a number of years if honeybees are not introduced by beekeepers.

A special problem can arise with the production of large fruits, especially those that are self-fruitful. Here there may be too much pollination where the fruit will be small and of low market value. Blossom thinning and thinning of developing fruit or the reduction of the number of pollinating insects are techniques that can be deployed.

1.8 Insect Pollinators for Rosaceous Fruit Crops

Rosaceous fruit crops are generally attractive to honeybees, **queen bumblebees** and many early season solitary bees, syrphid flies, bombyliid flies and to some extent blow flies. Certain fruit varieties such as pears are especially attractive to blow flies and syrphid flies, while plums are attractive to solitary bees. Syrphid flies may cover more trees and be especially valuable in cross-pollination. The efficiency of a particular insect species varies, however. Several genera of solitary bees are effective pollinators when orchards are adjacent to wild hilly terrain. Included are the genera *Andrena*, *Anthophora*, *Halictus* and *Tetralonia*. Other genera are more common under special conditions. Syrphid flies can be most abundant where there are large aphid infestations that serve as food for their offspring or shallow polluted water, which is the preferred medium of some syrphid fly larvae. Blowflies are common near refuse dumps, slaughterhouses or wherever there is abundant carrion. Honeybees may be successfully deployed in most fruit growing areas but it is sometimes difficult to obtain them. This is because most Rosaceae are not especially good honey plants and a beekeeper would rather place his hives in dandelion or some other early spring flour source. This is true especially if the orchards need to be heavily stocked. Beekeepers are also reluctant to place hives near orchards that will be treated with insecticides. Some areas are not suitable climatically for honeybees during pollination time. In certain areas of North America and northern Europe bumblebees are the only insects capable of foraging effectively during the blossoming season.

1.9 Pollination of Non Rosaceous Fruit Varieties

In grapes insects may assist in pollen transfer even though most commercial varieties are self-fruitful. Muscadine and Muscat grapes usually have sterile pollen so that they require interplantings of male plants or plants with good pollen. Honeybees and some solitary bees and many kinds of lies may be attracted to grapes, but in some areas pollinating insects are not very abundant. Blueberries are self-unfruitful and require bees for pollen transfer. They are frequently harvested from wild areas where solitary bees are plentiful. On the other hand, cranberries are self-fruitful, and although they may be auto-self pollinating, honeybees have been moved into cranberry areas with reported large increases in yields. There are few problems in citrus orchards with pollination. Navel oranges produce seedless fruit without pollination. Most commercial seeded types are self-fruitful but require pollination. Little pollination may be required because after one ovule is pollinated other seeds are produced apogamically from mother tissue and well-shaped fruits result. Orange blossoms are very attractive to bees and excellent quality honey crops can be produced. The common edible black mission fig produces parthenocarpic fruit with abortive seeds so it does not require pollination. The preferred Smyrna fig and several other varieties must be pollinated by another variety, however. The only insects that can pollinate figs are small wasps of the genus *Blastophaga*. These wasps develop in special gall flowers of a type of fig known as Capri. They emerge from these and accumulate pollen as they exit the fruit. They then enter the "eye" of a Smyrna fig and attempt to oviposit, unsuccessfully, in the long styles of the female blossoms. This activity effects pollination. In commercial plantings the Capri figs containing wasps have been gathered and hung on the branches of the Smyrna variety. Newer varieties of edible figs are available that avoid this procedure.

1.10 Pollination Status of Different Fruits and Nuts

Faegri and van der Pijl (1979) in their studies on pollination requirements in crops stated that some of the crops are highly dependent upon cross-pollinationg agents due to the presence of strong reproductive barriers such as self-incompatibility, different timing for floral maturity of reproductive organs and unisexuality. Under such circumstances a flower can not utilize its own pollen. Self-incompatibility flowers are genetically incapable of setting seeds in the same flower. Some of the self-incompatible fruit crops are apple (*Malus sylvestris* Mill.), almond (*Prunus amygdalus* Batsch), plum (*Prunus domestica* L.), pear (*Pyrus communis* L.), carambola

(*Averrhoa carambola* L.), almost all kinds of berries, sweet cherry (*Prunus avium* L.) and sour cherry (*Prunus cerasus* L.), passion fruit (*Possiflora* spp.), some cultivars of citrus (*Citrus* spp.) and grape (*Vitis vinifera* L.), and many cultivars of apricot (*Prunus armenica* L.). Guava (*Psidium guajava* L.) is partially incompatible. In case of protandrous crops, where the anthers mature before the receptivity of the stigma (Sihag 1997; Faegri and van der Pijl 1979). if pollination occurs due to self-pollen, the latter will go waste as it cannot germinate. This pollen can be useful only for the other flower(s) that have a receptive stigma. Some fruit crops like jujube (*Zizyphas jujuba* Mill. and *Z. mauritiana* Lamk.); are examples of protandry (Free 1993; McGregor 1976).

In yet another group of crops, contrary to the above, the plant bears flowers of one sex only i.e. male and female plants are separate – the male plant bears male flowers and the female plant bears female flowers, e.g. in papaya (*Carcia papaya* L.), date palm (*Phoenix dactylifera* L.) and Chinese gooseberry (*Actinidia chinensis* Planch.). In some plants, male and female flowers are borne on the same plant but maturity timings are different e.g. in all the cucurbits, coconut (*Cocos nucifera* L.), oil palm (*Elaeis guineensis* Jacq), etc. These are the cases of unisexuality. Under all of the above situations, pollen of the same flower cannot be utilized for pollination and seed/fruit set, and pollen from other flower(s) must be brought for this purpose. Besides these crops, there are others which are benefitted from cross-pollination in the enhancement of their seed/fruit production (Sihag 1997; Free 1993; McGregor 1976). Bees, due to their morphological adaptations such as branched hairs to collect pollen and nutritional requirements as they feed on pollen and nectar, can accomplish this task most efficiently. Among bees, honeybees are the most important group of pollinators due to their floral constancy and they can be deployed in desired numbers wherever their services are required. Two species of domesticated honeybees such as Asian honeybee *Apis cerana* and European honeybee *A. mellifera* can well be employed for the purpose of crop pollination. Inadequate pollination leads to low seed set and mishappen fruits in several cross-pollinated crops. Significant increase in seed/fruit yield has been reported in cross-pollinated crops by honey bees than those devoid of bee pollination (Free 1993; Sihag 1986). Therefore, honey bees play a significant role in seed/fruit production.

Interest in pollination is as ancient as civilized culture. Farmers have always shown a keen interest in the reproductive biology of plants and some mechanisms of fruit production (Real 1983) and while most of the studies have focused on the temperate zone, there is an expanding emphasis on investigations of tropical forests and plant ecosystems. Several studies of tropical plant pollination have been conducted in diverse natural habitats (Beach 1984; Valerio 1984; Gottsberger 1989) and this information has been applied to agricultural systems. Yet, it is necessary to remember that agroecosystems may lack some of the complex interactions observed in natural habitats. Young (1982) pointed out that for trees of tropical forests such as cacao, production of mature fruit is influenced by factors other than pollinator abundance. It is interesting to note that cacao originated as an understorey tree in Amazonian rainforests, where there was probably a well established equilibrium between pollinator abundance and abundance of flowers in the tree. Young (1982) concludes that successful pollination in a tropical crop depends in part upon the floristic complexity of the tree and the effects of surrounding habitats on the influx of insects influencing pollination within the agricultural habitat.

In conclusion there are two major and opposing forces that drive the dynamics of pest management from a worldwide perspective: trends toward globalization of markets versus trends toward sustainability of agricultural practices and conservation of biodiversity (Aluja 1994). In fact the market forces compel several countries to comply with quality standards established in other countries, that the fruit be without any cosmetic blemishes, and this can be only be achieved by use of agrochemicals or quarantine treatments (Sharp and Heather 2002).

Standard IPM methodologies, involving a number of control alternatives used alone or in combination have achieved some success in controlling pests such as fruit flies for individual growers (Hendrichs 1996). If the growers adopt a total pest population management strategy the chances of emergence of individual pests and their movement between neighboring orchards is sufficiently reduced. Under these area-wide conditions, which require an effective organization of growers, and in some instances increasing technical sophistication, IPM becomes much more effective. Malavasi et al. (1994) advocated the creation of pest free fields to facilitate quality production and increasing export potential of fruits. According to them creation of pest free fields require a very sensitive detection programm, strengthening of quarantine controls and preventing infestation during packing of fruits for exports.

In conclusion, for better management of tropical fruit, pests' habits, importance of their damage and action levels should be known as the foundation for any programme. More emphasis should be devoted to improved knowledge of pollinators and ways to improve their effectiveness; these will be determined by research and by practical experience. After these steps are taken, development of better monitoring techniques for pests and natural enemies should follow as well as the evaluation of a feasible biological control programme. Lastly, development of IPM packages for the whole pest complex should consider that all these arthropods inhabit the same universe. Some of them can be controlled with pesticides, others can be controlled by biological agents, cultural methods or by any means that are less harmful to the whole system (Fig. 1.2).

1.11 Pest Management During Pollination

Some of the precautions during application of pesticides for management of pests need to be undertaken to provide safety to bees and other pollinators on the crops. Monitor the pest levels regularly to minimize crop losses by adopting cultural, mechanical and biocontrol methods. If pesticide application is necessary, select the least toxic insecticides. Avoid application of pesticides during day time when bees are foraging in the field. Evening application is safer than the morning applications which could give a sufficient time lag to pollinators. Inform beekeepers at least

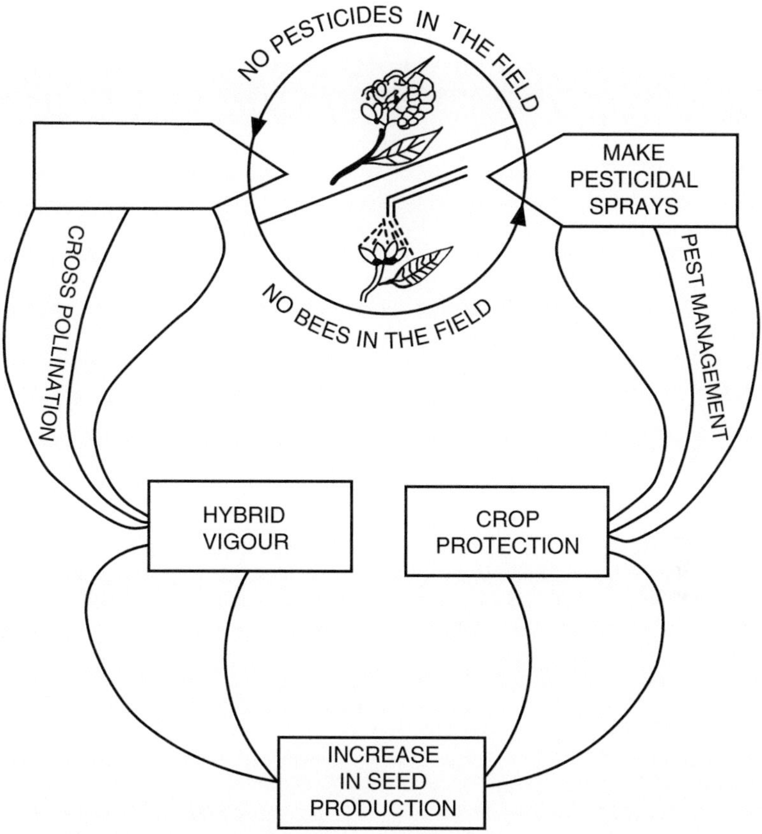

Fig. 1.2 Schematic model exhibiting safety of pollinators and pest management for improved fruit productivity (Source: Abrol 2009)

2–3 days before application, so that beekeepers could shift the colonies to safer places or restrict the bees inside the hives during application. Broad spectrum insecticides be avoided. Liquid applications are safer than the powdered forms.

Bibliography

Abrol DP (2009) Bees and beekeeping in India, 2nd edn. Kalyani Publishers, Ludhiana, 650 pp

Aguiar-Menezes EL, Menezes EB, Cassino PCR, Soares MA (2002) Passion fruit. In: Peña JE, Sharp JL, Wysoki M (eds) Tropical fruit pests and pollinators biology, economic importance, natural enemies and control. CABI Publishing, Wallingford, pp 361–390

Ahn SB, Kim IS, Cho WS, Choi KM (1989) Survey on the present situation of honeybee use as the pollinator in the plastic house of strawberry cultivation area by the questionnaire (Korean). Korean J Apicul 4:1–8

Akora RK (1998) Genetic resources of native tropical fruits in Asia: diversity, distribution and IPGRI's emphasis on their conservation and use. In: Akora RK, Ramanatha Rao V (eds)

Tropical fruits in Asia, diversity, maintenance, conservation and use. Proceedings of the IPGRI-ICAR-UTANET Regional Training Courses, Bangalore, 18–31 May 1997, pp 42–53
Alexander DME, Possingham JV (1984) Potential for fruit growing in tropical Australia. In: Page PE (ed) Tropical tree fruits of Australia. Queensland Department of Primary Industries, Brisbane, pp 1–19
Alford DV (1984) A colour atlas of fruit pests. Wolfe Publication, London
Aluja M (1994) Bionomics and management of *Anastrepha*. Annu Rev Entomol 39:155–178
Anonymous (1985) Integrated pest management for almonds. University of California, Division of Agriculture and Natural Resources, Oakland, p 156, Publication 3308
Anonymous (1991) Integrated pest management for citrus, 2nd edn. University of California, Division of Agriculture and Natural Resources, Oakland, p 144, Publication 3303
Anonymous (2005) Data-base of horticulture in Himachal Pradesh. Directorate of Horticulture, Shimla
Avery DT (1995) Saving the planet with pesticides and plastic. Hudson Institute, Indianapolis
Bailey J, Hoffman M, Olsen K (1988) Black light monitoring of two avocado insect pests. Calif Agric 42:26–27
Barfield BJ, Gerber JF (1979) Modification of the aerial environment of crops. Am Soc Agric Eng Monograph No 2. St. Joseph, Michigan
Baugher TA, Singh S (eds) (2003) Concise encyclopedia of temperate tree fruit. Haworth Press, New York
Beach JH (1984) The reproductive biology of the peach or 'pejibaya' palm (*Bactris gasiapaes*) and a wild congener (*B. porschiana*) in the Atlantic lowlands of Costa Rica. Principes 28:107–119
Beers E, Hull L, Jones V (1994) Sampling pest and beneficial arthropods of apple. In: Pedigo L, Buntin D (eds) Handbook of sampling methods for arthropods in agriculture. CRC Press, Boca Raton, pp 38–415
Bennett FD, Rosen D, Cochereau P, Wood B (1976) Biological control of pests of tropical fruits and nuts. In: Huffaker CB, Messenger PS (eds) Theory and practice of biological control. Academic Press, New York, pp 359–387
Boethel D, Eikenbary R (1979) Pest management programs for deciduous tree fruits and nuts. Plenum Press, New York, 256 pp
Borror DJ (1960) Dictionary of word roots and combining forms. Mayfield, Mountain View
Braga R, Cardoso JE, das Chagas F (1998) Pragas de fruteiras tropicais de importancia agroindustrial. Empresa Brasileira de Pesquisa Agropecuaria, Centro Nacional de Pesquisa Agrodindustria Tropical, Brasilia, p 209
Chagnon M, Gingras J, de Oliveira D (1989) Effect of honey bee (Hymenoptera: Apidae) visits on the pollination rate of strawberries. J Econ Entomol 82:1350–1353
Childers NF, Morris JR, Sibbett GS (1995) Modern fruit science, 10th edn. Norman F. Childers, Gainesville
Croft BA, Hoyt SC (1983) Integrated management of insect pests of pome and stone fruits. Wiley, New York
Day PR (1987) Crop movement: constraints and challenges. In: British crop protection conference on weeds. BCPC Publication Survey, Pulborough, pp 3–12
Diamond J (1997) Guns, germs, and steel: the fates of human societies. W.W. Norton & Co., New York
Downing C (1872) Downing's encyclopaedia of fruits and fruit trees of America: part I and part II. Wiley, New York
Duke JA (2001) CRC handbook of nuts. CRC Press, Boca Raton
Duke JA, DuCellier JL (1993) CRC handbook of alternative cash crops. CRC Press, Boca Raton
Ebeling W (1950) Subtropical entomology. Lithotype Process, San Francisco, 747 pp
Entwistle P (1972) Pests of cocoa. Longmans, London, 779 pp
Evans LT (1998) Feeding the ten billion: plants and population growth. Cambridge University Press, Cambridge
Facciola S (1990) Cornucopia: a source book of edible plants. Kampong Publications, Vista
Faegri K, van der Pijl L (1979) The principles of pollination ecology. Pergamon Press, Oxford, p 244

FAO (Food and Agriculture Organization) (2001) FAO statistical data base. http://appsl.fao.org/servlet/Xte.FAOStat Data Base
Faust M (1989a) Physiology of temperate zone fruit trees. Wiley, New York
Faust M (1989b) Physiology of temperate zone fruit trees. Wiley, New York
Flint ML (1998) Pests of the garden and small farm: a grower's guide to using less pesticide, 2nd edn. University of California Division of Agriculture and Natatural Resources, Oakland, Publication 3332. 276 pages
Free JB (1993) Insect pollination of crops, 2nd edn. Academic Press, London, p 684
Galletta GJ, Himelrick DG (eds) (1990) Small fruit crop management. Prentice-Hall, Eglewood Cliffs
Garner RJ (1988) The grafter's handbook, 5th edn. Cassell Publication, London
Garner RJ, Chaudhri SA (1976) The propagation of tropical fruit trees. Horticultural review No. 4, CAB International, Farnham Royal
Gilman EF (1997) An illustrated guide to pruning. Delmar Publication, Albany
Gold CS, Okech SH, Noko ES (2002) Evaluation of the pseudostem trapping as a control measure against banana weevil, *Cosmopolites sordidus* (Coleoptera: Curculionidae). Bull Entomol Res 92:35–44
Goodell G (1984) Challenges to international pest management research and extension in the third world: do we really want IPM to work? Am Entomol 30:18–26
Gottsberger G (1989) Beetle pollination and flowering rhythm of *Annona* spp. (Annonaceae) in Brazil. Plant Syst Evol 167:165–187
Gundia M et al (1995) The use of bumble bees, Bombus terrestirs, for pollination of green house strawberry. Hassadeh 75(9):49–50, 71–73
Gupta JK, Sharma PL (1998) Impact of number of bee visits on cauliflower pollination. In: Matsuka M, Verma LR, Wongsiri S, Shrestha KK, Pratap U (eds) Asian bees and beekeeping: progress of research and development. Apimondia Publication, Nepal, pp 183–185
Hansen JD, Armstrong JW (1990) The failure of field sanitation to reduce infestation by the mango nut weevil, *Cryptorhynchus mangiferae* (F.) (Coleoptera: Curculionidae), in Hawaii. Trop Pest Manag 36:359–361
Hansen RG, Wyse BW, Sorenson AN (1979) Nutritional quality index of foods. AVI Publication, Westport
Hare DJ (1994) Sampling arthropod pests in citrus. In: Pedigo L, Buntin D (eds) Handbook of sampling methods for arthropods in agriculture. CRC Press, Boca Raton, pp 417–432
Harris James G, Harris MW (1994) Plant identification terminology: an illustrated glossary. Spring Lake Publication, Spring Lake
Hartmann HT, Kester DE, Davies FT (1990) Plant Propagation: principles and practices, 5th edn. Prentice Hall Publication, Englewood Cliffs
Hedrick UP (1922) Cyclopedia of hardy fruits. MacMillan, New York
Helle W, Sabelis MW (1985a) Spidermites, their biology, natural enemies and control, vol 1A. Elsevier, Amsterdam, 405 pp
Helle W, Sabelis MW (1985b) Spidermites, their biology, natural enemies and control, vol 1B. Elsevier, Amsterdam, p 458
Hendrichs J (1996) Action programs against fruit flies of economic importance: session overview. In: McPheron B, Steck G (eds) Fruit fly pests: a world assessment of their biology and management. St Lucie Press, Delray Beach, pp 513–519
Hill D (1975) Agriculture pests of the tropics and their control. Cambridge University Press, New York, 516 pp
Hofmann S, Haufe M (1995) Effect of pollination by honey bees on yield and fruit quality of strawberry (*Fragaria x ananassa* Duch.). Erwerbsobstbau 37(5)
Houbaert D, Borremans G, Baetc W, Jacobs FJ (1992) Quality improvement of strawberries through optimal pollination. FruitteeLt 5(20):26–29
Hoyt SC, Burts EC (1974) Integrated control of fruit pests. Annu Rev Entomol 19:231–252
Indian horticulture data base (2009) National Horticulture Board, Ministry of Agriculture, Government of India, 85, Institutional Area, Sector-18, Gurgaon-122 015 INDIA
Jackson DI (1986) Temperate and subtropical fruit production. Butterworths of New Zealand, Wellington

Jackson D, Looney N, Morley-Bunker M (2007) The distribution of fruits. In: Jackson D, Looney N, Morley-Bunker M (eds) Temperate and subtropical fruit production. CABI Publishing, Netherlands, pp 3–10, 317 pp

Jain HK (2004) Scientific transformation of India horticulture: some strategic consideration. In: Chadha KL, Singh SK, Prasad KV, Som Dutt (eds) Souvenir first Indian horticulture congress 2004. Improving productivity, quality, post harvest management and trade in horticultural crops. HSI, New Delhi, pp 1–4

Jeppson LR, Keifer H, Baker E (1975) Mites injurious to economic plants. University of California Press, Berkeley, 614 pp

Johansen CA (1977) Pesticides and pollinators. Ann Rev Entomol 22:172–199

Keenan M (1997) Foreword. In: Smith D, Beattie G, Broadley R (eds) Citrus pests and their natural enemies. Department of Primary Industries, QI97030, Brisbane, 272 pp.

Klassen W (2000) Area-wide approaches to insect pest management: history and lessons. In: Tan KH (ed) *Proceedings of international conference in area-wide control of insect pests*, Penang, Malaysia, Penerbit Universiti Sains Malaysia, 28 May–2 June 1998, pp 21–38

Kremen C, Williams NM, Thorp RW (2002) Crop pollination from native bees at risk from agricultural intensification. Proc Natl Acad Sci U S A 99:16812–16816

Lamb KP (1974) Economic entomology in the tropics. Academic Press, New York, 1995 pp

Lapedes DN (ed) (1977) Mcgraw-Hill encyclopedia of food, agriculture, and nutrition. McGraw-Hill, New York

Lawrence GHM (1951) Taxonomy of vascular plants. MacMillan Publication, New York

Lewis Walter H, Elvin-Lewis Memory PF (1977) Medical botany: plants affecting man's health. Wiley, New York

Lind K, Lafer G, Schloffer K, Innerhofer G, Meister H (2003) Organic fruit growing. CABI Publication, Wallingford

Lindquist DA (2000) Pest management strategies: area-wide and conventional. In: Tan KH (ed) Proceedings of international conference in areawide control of insect pests, Penang, Penerbit Universiti Sains Malaysia, 28 May–2 June 1998, pp 13–19

Malavasi A, Rohwer G, Campbell DS (1994) Fruit fly free areas: strategies to develop them. In: Calkins CO, Klassen WA, Liedo P (eds) Fruit flies and the sterile insect technique. CRC Press, Boca Raton, pp 165–180

Mariu D (1999) Integrated pest management of tropical perennial crops. CIRAD and Science Publishers, France and Enfield, p 167

Masoodi AM (2003) Agriculture in Jammu and Kashmir. Mohisraw Book Series, Kashmir, 275 pp

Mayer DF (2003) Bee kills from pesticides and expert witness. Am Bee J 143(7):563

Mayer DF, Johansen CA (1983) Occurance of honey bee (Hymenoptera: Apidae) poisoning in eastern Washington. Environ Entomol 12(2):317–320

McGregor SE (1976) Insect pollination of cultivated crop plants. Agric. Handbook No. 496, USDA, Washington, p 411

Mitra SK (ed) (1997) Postharvest physiology and storage of tropical and subtropical fruits. CABI, Wallingford

Monselise SP (ed) (1986) CRC handbook of fruit set and development. CRC Press, Boca Raton

Moore JN, Ballington JR (eds) (1995) Genetic resources of temperate fruit and nut crops. *Acta Horticulturae* 290 1&2

Morton JF (1987) Fruits of warm climates. Julia F. Morton, Miami

Nyeki JA, Soltesz M (eds) (1996) Floral biology of temperate zone fruit trees and small fruits. Akademiai Kiado, Budapest

Ogawa JM, English H (1991) Diseases of temperate zone tree fruit and nut crops. University of California Division of Agriculture and Natatural Resources, Oakland, Publication 3345. 461 pages

Parker BL, Skinner M, Lewis T (1995) Thrips biology and management. Plenum Press, New York, 636 pp

Pedigo L (1994) Introduction to sampling arthropod populations. In: Pedigo L, Buntin D (eds) Handbook of sampling methods for arthropods in agriculture. CRS Press, Boca Raton, pp 1–14

Pena JE, Sharp JL, Wyoski M (2002) Tropical fruit pests and pollinators. CABI, Wallingford

Pimentel D (ed) (1981) CRC handbook of pest management in agriculture, vol 1. CRC Press, Boca Raton, 597 pp
Ploetz RC (ed) (2003) Diseases of tropical fruit crops. CABI, Wallingford
Popenoe W (1927) Manual of tropical and subtropical fruits. MacMillan, New York
Rana BS, Goyal NP (1991) Field toxicity of methyl demeton and dimethoate to the foragers of honeybee (*Apis cerana indica* F.) visitors to *Brassica chinensis* L. Indian Bee J 53(1–4):73–77
Real L (1983) Introduction. In: Real L (ed) Pollination biology. Academic Press, New York, pp 1–5
Rieger M (2006) Introduction to fruit crops. Food products Press, London Oxford
Ritchason J (1995) The little herb encyclopedia: the handbook of nature's remedies for a healthier life. Woodland Health Books, Pleasant Grove
Rom Roy C, Carlson RF (eds) (1987) Rootstocks for fruit crops. Wiley Interscience, New York
Rosen D (1990a) Armored scale insects: their biology, natural enemies and control, vol 1A. Elsevier, Amsterdam, 384 pp
Rosen D (1990b) Armored scale insects: their biology, natural enemies and control, vol 1A. Elsevier, Amsterdam, 687 pp
Ryugo K (1988) Fruit culture. Wiley, New York
Salunkhe DK, Kadam SS (eds) (1995) Handbook of fruit science and technology. Marcel Dekker, New York
Samson JA (1986) Tropical fruits, 2nd edn. Longman Science and Technical, Essex
Schaffer B, Andersen PC (eds) (1994) Handbook of environmental physiology of fruit crops, vol 2. CRC Press, Boca Raton
Schneider E (1986) Uncommon fruits and vegetables: a common sense guide. Harper & Row, New York
Sedgley MA, Griffin AP (1989) Sexual reproduction of tree crops. Academic press, London
Sharp JL, Heather NW (2002) Quarantine treatments for pests of tropical fruits. In: Peña JE, Sharp JL, Wysoki M (eds) Tropical fruit pests and pollinators biology, economic importance, natural enemies and control. CABI Publishing, Wallingford, pp 391–406
Sihag RC (1986) Insect pollination increases seed production in cruciferous and umbelliferous crops. J Apic Res 25(2):121–126
Sihag RC (1997) Pollination biology: basic and applied principles. Rajendra Scientific Publishers, Hisar, p 215
Singh Y (1979) Pollination activity on strawberry at Jeolikot, District, Nainital, India. Indian Bee J 41(1/2):17–19
Smirle MJ (1990) The influence of dextofying enzymes in insecticide tolerance in honeybee (Hymenoptera: Apidae). J Econ Entomol 3(3):715
Smith D, Beattie G, Broadley R (1997) Citrus pests and their natural enemies: integrated pest management in Australia. Queensland, Department of Primary Industries, Brisbane, 272 pp
Snowdon AL (1990) A color atlas of post-harvest diseases and disorders of fruits and vegetables. Vols. 1: general introduction and fruits. Wolfe Science Publication, London
Soule J (1985) Glossary for horticultural crops. Wiley, New York
SP-IPM (2010) Integrated pest management and crop health – bringing together sustainable agroecosystems and people's health. White Paper, SP-IPM Secretariat, International Institute of Tropical Agriculture (IITA), Ibadan, 17 pp
Swaine G, Ironside DA, Yarrow WH (1985) Insect pests of tropical fruit and vegetables. Queensland Department of Primary Industries, Information Series QI83021, Brisbane, 350 pp
Tandon PL (1998) Management of insect pests in tropical fruit crops. In: Akora RK, Ramanatha Rao V (eds) Tropical fruits in Asia, diversity, maintenance, conservation and use. Proceedings of the IPGRI-ICAR-UTANET Regional Training Courses, 18–31 May 1997, Bangalore, pp 235–245
Teskey BJE, Shoemaker JS (1978) Tree fruit production, 3rd edn. AVI Publication, Westport
The National Climate Data Center in Asheville (2014) NC provides detailed weather data on for North America. http://www.ncdc.noaa.gov/oa/ncdc.html
Thompson AK (2003) Fruit and vegetables: harvesting, handling, and storage. Blackwell, Oxford
Tukey HB (1964) Dwarfed fruit trees. Cornell University Press, Ithaca

Upshall WH (ed) (1976) History of fruit growing and handling in the United States and Canada 1860–1972. Regatta City Press, Kelowna

US Census Bureau (2001) Statistical abstract of the United States. http://www.census.gov/prod/www/statistical-abstract-us.html

Valerio C (1984) Insect visitors to the inflorescence of the aroid *Diffenbachia oerstedii* (Araceae) in Costa Rica. Brenesia 22:139–146

Verheij EW, Coronel RE (1992) Plant resources of South-East Asia. Edible fruits and nuts. Prosea Foundation, Bogor, 446 pp

Westwood MN (1993) Temperate zone pomology, 3rd edn. Timber Press, Portland

Whealy KA, Demuth S (eds) (1993) Fruit, berry, and nut inventory, 2nd edn. Seed Saver Publication, Decorah

White F, Elson-Harris M (1994) Fruit flies of economic significance: their identification and bionomics. CAB International, Wallingford, 601 pp

Wiersema JH, Leon B (1999) World economic plants: a standard reference. CRC Press, Boca Raton

Young A (1982) Population biology of tropical insects. Plenum Press, New York, 511 pp

Chapter 2
Pollination and Fruit Productivity

Asia has great diversity of tropical fruits than other continents. Asia has 500 different types of tropical fruit species. Major Asian countries producing tropical fruits include India, China, the Philippines, Indonesia, Thailand, and Pakistan. Indian subcontinent has over 300 species whereas Africa has about 1200. Tropical fruit production in Africa is mostly concentrated in Nigeria, Burundi, South Africa, Congo, Cameroon, Egypt, and Angola. India is the largest producer of tropical fruits followed by Brazil, China, Mexico, and the Philippines (FAO 2001). The most common tropical fruit producing areas are Central and South America (papaya, avocado, pineapple, guava), Asia (most citrus fruits, litchi), and South and South-east Asia (banana, mango, mangosteen, durian) (Gepts 2008). Intercontinental migrations have been responsible for spreading tropical fruits throughout the world which adapted differently depending upon different environments. For example mango which is native of Indo-Burma region has spread to South-east Asia to the east coast of Africa. The orange was moved, most likely by Arab traders, to the Mediterranean and southern Europe. Similarly, the Portuguese spread tropical fruits from their colony in Brazil around the Cape of Good Hope to Goa in India, Malacca in Malaysia, China and Japan (Table 2.1).

Fruit crops benefit greatly from insect pollination. There is an appreciable increase in fruit productivity and improvement in quality in self fertile, self incompatible and highly cross pollinated crops grown under different agroclimatic conditions. However, the pollinator communities are reported to vary across different environmental gradients and bio-geographic regions. All the flower visiting insects do not qualify to be the efficient pollinators as harmony is needed between floral visitors and the flowers. Different pollinating insects have been reported in different environments and different fruit crops (Verma and Chauhan 1985; Kumar 1988; Kumar et al. 1989; Mishra et al. 1976; Mattu et al. 1994). It is therefore essential to know which pollinator is most efficient (Sharma et al. 1974),based on its capacity to carry pollen, hairy body, floral fidelity, sufficient population and thoroughness in working. On the other hand, pollination requirements of different fruit crops/varieties must be evaluated. Entomophilous flowers

D.P. Abrol, *Pollination Biology, Vol. 1*, DOI 10.1007/978-3-319-21085-8_2

Table 2.1 World production of major tropical fruits from FAO Statistics Division (FAO 2009)

Fruit	Important producing countries
Avocado	Mexico, United States, Dominican Republic, Brazil, Colombia, Chile, South Africa, Indonesia, Israel, Spain
Banana Dessert	Burundi, Nigeria, Costa Rica, Mexico, Colombia, Ecuador, Brazil, India, Indonesia, Philippines, Papua New Guinea, Spain, Central America
Banana Plantain	Colombia, Ecuador, Peru, Venezuela, Ivory Coast, Cameroon, Sri Lanka, Myanmar
Citrus Oranges	Brazil, United States, India, Mexico, Spain, China, Italy, Egypt, Pakistan, Greece, South Africa
Citrus Tangerines and Mandarins	Brazil, United States, India, Mexico, Spain, China, Italy, Egypt, Pakistan, Greece, South Africa, Japan
Coconut	Indonesia, Philippines, India, Sri Lanka, Brazil, Thailand, Mexico, Vietnam, Malaysia, Papua New Guinea
Mango	India, Pakistan, Indonesia, Philippines, Thailand, Mexico, Haiti, Brazil, Nigeria
Papaya	Nigeria, Mexico, Brazil, China, India, Indonesia, Thailand, Sri Lanka
Pineapple	Philippines, Thailand, India, Indonesia, China, Brazil, United States, Mexico, Nigeria, Vietnam

producing nectar and pollen are variously coloured with scents to attract insect visitors. Cross pollination by insects may be a pre-requiste for fertilization (e.g. self-incompatible species) or may improve the quality opf the progeny (Free 1970). Adequate pollination may help to increase fruit size as in case of cucumbers or uniformity in fruit set (Williams et al. 1987).

2.1 Pollination Requirements

Pollination requirements vary for different fruit crops. If the flowers are not fertilized soon, they may drop early or inadequate pollination may lead to malformed or misshapen fruits. Even self pollinated varieties produce more and better quality fruit when cross pollinated than when self-pollinated. William (1970) has studied the effective pollination period of different fruit crops where the fertility of flower is more and early pollination could ensure proper pollen growth tube and maximum fruit set. Free (1993) has listed pollination requirements of different fruit crops based on their mode of reproduction. Most of the fruit varieties such apple (*Malus sylvestris* Mill.), almond (*Prunus amygdalus* Batsch), plum (*Prunus domestica* L.), pear (*Pyrus communis* L.), carambola (*Averrhoa carambola* L.), almost all kinds of berries, sweet cherry (*Prunus avium* L.) and sour cherry (*Prunus cerasus* L.), passion fruit (*Possiflora* spp.), are self-incompatible. Some cultivars of citrus (*Citrus* spp.) and grape (*Vitis vinifera* L.), and many cultivars of apricot (*Prunus armenica* L.). Guava (*Psidium guajava* L.) is partially incompatible. Jujube (*Zizyphas jujuba* Mill. and *Z. mauritiana* Lamk.); are examples of protandry (Free 1993; McGregor 1976).

2.2 Honeybees and Fruit Pollination

Honeybees and the flowering plants are mutually interrelated in evolution as reciprococally selective factors. Both benefit from each other, the bees get nectar and pollen and the plants get the pollen transferred between reproductive structures to initiate the process of seed setting. This mutual partnership ensures the reproductive success for the plants and food supply for the pollen vectors. Any deviation in this relationship is counter productive for both. This balance is at risk due to indiscriminate use of agrochemicals such as herbicides and insecticides used for protecting crops against insect pests. Even the feral populations have declined tremendously in recent years thereby risking the food security by insufficient pollination. The problem is more serious for horticultural crops which depend exclusively on honeybee colonies to meet their pollination needs (McGregor 1976). The practice of renting honeybee colonies for reported increased fruit yields through pollination in fruit orchards began in 1990s which is being followed at present (Todd and McGregor 1960).

Among all the pollinating insects, honeybees occupy a prime position as pollinating agents, because they be managed in artificial nests i.e. hives and can be placed in desired numbers whenever and wherever required. Foraging activities of honeybees are under the control of physical environment which comprises of several weather factors such as temperature, relative humidity, wind velocity and solar radiation which determines the initiation and cessation of foraging activities of honeybees. Foraging populations are directly related with the prevailing environmental conditions (Farrar 1931; Gooderham 1950).

Several factors influence the foraging activities of honeybee colonies. The plants with attractive flower colours and distinctive odours are more likely to be visited by honeybees than the others. Similarly, nectar and pollen availability can determine the foraging strength of honeybee foragers on different crops. If the nectar pollen supply is readily available, the foragers will continue to visit and recruit more and more foragers. Thus foraging strengths depend upon weather factors and availability of floral rewards.

2.3 Influence of Weather Factors on Foraging Activities

Weather plays an important role in determining the pollination process of plants. Weather affects foraging activities of bees as well as influence physiological processes of the plants. Foraging activities are hindered by cool temperature and wind velocity. Placing the honeybee colonies in sunny locations and protecting from wind could increase the foraging activities. In situations where no windbreak is available, provision of temporary wind breaks should be made. The foraging activities of honeybees are influenced by several factors which include temperature, relative humidity, light intensity, solar radiation, time of the day, dew drops on flowers, availability of nectar, pollen and concentration of sugars in nectar (Abrol 2007).

Each bee has a specific ecological threshold for commencement and cessation of foraging activity which differ inter and intra specifically depending upon the level of adaptation of a given species in an environment (Burrill and Dietz 1981; Abrol and Kapil 1986). Burrill and Dietz (1981) reported that the bee activity increased with temperature but was not effected by relative humidity and vapour pressure. Nunez (1977) found that in case of *Apis mellifera*, morning activity was related to nectar flow and in the evening it was correlated with the photoperiod. Iwama (1977) found that the interaction between temperature and light intensity was responsible for the flight activity of *Tetragonisca angustica*. Abrol and Kapil (1986) found that light intensity and solar radiations were important factors controlling flight activity of *Megachile lanata*.

Some of the investigators have placed more premium on temperature and solar radiation for initiation of flight activities of honeybees (Brittain 1933; Butler and Finney 1942; Burrill and Dietz 1981). They reported that honeybees donot fly below 9 °C. Burrill and Dietz 1981 reported that flight and temperature were linearly related in studies conducted between 14 °C and 22 °C. They further indicated that light is another limiting factor for flight activities of honeybees as the bees did not fly at low light intensities even when the suitable temperatures for flight were available. Phillips (1930a, b) also stated that no doubt some bees fly on cloudy overcast days, yet they prefer to remain indoor and close to the hive. Flight activity was negatively related to humidity and wind velocity (Brittain 1933; Szabo 1980; Williams and Sims 1977). Brittain (1933) indicated that forgaing activities of honeybee on apple declined when wind speed exceeded 7 mph. Heinrich (1996) found that honeybees can forage over 30 °C as they can regulate their body temperature using behavioural and physiological mechanisms. Gary (1999) concluded that European honeybee *Apis mellifera* initiated foraging activities at 13 °C and continued up to 43 °C for nectar and pollen collection. Thereafter, foraging activities continued for water collection only. Abrol (2010) studied the foraging behavior in relation to five environmental parameters; it was concluded that the foraging population correlated significantly and positively with air temperature, light intensity, solar radiation and nectar-sugar concentration and negatively with relative humidity.

Furthermore, colony's strength must be considered as a variable to determine the foraging populations in the field. 'During unfavorable weather, a lower percentage of the foraging population of larger colonies leaves the hive compared to smaller colonies (Taranov 1952; Free and Preece 1969)'.

2.4 Influence of Floral Attractants on Foraging Behaviour

Pollinators during foraging select the flowers depending upon several cues which may include colour of the flower, odour and nutritional value of nectar (Harborne 1982). The first stimulus the flower visiting insects receive is odour emanating from flowers from a considerable distance. The bees can detect odours with their antennae and newly recruited foragers during their first foraging visits rely solely on odours (Brett and Sullivan 1972; Wells and Wenner 1971, 1973; Wenner 1974) and are unable to locate food source that is uscented (Wenner et al. 1969; Wells and Wenner 1971; Friesen 1973) but once the pattern of foraging is established, then the bees can use visual landmarks to remember the location (Levin 1966; Wells and Wells 1985).

Once the pollinator is attracted to flower, the next cue directing them to nectar and pollen is colour of the flower. von Frisch (1950) has extensively studied on the colour preferences of bees. Bees can discriminate smallest differences in flower colours and are sensitive to the intensely UV-absorbing flavones and flavonols (Kevan and Baker 1983), however, insensitive to red hues, though they continue to visit red flowers if petals have UV-absorbing flavones (Harborne 1982; Kevan 1983). Experiments using artificial flowers have shown that when visual cues are limited, the bees may rely on odour cues (Wells et al. 1981, 1983; Wells and Wells 1985). Marking on petals also serve as cue to pollinators. Flowers with broken outlines are more stimulating than those with smooth outlines (Kevan and Baker 1983). Once the flower visited has alighted on the flower, nectar guides further direct the visitor to reach to nectar (Manning 1956; Free 1970; Jones and Buchmann 1974).

The availability of floral rewards determine the visiting pattern of insects on flowers. Nectar and pollen is required to meet the nutritional needs of the colony. Nectar containing sugar is the source of energy. It has mainly three sugars glucose, fructose and sucrose and some other sugars may also be present in small quantities (Baker and Baker 1983). Furthermore, based on dominance of sugars in nectar, they may be sucrose dominant (containing primarily glucose and fructose) or nectars containg equal amounts of glucose, fructose and sucrose. Honey bees prefer nectars containing sucrose to those with either glucose or fructose (Percival 1961). Wykes (1952a, b) reported that bees prefer equi-proportioned sugars characterized by approximately equal amounts of glucose, fructose, and sucrose. Whereas, Waller (1972) stated that honeybees prefer nectars containing 30–50 % sugars than those containing higher or lower than that concentrations. Later, Bachman and Waller (1977) reported that bees preferred nectars with dominant sugars than containing equal parts of sucrose, glucose, and fructose.

Amino acids, in nectar may affect plant–pollinator interactions and thereby exhibit a selective pressure on the flora in the honey bee habitat (Baker and Baker 1973a, b). Nectar secretion threshold varies among different plant species and even different cultivars of the same species (Beutler 1953). Wells et al. (1981) reported that mean yield of nectar per flower was negatively related to bee visits as all the flowers may not be rewarding.

Environmental factors have a pronounced impact on the amount of nectar and concentration of sugar in it. In case of open dish flowers, variations in nectar concentration are more depending upon the changes in humidity levels (Vansell 1934, 1942; Corbet et al. 1979a, b). The nectars may also become dilute and less attractive to bees if there is dew or rainfall. However, dry winds may result in more concentrated nectars increasing the attractiveness of bees (Kevan and Baker 1983). Shuel (1955a) reported that nectar secretion was more on sunny days compared to cloudy ones indicating that nectar sugars are direct product of photosynthesis.

Besides, soil moisture, composition of atomospheric pressure, size of nectar, age of flower, available carbohydrates and position of flowers on the plant also influence the amount of nectar secretion (Butler 1945a; Percival 1946; Wykes 1952c; Beutler 1953; Shuel 1955b, 1957; Free 1970; Loper et al. 1976). It is clearly indicated that nectar secretion is a complex process and the amount of nectar secreted depends upon biotic conditions, weather parameters and edaphic factors (Beutler 1953; Cruden and Hermann 1983).

Nectar being a rich source of carbohydrates provides energy to bees for various metabolic activities whereas pollen primarily meets the protein requirements of bees. Pollen also contains different types of other components such as starch, sugars, fatty acids, and trace amounts of inorganic salts (Todd and Bretherick 1942; Harborne 1982). Detailed accounts of pollen chemistry is available in Barbier (1970) or Stanley and Linskens (1974).

Anther dehiscence and shedding of pollen is also species specific and depends upon the prevailing weather conditions (Free 1970). For example, apple flowers require a minimum threshold of 10 °C for dehiscence (Percival 1955). Pollen collection by honeybees is also weather dependent. The pollen collection activities are more intense during days flooeing unfavourable weather conditions. This is partly due to the reason that pollen demands of the colonies increase following days of confinement (Free 1970). It also depends on the colony needs. If the colonies have sufficient eggs and larvae and an active queen, the need for pollen collection shall be more (Free 1967; Todd and Reed 1970; Al-Tikrity et al. 1972; Moeller 1977). Increased pollen foragers make colonies more valuable for pollinating crops where pollen collectors have a greater chance of contacting stigmata than nectar collectors (Roberts 1945; Free 1966c; Robinson 1979; Thorp 1979). The percentage of pollen collectors in the foraging population can also be increased by installing pollen traps that remove pollen loads from the legs of foragers as they enter the colony (Webster et al. 1985). Hence, pollen traps might increase the pollinating efficiency of colonies, but continuous pollen trapping can drastically reduce brood rearing (Moeller 1977).

Foraging range of different bee species varies depending upon their body size. *Apis cerana* has smaller foraging range (about 1 km) than *Apis mellifera* (3–4 km) (Goyal 1978; Punchihewa et al. 1985). The small foraging range of *A. cerana* in comparison to *A. mellifera* may be an additional advantage to breeders and seed growers because the foragers would tend to restrict themselves to smaller radii and, if desired, the breeders could establish the isolation yards for genetic purity of a cultivar easily by keeping *A. cerana* as a pollinator. Honeybees and bumblebees have large foraging distances and are known to forage over 10 km and occasionally may forage 20 km from the nests (Seeley 1985; Schwarz and Hurst 1997). The estimates of foraging distances of other bees are relatively less known but suugest that their foraging distances are generally few hundred metres to less than 1 or 2 km (Roubik and Aluja 1983; Roubik et al. 1986; Schwarz and Hurst 1997). Bees normally prefer to forage nearby resources but may move to long distances if resources are not available closer to the colony. Generally, the individual honeybee are floral constant (Grant 1950; Free 1963, 1970; Waddington and Holden 1979; Waddington and Heinrich 1981; Waddington 1983) but the colony may distribute their foragers among several plant species depending upon needs of the colony (Gary 1979).

Honeybees have a highly advanced communication system to locate food sources through various dances (von Frisch 1967, 1974; Lindauer 1971; Gould 1975, 1976; Gould et al. 1970). Abundance of nectar resources and concentration of sugars in nectar determines the size of foraging population on particular plant species (Butler 1945a, b; Visscher and Seeley 1982). As the quantity of nectar resources decline, the number of foragers that dance declines considerably (von Frisch 1967). Foragers that discover resources with high sugar concentration dance vigoursly and recruit new foragers more successfully (Lindauer 1948). Consequently, honeybee colonies adjust their allocation of foragers depending upon the availability of rewarding resources (Visscher and Seeley 1982).

2.5 Pollination and the Production of Horticultural Crops

Pollination requirements of fruit crops vary depending upon the reproductive system of each crop. Some of the crops are highly self-incompatible whereas others can set fruit with their own pollen. Detailed information is available in reviews by Free (1970), McGregor (1976), Crane and Walker (1984) and DeGrandi-Hoffman (1987). All these studies have highlighted the importance of honeybees in fruit crops pollination. Honeybee pollination not only affects the yield but improves the quality as well. Most of the crops require maximum ovules to be fertilized if not all for optimum fruit size and shape as in case of apples, berries and watermelon. Honeybee pollination also ensures uniform fruit setting with increased yield and high quality of produce if sufficient number of bees and pollen is available. DeGrandi-Hoffman (1987) listed number of fruit crops which depend or benefit from bee pollination which include: Kiwifruit (Chinese gooseberry), Onion, Cashew, Mango, Papaya, Brüssel sprouts, Rabbiteye blueberry Lowbush blueberry Highbush blueberry Sour-top bilberry Cranberry Guava Coconut Macadamia Strawberry Crab apple (dark crimson petals) Apple Crab apple (pink petals) Sweet cherry Plum (myrobolan or cherry) Tart cherry European plum Pear Almond Raspberries Coffee Lemon Sweet orange Grapefruit Litchi (lychee) and Tangerine (Tables 2.2, 2.3, and 2.4).

2.6 A Deciduous Fruit Crops

Most of the deciduous fruit crops exceptions to those which set fruit parthenocarpically depend upon insect pollination for production of seeds and fruit development. Almost all commercial cultivars of apple, plum and sweet cherry are generally self-incompatible and require cross-pollination. Though some of the cultivars of apple and peach are self-incompatible, yet they require bees to transfer pollen from anthers to stigma.

Table 2.2 Shows pollination requirements of crops, their dependence on animal pollinators, and the principal pollinators of major crops

Crop	Degree of cross-pollination	Total flowering period	Peak receptivity period of the stigma to pollen	Dependence on animal pollinators	Dependence ratio on animal pollinators			Honeybees, bumble-bees, halictid bees, *Eristalis* flies	Increase in yield from insect pollination (%)
					Min	Max	mean		
Fresh fruit									
Apple	All commercial varieties require cross-pollination	10–15 days	2–3 days	High	0.4	0.9	0.65	Honeybees, bumble-bees, halictid bees, *Eristalis* flies	180–6950
Apricot	Cross-pollination beneficial; for some cultivars it is essential	15–20 days	4–5 days	High	0.4	0.9	0.65	Honeybees, wild bees	5–10
Cherry	Cross-pollination essential	7–8 days	2 days	High	0.4	0.9	0.65	Honeybees, wild bees	56–1000
Citrus	Varies from self-fertile to self-sterile varieties	1 month	6–8 days	Low	0	0.1	0.05	Honeybees, bumble-bees, wild bees, flies	7–233
Gooseberry/ kiwi fruit	Cross-pollination essential	20–25 days	2–3 days	Essential	0.9	1.0	0.95	Honeybees	29–300
Grape	Generally self-fertile	20–25 days	3 days	0	0	0	0	Honeybees, halictid bees	23–54

Guava	Cross-pollination beneficial	20–25 days	1–2 days	Medium	0.1	0.4	0.25	Honeybees, bumble-bees, wild bees	–
Litchi	Cross-pollination beneficial	25–30 days	3 days	High	0.4	0.9	0.65	Honeybees, flies, ants	4,538–10,240
Mango	Cross-pollination highly beneficial	2–3 weeks	Few hours to 5 days	High	0.4	0.9	0.65	Flies, honeybees, butterflies, moths, beetles	–
Papaya	Cross-pollination essential	1 month	–	Low	0	0.1	0.05	Thrips, honeybees, butterflies, hawkmoths	–
Peach	Most varieties self-fertile; few self-sterile	20–25 days	3 days	High	0.4	0.9	0.65	Honeybees	7–3788
Pear	Partially or entirely self-sterile	7–10 days	4–5 days	High	0.4	0.9	0.65	Honeybees, flies, beetles	240–6014
Persimmon	Mainly self-fertile	25–30 days	3–4 days	High	0.4	0.9	0.65	Honeybees, bumble-bees	21
Plum	Varies from self-fertile to self-sterile varieties	1 week	2 days	High	0.4	0.9	0.65	Honeybees, bumble-bees, blow flies	5–10
Strawberry	Cross-pollination beneficial	30–35 days	3 days	High	0.4	0.9	0.65	Honeybees, wild bees	5–10

Sources: Compiled from Free (1993), Verma and Partap (1993), Klein et al. (2007)

Table 2.3 Floral biology and pollination requirements of important fruit crops

Crop	Extent of cross-pollination	Ratio of pollinizer variety to main variety	Total bloom period (days)	Peak receptivity period of stigma	Nectar or pollen potentials	Chief pollinators	Percent bloom require for good fruit set	Per cent increase in yield due to cross pollination	Pollination requirements (Number of hives/ha)
Almond	Cross pollination essential	One pollinizer row for two rows of main variety	30	3–4	N1P1	Honeybees	90–100	50–75	5–8
Apple	All commercial varieties require cross-pollination	Every third one in every third row should be a pollinizer	9	2–3	NP	Honeybees, Bumblebees, Halictus	50–95	180–6950	2
Apricot	Cross pollination beneficial and for some cultivars is essential	Same as for apple	15–20	4–5	NP	Honeybee	50–95	5–10	2.5
Banana	Pollination not essential	–	–	–	N2P2	Bats, birds	–	–	–
Black current	Cross pollination essential		21	5–6	N2P2	Honeybees	90–100	81–2200	–
Blueberry	Self-sterile to cross-sterile	One block to each variety to alternate with another variety	21	5–8	N2P2	Honeybees, bumblebees, solitary bees, flies	80–100	11–9800	3–12
Cherries	Cross-pollination is essential	One pollinizer for 10 trees cultivar	7–8	First two days after opening of flowers	N1P1	Honeybees	21–32	56–1000	2.5–3.0

Chestnut	Self sterile	1:1 or 1:2	–	45	NP	Honeybees, rose chafers, wild bees and wind	–	–	1–5
Citrus	Vary from self sterile to self fertile form	No specific recommendation	30	6–8	N2P2	Honeybees	80–90	7–233	1–2
Craneberry	Cross-pollination is essential	–	2–3 weeks	3	NP	Honeybees, bumblebees	90–100	9–2153	1hive/2.5 ha.
Grapes	Self sterile to self fertile	Interplanting of cultivars is beneficial	20–25	3	N3P3	Honeybees, other insects and wind	90–100	23–54 %	0.5–1.0
Guava	Self to Cross-pollination	1:2	25–45	4	P3N3	Honeybees, other insects	80–96	10–12	–
Hazel nut	Self sterile	12–20 % interplanting of 3–4 cultivars	–	30–90	P2	Wind	–	–	–
Litchi	Self-fruitful cross-pollination beneficial	–	26–36	3	N1P2	Honeybees, hoverflies, cvalli-phorids, *Musca* sp., wasps	50–90	3	4
Loquat	Self to cross-pollination	–	–	–	N3P	Honeybees and other insects	–	–	–
Mango	Self to cross-pollinated	–	–	5	N3P3	Honeybees, house flies, ants and syrphids	–	–	7–8

(continued)

Table 2.3 (continued)

Crop	Extent of cross-pollination	Ratio of pollinizer variety to main variety	Total bloom period (days)	Peak receptivity period of stigma	Nectar or pollen potentials	Chief pollinators	Percent bloom require for good fruit set	Per cent increase in yield due to cross pollination	Pollination requirements (Number of hives/ha)
Olive	Self sterile to self fertile	Interplanting of 2 cultivars in blocks	–	–	P2	Honeybees and wind	–	–	–
Papaya	Self to cross-pollinated	1:25	–	–	N3P2	Sphinax moth, honeybees, butterflies, humming bird, sunbird	–	–	–
Peach	Mostly self fertile and few self sterile	1:2	20–24	3	N1P1	Honeybees and other hymenop-terans, dipterans and wind	50–90	7–3788	2–3
Pears	Partly or entirely self sterile	Bartlett variety be interplanted with other varieties	7	4–5	N1P1	Honeybee, flies, beetles	–	–	–
persimmon	Cross-pollination not necessary	–	25–30	3–4	N3P3	Honeybees, bumblebees, wind	75	21	–

Pistachio	Cross-pollination	1: 8, males,female trees	–	–	P2	Wind, Honeybees			
Plum and prune	Vary from self-compatible to self-incompatible	Every fourth tree in very fourth row	5	Become receptive two days before the anthers dehisce	N1P1	Honeybees, wild bees, blow flies, bumblebees	50–90	5–10	2–3
Pomegranate	Self and Cross-pollination	–	–	–	N3P2	Honeybees, beetle and wind	–	–	–
Raspberry	Cross-pollination beneficial	–	3–6 weeks	2–3	N2P2	Honeybees	90–100	291–463	1–2
Sapota	Self sterile	–	–	–	–	Wind, Honeybees and other insects	–	–	–
Strawberry	Cross-pollination beneficial	No specific recommendation	45–60	3	N2P2	Honeybees, wild bees	90–100	5–10	1–10, 25 or even more

Source: Verma (1990)

N1 Major honey source; *N2* Medium honey source; *N3* Minor honey source; *P1* Major pollen source; *P2* Medium pollen source; *P3* Minor pollen source

Table 2.4 Increase in yield due to bee pollination (Source: Partap 1999a, b)

Crop	Increase (%)
Mango	3–5
Papaya	5–10
Pear	10–15
Plum	10–15
Apple	15–20
Citrus	10–20
Grapes	10–20
Guava	5–10
Litchi	20–25

In case of cross-pollinated fruit cultivars, provision of compatible pollen source (pollinizers) should be made for successful pollination. Pollinizers which are compatible with the main cultivars must bloom annually. The pollinizers should bloom 1–2 days before the main cultivar, however, this pattern may vary in different locations (Denis 1979). Blooming also depend upon prevailing weather conditions, cool temperature generally prolong the blooming whereas warmer temperature accelerates it (Morris 1921). Due to these reasons sometimes, there is an overlap between pollinizer and the main variety.

2.7 Species of Honeybees

There are nine species of honeybees found in Asia which include giant honeybees or rock bees (Apis dorsata and *Apis laboriosa*), the little honeybee (*Apis florea*), the small dark honeybees (*Apis andreniformis*). Among these, four species of honeybees such as *A. dorsata, A. laboriosa, A. florea* and *A. andreniformis* have not yet been domesticate and still exist in wild. Honey from these bees is harvested by traditional honey-hunting methods. Other two species such as indigenous Asiatic honeybee A. *cerana* and European honeybee A. *mellifera*, are domesticated and kept in modern hives for honey production and pollination. Apis dorsata is found throughout the Asian region up to 2000 m. It nests in open on tall trees, buildings and rocks. There may be clustering of more than 100 colonies at a place. Apis laboriosa is mostly found at 1200–3500 m in mountainous areas of Nepal, Bhutan, China and India (Sakagami et al. 1980; Underwood 1986; Batra 1995). *Apis dorsata* colonies are mostly found beneath rock on vertical cliff faces. Like, *A.dorsata*, they also occur in cluster of colonies. Both the species are good pollinators of variety of crops and other plants growing in nature.

Apis florea called dwarf honeybee is the smallest of all honeybee species. It also makes single combs on branches, bushes, hedges, small trees and chimneys. *Apis florea* is normally found in plains and up to 500 m in hilly areas. It is also migratory in nature and another important pollinator of agricultural crops. Sinilarly,

A. Andreniformis occurs at an elevation lower than 1,000 m. Its distribution range is south of Hengduan mountain range of China, Thailand and Malaysia where it builds its nests in small trees or bushes.

The Asian honeybee, A. cerana, and the European honeybee, *A. mellifera* are domesticated and can be kept in hives and managed for honey production and crop pollination. Both these species unlike *A. dorsata, A. laboriosa, A. andreniformis* and *A. florea* that build single comb nests in the open, are cavity nesting bees and makes multiple parallel combs inside a cavity. Recently, more species of *Apis* have been identified in Asia which include Apis koschevnikovi reported from Sabah, Malaysia, *Apis nigrocincta* reported from Philippines and *Apis binghami* reported from Sulawesi islands of Indonesia.

2.8 Role of Honeybees in Crop Pollination

Honeybees have special structural modifications for collecting nectar and pollen, such as pollen baskets and presence of hairs on its body and head which help them to pick up pollen grains. Being social insects, they continue to collect nectar and pollen throughout the year to feed the young ones. They have flower constancy and continue to vist the same plant species for as long as nectar and pollen are available. They have long working hours and adapted to different climates. Their colonies are very large 5000–80,000 individuals depending upon the species and have highly evolved communication system.

Honeybee colonies have been found to increase productivity levels of different crops such as fruits, vegetables, oilseeds, pulses and forage crops. Some of the studies have clearly established their role in improving quality and quantity of crops. Bee pollination was highly beneficial for apple (Dulta and Verma 1987), peach, plum, citrus, kiwi (Gupta et al. 2000) and strawberry (Partap et al. 2000a, b). Besides, there have been reports of reduced fruit drops and misshapen fruits following bee pollination in apple, peach, plum and citrus (Dulta and Verma 1987; Partap et al. 2000a, b). There was an increase in fruit juice and sugar content in citrus fruits following pollination in citrus fruits (Table 2.5).

The number of colonies required for pollination depends upon several factors which include duration of flowering, bee species used, amount of nectar pollen and number of plants where pollination is required (Kozin 1976; Verma 1990; Free 1993). In general, three strong colonies of *A. mellifera* and four to five strong colonies of *A. cerana* per hectare of crop are required (Kozin 1976; Verma 1990; Free 1993). The number of colonies required also varies from crop to crop and from season to season for the same crop. The colonies used for pollination purposes should have large amount of unsealed brood which will require more pollen resulting in receruitment of more pollen foragers. Weeds and other competing plants must be removed to ensure effective pollination.

Table 2.5 Percentage Increase in yield of some crops due to bee pollination (Abrol 2009)

Crop	Increase (%)
Fruit crops	
Apple	18.00–69.50
Almond	50.00–75.00
Apricot	5.00–10.0
Cherries	56.00–1000
Citrus	7.00–223.00
Grapes	23.00–54.00
Guava	12.00–30.00
Litchi	453.00–10,246
Plums	536–1655

2.9 Role of *Apis cerana* in Pollination of Fruit Trees

The association of A. *cerana* with fruit trees is well established. The trees for which clear documentation has been made include almond *Prunus dulcis* (Miller) D.A. Webb (Muttoo 1950; Bhalla et al. 1983a), apple *Malus domestica Borkh* (Sharma 1961; Rai and Gupta 1983; Verma and Dutta 1986), lime *Citrus aurantifotia* (Christm.) Swingle (Anonymous 1981), cranberry *Vaccinium macrocarpon* Alt. (Sharma 1961), litchi *Litchi chinensis* Sonner (Pandey and Yadav 1970; Dhaliwal et al. 1977; Phadke and Naim 1974; Dhoble and Shinde 1982), plum *Prunus domestica* L. (Sharma 1961), peach *Prunus persica* (L) Batsch. (Bhalla et al. 1983b; Kumar et al. 1984), phalsa *Grewia asiatica* L. (Parmar 1976), strawberry Fragaria vesca L. (Singh 1979), Chinese jujube *Ziziphus jujuba* Mill. (Ackerman 1961), durian *Durlo zibethinus* Murray (Crane and Walker 1983), *carambola Averrhoa carambota* L. (Nand 1971), coconut palm *Cocos nucifera* L. (Anonymous 1982), borassus palm *Borassus flabellifer* L. (Seethalakshmi and Percy 1979), and pear *Pyrus communis* L. (Sharma 1961). Muttoo (1950) cited the lack of fruit set in cultivated almonds resulted from failure in pollination, and he pleaded for the use of honeybees for the production of fruits in India. Sharma (1961) reported that 50–78% of the insect visitors to pear and cranberry were honeybees and on plum, apple, cherry, and peach the population of honeybees represented 33, 70, 45 and 63% respectively. A. *cerana* represented the largest number and was essential for fruit set. He further reported that pollinating insects greatly increased fruit set in four varieties of apple (Red-, Golden-, Red-delicious and American mother), pear, plum and cranberry. Persimmon, peach and cherry gave a commercial set even in the absence of pollinating insects but the set was higher on flowers receiving insect visits. Singh and Mishra (1986) observed the populations of bees and flies at different elevations in Himachal Pradesh (India), and flies were found to outnumber A. c. *indica* at all locations on all fruit blossoms. Mishra et al. (1976) reported that visits by honeybees on apple during 1200 and 1500 h were more than all other pollinators and red-delicious had higher fruit set near the bee colonies but that fruit set decreased with distance from the colonies. Verma and Dutta (1986) also compared the foraging behaviour of A. c. *indica* and A. mellifera in pollinating the flowers of apple. Without giving the relative abundance of different visitors, Bhalla et al. (1983b) reported that plum, peach, and almond were visited by honeybees A. C. *indica* along with other insects. Rai and Gupta (1983) also reported the role of honeybees in apple and pear pollination. Strawberry flowers are also visited by honeybees A. c. *indica* (Singh 1979) and among the insect visitors, their number was highest in a study in strawberry by Singh (1979). Pandey and Yadav (1970) studied pollination in litchi. They reported that 98–99 % of the visitors were Apoidea and A. *cerana* constituted 28 % of the total visitors. Phadke and Naim (1974) and Dhaliwal et al. (1977) also reported that A. *cerana* constituted 15 % of insect visitors to litchi blooms. Honeybees were the most important pollinating agents of phalsa and the plants of seedling origin were reported to be benefited by insect pollination (Parmar 1976). Fruit yield in lime was found to be increased from 15 to 17 times by the honeybees in Tamil Nadu, India (Anonymous 1981). Coconut was also found to double the fruit yield when honeybee colonies were moved near the trees. On durian (*Durlo zibethinus Murray*) in Singapore, honeybees collected pollen and A. c. *indica* visited the plants for nectar (Crane and Walker 1983) but their role in pollination is probably minimal.

2.10 Favourable Pollinating Characteristics of *Apis cerana*

In addition to its general pollinating ability, A. cerana possesses several positive characteristics which make it a superior pollinator, even compared to A. *mellifera.* Such characteristics include foraging behaviour, foraging rate, foraging range, flower constancy, and colony strength (Jhajj and Goyal 1979a, b). A. *cerana* foragers show the usual foraging behaviours on crops as is known for A. *mellffera*, i.e. the foragers include only pollen gathers, pollen as well as nectar gatherers, and only nectar gatherers. The different proportions of these categories of foragers vary during the course of the day and flowering span of the crop (unpublished). A. *cerana* has a higher wing-beat frequency (305 ± 16.2 per sec) than does A. *mellifera* (235.2 ± 7.5) (Goyal and Atwal 1977) and its foraging rate is also higher on 8. *juncea* (unpublished). These characteristics should make it a more efficient pollinator, at least in certain categories of crops where smaller numbers of pollen grains are required for pollination and flower size is small, e.g. cruciferous, umbelliferous, and many papilionaceous crops,

The small foraging range of A. *cerana* (about I km) in comparison to A. *mellffera* (3–4 km) (Goyal 1978; Punchihewa et al. 1985) may be an additional advantage to breeders and seed growers. This is because the foragers would tend to restrict themselves to smaller radii where pesticidal operations could be regulated more effectively by the growers and, if desired, the breeders could establish the isolation yards for genetic purity of a cultivar easily by keeping A. *cerana* as a pollinator for self-pollination. This would be especially beneficial in cruciferous crops where contamination of genotypes is common (Adlakha and Dhaliwal 1979). A. *cerana* has been reported to have a high floral fidelity (Chaudhary 1978) and out of 5,600 pollen loads analyzed, only 56 contained pollen from more than one plant species. Dhaliwal and Atwal (1986) reported that only 66.5 % of A. *mellifera* foraging on alfalfa carried pure pollen.

A. *cerana* colonies have smaller strength (not greater than 30,000) than those of A. *mellifera*, which can reach even 60–70,000 (Mishra and Sihag 1987). The larger size of A. *mellifera* colonies presents problems for management for pollination and smaller nuclei or packages are often used. This problem is overcome if A. *cerana* is used as a pollinator of crops. The colonies with smaller strength and small honey stores can be transported with little difficulty. However, for large crop fields, higher numbers of colonies of A. *cerana* would be required. A. *cerana* is also better adapted to higher altitudes as compared to A. *mellifera.*

Thus, it is clear that A. *cerana* is a pollinator of a large number of cultivated crops. Like A. *mellifera*, A. *cerana* can also be easily managed. In spite of its being less of a producer than A. *mellifera*, several of the favourable pollination attributes of A. *cerana* put this bee ahead in terms of its usefulness as a pollinator of crops not only possibly in areas of its natural distribution but also in new ones where it is not indigenous.

2.11 Management of Honey Bees for Pollination

Due to their morphological characters, social behaviour and management practices followed by our farmers, honeybees have become the most important and mostly the only pollinators of the crops. The uncertainty of populations and the difficulty in maintaining and using solitary bees, bumblebees and other pollinators' places all the onus of planned pollination on honeybees alone. The following points are most important to consider a colony for pollen foraging and getting desired pollination outputs.

Foraging Strength of Colonies In order to get maximum honey and pollination benefits from a colony, it must be full-sized and populous not a growing one-as the brood/bee ratio diminishes in it, so a greater proportion of bees is available for foraging. Matheson (1986, 1991) suggested that a colony used for pollination should contain at least 7 frames with 60 % brood (at least 25 % in egg or larval stage) in the brood chamber, a young prolific queen, at least 12 frames covered with bees, sufficient empty combs for expansion and enough honey and pollen stores to sustain it. Separate colony strength standards were also recommended for field colony and orchard colony by Roubik (1995).

Pollination Requirement and Concentration of Colonies The number of colonies needed per hectare of crop will depend on local conditions including the number of honeybees and other pollinators already present, the size of the crop and the presence of competing crops of the same and different species. It is generally recommended to have around 2.5 colonies/ha but it will depend on many factors like concentration of flowers, their attractiveness, competing insects and crops, species, place, percentage of flowers open at a time, etc. and the number of colonies required may be increased or decreased accordingly.

Foraging Efficiency of Colonies and Their Distance from the Crop Honeybees are capable of foraging at an enough honey and pollen stores to sustain it. Considerable distance from their hives but their efficiency is indirectly proportional to the distance covered. Generally foraging range is 2.5 km for *Apis mellifera*, 1.5 km for *A. cerana*, 3 km for *A. dorsata* and 1 km for *A. florea*. *Apis mellifera* have been observed to forage up to 11.3 km but foragers were concentrated within 0.6 km of their hives. The yields of the crops are more when the colonies are kept up to a distance of 0.5 km and decrease to almost half at a distance of 1.0 km and these impacts are even greater in poor season. The number of foraging bees on a crop diminishes with increase in the distance from the hive (Free 1993).

Moving Colonies to Crops It is advisable to take as short migrations as possible because of the risks involved in long distance migration which result in killing of brood and low foraging efficiency. The settling and arrangement of colonies after migration is very important. Formal arrangement with identical lives should be avoided as it may result in drifting of foragers especially during their first flight which may result in weakening of some colonies and strengthening of others, lowering the mean foraging potential, honey production and pollination efficiency. The colonies should be arranged irregularly in different directions and spacing. They should be put near landmarks and windbreaks and different colored boards may be placed above the entrance.

Time of Moving Colonies on Crops The origin of food stores of a colony plays important role in determining the species as they did before moving them. But the predominance of one species at the new site results in changing their foraging behavior. Therefore, colonies should be moved to the crop needing pollination only when it is sufficiently flowering. Care must be taken to avoid the blooming of too many flowers as an important part of the crop may fail to be pollinated. It is practical to move a few colonies to the crop at the beginning of flowering and the rest when more flowers have bloomed (Sale 1983).

Timing for placement of colonies is most important for effective pollination. It should concide with the blooming period of crops. Generally, the colonies should be placed in the orchards when 5–10 % of the crop is in bloom. If the colonies are brought earlier, they may fix their preference to alternate crops and ignore the target crops when it blooms and if late, may pollinate less vigorously resulting in poor yields and low quality of fruits. In case of short flowering period crops such as plum, bees must be moved immediately when flowering just begins as 50 % flowering is achieved within 3–4 days.

2.12 Arrangement of Colonies

Arrangement of colonies in crop is very important to ensure uniform distribution of foraging bees which will depend on the plant species, amount of nectar and pollen available/unit area, weather conditions and physical features of the area like topographic gradients, wind direction, shelter, etc. Placing colonies in the middle of the crop increases foraging area (Free 1993). Ideally colonies should be distributed singly over the crop which is always not feasible and practical. So, colonies should be kept in small groups of 4–5 at distances of about 200 in throughout the crop to provide sufficient overlap of foraging area between different groups. The colonies should face the direction of sun in the winter and should be sheltered during summer and rainy season. Windbreaks (natural and artificial) greatly benefit during chilly and windy conditions (Kumar et al. 1998). Honeybees prefer to forage near to their colonies if nectar and pollen resources are available. Colonies must be placed singly instead of groups and distributed evenly in the field to maximize pollination and good fruit set Gupta et al. (1993)

2.13 Directing Bees to Crops

Attempts to 'direct' the bees to the crops to be pollinated either by feeding them sugar syrup containing the scent of the crop or flowers immersed in the syrup or by spraying the crop with sugar syrup have met with a limited success. Recently the use of "Bee attractants" (Bee-Q and Bee-here) or attractive odors present in the pollen on high value crops met with varied success. Their application though increased forager population but they may be diverted from the flowers and spend much time in searching the plant leaves and sterns for pollen but of no avail (Patil et al. 2000).

2.14 Increasing the Attractiveness of Crops

The more appropriate approach for improved pollination is to increase attractiveness of the crop itself to the bees, by developing plant strains with more accessible nectar or pollen, or developing strains producing more nectar. Unfortunately, this factor is not a priority with the breeders and the recent hybrids (of berseem and sunflower) produce very low amount of nectar, further reducing their attractiveness to the bees (Kumar et al. 1998).

2.15 Increasing the Proportion of Pollen Gatherers

Increasing the proportion of pollen gatherers, which are more efficient pollinators than the nectar gatherers in a colony, is always beneficial. To increase the proportion of pollen gatherers Goodwin et al. (1991a, b) have suggested several ways. The foragers may be stimulated to gather more pollen by manipulating colony by placing brood frames opposite to the hive entrance or by removing pollen stores from the colonies or by using pollen traps. Another method is by feeding sugar syrup to colonies which will increase its pollination efficiency by greatly increasing the number and proportion of pollen-gatherers.

2.16 Using Man-made Devices to Increase Pollination

Devices like "pollen dispensers" or "pollen inserts" have been used in high valued crops like fruit trees. They are placed at the hive entrance containing hand collected pollen of required compatible cultivar. The out-going bees are forced to carry this pollen. In sunflower the male sterile plants caged with a honeybee colony and a dispenser with pollen had similar percentage of flower set, seed yield and seed oil to the male fertile plants. Newer and more effective dispensers have been developed now using bee collected pollen (Sasaki 1985). The simple "pollen enhancer" having row of soft nylon bristles at the entrance developed by Free et al. (1991) too are effective for the pollination of self in compatible tree fruits and hybrid seed production.

2.17 Breeding Honeybees for Pollination

Selection and use of Italian honeybees with long-tongue (6.8 mm) has been greatly reported for the pollination of red clover. Mackensen and Nye (1969), Nye and Mackensen (1970) and Hellmsich et al. (1985) have demonstrated the possibility of selecting honeybee colonies with high (2–15 times) or low levels of pollen collection for crops like alfalfa.

2.18 Use of Synthetic Pheromones

Use of synthetic queen pheromones to stimulate foraging (especially pollen collection) and Nasanov pheromone whose two components (citral and geraniol) have been reported to increase honeybee foraging in onion (Woyke 1981) and yield of apple (Ohe and Praagh 1983) hold a good promise. Use of synthetic alarm pheromones, a few minutes before insecticide application may help in repelling them from the crop thereby reducing bee loss, and some other pheromones need further refinement.

2.19 Diversification of Insect Pollinators

Amongst all the anthophilous insects, Hymenoptera are the only trustworthy pollinators of temperate fruit crops though dipterans also form an important group. The main problem with the latter group is that their pollination activity is irregular and unreliable. Even among Hymenoptera, all are not equally good pollinators and it is only Apoidea which are highly diverse structurally and behaviourally as well as taxonomically (Kevan and Baker 1983). Apoid bees which form a diverse group of Hymenoptera are well adaped for pollination. The solitary bees (*Megachile, Nomia* and *Osmia*) are also valuable pollinators of many crops but their usefulness is restricted because their population fluctuates greatly from year to year. Other bees like *Ceratina*, *Halictus* and *Andrena* are also important pollinators. However, bumble bee *Bombus* is one of the most efficient pollinator because its big size facilitates pollen transfer. Moreover, it can forage at temperature as low as 0oC when all other insect pollinators are immobilized. Bumble bees visit about twice as many as blossoms per unit time as other bees, and it works dawn to dark. All non-*Apis* bees have been named as "pollen bees" in 1992 which include all bees other than honeybees (Batra 1994).

On temperate fruit crops, the domesticated honeybees have been found to constitute high proportion of insect pollinators and there is no doubt that these are the only pollinators which can be relied upon for effective pollination and its population can be managed as per requirements of the crops easily. However, potential dangers have been found whenever one species has been used and relied upon for pollination. Outbreak of Thai sac brood viral disease of *A. cerana* during early eighties of the last century in this country resulted sudden collapse of managed honeybee population when more than 95 % of *A. cerana* colonies perished. Under such compelling circumstances need for diversification and conservation of pollinators is realized. Moreover, in cold desert areas like Spiti valley of Himachal Pradesh where honeybees do not occur in nature, the pollination is being accomplished in highly cross pollinated crop of apple by other native pollinators such as *Nomia* sp., *Halictus* sp., *Ceratina hieroglyphia, Xylocopa* sp., *Bombus* sp., Colletidae, *Prosopis* sp. (Hymenoptera); *Eristalis tenax, Scaeva pyrastis, Eupeodes frequens* and *Syrphus* sp. (Diptera).

2.20 Conservation of Pollinators

Clean and intensive cultivation has destroyed shelter locations and nesting sites of wild pollinators. This has resulted in disappearance of native pollinators in many areas where intensive agriculture was adopted. Indiscriminate application of insecticides has resulted in devastating setbacks to non-target useful fauna mainly pollinators and biocontrol agents. In Himachal Pradesh too during the apple bloom growers are known to spray highly toxic insecticides like methyl parathion which has resulted in loss of natural population of pollinators and now the growers have to make provision of bee colonies for proper pollination. For increasing the number of native pollinators development of habitat management programs has been suggested by many workers based on sound ecological principles. Further, there is a need for diversification bee forage plants due to excessive monoculture which can be achieved through plantation of multipurpose flora under social and agroforestry programmes. Artificial nesting sites have also been suggested for conservation and propagation of potential pollinators like bumble bees and alkali bees. The domesticated indigenous hive bee, *A. cerana* also needs conservation by encouraging traditional as well as beekeeping in modern hives.

2.21 Utilization of Honeybees for Pollination

Honeybee pollination is essential for the production of horticultural crops as most of the fruit plants are not only self incompatible but also cross incompatible. It has been found that the use of hive bees results in a manifold increase in the quality of produce. It is calculated that the value of honeybees as pollinators is many times greater than their value as honey and beeswax producers. In agro-ecosystem native pollinators are often too scarce to ensure adequate pollination as insecticides, herbicides and cultivation practices have reduced bee population in many areas to the point where they are insufficient for pollination of commercial plantings (DeGrandi-Hoffman et al. 1998). Pollinators must be recognized as an integral component of horticultural crop production system, since their presence strongly affects the crop size and potential value.

Utilization of honeybees as input in agriculture and horticulture needs to be recognised by policy makers, planners, development workers and growers. This is possible only if proper awareness on the role of honeybees in enhancing productivity of crops is created among the farmers, policy makers and extension workers. Partap (2000a, b, c) have suggested some of the steps for promoting honeybees for pollination as detailed below:

One of the most serious problems in the pollination of crops is the lack of knowledge about the usefulness of pollinating insects in crop production. While the technologies on crop pollination are available, their adoption so far has been very much lagging behind due to limited trained manpower and lack of extension programs on these aspects. Therefore, there is a critical need to strengthen the training capabilities of those involved in the crop production. Lack of awareness among farmers, extension workers and policy planners on the role of honeybees in increasing crop productivity is one of the most important factors for their underutilization as pollinators. Most of the farmers are not aware of the value of honeybees for agricultural production and they have been keeping honeybees for honey production only. There is a need to encourage beekeeping not for honey production or employment generation but for pollination for sustainable agricultural production to ensure food security. Capacity building at grass root level is intended to give theoretical as well as practical insight into different types of honeybees, other pollinating insects and their gainful utilization in crop production. Likely beneficiaries include farming community, orchidists/progressive beekeepers, and unemployed youth. Research on declining honeybees is necessary to conserve the bees for their utilization as crop pollinators. Pollinator decline throughout the world is a major problem in recent (Partap and Partap 1997, 2002; Partap et al. 2001; Eardley et al. 2006). Loss of habitat, changes in land use pattern, monocultures and negative impact of chemical fertilizers and insecticides are some the causes for their decline (Verma and Partap 1993; Aizen and Feinsinger 1994; Partap and Partap 1997, 2002; Allen-Wardell et al. 1998). Clearing of forests and grass lands for cultivation has adversely affected the nesting sites of wild pollinators (Partap and Partap 2002). In recent years, fluctuations in climate and abrupt changes thereof are possible reason of their decline (Abrol 2012). There is a need to generate human resource for conserving, rearing and promoting honeybees for pollination of crops and enhancing agricultural productivity.

Bibliography

Aalders LE, Hall IV (1961) Pollen incompatibility and fruit-set in lowbush blueberries. Can J Genet Cytol 3:300–307

Abrol DP (2007) Honeybee and rapeseed – a pollinator plant interaction. In: Gupta SK (ed) Advances in botanical research: rapeseed breeding, vol 45. Elsevier, London, pp 339–367

Abrol DP (2009) Bees and beekeeping in India, 2nd edn. Kalyani Publishers, Ludhiana, 719 pp

Abrol DP (2010) Foraging behaviour of *Apis florea* F., an important pollinator of *Allium cepa* L. J Apicul Res Bee World 49(4):318–325

Abrol DP, Kapil RP (1986) Factors affecting pollination activity of *Megachile lanata* Lepel. Proc Indian Acad Sci (AnimSci) 95:757–769

Ackerman WL (1961) Flowering, pollination, self-sterility and seed development of Chinese jujubes. J Am Soc Hortic Sci 77:265–269

Adlakha RL, Dhaliwal HS (1979) Insect pollination of seed cauliflower (*Brassica oleracea* var. *bottytis*) with particular reference to the role of honeybees. Indian Bee J 41(1/2):13–16

Ahmad F, Partap U, Joshi SR, Gurung MB (2002) Why the HKH region needs the Himalayan honeybee Apis cerana. ICIMOD, Kathmandu

Aizen MA, Feinsinger P (1994) Habitat fragmentation, native insect pollinators, and feral honey bees in Argentine "Chaco Serrano.". Ecol Appl 4:378–392

Al-Tikrity WS, Benton AW, Risius ML, Clarke WW (1972) The effect of length of stay of honey bee colony in a crownvetch field on its foraging behavior. J Apicult Res 11:51–57

Allen-Wardell G, Bernhardt P, Bitner R, Burquez A, Buchmann SL, Cane JH, Cox PA, Dalton V, Feinsinger P, Inouye D, Ingram M, Jones CE, Kennedy K, Kevan P, Koopowitz H, Medellin R, Medellin-Morales S, Nabhan GP, Pavlik B, Tepedino VJ, Torchio P, Walker S (1998) The potential consequences of pollinator declines on the conservation of biodiversity and stability of food crop yields. Conserv Biol 12:8–17

Ambrose JT, Mainland CM (1979) Rabbiteye blueberry *(Vaccinium ashei)* pollination tests. Proceedings 4th international symposium on pollination. Maryland Agricultural Experimental Station Special Miscellaneous Publication 1, pp 149–155

Anderson ED (1969) An appraisal of the beekeeping industry. U.S. Deptartment of Agricultural, Agr Res Serv (ARS) 42–150, 38 pp
Anonymous (1981) Work on bee pollination in India. Indian Bee J 43(4):140–145
Anonymous (1982) Does cocopalm benefit by bee pollination? Indian Bee J 44(1):24
Anonymous 1992. International conference on Asian honeybees and bee mites. Department of biology Chulalongkorn University Bangkok, Thailand and IBRA.
Bachman WW, Waller GD (1977) Honeybee responses to sugar solutions of different compositions. J Apic Res 16:165–173
Baker HG, Baker I (1973a) Some anthecological aspects of the evolution of nectar-producing flowers, particularly amino acid production in nectar. In: Heywood VH (ed) Taxonomy and ecology. Academic, London, pp 243–264
Baker HG, Baker I (1973b) Amino-acids in nectar and their evolutionary significance. Nature 241:543–545
Baker HG, Baker I (1983) A brief historical review of the chemistry of floral nectar. In: Bentley B, Elias T (eds) The biology of nectaries. Columbia University Press, New York, pp 126–152
Barbier M (1970) Chemistry and biochemistry of pollen. Prog Phytochem 2:1–34
Batra S (1994) Diversity with pollen bees. Am Bee J 134:591–594
Batra SWT (1995) Biology of Apis laboriosa Smith, a pollinator of apples at the high altitude in the Greater Himalaya of Garhwal. USDA Agricultural Research Series, Tektran
Benedek P (1985) Economic importance of honeybee pollination of crops at the national level in Hungary. In: Proceedings of the 29th international beekeeping congress, Budapest, Hungary, 25–31 Aug 1983, pp 286–289
Beutler R (1953) Nectar. Bee Wld 34:106–116, 128–136, 156–162
Bhalla OP, Verma AK, Dhaliwal HS (1983a) Insect visitors of mustard bloom (*Brassica campestris* var. *sarson*), their number and foraging behaviour under mid-hill conditions. J Entomol Res 7(1):15–17
Bhalla OP, Verma AK, Dhaliwal HS (1983b) Foraging activity of insect pollinators visiting stone fruits. J Entomol Res 7(2):91–94
Boller CA (1956) Growing *Blueberries* in Oregon. Oreg Agr Exp Stn Bull 499, 20 p
Brett CH, Sullivan MJ (1972) Bee attraction to cucurbit flowers and pollination. Bull. North Carolina Experi Stn 443
Brewer JW, Dobson RC (1969a) Pollen analysis of two highbush blueberry varieties *Vaccinium corymbosum*. J Am Soc Hort Sci 94:251–252
Brewer JW, Dobson RC (1969b) Seed count and berry size in relation to pollinator level and harvest date for the highbush blueberry, *Vaccinium corymbosum*. J Econ Ent 62:1353–1356
Brittain W H (1933) Apple Pollination Studies in the Annapolis Valley, N.S. Canada 1928–1932. Bull, Dept Agric Can New Ser No. 162
Burgett M, Fisher G (1979) An evaluation of Beeline™ as a pollinator attractant on red clover. Am Bee J 119:356–357
Burrill RM, Dietz A (1981) The response of honeybees to variations in solar radiation and temperature. Apidologie 12:319–328
Butler CG (1945a) The influence of various physical and biological factors of the environment on honeybee activity. An examination of the relationship between activity and nectar concentration and abundance. J Exp Biol 21:5–12
Butler CG (1945b) The behaviour of bees when foraging. J R Soc Arts 93:501
Butler CG, Finney DJ (1942) The influence of various physical and biological factors of ther environment on honeybeeactivity. An examination of the relationship between activity and solar radiation. Jexp Biol 18:206–212
Chaudhary RK (1978) Floral fidelity in the Indian honey bee (Apis cerana indica F.). Indian Bee J 40(2):33–35
Clark KK (1959) Honeybees in the orchard. Agriculture, Lond 65:573–575
Connor LJ (1970) Studies of strawberry pollination in Michigan. In: The indispensable pollinators, Ark Agr Ext Serv, Misc Pub 127, McGregor, 1976, pp 157–162

Connor LJ, Martin EC (1973) Components of pollination of commercial strawberries in Michigan. Hortscience 8:304–306

Corbet SA, Unwin DM, Prys-Jones OE (1979a) Humidity nectar and insect visits to flowers with special reference to Crataegus,Tilia and Echium. Ecol Entomol 4:9–22

Corbet SA, Willmer PG, Beament JWL, Unwin DM, PRYS-Jones OE (1979b) Post secretory determinants of sugar concentration nectar. Plant Cell Environ 2:293–208

Couston R (1966) Experiments on the influence of insect pollination of on soft fruits. Scot Beekeeper 43(3):39–40, (5):90–92

Crane MB, Lawrence WJC (1931) Inheritance of sex, colour and hairiness in the raspberry, *Rubus idaeus* L. J Genet 24:243–255

Crane E, Walker P (1983) The impact of pest management on bees and pollination. Tropical Development and Research Institute, London, 232 pp

Crane E, Walker P (1984) Pollination directory for world crops. IBRA, London

Cruden RW, Hermann SM (1983) Studying nectar? Some observations on the prt. In: Bentley B, Elias T (eds) The biology of nectaries. Columbia University Press, New York, pp 223–241

Darrow GM (1966) The strawberry. Holt, Rinehart and Winston, New York, Delaplanen ja Mayerin kirjassa 2000 sekä McGregor, 1976

Darrow GM, Moore JN (1962) Bluberry growing. U.S. Dept Agric Farm Bull 1951

DeGrandi-Hoffman G (1987) The honeybee pollination component of horticultural crop production systems. Hort Rev 9:237–249

De Grandi-Hoffman G, Collins AM, Martin JH, Schmidt JO, Spangler HG (1998) Nest defense behavior in colonies from crosses between Africanized and Eurpean honey bees *(Apis mellifera* L.) (Hymenoptera: Apidae). J Insect Behav 11:37–45

Denis FG Jr (1979) Factors affecting yield in apple with emphasis on Delicious'. Horticul Rev 1:395–422

Dhaliwal JS, Atwal AS (1986) Factors influencing tripping efficiency of bees on alfalfa (*Medicago sativa* L.). In: Kapil RP (ed) Pollination biology: an analysis. Inter-Indian Publication, New Delhi, pp 79–90

Dhaliwal HS, Scrivastava S, Adlakha RL (1977) Insect pollination of lychee, *Litchi chinensis Sonn* in the Valley areas in the Himalayas. In: Proceedings of XXVIth international apiculture congress, Adelaide, p 396

Dhoble SY, Shinde YM (1982) Role of insect pollinators for effective seed setting in sunflower. J Maharas Agricul Univ 7(1):103

Dulta PC, Verma LR (1987) Role of insect pollinators on yield and quality of apple fruit. Ind J Hortic 44:274–279

Eardley C, Roth D, Clarke J, Buchman S, Gemmill B (eds) (2006) Pollinators and pollination: a resource book for policy and practice. African Pollinator Initiative, Pretoria, 77 pp

Eaton GW, Daubeny H, Norman RC (1968) Pollination techniques for red raspberry breeding programs. Canad J Plant Sci 48(3):342–344

Eck P, Childers NF (1966) Blueberry culture. Rutgers University Press, New Brunswick and London, 378 pages

Emmett BJ (1971) Insect visitors to pear blossom. Plant Pathol 20:36–40

Faegri K, Van der Pijl L (1979) The principles of pollination ecology. Pergamon Press, London

FAO (Food and Agriculture Organization) (2001) FAO statistical data base. http://appsl.fao.org/servlet/Xte.FAOStat Data Base

FAO (2009) FAO crop production statistics. FAO, Rome. http://faostat.fao.org/site/567/default.aspx#ancor. Accessed 17 Nov 2009

Farrar CL (1931) A measure of some factors affecting the development of the honeybee colony. Unpublished doctoral thesis, Massachusetts State College, Amherst

Farrar CL, Bain HF (1947) Honeybees as pollinators of the cranberry. Cranberries 11

Fell RD (1995) Pollination. In: Hogmire H Jr (ed) Mid-Atlantic orchard monitoring guide. Natural Resource, Agriculture, and Engineering Service, Ithaca, pp 255–265

Filmer RS (1932) Brood and colony size as factors in activity of pollination units. J Econ Ent 25:336–343

Free JB (1960) The behaviour of honeybees visiting the flowers of fruit trees. J Anim Ecol 29:385–395
Free JB (1962a) The behaviour of honeybees visiting field beans *(Vicia faba)*. J Anim Ecol 31:497–502
Free JB (1962b) The effect of distance from pollinizer varieties on the fruit set on trees in plum and apple orchards. J Hon Sci 37:262–271
Free JB (1963) The flower constancy of honeybees. J Anim Ecol 32:119–131
Free JB (1965a) The behaviour of honeybee foragers when their colonies are fed with sugar syrup. J Apic Res 4:85–88
Free JB (1965b) The ability of bumblebees and honeybees to pollinate red clover. J Appl Ecol 2:289–294
Free JB (1966a) The pollinating efficacy of honeybee visits to apple flowers. J Hort Sci 41:91–94
Free JB (1966b) The pollination of the beans *Phaseolus multiflorus* and *Phaseolus vulgaris* by honeybees. J Apic Res 5:87–91
Free JB (1966c) The pollination requirements of broad beans and field beans *(Vicia faba)*. J Agric Sci Camb 66:395–397
Free JB (1966d) The pollinating efficiency of honeybee visits to apple flowers. J Hon Sci 41:91–94
Free JB (1967) Factor determining the collection of pollen by honeybee foragers. Anim Behav 15:134–144
Free JB (1968a) The behaviour of bees visiting runner beans *(Phaseolus multiflorus)*. J Appl Ecol 5:631–638
Free JB (1968b) The pollination of strawberries by honeybees. J Hon Sci 43:107–111
Free JB (1968c) The foraging behaviour of honeybees *(Apis mellifera)* and bumblebees *(Bombus* spp.) on blackcurrant *(Ribes nigrum)*, raspberry *(Rubus idaeus)* and strawberry *(Fragaria x Ananassa)* flowers. J Appl Ecol 5:157–168
Free JB (1968d) The pollination of black currants. J Hon Sci 43:69–73
Free JB (1968e) Dandelion as a competitor to fruit trees for bee visits. J Appl Ecol 5:169–178
Free JB (1970) Insect pollination of crops. Academic, London, 544 pp
Free JB (1993) Insect pollination of crops, 2nd edn. Academic, London
Free JB, Durrant AJ (1966) The transport of pollen by honeybees from one foraging trip to the next. J Hon Sci 41:87–89
Free JB, Preece DA (1969) The effect of the size of a honeybee colony on its foraging activity. Insects Soc 16:73–78
Free JB, Spencer-Booth Y (1964a) The effect of distance from pollinizer varieties on the fruit set of apple, pear and sweet cherry trees. J Hon Sci 39:54–60
Free JB, Spencer-Booth Y (1964b) The foraging behaviour of honeybees in an orchard of dwarf apple trees. J Hon Sci 39:78–83
Free JB, Paxton RJ, Waghwchoure ES (1991) Increased the amount of foreign pollen carried by honeybee foragers. J Apic Res 30:132–136
Friesen LJ (1973) The search dynamics of of recruited honey bees, *Apis mellifera ligustica* Spinola. Biol Bull 144:107–131
Gary NE (1967) A method for evaluating honeybee flight activity at the hive entrance. J Econ Entomol 60:102–105
Gary NE (1979) Factors that affect the foraging distribution of honeybees. Proceeding 4th international symposium Pollin Md Agric Exp Sta Misc Publ 1:353–358
Gary NE (1999) Activities and behavior of honey bees. In: Graham Joe M (ed) The hive and the honey bee. Dadant & Sons, Hamilton, pp 269–372
Gates BN (1911) The honey bees and craneberry growing. Cape Cod Craneberry Growers Assoc Annu Report, 24–28
Gepts P (2008) Tropical environments, biodiversity, and the origin of crops. In: Ming R, Moore P (eds) Genomics of tropical crop plants. Springer, New York, pp 1–20
Gooderham CB (1950) Overwintered colonies versus package bees for orchard pollination. Progr Rep Dom Apiarist Can Dep Agric (1937–1948), 7–8

Goodwin RM, ten Houten A, Perry JH (1991a) Effect of variations in sugar presentation to honey bees *(Apis mellifera)* on their collection of kiwifruit *(Actinidia deliciosa)* pollen. NZ J Crop Hort Sci 19:259–262

Goodwin RM, Houten A, ten and Perry JH (1991b) Feeding sugar syrup to honey bee colonies to improve kiwifruit pollen collection: a review. The 6th international symposium on pollination, Tilburg, Aug 1990. Acta Hortic 288, 265–269

Gould JL (1975) Honey bee recruitment: the dance language controversy. Science 189:685–693

Gould JL (1976) The dance language controversy. Quart Rev Biol 51:211–214

Gould JL, Henerey M, Macleod MC (1970) Communication of direction by the honeybee. Science 169:544–554

Goyal NP (1978) Performance of Apis mellifera and A. indica as observed in Punjab plains. Indian Bee J 40(1):3–5

Goyal NP, Atwal AS (1977) Wing beat frequency of *Apis indica* and *Apis mellifera*. J Apic Res 16(1):47–48

Grant V (1950) The flower constancy of bees. Bot Rev 16:379–398

Gupta JK, Goyal NP, Sharma JP, Gautam DR (1993) The effect of placement of varying numbers of Apis mellifera colonies on the fruit set in apple orchards having different proportions of pollinisers. In: Proceedings of international symposium on pollination in the tropics, Bangalore, pp 197–201

Gupta JK, Rana BS, Sharma HK (2000) Pollination of kiwifruit in Himachal Pradesh. In: Matsuka M, Verma LR, Wongsiri S, Shrestha KK, Partap U (eds) Asian bees and beekeeping: progress of research and development. Oxford and IBH, New Delhi, pp 169–170

Haragsimova-Neporasova L (1960) Measurement of nectar secretion in plants (in Czeech, English and other summaries). Ved Pr Vyzk Ustavu Vet CSAZV Brne 2:63–79

Harborne JB (1967) Comparative biochemistry of flavonoids. Academic, New York

Harborne JB (1976) Function of flavonoids in plants. In: Goodwin TW (ed) Chemistry and biochemistry of plant pigments, 2nd edn., pp 736–770

Harborne JB (1982) Introduction to ecological biochemistry, 2nd edn. Academic, New York

Haydak MH (1970) Honeybee nutrition. Annu Rev Entomo 15:143–156

Heinrich B (1996) How the honey bee regulates its body temperature. Bee World 77(3):130–137

Hellmisch RL, Kulincevic JM, Rothenbough WC (1985) Selection for high and low pollen hoarding honeybees. J Hered 76:155–158

Hughes HM (1961) Preliminary studies on the insect pollination of soft fruits. Expl Hort 6

Iwama S (1977) Influence of climatic factors on external flight activity of Tetragonisca angustica. Boletin de Zoologia Universidade Sao Paulo 2:189–201

Jaycox ER (1970a) Ecological relationships between honey bees and soybeans. Am Bee J 110:306–307, 343–345, 383–385

Jaycox ER (1970b) Pollination of strawberries. Am Bee J 110:176–177

Jhajj HS, Goyal NP (1979a) Comparative behaviour of pollen foragers of *Apis cerana indica* and *Apis mellifera*. J Apic Res 18(4):279–284

Jhajj HS, NP Goyal (1979b) Comparative studies on the flower constancy of *Apis cerana indica* F. and *Apis mellifera* L. In: Proceedings of 4th international symposium on pollination, Maryland, pp 333–339.

Johnston S (1929) Insects aid fruit setting of raspberry. Q Bull Mich Agric Exp Stn 9:105–110

Jones CE, Buchmann SL (1974) Ultraviolet floral patterns as functional orientation cues in hymenopterous pollination systems. Anim Behav 22:481–485

Kevan PG, Baker HG (1983) Insects as flower visitors and pollinators. Annu Rev Entomol 28:407–453

Kevan PG, Lane MA (1983) Texture sensitivity at the microscopic level in pollination and its significance in pollination. In: Proceeding of 5th international symposium pollination summaries, Centre National de la Recherche Agronomique, Versailles

Kevan PG, Lane MA (1985) Flower petal microtexture is a tactile cue for bees. Proc Natl Acad Sci USA 82:407–453

Kevan PG, Little RJ (1983) Floral colours through the insect eye: what they are and what they mean. In: Jones CE (ed) Handbook of experimental pollination. Van Nostrand Reinhold, New York

Klein AM, Vaissiére BE, Cane JH, Steffan-Dewenter I, Cunningham SA, Kremen C, Tscharntke T (2007) Importance of pollinators in changing landscapes for world crops. Proc R Soc 274, 303–313

Kozin RB (1976) Pollination of entomophilous agricultural crops by bees. Amerind, New Delhi

Kumar J (1988) Insect pollinators in temperate fruits during bloom. J Tree Sci 7:38–40

Kumar J, Gupta JK (1993) Nectar sugar production and honey bee foraging activity in three species of onion (Allium species). Apidologie 24(4):391–396

Kumar J, Mishra RC, Gupta JK, Dogra GS (1984) Pollination requirements of some peach cultivars. Indian Bee J 47:3–6

Kumar J, Rao KVK, Gupta PR, Dogra GS (1989) Temporal distribution of bees (Hymenoptera: Apoidea) on spring flowering crops in mid hills of Himachal Pradesh. Indian Bee J 51:55–58

Kumar J, Mishra RC, Thakur SS (1998) Use of honeybees for increasing agricultural yields. In: Mishra RC, Rajesh G (eds) Perspectives in Indian apiculture. Agro Botanica, Bikaner, pp 187–211

Levin MD (1966) Orientation of honeybees in alfalfa with respect to landmarks. J Apic Res 5:121–125

Lindauer M (1948) Uber die Einwirkung von Duft – und Geschmacksstoffen sowie anderer Faktoren auf die tunze der Bienen. Z Verlag Physiol 31(412):348

Lindauer M (1971) The functional significance of the honeybee waggle dance. Am Nat 105:89–96

Loper GM, Lapioli AM (1971) Photoperiodic effects on the emanation of volatiles from alfalfa *(Medicago sativa* L.) florets. Plant Physiol 49:729–732

Loper GM, Waller GD, Berdel RL (1976) Effect of flower age on sucrose content in nectar of citrus. Hortscience 11:416–417

Lundie AE (1925) The flight activities of the honeybee. U S Dep Agric Dep Bull 1328:1–38, Washington, DC

Mackensen O, Nye WP (1969) Selective breeding of honeybees for alfalfa pollen collection; sixth generation and outcrosses. J Apic Res 8:9–12

Manning A (1956) The effect of honey-guides. Behaviour 9:114–139

Margalith R, Lensky Y, Robinowitch H (1984) An evaluation of bee line as a honeybee attractant to cucumbers and its effect on hybrid seed production. J Apic Res 23:5054

Martin EC (1966) Honeybee pollination of the highbush blueberry. Am Bee J 106:366–367

Marucci PE (1966) Blueberry pollination. Am Bee J 106:250–251, 264

Marucci PE, Moulter HJ (1977) Craneberry pollination in New Jersey. Acta Hortic 61:217–222

Matheson AG (1986) Colony standards for Kiwifruit pollination. NZ Beek 189:20–22

Matheson AG (1991) Managing honeybee pollination of Kiwifruit (Actinidia deliciosa) in New Zealand – a review. Acta Hort 288:213–219

Mattu VK, Chaudhary DK, Kumar L (1994) Foraging ecology of *Apis cerana* F. and *A. mellifera* L. in pollinating stone fruits. Pest Manag Econ Zool 2:35–39

Mayer DF, Johansen CA (1982) Field evaluation of chemical attractants on tree fruits. Am Bee J 122:287–289

McGregor SE (1976) Insect pollination of cultivated crop plants. Agricultural Handbook Number 96. Agricultural Research Service, United States Department of Agriculture, Washington, DC, 411 p

Mishra RC, Sihag RC (1987) Apicultural Research in India. All Indian co-ordinated project on honeybee research and training. Haryana Agricultural University Press, Hissar, 120 pp

Mishra RC, Dogra GS, Gupta PR (1976) Some observations on insect pollination of apple. Ind Bee J 38:20–22

Moeller PE (1977) Managing colonies for pollen production. In: Proceedings of 26th international Agriculture of Congress Adelaidae, Apimondia Publication Buchaarest, pp 232–239

Mommers J (1961) De bestuiving van aardbeien onder glas. Meded Dir Tuinb 24:353–355

Mommers J (1972) De bestuiving van arrdbeien onder glas. Meded Dir Tidnb 24:353–355
Moore JN (1969) Insect pollination of strawberries. J Am Soc Hort Soc 94:367–364
Morris OM (1921) Studies in apple pollination. Wash Agric Exp Sta Bull 163
Morrow EB (1943) Some effects of cross-pollination versus self-pollination in the cultivated blueberry. Proc Am Soc Hort Sci 42:469–472
Muttoo RN (1950) Honeybees and fruit crops. Indian J Horti 7(3/4):17–20
Nand D (1971) Pollination, fruit set and fruit development in carambola (*Averrhoa carambola* L.). Indian J Hortic 28(4):278–284
Noro K, Yago M (1934) Studies on sterility of Japanese pear with special reference to hand pollination, dehiscence of anthers and insect visitors. Bull Schizuokaken Agric Expt Stn 29:13
Nunez JA (1977) Nectar flow by melliferous flora and gathering flow by *Apis mellifera ligustica*. J Insect Physiol 23:265–276
Nunez JA (1982) Honeybee foraging strategies at a food source in relation to its distance from the hive and the rate of sugar flow. J Apic Res 21:139–150
Nye WP, Anderson JL (1974) Insect pollinators frequenting strawberry blossoms and the effect of honey bees on yield and fruit quality. J Am Soc Hon Sci 99:40–44
Nye WP, Machensen O (1970) Selective breeding of honeybees for alfalfa pollen collection: with tests in high and low alfalfa pollen collection regions. J Apic Res 9:61–64
Ohe W, Praagh JP (1983) Versuche zur Duftlenkung in Obstanlagen. Nordwestdeustsche Imkerzeitung 35:100
Pandey RS, Yadav RPS (1970) Pollination of litchi (*Litchi chinensis*) by insects with special reference of honeybees. J Apic Res 9(2):103–105
Parmar C (1976) Pollination and seed set in phalsa (*Grewia asiatica* L.). Agric and Agro-Indust J 9(6):12–14
Partap U (1999a) Conservation of endangered Himalayan honeybee, Apis cerana for crop pollination. Asian Bee J 1:44–49
Partap U (1999b) Pollination management of mountain crops through beekeeping: trainers' resource book. ICIMOD, Kathmandu
Partap U (2000a) Foraging behaviour of Apis cerana on sweet orange (Citrus sinensis var. Red Junar) and its impact on fruit production. In: Matsuka M, Verma LR, Wongsiri S, Shrestha KK, Partap U (eds) Asian bees and beekeeping: progress of research and development. Proceedings of 4th Asian Apicultural Association International Conference, Kathmandu, pp 174–177
Partap U (2000b) Pollination of strawberry by the Asian hive bee, Apis cerana F. In: Matsuka M, Verma LR, Wongsiri S, Shrestha KK, Partap U (eds) Asian bees and beekeeping: progress of research and development. Oxford and IBH, New Delhi, pp 178–182
Partap U (2000c) Foraging competition between Apis cerana and Apis mellifera and its impact on crop pollination. In: Matsuka M, Verma LR,Wongsiri S, Shrestha KK, Partap U (eds) Asian bees and beekeeping: progress of research and development. Oxford and IBH, New Delhi, pp 206–208
Partap U, Partap T (1997) Managed crop pollination. The missing dimension of mountain agricultural productivity. International Centre for Integrated Mountain, Kathmandu
Partap U, Partap T (2002) Warning signals from apple valleys of the HKH region: pollination problems and farmers' management efforts. ICIMOD, Kathmandu
Partap U, Verma LR (1992) Floral biology and foraging behaviour of Apis cerana on lettuce crop and its impact on seed production. Prog Hortic 24:42–47
Partap U, Verma LR (1994) Pollination of radish by Apis cerana. J Apic Res 33:237–241
Partap U, Shukla AN, Verma LR (2000a) Pollination of peach and plum by Apis cerana. In: Matsuka M, Verma LR, Wongsiri S, Shrestha KK, Partap U (eds) Asian bees and beekeeping: progress of research and development. Oxford and IBH, New Delhi, pp 171–173
Partap U, Shukla AN, Verma LR (2000b) Comparative attractiveness of broad-leaf mustard, cauliflower and radish to Apis cerana in Kathmandu Valley of Nepal. In: Proceedings of the 4th Asian apicultural association international conference, Kathmandu, pp 203–205
Partap U, Partap T, Yoghua H (2001) Pollination failure in apple crop and farmers' management strategies in Hengduan, China. Acta Hortic 561:225–229

Partap U, Partap T, Sharma HK, Phartiyal P, Marma A, Tamang NB, Ken T, Munawar MS (2012) Value of insect pollinators to Himalayan agricultural economies. ICIMOD, Kathmandu
Patil BS, Viraktamath S, Lingappa S, Giraddi RS, Parameshwarappa K, Bhat ARS (2000) Effect of Bee-Q and Bee- here on pollination and yield of sesamum. Insect Environ 5(4):151
Percival MS (1946) Observations on the flowering and nectar secretion of *Rubus fructicosus* (Agg.). New Phytol 45:111–123
Percival MS (1955) The presentation of pollen in certain angiosperms and its collection by *Apis mellifera*. New Phytol 54:353–368
Percival MS (1961) Types of nectar in angiosperms. New Phytol 60:235–281
Petkov V (1963) Nectar production in cultivated raspberry. Sel Nauk 2:201–207
Petkov V (1965) Contribution of honeybees to the pollination of strawberries. Gradinar lozar Nauk 2:421–431
Phadke KG, Naim M (1974) Observations on the honeybee visitation to the litchi (*Nepheliurn litchi*) blossom at Pusa (Bihar, India). Ind Bee J 36(1–4):9–12
Phillips EF (1930a) Bees for the orchard. J Econ Entomo 23:218–223
Phillips EF (1930b) Honey bees for the orchard. Cornell Agric Exp Sta, Bull 190
Prasad R, Chand H, Singh R (2000) Effect of honeybee, *A.mellifera* pollination on the yield and germination of onion, *Allium cepa* seeds. Shashpa 7(1):53–55
Punchihewa RWK, Koeniger N, Kevan PG, Gadawski RM (1985) Observation on the dance communication and natural foraging ranges *of Apis cerana*, *Apis dorsata* and *Apis florea* in Sri Lanka. J Apicul Res 24:168–175
Rai KM, Gupta BP (1983) Role of honeybees and other insects as pollinators in apple and pear. Indian Bee J 45(2/3):56–57
Rajotte EG, Roberts RB (1979) Nectar sugar dynamics of high bush blueberry cultivars (Vaccinium corymbossum L.). Proceeding 4th international symposium. Pollin Md Agric Exp Sta Misc Publ 1:157–164
Rejput CBS, Singh JB (1967) Pollination and fruit setting studies in strawberries. Indian J Hortic 24:30–32
Rigby R, Dana MN (1971) Seed number and berry volume in cranberry. Hort Science 6:495–496
Roberts RH (1945) Blossom structure and setting of delicious and other apple varieties. Proc Am Soc Hortic Sci 46:87–90
Roberts RB (1979) Energetics of cranberry pollination. In: Proceedings of 4th International symposium pollination. Maryland Agricultural Experimental Station, Spec Misc Pubi 1, pp 431–440
Robinson WS (1979) Effect of apple cultivar on foraging behaviour and pollen transfer by honey bees. J Am Soci Hortic Sci 104:596–598
Roubik DW (1995) Pollination of cultivated plants in the tropics. FAO agricultural services Bulletin No. 118, FAO, Rome, 196p
Roubik DW, Aluja M (1983) Flight ranges of *Melipona* and *Trigona* in tropical forests. J Kans Entomol Soc 56:217–222
Roubik DW, Moreno JE, Vergara C, Wittmann D (1986) Sporadic food competition with the African honey bee: projected impact on neotropical social bees. J Trop Ecol 2:97–111
Sakagami SF, Matsumura T, Ito K (1980) Apis laboriosa in the Himalaya, the little known world's largest honeybee (Hymenoptera: Apidae). Insecta Matsumurana NS 19:47–77
Sale PR (1983) Kiwifruit culture. GPO, Wellington, New Zealand
Sasaki M (1985) Utilization of pollen loads for artificial pollination of fruit trees. In: The 30th international apicultural congress of Apimondia, Nagoya, pp 390–393
Schwarz MP, Hurst PS (1997) Effects of introduced honey bees on Australia's native bee fauna. Vic Nat 114:7–12
Seeley TD (1983) Division of labor between scouts and recruits in honey bee foraging. Behav Ecol Sociobiol 12:253–259
Seeley TD (1985) The information-center strategy of honeybee foraging. Fortschr Zool 31:75–90
Seethalakshmi TS, Percy AP (1979) *Borassus flabellifer* (palmyrah palm) a good pollen source. Indian Bee Journal 41(1/2):20–21

Shanks CH (1969) pollinationof raspberries by honeybees. J Apic Res 8:19–21
Sharma PL (1961) The honeybee population among insects visiting temperate zone fruit flowers and their role in setting fruit. Bee World 42(1):6–8
Sharma PL, Dhaliwal HS, Kakar KL (1974) Insect visitors and pollinators of cauliflower (*Brassica oleracea* var. *botrytis*) seed crop bloom. Himachal J. Agric Res 2:74–78
Shaw FR, Bailey JS, Bourne AI (1939) The comparative value of honeybees in the pollinationof cultivated blueberries. J Econ Entomol 33:872–876
Shaw FR, Savos M, Shaw WM (1954) Some observations on the collection habits of bees. Am Bee J 94:422, Free, 1970
Shuel R (1955a) Nectar Secr Am Bee J 95:229–234
Shuel R (1955b) Nectar secretion in relation to nitrogen supply, nutritional status, and growth of plant. Can J Agrie Sci 35:124–138
Shuel R (1957) Some aspects of the relation between nectar secretion and nitrogen, phosphorus, and potassium nutrition. Can J Plant Sci 37:220–236
Singh Y (1979) Pollination activity on strawberry at Jeolikot, Dist. Nainital, India. Indian Bee J 41(1/2):17
Singh R, Mishra RC (1986) Flower visiting flies of fruits in Himachal Pradesh. Proc Indian Nat Sci Acad 52B(4):451–453
Skrebtsova ND (1957) The role of bees in pollinating strawberries. Pchelovodstvo 34:34–36 (in Russian)
Skrebtsova ND (1958) Effectiveness of the saturatedpollinationof straw-berries by bées. Byull Nauchno-Tekh Infm Nauchno-Issled Inst Pchelovod 3–4:25–26 (in Russian)
Stanley G, Linskins HF (1974) Pollen: biology biochemistry and management. Springer, Berlin
Stephen WP (1958) Pear pollination studies in Oregon. Bull Ore Agric Exp Stn 43
Szabo TI (1980) Effects of weather factors on honeybee flight activity and colony weight gain. /. Apic Res 19:164–171
Taranov GF (1952) Rules governing the flight activities of bees. Zool Z 31:61–71
Thorp RW (1979). Honey bee foraging behavior in California almond orchards. In: Proceedings of 4th International symposium pollination. Maryland Agricultural Experimental Station, Spec Misce Pubi 1, pp 385–392
Todd FE, Bretherick O (1942) The composition of pollens. J Econ Entomol 35:312–317
Todd FE, Mcgregor SE (1960) The use of honeybees in the production of crops. Annu Rev Entomol 5:265–278
Todd FE, Reed CB (1970) Brood measurement as a valid index to the value of honeybees as pollinators. J Econ Entomol 63:148–149
Underwood BA (1986) The natural history of Apis laboriosa Smith in Nepal. Thesis, Cornell University, Ithaca
Vansell GH (1934) Relations between the nectar concentration in fruit blossoms and the visits of honey bees. J Econ Entomol 27:943–945
Vansell GH (1942) Factors affecting the usefulness of honeybees in pollination, U.S. Department Agriculture Circular No. 650. U.S. Department of Agriculture, Washington, DC
Vansell GH, Reinhardt JF (1948) Do honey bees help pollinate baby lima beans? Gleanings Bee Cult 76:678–679
Vansell GH, Watkins WG, Bishop RK (1942) Orange nectar and pollen in relation to bee activity. J Econ Entomol 35:321–232
Verma LR (1990) Beekeeping in integrated mountain development: economic and scientific perspectives. Oxford and IBH, New Delhi
Verma LR, Chauhan P (1985) Distribution, abundance and diversity of insect pollinators in apple orchards of Shimla hills. Ind J Ecol 12:286–292
Verma LR, Dutta PC (1986) Foraging behaviour of *Apis cerana indica* and *Apis mellifera* En pollinating apple flowers. J Apic Res 25(4):197–201
Verma LR, Partap U (1993) The Asian hive bee, Apis cerana, as a pollinator in vegetable seed production. ICIMOD, Kathmandu

Verma LR, Rana RS (1994) Further studies on the behaviour of Apis cerana and Apis mellifera foraging on apple flowers. J Apic Res 33:175–179
Visscher PK, Seeley TD (1982) Foraging strategy of honeybee colonies in a temperate deciduous forest. Ecology 63:1790–1801
von Frisch K (1950) Bees, their vision, chemical senses and language. Cornell, Ithaca
Von Frisch K (1965) Tanzsprache and Orientierung der Bienen. Springer, Berlin/Heidelberg
von Frisch K (1967) The dance language and orientation of bees. Harvard University Press, Cambridge
Von Frisch K (1974) Decoding the language of the bee. Science 185:663–668
Waddington KD (1983) Floral-visitation-sequences by bees: models and experiments. In: Jones CE, Little RJ (eds) Handbook of experimental pollination biology. Van Nostrand Reinhold, New York, pp 461–473
Waddington KD, Heinrich R (1981) Patterns of movement and floral choice by foraging in bees. In: Kami A, Sargent T (eds) Foraging behavior: ecological, ethological, and psychological approaches. Garland Press, New York
Waddington KD, Holden LR (1979) Optimal foraging: on flower selection by bees. Am Nat 114:179–196
Waller GD (1970) Attracting honey bees to alfalfa with citral, geraniol, and anise. J Apic Res 9:9–12
Waller GD (1972) Evaluating responses of honey bees to sugar solutions using an artificial flower feeder. Ann Entomol Soc Am 65:857–862
Webster TC, Thorp RW, Briggs D, Skinner J, Parisian T (1985) Effects of pollen traps on honey bee (Hymenoptera: Apidae) foraging and brood rearing during almond and prune pollination. Environ Entomol 14:683–686
Wellington RA (1913) Raspberry breeding. American Society Horticulture Science Proceedings 10th Annual Meeting, Washington, DC, pp 155-159
Wells H (1983) Population equilibria and stability in plant-animal pollination systems. Theor Biol 100:685–699
Wells PH, Wells H (1985) Ethological isolation of plants, 2. Odour selection by honeybees. J Apic Res 24:86–92
Wells PH, Wenner AM (1971) The influence of food scent on behavior of foraging honey bees. Physiol Zool 44:191–209
Wells PH, Wenner AM (1973) Do honey bees have language? Nature 141:171–175
Wells H, Wells PH, Smith DM (1981) Honey bee responses to reward size and colour in an artificial flower patch. Apic Res 20:172–179
Wells PH, Wells H, Smith DM (1983) Ethological isolation of plants, 1. Colour selection by honeybees. J Apic Res 22:33–34
Wenner AM (1974) Information transfer in honey bees: a population approach. In: Krames L, Pliner P, Alloway T (eds) Nonverbal communication. Plenum Press, New York, pp 139–169
Wenner AM, Wells PH, Johnson DL (1969) Honey bee recruitment to food sources: olfaction or language? Science 169:84–86
Whatley BT, Lackett JL (1979) Effects of honey bee pollination on fruit set, yield, and quality of Rabbiteye blueberry (*Vaccinium ashei)* Reade. In: Proceedings 4th international symposium pollination, Maryland Agricultural Experimental Station. Spec Miscell Publ 1, pp 143–148
Wilhelm S (1974) The garden strawberry: a study of its orgin. Am Sci 62:264–271
Williams RR (1970) Factors affecting pollination in fruit trees. In: Luckwill LC, Cutting CV (eds) Physiology of tree crops. Academic, London, pp 193–207
Williams NH (1983) Floral fragrances as cues in animal behavior. In: Jones CE, Little RJ (eds) Hand-book of experimental pollination. Van Nostrand Reinhold, New York, pp 50–72
Williams IH (1991) Crop pollination and sustainable agriculture. Bee World 72:45–46
Williams RR, Sims FP (1977) The importance of weather and variability in flowering time when decidingpollinationscheme for Cox's Orange Pippin. Exp Hortic 29:15–26
Williams IH, Martin AP, White RP (1987) The effect of insect pollination onplant development and seed production in winter oil-seed rape (*Brassica napus* L.). J Agric Sci Camb 115:67–73

Wood GW (1968) Self-fertility in the lowbush blueberry. Can J Plant Sci 48:431–433
Woyke HW (1981) Some aspects of the role of the honeybee in onion seed production in Poland. Acta Hortic 111:91–98
Wu YR, Kuang BY (1987) Two small species of honeybees: a study of genus Micrapis. Bee World 68:152–155
Wykes GR (1952a) An investigation of the sugars present in the nectar of flowers of various species. New Phytol 51:210–215
Wykes GR (1952b) The preference of honey bees for solutions of various sugars which occur in nectar. J Exp Biol 29:511–519
Wykes GR (1952c) The influence of variations in the supply of carbohydrate on the process of nectar secretion. New Phytol 51:294–300

Chapter 3
Pest Problems in Fruit Crops

3.1 Introduction

In the green revolution, most of the emphasis was on high-yielding cereal grains such as rice, wheat, and corn, while horticultural crops received little attention for their role in food security and reduction of poverty and malnutrition. Sustainable horticultural crop production should be considered in association with other agricultural crop production programs as a long-term approach to end poverty, improve nutrition, and sustain economic growth. These crops are well-suited for commercial, small scale and subsistence farming and constitute over 90 % of the products grown in home gardens. The population of world is already over seven billion and is expected to increase by another two billions or more by mid–century thereby exerting huge pressing to meet the growing need for food worldwide. The demand for food from horticultural crops is expected to increase even more as many Asian countries become more developed. Agriculture will be required to meet this food demand with less land, less water, and less labor. Although staple food is provided by the two main cereals, rice and wheat, food from horticultural crops such as beans, vegetables, and fruits is also significant for humans. The huge variety of food produced from horticultural crops will provide not only diversity in taste but also sources of nutrients, minerals, and vitamins for humankind to enjoy healthy food. Nonetheless, insect pests are constant threats to horticultural crop production. Insects are the principal causes of yield losses in cabbage, mango, and bean crops. Currently, pests such as insects, mites, bacteria, fungi, viruses, nematodes, and weeds often cause up to 40 % reduction in crop yields (SP-IPM 2010). Insect pests are responsible for a major part of these losses due to direct feeding damage and/or vectoring of viral, bacterial, and fungal diseases.

Development of effective pest management programs cannot move forward unless the pests are correctly identified and their distribution, biology, nature of damage, and important natural enemies are known. Understanding the center of origin for both the crops and their pests can be important for developing effective pest management programs. As expected, the crops that originated in Asia have

D.P. Abrol, *Pollination Biology, Vol. 1*, DOI 10.1007/978-3-319-21085-8_3

more arthropods associated with them than crops introduced from the neotropical, European, and Afro-tropical regions. This is especially true for invasive pests that originated in one region and were accidentally introduced into another without their natural enemies to keep them in check. Therefore, it would be absolutely necessary to determine the origin of invasive pests such as papaya mealy bugs *Paracoccus marginatus* Williams and Granara de Willink. It spread from Mexico in the early 1990s, but parasitoids were identified in its center of origin and were introduced to several countries in the Caribbean, South America, Pacific Asia and Africa where they effectively controlled this pest (Muniappan et al. 2009, 2010). These parasitoid introductions resulted in billions of dollars in savings and averted the collapse of the papaya industry.

3.2 Ecology of Pests

Strategies for development of successful IPM programmes require proper understanding of ecology of the pests and their natural enemies (Peña et al. 2002). India is a major contributor of fruits and vegetables ranking second in the world. The area under fruits alone is 32.05 lakh hectares with a production of 329.55 lakh tonnes. One of the limiting factors in fruit production is severe attack of arthropod pests and diseases affecting yield and quality. Unlike agricultural crops, horticultural crops are grown as perennial monocultures, therefore, harbor the insect pest and disease problems that are entirely different and complex in nature. These two factors impose huge losses upon horticulture in terms of alleviated production, reduced quality and cost of management. A large number of pests have been observed on different horticultural crops with the effects ranging from mild damages to serious losses in production and quality. Various management practices for protection of fruit crops are discussed below:

3.3 Management of Pest Problems in Fruits

Struggle for existence between pests and plants have been going on right from the time of their origin. Insects became pests when confronted with the needs of human beings. Farmers in order to manage these pests adapted hit and trial methods which included cultural and mechanical practices. Later, these practices became part of their strategies to control pests. The earliest known example of utilization of biological control is use of predatory ants, *Oecophylla smaragdina* Fabricius by Chinese for the control of leaf feeding insects in citrus orchards. They used bamboo stems to facilitate the movement of ants between different citrus trees (McCook 1882, in DeBach 1964). The use of chemical control came much later around 2500 years ago. But the management practices began to change from late 1800 to 1940s when use of soaps, oils, resins and plant derived compounds are used. Dichloro diphenyl trichloroethane (insecticide), ferbam (fungicide), and 2, 4-D (herbicide) were the first to be discovered in 1940s marling the beginning of era of chemical age (Arneson and Losey 1997). During 1940 to 1950, they became more popular to control crop pests and enhance agricultural productivity. It was only in 1962 when Rachel Carson in her book *Silent Spring* made aware of the deleterious effects of these pesticides on wild life, human health and the environment.

The repeated use of pesticides for control of crop pests has lead to development of resistance in pests, contamination of ground and surface water, atmospheric pollution and negative effects on non target organisms (Adams 1990; Beaument 1993; Howard et al. 1991; Mullen 1995). The realization of negative effects of these pesticides led to development of new concept called integrated pest control which subsequently came to known as Integrated Pest Management (IPM) (FAO 1966). Later, concepts of economic threshold (ETL) and economic injury levels (EIL) came into existence for making control measures. Different investigators have proposed more than 70 definitions of integrated pest management (Bajwa and Kogan 1998) between the period 1950 and 1998. The IPM emphasized the use of multiple tactics in compatible manner which are safe, cheap and sustainable with minimum harmful effects as compared to single control tactics (Dent 2000). Consequently an economically efficient and ecofriendly, more integrated approach was advocated to keep the pest populations below economic threshold levels (Norton and Mullen 1994).

3.3.1 Biological Control of Fruit Pests

There are several examples where biological control has proved successful. In 1762, Indian bird Mynah was introduced in Mauritius for the control of red locust *Nomadacris septemfaciata* Serv., The cottony cushion scale was controlled in California by introducing predatory insects from Australia during 1880s. Similarly, woolly aphid *Eriosoma lanigerum* (Hausm.) of apple in India was successfully controlled by introduction of hymenopterous parasite, *Aphelinus mali* (Hald.) in 1937 from North America. Australian lady bird beetle predator *Rodolia cardinalis* (Muls) obtained from the USA in 1929 and from Egypt in 1930 was used against *Icerya purchasi* in India which gave a spectacular control. *Trichogrammatid* egg parasitoid alone or in combination with *Bacillus thurigiensis* var. kurstaki or baculoviruses are some of the promising biocontrol agents used in horticultural ecosystems for control of coccinellid beetles or mealy bugs and scale insects. Similarly, parasitoid and predators for san jose scale and wolly aphis on apple and phytoseiid mites for tetranychid mites, entomopathogenic nematodes for soil borne pests and entomofungal pathogens for several hoppers and mites are found effective.

3.3.2 *Pre-requisites for Biocontrol*

3.3.2.1 Survey and Surveillance for Base Line Data

To understand, the real nature of problem and devise pest management strategies, 1 Survey and surveillance of pests is a prerequisite for decision making process. This should be followed by correct identification of the pest or disease. An incorrect identification may lead to unnecessary sprays and wastage of resources.

3.3.2.2 Pest Forewarning in Fruit Crops in Relation to Weather Factors

Congenial weather conditions encourage dises and pest outbreaks. Pandey et al. (2003) found that day and night temperature greater than 35° C and 23° C, respectively in conjunction with relative humidity in the range of 50–80 % and vapour pressure of 20–24 Hg were found conducive for breeding and population build up of mango hoppers. Most of the growers depend upon phenological schedule of crop to adopt pest control measures. However, this method may not prove effective in different years as the pest populations may not peak at the same stage of development each year. Evidently, forecasting of pest development based on day degrees is the most accurate method and disease and pests propensity can be predicted in space and time with reasonable accuracy on the basis of weather variables (Huda and Luck 2008). Weather forecasting models have been developed for apple scab, fire blight, and powdery mildew using weather data of specific field. Information on weather parameters at micro-level not only helps to forewarn the pest and disease attack but also suitable time for application of control measures.

3.3.2.3 Field Monitoring and Scouting of Pest Population

Insect pest populations fluctuate from 1 year to another, therefore, monitoring of their populations is essential to undertake management decisions (Barnes 1990). Survey and surveillance requires monitoring of injury caused by different insect pests. Survey routes at block and village levels be identified involving extension functionaries and farmers to initiate field scouting. The pests and disease need to be observed at intervals and control measurers only when pests and diseases cross ETL. Field monitoring of insect pests and beneficial organisms is needed to design, evaluate and proper execution of IPM practices (Peña 2004).

3.3.2.4 Threshold Level Determination

When spray is necessary as a control tactic, consider all the monitoring information and other factors such as tree stage, fruit age, pest stages, climatic factors, fruit variety, etc. for taking the right decision. The density of a pest, given all these factors, that will result in an economic loss of the crop, which outweighs the cost of the control measure, is termed the economic injury level (EIL). EIL have been scientifically determined for only some pests, thus for many pests and fruit varieties thresholds have been devised by the scientists based on experience, level of aversion to perceived risk of crop loss, etc.

3.4 Cultural Methods

Cultural methods have proved useful in controlling insect pests. Clean cultivation and ploughing creates condions unfavourable for development of insect pests by exposing dormant stages or burying them deep into the soil such as mealy bugs on citrus, guava and mango, san jose scale on apples and peach leaf curl aphid on peach. Pruning also helps to reduce the pest infestation pressure and increase the number of beneficial insects. planting of trap/cover crops in fruit orchards help to sustain the diversity and population of beneficial insects including natural enemies Hoyt and Burts (1974).

3.5 Mechanical and Physical Methods

Removal or destruction of pests is most beneficial methods especially for fruit growers in home gardens. Collection and destruction of insect stages helps in checking the outbreaks. Bagging of fruits is yet another method to prevent damage of butterfly and fruit sucking moths in pomegranate and citrus. Use of slippery alkathene sheets around tree trunks prevents climbing of certain stages of insects to the foliage. This method is widely used for control of mango mealy bugs (Singh et al. 2001a). Use of physical factors like heat, cold, sound, radiation etc. has also proved promising in certain situations.

3.6 Host Plant Resistance

Despite great achievements in host plant resistance in field crops, similar gains are still lacking in fruit crops. Sharma and Singh (2006) reviewed the host plant resistance in fruit crops in India and reported that information on insect damage is available in citrus, mango, apple, guava, peach, plum, banana, grapes, date palm, ber, sapota etc. but no efforts has been made to incorporate the resistant and or tolerant varieties in breeding programmes. Jothi et al. 1994 reported that most of the mango varieties are susceptible to fly attack except Langra, Dashehari and Bombay Green. Angeles (1991) reported that mango *Mangifera altissima is* not affected by mango pests such as leafhoppers, tip borers and seed borers, in the Philippines

3.7 Genetic Basis of Pest Control

The pest population can be controlled by disrupting their normal reproduction through use of substances that induce sterility. The sterilization of males using chemosterlents like TEPA, METEPA and apholate or with direct application of γ-radiation and releasing them has been used to control oriental fruit fly, melon fruit fly and codling moth. The females mated with such sterile males produce unfertilized and non viable eggs thereby suppressing the emerging pest populations. The use of sterile insect technique could considerably reduce insecticide usage on grapes and codling moth (Barnes and Eyles 2000).

3.8 Use of Novel Insecticides

In case of insect pests which are resistant to pesticides could be effectively managed by adopting insect growth regulators and less hazardous encapsulated pesticides (Blomefield 1997). These new insecticides have an added advantage of acting on biological processes of insects such as moulting and they have greater selectivity to target species specific and are less harmful to natural enemies. They are most commonly effective against lepidopteran pests, sucking insects, dipteran, leafminers lepidopteran pests, sucking insects, dipteran, leaf miners (Barnes 1999).

3.9 Use of Pheromone Traps

Mating disruption in codling moths and oriental fruit fly using pheromone traps has been found quite effective. Feeding sex attractants against fruit flies have given good results. Few traps are species specific like weevils and lepidopterans (Bailey et al. 1988; Gold et al. 2003). Several other traps like pheromone traps, yellow pan traps and sticky traps help to monitor the initial build up of pest populations. Florescent lights help to repel nocturnal fruit piercing moths (Whitehead and Rust 1972). Use of pheromone and other traps needs to be popularized among the farmers. Pheromone traps with specific lure should be installed at a distance of 50 m at 5 per trap/ha for each insect species. Similarly, yellow sticky traps for whitefly @10 yellow pans/sticky trap per ha are recommended. For monitoring of black fly Yellow colour traps reflecting light at the wave length of 550 nm may be installed and operated for 2 h in the evening.

3.10 Farmers Field School (FFS)

Farmer field schools help the farmers to interact and discuss the problems in farming. UN FAO organised first farmers field schools in Indonesia, since then more than two million farmers have participated in more than a dozen countries. It is good to have programme facilitators to get materials organised and solve problems in a participatory ways (Roling and van de Fliert 1994; Nelson 1994).. In recent years emphasis on participatory approaches to IPM has intensified. The adoption of fruit and vegetable IPM practices by farmers in India is the success story validated by World Bank funded National Agriculture Technology Project (Started in 1994) and Horticultural Technology Mission Mode Projects (NATP/HTMM projects). These projects resulted into creation of more awareness among the farming communities about the uses and abuses of pesticide application and their pressure on ecosystem. Because of this method, many undesirable effects associated with pesticide use may either be reduced or eliminated. These effects would include resistance development in pests to pesticides, pest resurgence and secondary pest outbreaks, excessive pesticide residue in table fruits and vegetables, cases of poisonings (human, domestic animals and wildlife), adverse impacts on non-target organisms, and environmental pollution. Additionally, lower imports of pesticides leads to savings in foreign exchange.

3.11 Pest Management Policy for India

Indiscriminate and injudicious use of chemical pesticides in agriculture has resulted in several associated adverse effects such as environmental pollution, ecological imbalances, pesticides residues in food, fruits and vegetables, fodder, soil and water, pest resurgence, human and animal health hazards, destruction of biocontrol agents, development of resistance in pests etc. Therefore, Govt. of India has adopted Integrated Pest Management (IPM) as cardinal principle and main plank of plant protection in the overall Crop Production Programme which could be interpreted as policy formulations incorporated into the Agenda 21 declarations of this country.

To alleviate the ill effects of pesticides, India has officially adopted IPM as its policy and is a prominent feature in recent Five Year Plans. On a broader scale, IPM is defined and explained in terms that encompass the farm families and their environment, and regional food security. The essential element for IPM includes one or more management activities that are carried out by farmers that result in the density of potential pest populations being maintained below levels at which they become pests, without endangering the productivity and profitability of the farming system as a whole, the health of the farm family and its livestock, and the quality of the adjacent and downstream environments.

Pests of economic importance in temperate and tropical areas and their management are discussed below:

3.12 Pests of Economic Importance in Temperate Fruits

A large number of pests have been observed on different horticultural crops with the effects ranging from mild damages to serious losses in production and quality. The arthropod pest spectrum of various crops in temperate fruits is very wide, they include:

3.12.1 San Jose Scale

San Jose scale, *Quadraspidiotus perniciosus* is one of the most serious pests of apple. Its first peak commences in the second or third week of May and second peak during second or third week of August and the third generation overwinters. The pest is a sucking type, feeding on the young and old branches and twigs, reducing the vigour and vitality of trees, the surfaces of the branches appear ash grey, fruit is severe damaged rendering it unmarketable.

3.12.2 Woolly Apple Aphid

Woolly apple aphid (*Eriosoma lanigerum*), is another most serious pest on apple. It appears during attack 2nd week of March and continues up to September. The pest migrates from roots to aerial portion and vice versa during spring and fall, with maximum dispersal during day between 4 and 6 p.m. These form colonies on the bark of the tree trunk and the branches characterized by white cottony cushion and suck out the plant sap, weakening the tree and make it prone to the attack of other diseases.

3.12.3 Apple Green Aphid

Apple green aphid (*Aphis pomi*) drains sap from the leaves, flowers, buds and terminals of apple trees reducing their vitalities. It remains active from 2nd week of April to last week of May.

3.12.4 Mites

Several species of phytophagous mites have been reported attacking apple in Kashmir (Rather and Masoodi 1988). However, the European red mite (*Panonychus ulmi*) and two spotted spider mite (*Tetranychus urticae*) are the most serious one. European Red Mite has emerged as a limiting factor in apple production in Kashmir. European red mite completes six generations with high biotic potential, starting its activity in mid spring and continuing till late autumn. The infestation of the mites causes yellowing and bronzing of the foliage, affecting the photosynthetic efficiency of the leaves and results in the premature leaf drop. The situation is worsened by the presence of *Alternaria* causing complete leaf blotch rendering the apple tree denuded (Zaki et al. 2004).

3.12.5 Stem Borer and Bark Beetles

Stem borer (*Aeolesthes sarta*) and bark beetles (*Scolytus* sp.), the pests of both fruit as well as forest trees, have emerged as a serious problem for apple and walnut.

3.12.6 Codling Moth

Codling moth (*Cydia pomonella*) is also one of the serious pests of apple, pear, walnut and apricot etc. However, in Jammu and Kashmir its incidence is confined to Leh and Kargil. The pest inflicts direct damage to the apple and apricot rendering the fruit unfit for the human consumption and market. The extent of losses caused to apple crop alone is the tune of 50 % in Ladakh.

3.12.7 Indian Gypsy Moth

Indian gypsy moth (*Lymantria obfuscata*) is a voracious polyphagous pest. It attacks apple, walnut and other fruit trees and plantations.

3.12.8 Chaffer Beetles

Chaffer beetles (*Adoretus simplex, Brahmina coriacea, Holotrichia longipennis*) damage fruit crops especially High-density plantation of apple and walnut. Among 70 species of leaf feeding beetles recorded in temperate regions of India, scarabaeids, cetoniids, curculionids and chrysomelids *in* that order cause significant damage to pome and stone fruits.

3.12.9 Walnut Weevil

Walnut weevil (*Alcides porrectirostris*) prefers thin shelled varieties to the less valuable thick shelled walnut types. The kernel of the nut is damaged and the attacked fruit usually drops or is unmarketable. The dropping of the fruit starts from 2nd week of May till end of August. Another important pest associated with walnut is **grey weevil** (*Myllocerus* spp.). It feeds on foliage by biting holes in the lamina or by peripheral feeding on margins. The maximum damage is caused from mid May to mid June.

3.12.10 Leaf Rollers

Among leaf rollers, folders *Archips pomivora* is the most important. It causes maximum damage in June. It attacks foliage but injury on fruits and their rotting cause economic losses. An important pest of Red cherry and Black cherry is **cherry fruit worm**. The extent of damage if caused by this pest is predominantly visible when the cherry starts to ripen. The cherry is also infested by Cherry fruit fly if the fruit harvest is delayed in certain localities.

3.12.11 Pomegranate Fruit Borer

Pomegranate fruit borer (*Deudorix epijarbas* Moor) a widely distributed and predominant pest of pomegranate limiting its cultivation and causing considerable loss to the crop.

3.12.12 Almond Mealy Bug

Almond mealy bug (*Drosicha dalbergia*) is a serious pest on almond. It feeds on both aerial and underground parts of the plant. The infested plants exhibit sickly appearance, thereby, reducing the quality and quantity of the fruit.

3.13 Pome Fruits

Pome fruits comprising of apple and pear are attacked by a variety of insect pests which cause considerable damage. More than 70 insects and mite pests have been reported attacking pome fruits. Among several pests attacking apple and pear San Jose scale, *Quadraspidiotus pernicious* (Comstock) is one of the most serious pests in the world. It has a very wide range and can be disseminated through infested nursery stocks to different countries of the world (Kozar et al. 1994). Aphids are brown or greyish in appearance, suck sap from the plants and may result in gall formation on aerial and underground parts. The other pests include Tent caterpillar, *Malacosoma indica* Walker, Indian gypsy moth, *Lymantria obfuscata* Walker infesting temperate fruits in different parts of India (Masoodi 1985; Srivastava and Masoodi 1985). The important predators and parasites include ectoparasitoid *Aphytis* sp., endoparasitoid *Encarsia perniciosi* and coccinellids predator, *Chilocorus bijugus*, *C. circumdatus* and *Coccinella septumpunctata* (Bull et al. 1993). Peach is the major stone fruit grown commercially in temperate to subtropical conditions. It is attacked by leaf curl aphid, peach mealy aphid, peach green aphid, san jose scale, fruit moth, peach fruit flies, flat headed stem borer, bark eating caterpillar, peach twig borer, grey weevils and cockchafer beetles. The rapid expansion of area under cultivation has led to the emergence of many indigenous insect pests of secondary importance as major pests (Tables 3.1, 3.2, 3.3, 3.4 and 3.5).

Table 3.1 Major arthropod pests of pome fruits under Kashmir agro-climatic conditions (Source: Zaki et al. 2008b)

Pest	Scientific name	Active period	Remarks
San Jose scale	Quadraspidiotus perniciosus	Mar.–Oct.	Serious key pest
Wooly apple aphid	Eriosoma lanigerum	Feb.–Nov.	Perennial pest
Gypsy moth	Lymantria obfuscata	Apr.–June	Attaining a serious status
Apple green aphid	Aphis pomi	May–Aug.	Perennial pest
Short hole borer	Scolytoplatipus sp.	Apr.–Oct.	Moderately serious pest on apple but most serious on walnut
Stem borer	Aeolesthus sarta	Apr.–Nov.	Serious pest
Pear psylla	Psylla pyricola	May–Sep.	Minor pest
Leaf minor	Lyonetia prunifoliella	Apr.–June	Moderately serious
Chaffer beetles	Protacta neglecta, Adoretus spp.	Grub attacks throughout the year. Adult from June–July	Serious, and moderately serious, respectively
Tent caterpillar	Malacosoma kashmiriensis	Apr.–Nov.	Moderately serious
Walnut weevil	Alcides porrectirostris	Apr.–Sep.	Most serious
Pin hole borer	Scolytoplatipus sp.	Apr.–Oct.	Most serious
Cabbage butterfly	Pieris brassicae	Mar.–Oct.	Serious
Mustard saw-fly	Athalia proxima	Feb.–May	Sporadic pest
Aphid	Myzus persicae, Lipaphis erysimae	Throughout the year	Most serious
Diamond backmoth	Plutella xylostella	June–Oct.	Moderately serious
Bark beetle	Scolytus nitidus	Apr.–Oct.	Moderately serious
Cut worm	Agrotis ipsilon	May–June	Serious
Army worm	Mythimna separata	Mar.–Sep.	Sporadic pest
Cabbage aphid	Brevicoryne brassicae	Throughout the year	Serious
European red mite	Panonychus ulmi	Apr.–Nov.	Key and serious pest
Two spotted spider mite	Tetranychus urticae	Apr.–Nov.	Key and serious pest

Table 3.2 Insect pests attacking walnut in Kashmir (Source: Zaki et al. 2008b)

S. No.	Name of the pest	Scientific name	Order	Family
1	Walnut weevil	Alcides porrectirostris (Marshall)	Coleoptera	Curculionidae
2	Stem borer	Aeolesthes sarta (Solsky)	Coleoptera	Cerambycidae
3	Pin hole borer or shot hole borer	Scolytoplatipus spp.	Coleoptera	Scolytidae
4	Chaffer beetle	Protecta neglecta Adoretus simplex	Coleoptera	scarabaidae
5	Indian gypsy moth or hairy caterpillar	Lymantria obfuscata (Walker)	Lepidoptera	Lymantriidae
6	Walnut leaf defoliater	Chaetoprocta odata (Hewitson)	Lepidoptera	Lycaenidae

Table 3.3 Comparative bioefficacy of different summer spray oils against European red mite (ERM) in Kashmir (Source: Zaki et al. 2008b)

		Per cent mortality			
		Arabal (Srinagar)	Hanjiwara (Baramulla)		
Treatment	Concentration (%)	2004	2003	2004	Pooled mean
PD spray oil	0.75	83.53	92.42	82.00	86.83
	1.0	92.99	95.67	91.44	94.15
ATSO spray oil	0.75	83.21	90.45	84.05	83.34
	1.0	89.20	96.74	88.18	92.45
Orchex-796 spray oil	0.75	83.01	92.12	82.75	85.96
	1.0	93.35	94.69	87.09	91.71
Arbofine spray oil	0.75	84.53	92.94	82.64	87.75
	1.0	92.35	95.24	91.08	94.01
HP spray oil E	0.75	81.43		87.63	84.53
	1.0	90.18		88.40	
HP spray oil	0.75	77.98		81.21	79.59
	1.0	88.18		87.38	87.78
Sparrow 888 plus spray oil	0.75	76.88		71.89	74.38
	1.0	88.69		80.20	84.44
Control		21.61		25.87	23.74

Each figure is mean of 15 replicates and each replicate is mean of 12 observations

3.14 IPM for Major Insect Pests on Temperate Fruit Crops

3.14.1 Mineral Oil Sprays

Horticulture mineral oils that are used for the management of important fruit pests are physical toxicants and result in the death of insect pests due to asphyxiation. They are eco-friendly and bio-safe as they come from earth and are formed as a

Table 3.4 Showing plant protection schedule for the management of pests and diseases of apple (Source: Zaki et al. 2008b)

Spray	Tree stage	Fungicide/100 lit of water	Insecticide/Acaricide/HMO per 100 lit of water
I	Green tip (delayed dormancy) (Mid March-Mid April)	–	Horticultural Mineral Oil (HMO) @ 2 litres (2 %)
II	Pink bud (Mid April-End April)	Myclobutanil 10 WP (30 g) or Dodine 65 WP (60 g) or Bitertanol 25 WP (50 g) or Fenarimol 12 EC (40 ml)	[c]3–4 days after II fungicidal spray in case spray I is missed. Chlorpyriphos 20 EC (100 ml) or Summer spray oil (750 ml) or Fenazaquin 10 EC (40 ml) or Herbal (200 ml)
III	Petal fall (80–100%)	14–18 days after II spray Hexaconazole 5 EC (30 ml) or Difenaconazole 25 EC (30 ml) or Triadimefon 25 WP (50 g) or Diniconazole 25 WP (40 g)	3–4 days after III fungicidal spray Quinalphos 25 EC (100 ml) or Methyl-o -Demeton 25 EC (80 ml) or Phosalone 35 EC (140 ml) or Summer Spray Oil (750 ml) or Fenazaquin 10 EC (40 ml)
IV	Fruit let (pea size)	14–18 days after III spray Ziram 80 WP (200 g) or Mancozeb 75 WP (300 g) or Ziram 27 W/V (600 ml) or Captan 50 WP (300 g) or Propineb 75 (300 g) or Mancozeb Flowable 35SL (300 ml) or Zineb 75 WP (300 g)	[c]3–4 days after IV fungicide spray Chlorpyriphos 20 EC (100 ml) or Methyl-o- Demeton 25 EC (80 ml) or Phosalone 35 EC (140 ml) Or Dicofal 18.5 EC (108 ml) or Summer Spray Oil (750 ml) or Fenazaquin 10 EC (40 ml)
V	Fruit development I	14–18 days after IV spray Bitertanol 25 WP (50 g) or Dithionon 75 WP (75 g) or Penconazole 10 EC (50 ml) or Fenarimol 12 EC (40 ml)	3–4 days after V fungicide spray Chlorpyriphos 20 EC (100 ml) or Methyl-o-Demeton 25EC (80 ml) or Dimethoate 30 EC (100 ml) or Docofol 18.5 EC (108 ml) or Summer Spray Oil (750 ml) or Fenazaquin 10 EC (40 ml) or Abamectin 1.8 EC (55.5 ml)

(continued)

Table 3.4 (continued)

Spray	Tree stage	Fungicide/100 lit of water	Insecticide/Acaricide/HMO per 100 lit of water
VI	Fruit development II	14–18 days after V spray Hexaconazole 5 EC (30 ml) or Myclobutanil 10 WP (30 g) or Triadimefon 25 WP (50 g) or Diniconazole 25 WP (40 g)	–
VII	Fruit development III	[d]14–18 days after VI spray Difenoconazole 25 EC (30 ml) or Dodine 65 WP (60 g) or Dithionon 75 WP (75 g) or Penconazole 10 EC (50 g)	3–4 days after VI fungicidal spray Chlorpyriphos 20EC (100 ml) or Dimethoate 30EC (100 ml) or Endosulfan 35EC (140 ml) or Methyl-o-Demeton 75 EC (80 ml) or Phosalone 35 EC (140 ml) or Fenazaquin 10EC (40 ml) or Abamectin 1.8 EC (55.5 ml)
VIII	Fruit development IV	[a]14–18 days after VII spray Hexaconazole 5 EC (30 ml) or Bitertanol 25 WP (50 g) or Fenarimol 12 EC (40 ml)	–
IX	Pre-harvest	[b]25 days before harvest Ziram 80 WP (200 g) or Mancozeb 75 WP (300 g) or Ziram 27 W/V (600 ml) or Captan 50 WP (300 g) or Propineb 75 WP (300 g) or Mancozeb Flowable 35 SL (300 ml) or Zineb 75 WP (300 g)	–
X	Post harvest	–	[e]Ethion 50 EC (100 ml) or Fenazaquin 10 EC (40 ml) or HMO (75 ml) or Chlorpyriphos 20 EC (100 ml)

[a]Need based spray if leaf spotting incidence is more than20 %
[b]Need based spray for long term storage of fruits
[c]Need based spray when 5–8 mites/leaf are observed
[d]Need based spray when 10 or more mites/leaf are observed
[e]Need based spray when 20 or more mites/leaf are observed during late August and spray at Fruit Development III (during August) has been missed

Table 3.5 Biological control agents reported from Jammu and Kashmir (Zaki et al. 2008b)

Pest	Pest status	Biological control agents
San Jose scale (Quadraspidiotus perniciosus)	Serious key horticulture pest	Aphytis proclia, A. paramculicornis, Azotia kashmirensis, Maristta camesi, Encarsia perniciosi (Parasitoids); Chilocorus bijugus, Pharoscymnus flexibilis, Cybocephalus sp., Sticholitis marqinalia, Adalia tetraspilota, Melachius sp., Harmonia sp., Coccinella septempunctata, Chrysoperla carnae (Predators)
Apple aphid (*Aphis pomi*)	Perennial pest	Coccinella septempunctata Linnus, Adalia tetraspilots Hoe, Hippodamia variegate Goeze, Harmonia breiti Madar, Pharoscymnus sp., Symnus grsoilis, Exochomus uropygialia, Plantynaspis sp. Mulsant, Propyles japonica Thunberg, Dropephylla almorensis Cham, Chrasoperis carnae Stephen, C. prestes Banks, Tomnoatethus (Ectemnus) paradoxus Hetchinson, Lygus sp., Episyrphus balteatus DeGreer, Eumerus elbirfrons, Metasyrphus confrator Weid, Eristalis tenax Linnaeus, E. arbustorum Linnaeus, E. correalis Fabricius, Scaeva pyrastri Linnaeus, Sphaerospheria ecripta Linnaeus, Aphidoletes aphidomyza Rondani, Pachyneuron sp.
Wooly apple aphid (*Eriosoma lanigerum*)	Key pest	C. septempunctata, C. bijugus (Predator); Aphelinus mali (Parasitoid)
European red mite (*Panonychus ulmi*)	Key and serious pest of apple	Anystis baccurum Linnaeus, Eusius vignus Rishi and Rather, Phytoseius domesticus Rather, Euseius insanus Khan and Chaudhri, Amblyseius finlandicus Oudemans, Phytoseius mixtus Chaudhri, Euseius sp., Cheyletus sp., Agistemus sp. (Acarine predators); Hippodamia sp., Stethorus punctillum, Stethorus punctum, Chrysoperla carnea, Oligota sp., Anthocoris sp., Scolothrips sp. (Arthropod predators)
Mealy plum aphid (Hyalopterus arundinis)	Moderately serious	Coccinella undecimpunctata, Adalia tetraspilota, Hippodamia variegata, Scymenus gracilis, Episyrphus balteatus, Eumerus nepalensis, Sphaerophora indica, Syrphus sp., Anthocoris sp., Chrysoperla carnia, Trioxys rishii
Peach curl aphid (Anuraphis helichrysii)	Serious pest	Scymenus gracilis
Walnut aphid (Callaphis juglandis)	Serious pest	Metasyrphus corollae
Oystershell scale (Lepidosaphes kirgisica)	minor	Aphytis paramaculicornis

result of transformation of plankton due to the action of high pressure and high temperature over millions of years. Comparative bioefficacy of different summer spray oils against San Jose scale and European red mite (ERM) in Kashmir during 2004–2008 resulted in a maximum mortality per cent of about 87.71 and 94.15, respectively (Table 3.7). A number of such mineral oils, viz., ATSO spray oil, HP spray oil, Arbofine spray oil, Mak all season spray oil, etc. at 2 % and 0.75 % concentrations during winter and summer, respectively, have been found effective against the scale and mites (Sofi et al. 2008; Zaki et al. 2008a, b).

3.14.2 Chemical Control

The most critical issue on the use of chemical control is the precise timing of insecticidal application, which is important for achieving the desirable control of important insect pests. Every year a spray schedule is prepared where the proper timing and application of particular pesticides is provided to farmers. By and large, two summer sprays of an acaricide/insecticide or insecticide-cum-acaricide is advocated during the third to fourth week of May and first to second week of July, respectively, to control and manage most of the arthropod pest of apple crop in Temperate Horticulture. Normally no acaricide/insecticide is recommended beyond that period, thereby ensuring that no acaricide/insecticide residues are in the fruit at the harvest. Need based sprays are recommended only if the mite and /or San Jose scale populations exceed economic threshold level during August or after harvesting.

3.14.3 Biological Control

The biological control agents associated with the important pests of horticultural crops. The important biocontrol agents include *Aphytis proclia, Encarsia pernisiosi, Chilocorus bijugus, Adalia tetraspilota, Coccinella septumpunctata* and *Chrysoperla carnae*, against San Jose scale (Anonymous 1988); *C. septumpunctata, A. tetraspilota* and *Symnus grsoilis* (against Apple aphid); *Aphelinus mali*, *C. septumpunctata*, *C. bijugus, Amblyseius finlandicus, Phytoseius mixtus* and *Stethorus* sp. (against European red mite and Two spotted spider mite); *Scymenus gracilis* (against Peach leaf curl aphid). During 1989–1990 studies on the population record of predators associated with woolly apple aphid in Kashmir revealed presence of *Aphelinus mali* an eulophid endoparasite and as many as seven predators. During the year 2000, observations in respect of predators were same. However, during 2002 only three predators including *Chilocorus* sp., *Coccinella* sp. and *Phasoscymnus balteatus* were found. The prevalence of pests has been observed to directly influence the population status of its natural enemies. Appearance and population built-up as well as disappearance of coccinellid beetle and syrphid fly was found to follow the population built-up pattern of their host aphid.

3.14.4 Cultural Control

3.14.4.1 Burlapping

Burlapping is a technique of placement and wrapping of a gunny bag band or corrugated cardboard band around the tree trunk at 4–6 ft above ground level to provide artificial hibernating/nesting sites. Gunny bag burlaps have been found to be more effective and economic because these can be re-used. Burlapping in integration with the insecticidal spray at Kargil under moderate infestation showed that the number of larvae trapped in corrugated cardboard in different insecticidal spray ranged between 2.3 larvae in fenvalerate treatment to 12.3 larvae in endosulfan as compared to 17.3 larvae in unsprayed control. The results indicated that even without insecticidal application the efficacy of this practice is quite satisfactory. Gunny bag burlaps have been found to be more effective and economic because these can be re-used. The practice of burlapping have also been found effective in the trapping, collection and destruction of caterpillars and eggs of Indian gypsy moth, which otherwise take shelter under the bark of the trees.

3.14.4.2 Fruit Destruction

Destruction of infested fruits is one of the important cultural practices for the management of various insect pests. This method can bring up the effective control of fruit borers eg. Cherry fruit fly. The practice of destroying or burying the apple fruit infested with codling moth in farmer's field could reduce larval population and subsequent incidence for the next year.

3.14.4.3 Collection and Destruction of Wood

Wood is more susceptible to the infestation of stem borers, pin hole borers and short hole borers. Therefore, placing some logs of dry hard wood in the orchards may considerably prevent these orchards from infestation of such insect pests. The wood thus placed act as a trap strategy to these pests, which need to be collected and destroyed.

3.14.4.4 Training and Pruning

Proper training and pruning of fruit trees is very important cultural practice as it helps in the management of economically important insect and non-insect pests eg. Aphids, San Jose scale, European red mite, bark beetle, etc. Off season biology of the important pests is carried over on the tree branches and trunk/bark, therefore, pruning of infested parts reduces the initial pest load in the fruit orchards. The borer and bark beetle infested branches and twigs should be removed.

3.14.4.5 Fertilizer Management

It has been found that the use of fertilizers help in improving the fertility of soil, however, the balanced dosage of fertilizers is equally important. When the nitrogen is applied in higher amounts, it increases the quantity of sap but at the same time increases the infestation of sap sucking pests eg. Green aphids, black aphids, wooly aphid, European red mite, peach leaf curl aphid, San Jose scale, two spotted mite, almond mealy bug, etc. Balanced doses of fertilizers, micronutrients and higher doses of potassium were found to reduce the infestation of ERM and aphids.

3.14.4.6 Habitat Management

The ground cover management around fruit trees is another important cultural practice as it provides habitation to various predators eg. *Amblyseius* sp. during application of pesticides and dormancy. It also harbours the prey for such predators during the off season. Therefore, in order to enhance the efficiency of biological control of European red mite, it is advised to allow some flora around the apple tree trunks.

3.14.4.7 Scrapping

Scrapping of San Jose scale infested twigs with khurpi is another technique, yet uncommon among the farmers but can help in reducing the infestation and further increment of scale population considerably.

3.14.4.8 Quarantine Control

The distribution of codling moth in India is restricted to Ladakh only. In order to prevent its accidental entry to other localities and particularly the temperate fruit growing states of Himachal Pradesh, Uttaranchal, Arunchal Pradesh and Uttar Pradesh, the pest has been brought under domestic quarantine through an SRO promulgated by the Government of Jammu & Kashmir under "Destructive Insect Pest and Weeds Act, 1914 amended 1968" Clause-II of Government of India. The export of fresh fruit from Ladakh to other places is banned which is regulated by the Jammu and Kashmir State Department of Horticulture, at the Sonamarg post. The domestic quarantine has, therefore, checked its entry form Ladakh to other places. However, with the operating of Leh-Manali road and no quarantine check post on the highway there are all apprehensions of this pest getting entry into Himachal Pradesh through the fresh fruit transport by the tourists and exchange of plant materials. The issue needs to be reviewed seriously by the Government of Himachal Pradesh and the Government of India for Northern and Noth eastern states. Import of plant materials from other countries in the past without following proper regulations have introduced a number of pests and diseases not reported earlier. Root rot of apple in epidemic form in Kashmir and the report of dreaded blackline virus in walnut nursery at Zainpora, Kashmir which prompted the Department of Horticulture to destroy the plantations, are two examples of many unwanted troubles brought in due to poor quarantine. Fire blight caused by *Erwinia amylovora* reported infecting roses by Indian Institute of Horticulture, Bangalore (Personal communication), has probably been introduced in the country along with the rose plants imported for the new varietal introductions. The apple growing states of India should be vigilant on the introduction of rose plants infected with the bacteria through All India co-ordinated research project on Ornamental Crops, which may destroy apple crop on which the disease is not reported so far. The domestic quarantine measures need to be enforced on the movement of the rose plants in the apple growing states.

3.15 Constraints in Management of Insect Pests of Fruit Crops

Despite several technological innovations and expansion of area under cultivation of fruits, the pace of development has not been up to the mark due to devastating emerging challenges in management of insect pests and diseases causing poor productivity and quality in apple/walnut orchards in Kashmir, thereby, putting enormous pressure on the sustainability of horticulture and export of fruits to international markets. European red mite (*Panonychus ulmi*), stem borer and bark beetles on fruit plantation have emerged as most serious pest problems, due to drought like conditions in the valley during last few years. European red mite has become one of the limiting factors in apple orchards in recent past, posing a serious challenge in its management.

3.15.1 Mineral Oil Sprays

Dormant spray oils which were earlier used regularly against San Jose scale and would also kill over wintering eggs of European red mite, had been partially abandoned due to ease of application of synthetic insecticides, high input and labour cost, hostile weather conditions during dormancy, shift in priorities from pests to apple scab disease control when it appeared in epiphytotic form, replacement of diesel oil by summer pesticide spray. Such an omission in some orchards has resulted in population explosion of the ERM to damaging proportions.

3.15.2 Disruptive Use of Pesticides

Out break of ERM, San Jose scale and wooly aphid has largely been due to indiscriminate use of broad-spectrum pesticides killing natural enemies. Consequently in the absence of natural controlling factors the mite population and San Jose scale increased to alarming proportions. The predator activity in the pesticide managed and unmanaged orchards showed higher activity of predators with lower mite infestation in the latter in comparison to the former (Zaki and Chan 2003).

3.15.3 Insect Resistance to Insecticides

Development of resistance of mites to pesticides is a major challenge influencing the changing pattern of chemical pest control strategies and is, therefore, the most cogent reason for an integrated approach to pest control in apple orchards. ERM has developed resistance to almost the entire-spectrum of acaricides and insecticides-cum-acaricides used in tree fruit production (Zaki et al. 2005) in Kashmir. European red mite has developed 7–80 fold resistance to endosulfan, dimethoate, chlorpyriphos, methyl-o-demeton and fenazaquin.

3.15.4 Non-Standard Pesticides

The sale of spurious pesticides has resulted in loading the orchards with disastrous chemicals, which besides their toxic hazards have changed the balance between pests and natural controlling factors. Such spurious pesticides besides being ineffective impose a sub lethal dose selection pressure on the key pests.

3.15.5 Nutritional Imbalance

There exists a direct relationship between levels of nitrogen and mite population. Most of the orchards do not receive balanced dose of fertilizers and urea seems to be predominant fertilizer applied to apple trees allowing succulence and predisposition to sucking pests.

3.15.6 Adverse Effect on Natural Enemies

The repeated applications of insecticides to temperate fruits with broad-spectrum pesticides have a destructive effect on beneficial insect populations. Fields and orchards under heavy treatment schedules have become veritable “biological deserts”. The re-establishment of adequate complexes of natural enemies in such biological deserts may require few years after the use of insecticides/acaricides is abandoned.

3.15.7 *Pesticidal Poisoning of Pollinators*

The poisoning of honey bees by the application of insecticides is a major problem of beekeepers especially where applications are made to crops during bloom period. Because of domestication honey bees have not developed the wide variety of detoxification enzymes as found in many other insects, therefore, are highly susceptible to pesticides. Wild bees such as bumble bees, *Megachile* and other bees are also important for pollination and are highly susceptible to insecticides.

3.15.8 *Pre-mature Leaf Fall*

The premature leaf fall in apple orchards has become an alarming syndrome in recent years. The problem is enhanced and further aggravated by *Alternaria* leaf blotch particularly under high temperature and higher humidity regimes. Higher mite population in the *Alternaria* affected orchards and concurrent presence of the two in the affected orchards lead to higher disease incidence, intensity and premature leaf fall.

3.15.9 *Socioeconomic Constraints*

Participation of people/farmers in taking up the management strategies seriously is very poor.

3.15.9.1 Kashmir as Agro-export Zone

Declaration of the state as Agro-Export Zone for the apple and walnut fruits was one of the major challenge before growers, development officers and the agricultural scientists to figure in export map. The fruits for export have to be qualitative, competitive, standard grade and produced either organically or free of all pesticide residues. Apple concentrate being exported have been found free from objectionable pesticide residues. There is strong need to promote biological and genetic methods of pest control as vital components of IPM to further enhance the potential of apple and walnut export.

3.15.9.2 High-Density Plantation

High-density plantation with dwarf cultivars is an important strategy for enhancing productivity manifold but it has been found to be highly susceptible to pests. Therefore, new management strategies have to be developed for combating the pest in high-density plantation in order to enhance productivity. High density plantation have been found to be infested by a different pest spectrum in addition to the pest recorded in traditional orchards, therefore the pest and disease management in this system is more management, labour and cost intensive.

3.16 Changing Pest Scenario in Fruits

In recent past, many new varieties have been introduced to increase the production growth rate of fruits. These input intensive high yielding varieties/hybrids with good agronomic traits and yield attributes have gradually changed the pest situation. Consequently, a shift in pest status has been realized in time and space. Many pests have adapted new hosts, developed resistance to pesticide and often there are secondary out breaks. New pests like mango stone weevil showed its severity in mango. Earlier shoot gall maker and *Deudorix isocrates* of aonla known to be minor pest is gradually becoming a regular problem in aonla, whereas, fruit fly have intensified the severity of occurrence.

3.17 Current Status of Pest Control in Fruits

Among the various methods of pest control, chemical method still enjoys first choice because of its quick and certain action. Unlike cereals, pulses and other crops, fruits harvested at shorter intervals used, as fruits are prone to retain pesticide residue. The persistent residue of major groups of insecticides has been detectable to considerable extent. India being the home of maximum fruit crops is a large reservoir of the natural enemies of pests and pollinators involved with them. But careful studies are required to use the predators/parasitoids/pollinators as a tool of pest management and increasing the fruit yield. Release of *Trichogramma* spp. (egg parasitoid), *Chrysoperla carnea* (predator), Syrphid fly and coccinellid beetle has shown positive results. The feasibility of biocontrol agents in fruit is comparatively less because of the annual nature of crop with short growing period.

3.18 Non Chemical Approaches

3.18.1 Cultural Practices

This component of Integrated Pest Management requires no additional budget for the farmers. Rather, these aspects are more sustainable and eco-safe. But it needs careful consideration with sufficient planning in advance. Crop husbandry of fruit crops is radically different to that of field crops and under many situations the pest problems are permanent nature. Several cultural practices if performed regularly in orchard ecosystem would be helpful in either elimination or at least reducing the carryover of pest population. Routine agronomic practices like selection of varieties, intercultural operation like sanitation, tillage, pruning and thining and raking of the soil etc., trap cropping/intercropping should be suitably modified to make uncongenial condition for pests and to enhance the activity of natural enemies.

Keeping in mind the diversity and intensity of pests in particular place, selection of resistant/less susceptible varieties, holds well in pest management. Plant resistance provides a built in ability to allow less pest and cut off the extra load of insecticides in the crop. Being completely safe, host plant resistance fits well with all other components. Unlike cereals, pulses, oilseeds and vegetables in fruits very less number of resistant/less susceptible varieties have been developed.

3.18.2 *Tillage*

Many insects like mango mealy bug, gall makers, leaf-eating weevil, litchi nut borer, fruit fly of guava, pear and peach *Bactrocera dorsalis* and ber *Carpomyia vesuviana* Costa are found to be controlled by ploughing the soil during May-June as the gorged female lay eggs in the soil. The eggs are exposed to either intense heat or natural enemies. Similarly, the pre- pupal and pupal populations of Bactrocera dorsalis occurred in the soil and shallow ploughing with cultivator.

3.18.3 *Pruining*

Pruning the citrus plants in December–January will effectively check the incidence of citrus leaf minor, *Phyllocnistis citrella.* Pruning of branches in mango reduces the prevalance of leaf webber and gall insects (Table 3.6).

Table 3.6 Resistant/tolerant varieties, germplasm of some fruit pests

Crop/Pest	Varieties/germplasm
Ber fruit fly	Katha & Tikadi, Dodhia and Meharun (Pareek 1983)
Ber fruit borer	Gola, Benarasi, Khatti, Sua (Batra 1991)
Citrus leaf minor	Citrumelo, Morton, Sacoton and Savage (Least susceptible) Batra and Sandhu 1981, Batra and Sharma 1993
Pomegrannate fruit borer	Speen, Sakar, G- 137 and Muskat (Least susceptible)

3.18.4 Sanitation (Collection and Destruction)

Hand picking and destruction of insect stages e.g. hand picking of lemon butterfly caterpillar, *Papilio demolius* Linn., brushing of scales of mango, dropped guava, ber, pear, peach etc., fruits infested by fruit fly maggot *Bactrocera* spp., dropped mango leaves mined by grubs of leaf cutting weevil, mummies of litchi for litchi nut borer is most effective method in checking the pest outbreaks. Removal and destruction of infested plant parts is also effective in regulating the pest populations of citrus leaf minor, mango stem borer, *Bactrocera rufomaculata* de Geer and guava (*Microcolona technographa* Meyrick).

3.19 Mechanical Methods

3.19.1 Exclusion Through Screens and Barriers

Application of fluppy band of 15 cm wide or a band of a folded slippery sheet like alkathene around the trunk of mango against mealy bug. Wrapping individual fruits of pomegrannate and citrus with butter paper envelops to save them from the attack of anar butterfly and fruit sucking moths. Covering the fruits with polythene bags for controlling fruit flies in case of guava and mango, squirrels in guava.

3.19.2 Traps

Different types of traps is also recommended for the monitoring, reducing the pest population and will also help in application of in the orchard. Bait trap Banana or other over ripe waste fruit pulp (1 kg) + yeast (10 g) + citric acid (5 g) @ 100 g/ plastic made flat tray. Monitoring may also be made with methyl eugenol (2 ml) + malathion (2 ml) + water (1lit) by keeping in a jar @ 500 ml per jar and No mate® cu-lure block for the fruit fly. A circular whole should be made in the jar to facilitate the entry of fruit fly.

Pheromone trap and light traps is also effective for the many lepidopterous insects.

3.19.3 Preparation Procedure for Bait Traps

The plastic tray should be covered with another inverted tray of the same type. The gap between the rims of the both the trays should be kept 4 cm to avoid the problem of bird poisoning and deposition of dust and the rain water.

3.19.4 *Jarring*

Placing cut pieces of banana pseudostem to attract banana pseudostem weevil, *Odoiporus longicollis*. Installing jam jars suitably baited with sugar, yeast and fruit juice and sunk to the rim in soil to trap crickets or hung on tree to trap fruit sucking moths.

3.20 Natural Enemies

Huge diversity of natural enemies consisting of predators, parasitoids and parasites are found in fruit ecosystem which are rersponsible for causing damage to insect pests, thus sustaining the natural balance of keeping the insect pests under check. Till now except a few, most of these natural enemies are under exploited for pest management in vegetables. Among the egg parasites *Trichogramma chilonis* Nagraja has been utilized to check the anar butter fly @ 250,000 parasitised eggs/ha at intervals of 10 days to reduce infestation.. Release of *Cryptolaemus montrouzeri* @ 20–30/tree gives excellent control of *Ferrisia virgata* on guava. *Verticillium lecanii* @ 1×10^8 spore/ml against mango leaf hopper. In citrus release of *Cotesia nigita* @ 10–15/tree and *Aphytis melinus* De Bach @ 2000/tree for red scale. *Verticillium lecanii* Zimmemlan @ 1×10^8 spore/ml for green scale. Another predator, which needs immediate attention, is Coccinellds, syrphid fly and *Chrysoperla carnea*. These are effective predators for the controlling of soft bodied insects i.e. aphid, jassid and eggs as well as early instar larvae of some lepidopterous borers. Mass rearing of these predator needs to be streamlined for economising its use as biocontrol agent.

Development of efficient and economic rearing technique of these bioagents will be the first step towards successful biological control. Besides, improvement of field release technique to maintain optimum potential is most important. The strains having potency to survive in toxic environment will fit well into the Integrated Pest Management system. Careful consideration of control methods to maintain and enhance the resident natural enemies is the most useful way, if the area is endowed with sufficient population of particular natural enemy (Table 3.7).

3.21 Microbial Control

Through this method, entomopathogenic bacteria, fungi and viruses are utilized for control of insect pests. These disease-causing microorganisms are produced in the laboratory and sprayed in the field to spread the disease in the host insect. They are cheaper than the pesticides, ecofriendly and compatible with natural enemies existing in the ecosystem. Fruit pest control through microbial intervention is so far limited

Table 3.7 Natural enemies of insect pests of some important fruit crops (Source: Kaul et al. 2009)

Pest	Parasitoid/Predator
Citrus	
Leaf minor	Citrostichus quadristriatus
	Citrostichus phyllocnisttoids
	Ageniapsis spp.
Butterfly	Trichogramma chilonis
	Cotesia spp.
Semilooper	Trichogramma chilonis
Psylla	Tamarixia radiata
	Episyrphus balteatus
Black fly	Encarsia divergens
	Encarsia merceti
Mango	
Leaf hopper	Coccinella spp.
	Chrysoperala carnea
	Verticillium lecanii
Mealy bug	Coccinella spp.
	Chrysoperala carnea
Fruit borer (citrus,guava and aonla)	Oenocyrtus papilionis
	Telonomus cyrus
	Trichogramma chilonis
Aphid (citrus and aonla)	Chrysoperala carnea
	Coccinella spp.

to few pests only. Considering the wide spread use of chemicals and consequent residue problem in fruits, biological control of fruit pests became more viable. Large-scale cultural operations, short cropping period and chemical intervention did not allow the natural enemies to grow to their potential. Thus conservation and periodical release of natural enemies in fruit ecosystem should be taken up. Besides host specific insect viruses and other entomopathogenic microorganisms when available in epizootic form or used through artificial application to the target site, provides good control of several pests of fruits. Some important parasitoids, microbial agents recommended against fruit pests.

3.22 Judicious Selection of Chemicals

There is large number of reports of successful chemical control of fruit pests. But various ill effecrs by the insecticides like, insecticide resistance, secondary pest outbreak, pest resurgence and killing of non target organisms etc. Have now threatens the widespread use of synthetic insecticides.

3.23 Choice of Chemical

Residue of major group of insecticides has been detected to considerable extent in fruit ecosystem. To avoid the persistent residues, choice of chemical plays a crucial role. Considering the picking interval and waiting period, the insecticides should be chosen.

3.23.1 Economic Threshold Level (ETL)

Instead of going for routine schedule application of insecticide, need based spraying on the basis of economic threshold level will help to reduce pesticide consumption and environmental abuse. ETL based applications help to keep the pest under check on one hand and maintain pest residue for natural enemies on the other.

3.23.2 Selection of Safe Chemicals

Chemical insecticides cannot be out rightly rejected from fruit pest control schedule. But selection of insecticides, which are comparatively safe to the insect natural enemies, should be taken into consideration.

3.24 Future Thrust

Research, education and demonstration of biologically based technologies for pest management in fruit crops are still needed. Emphasis should be given on the following aspects in the coming years.

1. Bioagents particularly predators, parasites and entomopathogens regulate the success of IPM programme. Research should be intensified to economise the production of bioagents with maximum quality assurance.
2. Quality control is an important pre-requisite in bio-pesticides. Besides, procedure for registration and quality control should be made simpler. Pesticide resistant natural enemies through artificial selection may sustain well in toxic environment also.
3. Development of chemical molecules having lower quantity of application rate in ecosystem, safer to natural enemies, lower mammalian toxicity and higher insecticide activity will strengthen the success of IPM.
4. Studies on Economic Threshold level and waiting period of insecticides is needed to the application of pesticide and use of fruits for consumption after pesticide application.

5. Government policies to provide incentives for the industries engaged in the production of bioagents. Strict monitoring in the pesticide market is needed to cut down the banned insecticides and support the bio rational pesticides.
6. Successful IPM technologies should be implemented and promoted on wide area basis in the farmer's field. Region specific IPM modules should be developed and demonstrated in farmer's field.

Bibliography

Adams WM (1990) Green development: environment and sustainability in the third world. Routledge, London

Aluja M (1994) Bionomics and management of *anastrepha*. Annu Rev Entomol 39:155–178

Angeles DE (1991) *Mangifera altissima*. In: Coronel RE, Verheij EWM (eds) Edible fruits and nuts. Plant Res. S.E. Asia 2 (PROSEA), Pudoc, Wageningen, pp 206–207

Anonymous (1988) Annual report, division of entomology, SKUAST-K Shalimar, technical document No 3, pp 45–56

Anonymous (2007) Economic survey 2007–2008, II ed. Directorate of Economics and Statistics, Planning and Development Department, Jammu and Kashmir Government

Anonymous (2008) Apple spray schedule for the year 2008. Department of Horticulture, Government of Jammu and Kashmir, Srinagar, 6 p

Arneson PA, Losey JE (1997) Integrated pest mangement-lecture notes. Cornell University, Tirana, Albania

Bailey J, Hoffman M, Olsen K (1988) Black light monitoring of two avocado insect pests. Calif Agr 42:26–27

Bajwa WI, Kogan M (1998) Compendium of IPM definitions (CID): a collection of IPM definitions and their citations in worldwide IPM literature. Integrated Plant Protection Center (IPPC), Oregon State University

Barnes BN (1990) FFTRI monitoring manual for orchard research institute, Stellenbosch, 21 pp

Barnes BN (1999) Fruit fly baiting in deciduous fruit doing the right things right. Deciduous Fruit Grow 49:5–7

Barnes BN, Blomefield TL (1997) Goading growers towards mating disruption: the south African experience with graphotita molesta and cydia pomonella (Lepidoptera: tortricidae). IOBC WPRS Bull 20:45–46

Barnes BN, Eyles DK (2000) Feasibility of eradicating *ceratitis* spp. Fruit flies from the western cape of South Africa by the sterile insect technique. In: Tan KH (ed) Area wide control of fruit flies and other insect pests. Penerbit Universiti Sains, Pulau Pinang, pp 449–455

Batra RC, Sandhu GS (1981) Differential population of citrus leaf miner and its parasiteson some commercial citrus cultivars. J Res Punjab Agric Univer 18:170–176

Batra RC, Sharma DR (1993) Current status of insect resistance in fruit crops. In: Dhaliwal GS, Dilawari VK (eds) *Advances in host plant resistance* to insects. Kalyani Publishers, Ludhiana, p 443

Beaument P (1993) Pesticides, policy and people. Pesticide Trust, London

Beers E, Hull L, Jones V (1994) Sampling pest and beneficial arthropods of apple. In: Buntin D, Pedigo L (eds) Handbook of sampling methods for arthropods in agriculture. CRC Press, Boca Raton, pp 383–415

Bentley JW, O'Neil RJ (1997) On the ethics of biological control of insect pests. Agric Hum Values 14(3):283–289

Bhagat KC, Masoodi MA, Koul VK (1988) A note on the occurrence of a coccinellid predator in Kashmir. Curr Res, Univ Agric Sci, Bangalore 17:49–51

Blomefield TL (1997) Managing resistance of codling moth, Cydia pomonella (L.) in South African pome fruit orchards, Combating resistance. IARC-Rothamsted, Harpenden, pp 14–16

Boavida C, Neuenschwander P, Schulthess F (1992) Spatial distribution of *Rastrococcus invadens* Williams (Homoptera: pseudococcidae) in mango trees. J Appl Entomol 114:381–391

Bull BC, Raupp MJ, Hardin MR, Sadof CS (1993) Suitability of five horticulturally important armored scale insects as hosts for an exotic predaceous lady beetle. J Environ Hortic 11:28–30

Campbell CW, Knight AJ (1983) Production de litchis. Communication at XIII Congreso de N.O.R.C.O.F.E.L Puerto de la Cruz. Ministerio de Agricultura Espana (En Prensa)

Chadha KL, Choudhary ML (2007) Report of the working group on horticulture, plantation crops and organic farming for the XI Five Year Plan (2007–12), GOI, Planning Commission, Jan 2007

Chang KP (1970) A study of Anoplophora macularia, aserious insect pest of litchi. Taiwan Agr Q 6(3):133–143

Cochran WG (1977) Sampling techniques. Wiley, New York 3 rd ed

Cunningham IC (1984) Mango insect pests. Australian mango research workshop. CSIRO, Melbourne, pp 211–224

Daku L, Norton GW, Pfeiffer DG, Luther GC, Pitts CW, Taylor DB, Tedeschini J, Uka R (2000) Farmers' knowledge and attitudes towards pesticide use and olive pest management practices in Vlora, Albania: A Baseline Survey. IPM CRSP working paper 001–1, Virginia Polytechnic Institute and State University

de Laroussilhe F (1980) Le Manguier. G.P. Masoneuve & Larose, Paris

DeBach P (1964) Biological control of insect pests and weeds. Chapman & Hall, London, 844 p

Dent D (2000) Insect pest management, 2nd edn. CABI Publishing, Ascot, 399 p

DPPQS, Directorate of Plant Protection, Quarantine and Storage (2001) Integrated pest management package, Government of India. (IPM package no: 28, 31, 32, 34, 37)

Drew RAI, Hancock DL (1994) The *bactrocera dorsalis* complex of fruitflies (diptera: tephritidae: dacinae) in Asia. Bull Entomol Res Suppl Ser 2:68

Falcon LA, Huber J (1991) Biological control of codling moth. In: van der Geest LPS, Evenhuis HH (eds) Tortricid pests, their biology, natural enemies and control. World Crop Pests. 5: 355–369.

Farahbakhch G, Moini M (1975) Olive pests in Iran, Report of plant pests and diseases research institute. National Agricultural Research Organization, Ministry of Agriculture and Natural Resources, Tehran, pp 1–14

Flint ML, van den Bosch R (1981) Introduction to pest management. Plenum Press, New York

Galan-Sauco V (1990) Los frutales tropicales en los subtropicos. Ed. Mundi Prensa, Madrid, pp 83–84

Gold CS, Nankinga C, Niere B, Godonou I (2003) IPM of banana weevil in Africa with emphasis on microbial control. In: Neuenschwander P, Borgemeister C, Langewald J (eds) Biological control in IPM systems in Africa. CABI Publishing, Wallingford, pp 243–257, 400 p

Golez HG (1991) Bionomics and control of the mango seed borer, *Noorda albizonalis* Hampson (Pyralidae: Lepidoptera). Acta Hort 291:418–424

Haji FNP, Barbosa FR, Lopes PRC, Moreira AN, de Alencar JA, Ferreira RCF (2004) Monitoring mango pests within an integrated production programme in Brazil. Acta Hort 645:163–165

Hameed SF, Sharma DD, Agarwal ML (1992) Integrated pest management in litchi. National seminar on recent developments in litchi production held at Rajindra Agricultural University, Bihar, Pusa, 30–31 May 1992, Abstract of papers, pp 38

Hansen JD, Amstrong JW (1990) The failure of field sanitation to reduce infestation by mango nut weevil, cryto mangiferae in Hawaii. Trop Pest Manag 36:359–361

Hare DJ (1994) Sampling arthropod pest in citrus. In: Pedigo L, Buntin D (eds) Handbook of sampling methods for arthropods in agriculture. CRC press, Boca Raton, pp 417–432

Haseeb M (2007) Current status of insect pest problems in guava. Acta Hort 735:453–467

He DP (2001) An overview of integrated management of insect pests in litchi orchards of Guangdong. Acta Hort 558:401–405

Hill D (1975) Agricultural insect pests of the tropics and their control. Cambridge University Press, New York

Howard PA, Michalenko EM, Sage GW, Jarvis WF, Meylan WM, Basu DK, Beauman JA, Gray DA (1991) Handbook of environmental fate and exposure data for organic chemicals. Lewis Publishers, Chelsea

Hoyt SC, Burts EC (1974) Integrated control of fruit pests. Ann Rev Ento 19:231–252

Huda AKS, Luck J (2008) Early warning of pest/diseases for selected crops using climate information and crop simulation modeling approach under climate change scenarios. Proceeding of the international symposium on agrometeorology and food security, 18–21 Feb 2008, Hyderabad, pp 11

Jothi DB, Tandon PL, Verghese A (1994) Hot water immersion as a quarantine treatment for Indian mangoes infested with the oriental fruitfly, Bactrocera dorsalis (Hendel) (Diptera:Tephritidae). FAO Plant Protect Bull 42(3):158–159

Kaul V, Kesar YK (2003) Incidence and management of lepidopteran fruit borers of guava (Psidium guajava L.) in Jammu, India. J Asia-Pacific Entomol 6(2):201–205

Kaul V, Malik GH, Shankar U, Monobrullah M (2007a) Incidence and management of olive psylla, *Euphyllura pakistanica*. Indian J Entomol 69(4):331–340

Kaul V, Tiku AK, Shankar U, Monobrullah M (2007b) Green stink bug (hemiptera: pentatomidae) recorded as a new pest of olive in India. J Asia-Pacific Entomol 10(1):81–83

Kaul V, Shankar U, Khushu MK (2009) Bio-intensive integrated pest management in fruit crop ecosystem. In: Rajinder P, Dhawan AK (eds) Integrated pest management: innovation development process, vol 1. Springer, Dordrecht/London, pp 631–666

Keenan M (1997) In: Smith D, Beattie G, Broadley R (eds) Citrus pests and their natural enemies. Deptt. of Primary Industries, Q 197030, Brisbane, 272 p

Khangura JS, Kanwar JS, Singh H, Singh D (1992) Pest management in litchi (litchi chinensis Sonn.). National seminar on recent developments in litchi production held at Rajindra Agricultural University, Bihar, Pusa, 30–31 May 1992, Abstract of papers, pp 40–41

Kozar F, Brown MW, Lightner G (1994) Spatial distribution of homopteran pests and beneficial insects in an orchard and its connection with ecological plant protection. J Appl Entomol 117:519–529

Kumar N, Suresh J, Anbu S (2007) Problems and prospects of guava cultivation under Tamil Nadu conditions. Acta Hort 735:329–334

Lichtenberg E, Zilberman D, Archibald SO (1990) Economics and pesticides. Chapter 5. In: Zilberman D, Lichtenberg E (eds) Economic perspectives on pesticide use in California: a collection of papers. Berkley University, California.

Malo SE, Campbell CW (1994) The guava. Horticultural sciences department fact sheet HS-4. Florida Cooperative Extension Service, Institute of Food and Agricultural Sciences, University of Florida

Mani M, Krishnamoorthy A, Singh SP (1989) The impact of the predator, *Cryptolaemus montrouzieri* Muls. on the pesticide resistant populations of the striped mealybug *Ferrisia virgata* (Ckll.) on guava in India. Insect Sci Appl 11:167–170

Masoodi MA (1985) Growth response of *Lymantria obfuscate* Walker in relation to tannin content in different host foliage. Indian J Entomol 47(4):422–426

Mathur RN (1975) Psyllidae of the Indian subcontinent. Indian Council of Agricultural Research, New Delhi, 429 p

Mathur AC, Tondon PL (1974) Litchi mite can be controlled. Indian Hort 19:11–12

McCook HC (1882) The honey ants of the garden of the gods, and the occident ants of the American plains. J. B. Lippincott & Co., Philadelphia

Miller JC, Hanson PE, Dowell RV (1987) The potential of gypsy moth as a pest of fruit and nut crops. Calif Agr 41:10–12

Morin C (1967) Cultivo de Frutales Tropicales. Librerias ABC, Lima, pp 225–229

Mullen JD (1995) Estimating environmental and human health benefits of reducing pesticide use through integrated pest management programs. Master thesis, Virginia Polytechnic Institute and State University, USA

Muniappan R, Shepard BM, Watson GW, Carner GR, Sartiami D, Rauf A, Hammig MD (2009) First report of the papaya mealybug, *Paracoccus marginatus* (Hemiptera: Pseudococcidae), in Indonesia and India (2008). J Agric Urban Entomol 25:37–40

Muniappan R, Shepard BM, Watson GW, Carner GR, Rauf A, Sartiami D, Hidayat P, Afun JVK, Goergen G, Ziaur Rahman AKM (2010) New records of invasive insects (Hemiptera: Sternorrhyncha) in Southeast Asia and West Africa (2009). J Agric Urban Entomol 26:167–173

Murray RC (1991) Arthropod pests of mango in the Caribbean Islands, Review of crop protection. University of West Indies, Kingston

Navon A, Cohen M, Keren S, Levski S, Harel M, Dunkelblum E (1997) Trapping of the leopard moth, *Zeuzera pyrina* L. (Cossidae: Lepidoptera) with sex pheromone traps for monitoring the pest experiments in apple and pear orchards. Alon Hanotea 51(8):358–366

Nelson KC (1994) Participation, empowerment, and farmer evaluation: a comparative analysis of IPM technology generation in Nicaragua. Agr Hum Values 11:109–125, (Spring-Summer 1994)

Nijjar GS (1972) Litchi cultivation. Punjab Agriculture University, Ludhiana

Norton GW, Mullen J (1994) Economic evaluation of IPM programs: a literature review. Report 448–120. Cooperative Extension, Blacksburg

Norton GW, Rajotte EG, Gapud V (1999) Participatory research in integrated pest management: lessons from the IPM CRSP. Agr Hum Values 16:431–439

Pandey V, Patel MG, Chaudhari GB, Patel JR, Bhatt BK, Vadodaria RP, Shekh AM (2003) Influence of weather parameters on the population dynamics of mango hopper. J Agrometeorol 5(1):51–59

Pareek OP (1983) The Ber. ICAR, New Delhi, p 5

Pedigo LP (2001) Entomology and pest management, 4th edn. Prentice Hall, New Jersey

Peña JE (2004) Integrated pest management and monitoring techniques for mango pests. Acta Hort 645:151–161

Pena J, Johnson F (2001) Insect management in guava. Document ENY- 412. Florida Cooperative Extension Service, Institute of Food and Agricultural Sciences, University of Florida

Peña JE, Sharp JL, Wysoki M (2002) *Tropical fruit pests and pollinators: biology, economic importance,* natural enemies and control. CAB International, Wallingford

Pimental D, Acquay H, Biltonen M, Rice P, Silva M, Nelson J, Lipner V, Giordano S, Horowitz A, D'Amore M (1992) Environmental and economic costs of pesticide Use. BioScience 42(10):750–760

Prasada Rao GSLHV, Beevi SP (2008) Forewarning tea mosquito bug helopeltis antonii sign. (miridae:hemiptera) in cashew. Proceeding of the international symposium on agrometeorology and food security, Hyderabad, 18–21 Feb 2008, pp 203

Rather AQ, Masoodi MA (1988) Survey of mite pests and their natural enemies. In: Annual report, division of entomology, SKUAST-K Shalimar, pp 71–75

Roling N, van de Fliert E (1994) Transforming extension for sustainable agriculture: the case of integrated pest management in rice in Indonesia. Agric Hum Values, Spring-Summer 11:96–108

Roy RS, De BN (1950) Control of litchi mite. Indian J Hort 7(1):16

Sauco GV, Menini UG (1989) Litchi cultivation. FAO plant production and protection paper. 83

Shankar U, Kaul V, Rai S (2007) New recode of anar fruit borer, Deudorix isocrates (Fabricius) on aonla. J Res, SKUAST-J 6(2):292–295

Sharma DD (1985) Major pests of litchi in Bihar. Indian Farming 35(2):25–26

Sharma DR (2006) Management of insect pests of tropical and sub-tropical fruit crops. In: Proceedings of advance training course on emerging trends in economic entomology, 21st Feb–13th Mar 2006, Hisar, India, pp 123–135

Sharma DR, Singh S (2006) Management of insect pest of temperate fruits. In: Proceedings of advance training course on Emerging trends in economic entomology, 21st Feb – 13th Mar 2006, Hisar, pp 111–122

Singh SP (1989) Achievements of AICRP on biological control. Tech Bull No 2. Biological Control Centre, NCIPM, Faridabad, 20 pp

Singh G (1991) Loss assessment, ecology and management of mango fruitfly, *Dacus* sp. Acta Hort 291:425–436

Singh A, Trivedi TP, Sardana HR, Dhandapani A, Sabir N (2001a) Resource manual on validation and promotion of IPM. National Centre for IPM, New Delhi, p 119

Singh SP, Rao NS, Ramani S, Poorani J (eds) (2001b) Annual report 2000–01, project directorate of biological control, Bangalore. PDBC, Bangalore, 218 p

Smith D, Beattie G, Broadley R (1997) Citrus pests and their natural enemies: IPM in Australia. Queensland, Department of Primary Industries, Brisbane, 272 p

Sofi MA, Zaki FA, Dar MA, Khan ZH (2008) New approaches to San Jose scale control by newer oils alone and in combination with insecticides and their effect on grading of apple in Kashmir. Environ Ecol 26(3A):1369–1372

SP-IPM (2010) Integrated pest management and crop health – bringing together sustainable agro-ecosystems and people's health. White Paper, SP-IPM Secretariat, International Institute of Tropical Agriculture (IITA), Ibadan, 17 pp

Srivastava AS, Masoodi MA (1985) Influence of host plants on the development and survival of *Lymantria obfuscate*. FAO Plant Protect Bull 33(2):67–69

Syed RA, Ghani MA, Murtaza M (1970) Studies on trypetids and their natural enemies in West Pakistan. III. *Dacus (Strumeta) zonatus* (Saunders). Tech Bull Commonw Inst Biol Control 13:1–16

Tandon PL, Verghese A (1985) World list of insect, mite and other pests of mango. Technical document no 5. IIHR, Bangalore, p 22

Thakur JN, Rawat US, Pawar AD (1989a) Investigations on the occurrence of natural enemies of San Jose scale, *quadraspidiotus perniciosus* Comstock (hemiptera: coccidae) in Jammu & Kashmir and Himachal Pradesh. Entomon 14:143–146

Thakur JR, Sharma PC, Gupta PR (1989b) Occurrence of olive psylla, *euphyllura pakistanica* on olive in India: a new record. Trop Pest Manag 35(3):331

Veerish GK (1989) Pest problems in mango – world situation. Acta Hort 231:551–565

Villeirs EADE, Matthee WA (1973) The occurrence of bark borers on litchi and other fruit tree. Citrus Grow Sub-Tropical Fruit J 477:17–19

Wearing CH (1975) Integrated control of apple pests in New Zealand. 2. Field estimation of fifth-instar larval and pupal mortalities of codling moth by tagging with cobalt-58. New Zealand J Zool 2:135–149

Whalon ME, Croft BA (1984) Apple IPM implementation in North America. Annu Rev Ent 29:435–70

Whitehead VB, Rust DJ (1972) Control of fruit piercing moth, *Sorrodes partita* (Fabr.) (Lepidoptera: Noctuidae). Phytophylactica 4:9–12

Zaki FA, Chan SA (2003) Management of European red mite (*Panonychus ulmi*) on apple in Kashmir. SKUAST J Res 5:198–203

Zaki FA, Dhobi GM, Mantoo MA (2004) Interaction of European red mite (*Panonychus ulmi* Koch) infestation with *Alternaria* sp. in premature leaf fall in apple (*Malus domestica* Borkh). In: Kamili AN, Yousuf AR (eds) Bioresources: concerns and conservation. Centre of Research for Development, University of Kashmir, Srinagar, pp 119–126

Zaki FA, Lone AH, Mantoo MA (2005) Resistance in European red mite (*Panonychus ulmi* Koch) to some pesticides on apple in Kashmir. In: Dilawari VK, Doel GS, Joia BS, Chhuneja PK (eds) Proceedings of the 1st congress on insect science: contributed papers, INSAIS, PAU, Ludhiana, 15–17 Dec, pp 161–163

Zaki FA, Wani AR, Wani NA, Parray MA, Mantoo MA, Rather AQ (2008a) Efficacy of newer acaricides against European red mite Panonychus ulmi (Koch) on apple Malus domestica (Borkh) in Kashmir. Paper presented in IUPAC sponsored 1st international conference on agro-chemicals protecting crop, health and natural environment, abstrat. no. IL-29, 08–11 January 2008, New Delhi, India

Zaki FA, Mantoo MA, Parray MA, Khan AA, Rather AQ, Wani NA, Wani AR (2008b) Paraffinic mineral oils – a new eco-friendly agrochemical strategy for agricultural pest management. IUPAC sponsored first international conference on agrochemicals protecting crop, health and natural environment, New Delhi, 08–11 Jan 2008, Abstract No P-77

Zouiten N, Hadrami IE (2001) Olive tree psyllid: state of knowledge and perspective of control. Agricultures 4(10):225–232

Chapter 4
Pome Fruits

Pollination is one of the most essential and critical stages in the production of fruits and vegetables. Most of the apple varieties are self incompatible and require a pollinating agent for transfer of pollen to set fruits. Even the self-compatible varieties produce quantitatively and qualitatively better fruits if receive pollination. Waite (1895, 1899) provided first scientific evidence that apple and other pome fruits benefit from the interplanting and cross pollination between cultivars. Since then several investigators have worked on the pollination requirements of apples and pears in the world (Kendall 1973; Kendall and Solomon 1973; DeGrandi-Hoffman et al. 1985, 1995; Kuhn and Ambrose 1982; Free 1993; Gupta et al. 1993; Gardner and Ascher 2006; Klein et al. 2007; Thomson and Goodell 2001).

4.1 Pollination Studies

4.1.1 Apple

The apple flowers occur in a cluster at the tip of woody shoots called spur in the axils of leaves. The primary or king bud opens first which produces the best fruit and if the king bloom fails lateral blooms which open after a day or two can produce fruit. Each flower has five stigmas which join into a style that leads to the ovary. Each ovary is divided into five compartments with two ovules each. The style is surrounded by 20–25 pollen bearing stamens. Five pinkish petals surround the sexual parts (Fig. 4.1). Bhartiya et al. (1983) found that number of sepals and petals did not differ but number of stamens varied in different cultivars. Nectar is secreted between bases of stamens and styles. Apple flowers produce both nectar and pollen.

D.P. Abrol, *Pollination Biology, Vol. 1*, DOI 10.1007/978-3-319-21085-8_4

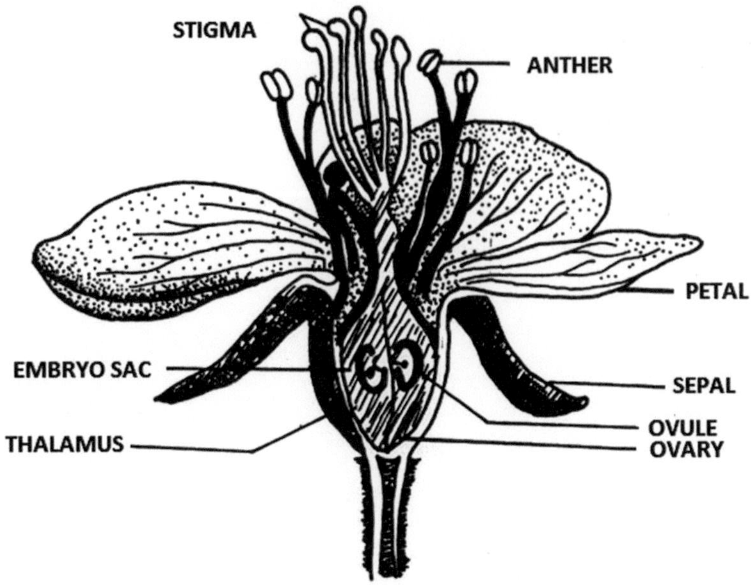

Fig. 4.1 Flower structure of apple

4.2 Apple Pollination Requirements

The pollination requirements vary in different cultivars of apple. Some of the cultivars are completely self-incompatible (Griggs 1970). Some set no fruit when self pollinated while as others set various proportions of fruit (Brittain 1933; Burrell and Parker 1932; Latimer 1931; MacDaniels and Heinicke 1929; Overholser 1927). Interplanting of polenizer cultivars is necessary which has the most compatible pollen and blooms simultaneously with the main cultivar.

Apple pollen is heavy and not carried readily by the wind as the pollen of some tree species, such as conifers and nuts. The pollen is transferred primarily by insects, especially honey and bumble (*Bombus* sp.) bees. During bloom, prolonged periods of cool weather or rain, which limit bee flight, can be detrimental to fruit set. Fruit growers rent honey bees from apiculturists during the bloom period, a minimum of four or five strong colonies per hectare being recommended in mature orchards. The bees must be removed from the orchard prior to application of insecticides. For maximum effectiveness, the orchard floor should be mowed to remove competing weed flowers. Timing is important in order for pollination to be effective. Williams (1966) proposed the term 'effective pollination period' (EPP) for the interval during which pollination will result in fertilization. Pollination must occur within a given 'window' (the EPP) in order for the pollen-tube to reach the ovule and for fertilization to occur while the ovule is still receptive. The length of the EPP varies with cultivar, tree condition and temperature.

4.3 Apple Pollinators

The role of insect pollination in apple was first established by Waite 1895. In apple orchards cross pollination is done by insects, especially bees and due to sticky nature of pollen, role of wind has been considered of little significance. Honeybees have been recognised as the most important and efficient pollinators of apple (Free 1964). Besides various species of wild bees such as *Andrena, Bombus, Halictus,* and *Osmia* (Brittain 1933; Free 1964; Glukhov 1955; Hutson 1926; Kitamura and Maeta 1969; Loken 1958; Phillips 1933). The major problem with the use of wild bees is that their number fluctuates between years and different locations and cannot be relied upon. Honeybees which are available in sufficient numbers and can be concentrated in desired numbers in orchards where ever their services are required. However, there are several factors which may be considered while planning pollination in apple orchards which include the strength, placement, and manipulation of colonies, the effects of competing plants, soil, and weather, and other factors both within the colony and in the environment which contribute to the effectiveness of honey bees.

Fruit set in three varieties of apple (*Malus sylvestris*) viz. Red-golden, Red-delicious and American mother was found to be greatly increased by insect pollination (Sharma 1961). Honeybees, *Eristalis sp. and Syrphus sp.* were the main visitors. Red-delicious had higher fruit set near the bee colony and decreased with distance. Flies were found to outnumber *Apis cerana indica* at all locations on all fruit blossoms at different elevations in Himachal Pradesh (India). Rai and Gupta (1983) also emphasized the role of bee pollination in the fruit set of apples. However, exact pollination requirements of the crop in Asia have not yet been studied. On apple bloom besides honeybees frequent insect visitors are species of *Eristalis* at different places on Shimla hills Himachal Pradesh, India (Verma and Chauan 1985). Singh and Mishra (1986) recoded nine species of insects visiting apple flowers in different areas of Himachal Pradesh. Kanwar 1987 reported a species of wild pollinating fly playing an important role in apple pollination. Verma and Chauhan recorded 44 species of insects pollinating apple flowers in Shimla hills Himachal Pradesh. Abrol (1989) presented a detailed list of pollinators of different crops under Jammu and Kashmir (India) conditions.

Studies on comparative foraging behaviour of A. cerana and A. mellifera on apple flowers revealed that foraging activities of A. cerana started at lower temperature and much earlier than A. mellifera (Verma and Dulta 1986) and A. mellifera visited significantly more flowers per foraging trip than A. cerana (Verma and Rana 1994).

Dashad and Sharma (1994) reported that foragers of *Apis dorsata* had higher foraging rates as compared to their counterparts of *A. cerana* and *A. mellifera* on apple flowers. Dipteran flies had very low foraging rate as compared to honeybees. Later Dashad et al. (1994) conducted studies on loose pollen grains carrying capacity of insect-visitors of apple bloom and found that the amount of loose pollen on the body varies with the plant species and the varieties on which the insects are working.

Honeybees have been recognised as the most important pollinators of apple, but sometimes they rob the flower of its nectar without coming in contact with sexual parts of the flower. Solitary bees are more efficient as apple pollinators but their availability in sufficient number is a major constraint. Several investigators have used pheromone based attractants to increase bee visitation to apple flowers (Mayer et al. 1989a; Currie et al. 1992). Honeybee efficiency further depends upon the presence of large populations of adult population with plenty of brood. Colonies placed for apple pollination should have a minimum strength of 20,000 bees on six frames (Mayer et al. 1986; Ambrose 1990). Placement of colonies in apple orchards is another important factor. Colonies should be moved in apple orchards after about 5 % of the orchard is in bloom so that bees focus their attention on target crop rather than visiting competing plants (Mayer et al. 1986). Batra (1984) reported that in Japan in comparison to honeybees, orchard mason bees visited more apple flowers per minute and contacted the sexual columns 26 times more frequently. Torchio (1985) observed that orchard mason bees such as *O. cornifrons, O. lignaria lignaria,* and *O. Lignaria propinqua* land directly on the anthers and stigma of the apple blossom, thereby acting as successful apple pollinators. Inspite of their effectiveness as most efficient pollinators of apple, their management has not reached to desired levels due to problems related to their timing of emergence and apple bloom (Kuhn and Ambrose 1984). Different investigators have recommended have recommended variable densities of bee hives per acre for apple pollination in different geographic areas with an average of 3.7 hives/acre (Humphry-Baker 1975; Crane and Walker 1984; Ambrose 1990; Kevan 1988; Mayer et al. 1986; McGregor 1976; Levin 1986; Kevan 1988; British Columbia Ministry of Agriculture Fisheries and Food 1994; Scott-Dupree et al. 1995). Mayer et al. (1986) recommended 20–25 honey bees per tree per observation min for effective pollination in apples. In case of orchard bees, Torchio (1985) recommended 250 orchard bees per acre (618 bees for 1 ha). For almond, Bosch (1994a, b) recommends three *O. cornuta* females per tree. For a variety of orchard fruits, Batra (1982) recommends 2834 *O. Cornifrons* acre_[1] (7000 [ha]). For highbush blueberry, Torchio (1990a, b) recommends 300 nesting female *O. ribifloris* acre 21 (741 ha21).

Soil nesting (non-Nomia) bees such as *A. pilipes villosula* has also been reported as apple pollinator of minor significance (Batra 1994). However, this species has the capacity to work under cool, damp conditions from early morning to late evening.

Although honey bees are generally good pollinators of apple, they tend to rob flowers of 'Delicious' apple varieties. Some species of *Andrena* help compensate for this because they legitimately visit apple flowers (including 'Delicious'), specialize on apple, and work in cool temperatures (Parker et al. 1987).

Leaf cutting bees visit apple blossoms, but they tend to fly only when temperatures exceed 75 °F (24 °C); moreover, it is difficult to synchronize bee incubation and emergence with apple bloom. Bees in the genus *Osmia* (family Megachilidae) have proven themselves to be effective pollinators of apples and other orchard fruits. These solitary bees nest in hollow reeds or pre-existing holes in wood such as abandoned beetle burrows or nail holes. They may nest in large aggregations if nest holes are abundant. *Osmia* bees partition cells and seal their nests with mud, chewed leaf material, or a mixture of both – hence, they are sometimes called orchard mason bees. The eastern subspecies (*O. lignaria lignaria*) occurs from the eastern slopes of the Rocky Mountains to the Atlantic. The western subspecies (*O. Lignaria propinqua*) occurs from the western slopes of the Rockies to the Pacific Ocean. Females of the blue orchard bee have a pair of hornlike projections extending from the lower face. The blue orchard bee is shiny blue/black and about two-thirds the size of a honey bee. The male is about a third smaller than the female and has a white patch of hair on the face and long curved antennae. Females have no white on the face and their antennae are about half as long as those of males. The horned-face bee (*O. cornifrons*) was introduced from Japan into Utah in the 1960s, and from Utah to Maryland by 1978 (Batra 1989); it has since become established in many areas of the eastern US and Canada. It has a pair of horn-like projections on the lower face. The orange orchard bee (*O. cornuta*) was introduced from Spain to California almond orchards in the 1980s (Torchio 1987). The female is slightly larger than the female blue orchard bee. Its most distinguishing characteristic is an abdomen coated with beautiful, bright orange hair. It also has a pair of horns on the lower face.

Apples grown from areas near *Osmia* nests had comparatively more seeds and better fruit shape (Kuhn and Ambrose 1984). However, these introduced populations of *Osmia* did not establish well in the test orchards, so the long-term benefit of these introductions was doubtful.

4.4 The Role of Native Bees in Apple Pollination

While honey bees are generally viewed as essential pollinators in apple orchards (McGregor 1976; Free 1993), there is evidence that other bee species are contributing significantly to apple pollination. Kendall and Solomon (1973) analyzed the amount and composition of pollen carried on a variety of insects collected visiting apple blossoms in the UK. They counted (1) the total number of pollen grains present on the body, and (2) the proportion of pollen grains that were from Rosacaeae (they lumped *Malus*, *Prunus*, *Crataegus* and *Ameliantchier* together because they could not be separated morphologically). Honey bees ranked 11th out of the 20 bees tested in the quantity of Rosaceae pollen carried (roughly 4000 grains/worker). A number of other bee species, mostly in the genera *Andrena*, *Bombus*, and *Osmia* carried significantly more Rosaceae pollen than honey bees. Among the top three species were *Andrena pubescens*, *Andrena haemorrhoa* and *Andrena coitana*, with between 16,000 and 24,000 grains of Rosaceae pollen per bee. These bees are either equal to, or smaller than, worker honey bees, so this effect is not due simply to differences in body size. Most bee species in the survey carried between 60 % and 90 % Rosaceae pollen and the above three *Andrena* species carried between 81 % and 97 % Rosaceae pollen, suggesting a high level of Rosaceae specialization. The results of this study suggest that, while honey bees are certainly capable of effective apple pollination, there were ten species of native bees that carried more Rosaceae pollen, on a per-bee basis, than honey bees.

In a related study, Kendall (1973) examine pollination effectiveness of various bee species in apple orchards. Bees were collected, killed, and the venter of the bee was brushed against a test (virgin) stigma. He measured the percentage of ovules fertilized by honey bees, six species of bumblebees (*Bombus*), eight species of *Andrena*, and one *Osmia*. As controls he used stigmas cross-pollinated by hand, self-pollinated by hand, and un-pollinated. The highest pollination rates were achieved by two species of *Andrena* (*A. haemorrhoa* and *A. jacobi*), which proved to be significantly better than honey bees. In addition, species of *Andrena*, *Halictus*, and *Osmia* showed less variation *among orchards* in pollinator effectiveness than honey bees.

Kendall concluded that "when abundant, female solitary bees must be valuable as cross-pollinators, and some species such as *Andrena haemorrhoa* and *A. jacobi* are consistently better pollinators of the flowers they visit than are similar numbers of honey bees." While the methods used in this study are simple, they provide evidence that native bees are no worse and, in some cases better, as fruit pollinators than honey bees.

Thomson and Goodell (2001) performed a much more rigorous analysis of pollinator effectiveness by examining both pollen removal and pollen deposition *per visit* by live honey bees and bumblebees. While *Apis* and *Bombus* removed similar amounts of pollen per visit, *Bombus* deposited slightly more pollen on stigmas per visit than *Apis*. Pollen collecting bees of both species removed more pollen per visit than nectar collecting bees. *Apis* showed a fairly high frequency (up to 30 %) of "sideworking" (approaching nectaries laterally without contacting either stigma or anthers). These visits resulted in significantly lower levels of pollen deposition than regular visits.

Finally, detailed studies of non-*Apis* managed apple pollinators, such as *Osmia*, suggest that on a per-bee basis, *Osmia* females are significantly more effective than honey bees in apple pollination. Vicens & Bosch (2000a) compared per-visit fruit set by foraging female *Osmia cornuta* and honey bees in commercial orchards in northeastern Spain. One third of the *Osmia* visits resulted in commercial fruit set – a fivefold higher rate than for *Apis* visits. *Apis* foragers made little contact with the stigma, primarily because the majority of *Apis* floral visits (97 %) were for nectar rather than pollen. In addition, *Osmia* foragers visited more flowers per minute and showed a strong preference for *Malus* pollen. Related studies (Vicens and Bosch 2000b) showed that *Osmia* females forage at lower ambient temperature than *Apis*. Other *Osmia* species have been shown to be better apple pollinators than *Apis* in Japan (*O. cornifrons*; Maeta and Kitamura 1974) as well as in the US (*O. lignaria*; Torchio 1985).

4.5 Number of Hives Per Acre

The scientific literature average is 1.5 hives per acre, and this is the recommended number for apple growers. However, different factors can affect the number of hives needed to ensure optimal pollination. The attractiveness of the crop has a large effect on the foraging activity of the pollinators. If the crop is not appealing to the pollinator, or if there is a more rewarding crop in bloom nearby, more number of hives per acre may be required. Therefore, it is important to remove (if possible and legal) any non-target forage that may entice the bees from the apple blossoms, including flowering weed on the orchard floor. If it is not possible to remove these plants, then more hives may be needed to ensure that the apple blossoms are visited. The local population density of wild bees can also affect the number of hives necessary for pollination services. If the wild bees are available in sufficient numbers, even lower number of hives per acre may be result in successful pollination.

4.6 Chemical Attractants

There are several chemical attractants available, most of which are composed of synthetic honey bee pheromones. These chemicals can stimulate increased bee visitation and recruitment, and in some cases they can promote the earlier onset of daily foraging activities. These compounds are particularly helpful to use when there are suboptimal pollination conditions. However, while these attractants may increase bee visitation, they may not necessarily increase pollination.

4.7 Pesticides

In agriculture, it is often necessary to use chemical insecticides and herbicides to remove unwanted pests and plants. Unfortunately, these chemicals can have adverse effects on the pollinator community, especially if they are applied while the target crop is in bloom. Sometimes insecticides do not kill bees but instead repel them from a crop and interfere with pollination. Honey bees were shown to be repelled from apple flowers sprayed with the insecticide dimethoate for at least 2 days after treatment (Danka et al. 1985). Vandame et al. (1995) reported that sublethal doses of insecticides can change the homing behaviour of bees. They placed a colony of honeybees in an insect proof cage and trained the workers to visit feeder at a distance of 8 m from the hive. Selected honey bees were then treated with sub-lethal doses of the insecticide deltamethrin and subsequently released. Of the treated bees, 81 % failed to return to their nest within 30s, a figure three times longer than the average time for non-treated bees. Considering the widespread use of insecticides in a typical agricultural landscape, it is possible that numerous similar sub-lethal effects are manifest and contributing to pollination problems.

First, it is important to underscore that not all insecticides are equally hazardous to bees, and bee species vary in their susceptibilities (Mayer et al. 1994b). The formulation of an insecticide affects its toxicity to bees. Granules and solutions are generally less hazardous than wettable powders and dusts. Non-sugary bait formulations are also relatively safe. For example, carbaryl-laced wheat bran flakes used to control rangeland grasshoppers are relatively safe to alfalfa leafcutting bees (Peach et al. 1994, 1995).

Many insecticides are deadly to bees when they are first applied, but they degrade within hours to relatively non-hazardous levels. This means that certain acutely toxic insecticides can be applied in the late evening, and by morning the pest will be controlled and the residues will be sufficiently degraded so that dayflying bees can forage with relative safety. One exception to this scenario will be of concern to growers who rely on bumble bees or squash bees for pollination. These bees frequently spend the night in cucurbit flowers, and evening insecticide applications are no guarantee for their safety.

Finally, an emphasis on pest scouting, treatment thresholds, precision sprays, can help to reduce the number of sprays needed per season. IPM practices since the 1960s and 1970s in the western US have reduced the number of lucerne sprays per season from 6–8 to 1–3 with a noticeable benefit to pollinators (Peterson et al. 1992).

If chemical control is needed during the pollination period, there are a few things to consider so that the pollinator community is minimally impacted. First, do not spray during the flowering period if at all possible. This will help minimize the exposure of the pollinator to the potentially harmful chemicals. Second, use chemicals with low bee toxicity, particularly those with short residual times and moderate to low LD50 ratings as outlined on the chemical label. In general, granule and liquid formulations are safer than powder and dust applications. The granule and liquid formulations minimize drift onto non-target flowering plants. Finally, late-afternoon or evening application of chemicals is recommended to minimize the exposure of foraging bees to potentially harmful chemicals. The best approach is to anticipate and manage pest problems before bees are placed in the orchard.

It is a general principle that insecticides should never be applied to plants while they are flowering. However, there are occasions in which a crop must be treated for a serious pest even while the crop is in bloom. Fortunately, there are ways to reduce bee kill even in these situations.

Research in biological and botanical insecticides is yielding products that are safer for bees. Some of these products are with the active ingredients *Bacillus thuringiensis* and diflubenzuron. Many of these products are specific to particular pest species and are relatively innocuous to non-target organisms.

Nevertheless, many biological insecticides are relatively safe for honey bees and may, in some cases, allow growers to spray crops during bloom, something inadvisable with 'harder' chemicals. For example, treating blooming canola with azadirachtin (an extract of the neem tree) at concentrations of up to 150 ppm did not repel honey bees or other pollinators, and no residues of azadirachtin were found on the bodies of bees foraging on treated plants (Naumann et al. 1994a). Certain doses of the insecticidal fungus *Metarhizium flavoviride* control pest locusts but kill relatively few honey bees (Ball et al. 1994). These kinds of pesticides may be promising alternatives when growers are faced with emergency pest outbreaks during crop bloom. One should always avoid untested mixes of two or more pesticides because the combined activity can be unpredictable. Sometimes a blend of two or more pesticides can be more toxic than any one of the ingredients by itself. This interactivity is called *synergism*. As an example, the insecticide Karate® is already highly toxic to honey bees, but when blended with the fungicides Impact® or Sportak® its toxicity increases up to 16 times (Pilling and Jepson 1993). Herbicides, plant growth regulators, and fungicides are relatively safe for bees but there are exceptions such as the herbicides 2,4-D and MSMA, the growth regulator carbaryl, and the fungicide binapacryl (Johansen and Mayer 1990).

Sometimes insecticide-damaged honey bee colonies can be rehabilitated. The first step is to move hives away from the hazard. The colony may recover on its own if only the older adult population was affected and if it has plenty of honey and pollen. However, if brood and nurse bees continue dying then this means that the pollen is contaminated. In this case all combs with pollen must be removed. It may be possible to salvage these combs by soaking them in water for several hours, washing the pollen out of the cells, and air-drying, but it is safer simply to discard them. Weakened colonies can be fed to stimulate brood production. They can be strengthened with the addition of queenless packages of bees or combined together to make up stronger colonies. It is important to quickly ascertain if the queen is still functioning normally and replace her if there are signs of reduced egg-laying capacity. Obviously, pesticide applications must be clearly worked out between the grower and the bee-keeper. Agricultural universities, extension services, crop consultants, and government departments of agriculture are good sources of research-based, environmentally sound pest control recommendations which can help minimize bee losses.

4.8 Renting a Pollinator Hive and Setting up a Pollination Contract

A 'pollination fee' is the cost to rent a hive of bees during the bloom of a particular crop. Pollination contracts are made between the grower and the beekeeper to help ensure that a sufficient number of bees are present in the during apple bloom to produce better quality and increase in fruit production.

4.9 Pollination in Pear

The problem of self sterility and need for insect pollination in pears was first discovered by Waite (1895). This discovery led to surge of further studies in fruit pollination (Crane 1876; Hutchinson 1886; Muller 1883). Most of the pear varieties are self unfruitful and need cross pollination by insects. High densities of bees in pear orchards not only ensure good fruit set but increase seed numbers in each fruit. The fruits thus produced are of better quality and even shaped. The pear flowers produce small amounts of nectar, low in sugar mostly unattractive to bees, thus requiring more pollinizers and bees for effective pollination.

The pear (*Pyrus communis*) flower is about 1 in (2.54 cm) wide, has five white petals, and occurs in clusters along the branches. There are five styles, and the stig-

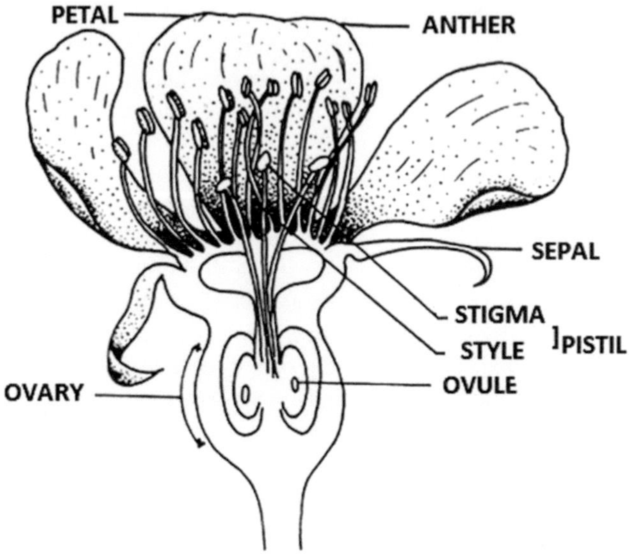

Fig. 4.2 Flower structure of pear

mas mature earlier than the anthers. The style stands erect upon opening and the stamens bend so that the anthers are well below the receptive stigma (Fig. 4.2). Later, the stamens stretch to their full length and the fully-mature anthers release pollen. Thus, the flower's behaviour discourages selfing. The ovary is made up of five compartments with two ovules each. At first, only a few flowers open on a tree and, depending on the weather, more flowers open each day for 3–7 days. Pear produces abundant pollen but little nectar. The sugar content of pear nectar is low, 25 % or less, depending on the variety. Therefore, bees visit other plants that are more attractive, and most bee visitors on pear are pollen collectors.

4.10 Pollination Requirements

Pear varieties range from self-fertile to self-sterile. Some are only partially self-fertile and benefit measurably from cross-pollination. Some varieties show different degrees of self-fertility in different regions and growing practices (Free 1993).

4.11 Pollinators

Many types of bees and flies visit pear, but honey bees are the only practical pollinator. Honey bees do not prefer pear, and honey bee management centres on overcoming this liability. High honey bee numbers can help increase bee visitation to flowers through increased competition (Plates 4.1 and 4.2). Humphry-Baker

Plate 4.1 (**a**) Severely infested mango fruits by mealybugs (**b**) Mango fruit fly ovipositing on mango fruit (**c**) Hoppers infestation on mango leaves (**d**) Mango inflorescence damaged by hoppers (**e**) Mummified fruits to guava due to infestation of Deudorix isocrates (**f**) Guava aphids infestation on newly emerged shoots (**g**) Mealybugs infestation on guava twigs (**h**) Parrot damaging the guava fruit

Plate 4.2 A view of Apis mellifera apiary in fruit orchard. A cherry tree in bloom, Apis cerana on apple flowers. Apis dorsata on peach flowers, Apis mellifera on peach flowers, A. dorsata on guava flowers, A view of Apis mellifera apiary in citrus orchard, Bombus haemorrhodalis on citrus flowers, Apis cerana and Apis dorsata on citrus flowers, A. view of apiary in litchi orchard, litchi bloom with insect pollinators, Apis dorsata on litchi flowers, Apis florea on ber flowers, Apis dorsata on guava, A.dorsata on litchi flowers. Apis cerana on strawberry flowers, Apis florea, flies and other pollinators on strawberry flowers, well formed and malformed fruits in strawberry and aphids attacking strawberry plants

Plate 4.2 (continued)

Plate 4.2 (continued)

Plate 4.2 (continued)

(1975) recommended, 'twice the bee hives per acre in pears than in other fruits'. A second way to increase honey bee efficiency is to use inexperienced bees (Mayer 1994).

Honey bee attractants may be helpful in a crop like pear that is not attractive to honey bees. Bee-Scent®, a Nasonov pheromone-based attractant, increased honey bee visitation in Washington, USA for 24 h post-treatment in 'Bartlett' and 'Bosc', but not in 'Anjou'. Bee-Scent® increased fruit-set by 23 % in 'Bartlett' and by 44 % in 'Anjou'. Bee-Scent Plus® increased fruit-set by 44 % in 'Bartlett' (Mayer et al. 1989a). Queen mandibular pheromone (QMP)-based attractants increased honey bee visitation in 'Anjou' and 'Bartlett' in Washington and British Columbia and increased fruit diameter, all of which translated to a US\$427 acre^{-1} (US\$1055 ha^{-1}) increase in farm gate revenue (Currie et al. 1992). In another study, QMP-based attractant did not increase honey bee numbers nor fruit-set in 'Anjou', but did increase fruit size by 7 % which translated to a US\$162 acre^{-1} (US\$400 ha^{-1}) increase in farmgate revenue (Naumann et al. 1994b). A number of investigators have recommended variable number of honeybee colonies for pollination in pears. For example, Humphry-Baker (1975) recommended two colonies per acre and five colonies per ha. Similarly, Kevan (1988) recommended 0.2–4 colonies per acre with an average of 1–5 colonies per ha. Mayer et al. (1990) recommended 10–15 honey bees per tree min^{-1}.

There has been keen interest in identifying other bee species that do not share honey bees' disinterest in pear. However, the search has not been fruitful. A solitary bee, *Andrena nivalis*, in western North America visits pear and probably is an effective pollinator when it occurs in large numbers, but its active season does not always coincide with pear bloom (Miliczky et al. 1990). Spring-introduced horned-face bees, *Osmia cornifrons,* and blue orchard bees, *O. Lignaria propinqua,* rarely visit pear flowers in south central Washington and instead visit other flowering plants. The incidence of flower visitation by commercially reared and hived bumble bees totalled only three bees over 3 years in an orchard of 'Anjou', 'Bartlett', and 'Bosc' pears in Washington (Mayer et al. 1994a, b). Some speculate that flies are good pollinators of 'Anjou', but there is no research to support this.

4.12 The Role of Insects in Pollination

Some plants could produce crop without pollination. Nevertheless, most of our fruits – like pear – need the pollen from the suitable pollinator and also the transfer by the pollinating insects to be grown economically. There are self-infertile ones among pear species, which need mediators (vectors) taking the pollen from each tree or species to the other ones. Honeybees with specialized body parts and availability in sufficient numbers make the important pollinators (Benedek and Finta 2005).

4.13 Facts Determining the Need of Bee-Colonies

Benedek et al. (1976) refers to the point, that a pear produces less pollen than an apple, however its pollen is more valuable for bees as of resulting favourable biological effects. The disadvantageous weather and the lack of adequate bee-colonies in the nearby make a decrease in the period and efficiency of pollination (Free 1970, 1993; Benedek et al. 1989; Benedek and Nyéki 1995, 1996a, 1996b, 1997; Benedek, Szabó and Nyéki 2000; Benedek et al. 2000; Roversi and Ughini 1986). It is more than challenging to determine the number of bee colonies in an orchard on a per hectare basis. In the very beginning a single family was told to be enough on 1–2 ha. Benedek et al. (1976) identified, that 2.5–3 colonies should be sufficient per hectare taken that no negative effects occur.

Mason bees, *Osmia* spp., are recognized as excellent pollinators, important in North America and many other parts of the world primarily for pollination of apples, pears, almonds, cherries, caneberries, blueberries, and other tree fruits, but also for rape pollination (Torchio 1988; Gathmann et al. 1994; Bosch and Kemp 1999;

Delaplane and Mayer 2000; Havenith 2000; Sekita 2000; Kemp and Bosch 2002; Abel et al. 2003; Li et al. 2004; Monzon et al. 2004; Cane 2005a, b; Bosch et al. 2006; Gardner and Ascher 2006; Krunić and Stanisavljević 2006a, b; Sheffield et al. 2008; Steffan-Dewenter and Schiele 2008; Matsumoto et al. 2009; Sampson et al. 2009). Bosch and Kemp (2002) provide a review of *Osmia* research, evaluating them as crop pollinators. Fruit crops cultivated in Virginia that are bee pollinated include apple, pear, peach, almond, nectarine, apricot, plum, cherry, blueberry, caneberry (raspberry, blackberry, black raspberry, dewberry), strawberry, loganberry, elderberry, gooseberry, persimmon, serviceberry, passion flower (Maypops), and hardy kiwi. Many of these require cross-pollination for optimum fruit and seed set (McGregor 1976; Free 1993; Fell 1995).

4.14 IPM Pome Fruits

Apple and pear commonly called as Pome fruits are attacked by more than 70 insect pests and mites which cause considerable damage.

4.15 IPM in Apples

Integrated pest management (IPM) of apples is very complex due to their perennial growth habit and physical complexity. The different parts of apple tree provide multiple habitats suitable for arthropod colonization. Oatman et al. (1964) recorded as many as 763 species of arthropods associated with apple trees. Of these more than 100 species or so were considered as pests. In general, only a dozen or so arthropods in any given region are considered serious or chronic pests. Some of the pests are most common wherever apples are grown which include codling moth, the European red mite and, to a lesser extent, the two spotted spider mite whereas some others are strictly regional pests. The tetranychid mite (*Tetranychus urticae* Koch, *Panonychus ulmi* (Koch) and *Tetranychus mcdanieli* McGregor) all feed in the same manner and cause a similar type of foliar damage (Beers et al. 1993). The leaf rollers feed on the leaf and surface of apple fruits. Weevils and thrips are examples of two unrelated taxa but cause similar damage such as surface feeding and oviposition, thereby leaving a superficial scar. A number of pest species are strictly monophagous on apple (e.g. *Aphis pomi* De Geer), while others are oligophagous or even highly polyphagous (e.g. *T. urticae*). However, many species exhibit a certain degree of plasticity in their feeding behaviour and are capable of shifting hosts or expanding their host range over time. An example is the apple maggot, *Rhagoletis pomonella* Walsh which recently shifted its host range from apple to cherry.

The introduction of synthetic organic pesticides in later part of twentieth century resulted in a qualitative change in pest management. They had a very broad range, highly effective and relatively inexpensive. They were very easy to use and had long residual toxicity to pests which made them very popular among the pest control agencies. However, these benefits were overshadowed by the resistance problems developing in pests during very short period of their regular use. The problems associated with the use of these products became apparent after a relatively short time, including environmental persistence and damage, mammalian toxicity, possible consumer effects from residues on foods such as carcinogenicity, teratogenicity, mutagenicity or chronic neural effects, and destruction of pests' natural enemies. These problems resulting from so called 'pesticide treadmill' provided a new initiative to reconsider the pest management tactics in the form of a pest management system later known as 'integrated pest management', or IPM (Stern et al. 1959). The basic philosophy of this system was to optimize and harmonize the management tactics in a econonomically feasible, environmentally safe and socially acceptable (Rabb 1972). The objective of this system was to utilize all the available options to keep the pest population below economic levels with minimum use of pesticides.

Conservation biological control uses a natural enemy species that already occurs in the region and makes the environment more favourable for its growth and development. This can include cultivating plants in the vicinity of the orchard that provide an alternative insect host or habitat or avoiding pesticides that are toxic to one or more life stages. This type of system was refereed as 'integrated control'.

However, there is still an interest in the importation of natural enemies, which, if established, become candidates for conservation biological control. The last methods involve ongoing releases of artificially reared natural enemies; these can occur either occasionally (augmentative) or in the form of a 'biological pesticide' (inundative). Because the expense of rearing natural enemies can be considerable, the latter two methods have been little implemented. Cultural control involves manipulating the orchard or the immediate environment to reduce pest numbers or mitigate pest damage. Irrigation may reduce water stress and allow arthropod-stressed trees to produce better than they could otherwise. Orchard floor management may allow more natural enemies to build up in the cover crop and be available to reduce arboreal pest populations. Reducing fertilization so that vegetative growth is minimized may slow the population growth of flush-feeding insects, such as aphids. Cultivars or strains that have reduced terminal growth, such as the spurtype cultivars, may play the same role. In general, however, producers prioritize plant growth and productivity in their orchard management practices, which may conflict with the optimal pest-control practice. Host-plant resistance, while frequently used in field crops, has been of little significance in orchard crops. The horticultural characteristics, especially precocity, productivity, flavour and storability, are the primary drivers of cultivar choice. One notable exception is the use of resistant rootstocks for woolly apple aphid.

Ultimately, IPM can be viewed as just another evolutionary step in our overall problem-solving process in agriculture. More recently, theories have emerged that take the next logical step of integration to the entire production system – integrated fruit production, or IFP (Boller et al. 1998). To an extent, this may be viewed as a reincarnation of the organic-production philosophy, which also encompasses all aspects of the production system but with the additional caveat of restricting the materials used to only naturally occurring, minimally processed products (in terms of pesticides, plant-growth regulators and fertilizers).

Different group of insects which attack apple trees at different stages of development and at different levels are given below:

4.16 Fruit Feeders

This group of insects attacks the fruit directly, leaving either feeding scars or deep entries, potentially serving as an infection site for pathogens. The EILs for these pests are relatively straightforward for fresh-market fruit, because virtually all defects are removed during packing. The issue is somewhat clouded for processing fruit, where some level of damage, especially healed surface damage, does not detract from the utility or quality of the fruit. Overall, pest-management programmes have focused most intensely on this group of pests because of their clear and apparent effect on usable yield.

The fruit may be attacked at almost any point during the growing season, from early in the bud stage to harvest. Fruit attacked early in the season is more likely to abscise naturally, or it can be selectively thinned during hand-thinning. Fruit attacked during the mid-season is more likely to stay on the tree and thus has a higher likelihood of being harvested. Fruit attacked very late may generate sufficient ethylene to abscise prematurely and has a slightly reduced chance of entering the packing or processing plant. Clearly, excessive amounts of fruit drop just before harvest will have a detrimental effect on yield.

4.17 Direct Pests of Buds and Fruitlets

4.17.1 Noctuids (Lepidoptera: Noctuidae)

This group, called the green fruitworms in North America, include *Orthosia hibisci* (Guenée), *Amphipyra pyrimadoides* (Guenée) and *Lithophane antennata* (Walker) in which the young larvae feed on developing buds and fruitlets. The feeding damage prevents, causes premature abscission and distortion of the fruit. Several species, such as *Orthosia incerta* (Hufnagel), may be found in Europe, depending on the region (Carter 1984). These pests may be regionally important, but are generally considered minor. Pheromones may be used to monitor their flight to help predict phenology and relative abundance, e.g. *Graphania mutans* in New Zealand (Burnip et al. 1995).

4.18 Weevils (Coleoptera: Curculionidae, Attelabidae)

Several species of weevils have been reported damaging buds, fruits, foliage and woody tissues of apple trees. Two species of weevils which are considered to be major pests include apple blossom weevil, *Anthonomus pomorum* (L.), and the plum curculio, *C. nenuphar* (Herbst) (Toepfer et al. 1999). Apple-blossom weevil adults feed on developing apple buds in spring. Feeding is followed by oviposition and larval feeding on the bases of flower petals, resulting in sterility and a brown-capped appearance of the flowers. Low to moderate populations may act as natural blossom thinners. Large populations, more common in recent years, can overthin the crop. Plum-curculio adults likewise feed on developing apple buds in spring but also feed upon and then oviposit into young fruitlets, where larvae tunnel and cause most injured fruitlets to drop. Injured fruit remaining on trees are scarred by the feeding and ovipositional wounds, which usually render injured fruit unmarketable. Whereas apple-blossom weevils and northern populations of plum curculios have one generation per year, more southern populations of plum curculio have an additional generation and threaten not only fruitlets but also apples approaching maturity.

An understanding of the ecology of these weevil species is the key to successful management (Vincent et al. 1999). Both species can build into large populations on unmanaged host trees. In some locales, plum curculio annually infests 90 % of the fruit on unmanaged trees. Although resident vertebrate and invertebrate predators, parasitoids and pathogens do have some impact, the degree of population suppression by these biocontrol agents has generally been insufficient to maintain infestations below levels that threaten the quality of buds or fruitlets. Application of organophosphate or other insecticides timed to coincide with pulses of adult immigration continues to be the main approach to managing both of these pests. Because there still exists no truly effective trap for monitoring immigrant adults (Prokopy et al. 1999), timing of application is based on degree-day models that predict periods of immigration (Reissig et al. 1998). Improved understanding of the ecology of these species has facilitated excellent orchard-wide control using a much-reduced amount of material through restricting application to only those orchard trees most likely to become infested, i.e. trees within 20 m or less of the perimeter (Vincent et al. 1997).

4.19 Mirids (Hemiptera: Miridae)

Like the weevils, the serious mirid pests of apple are orchard invaders, completing the majority of their life cycle outside the orchard and immigrating only during brief periods to feed on fruit. This presents an additional challenge to pest management in that the grower is forced to respond reactively, rather than being able to take proactive steps in management. The tarnished plant bug *Lygus lineolaris* Palisot de

Beauvois is a sporadic pest of apple. It pierces the developing fruit let with its piercing–sucking mouth-parts, leaving a deep, inverted dimple on the mature fruit. Although the mullein plant bug (*Campylomma verbasci* (Meyer)) feeds in a similar way, it leaves a raised corky wart on the fruit. Several other pests in the same group occur in different areas of Europe and North America, including the genera *Lygocoris*, *Lygidea*, *Heterocordylus* (Boivin and Stewart 1982), *Campyloneura*, *Plesiocoris*, *Blepharidopterus* (Alford 1984) and *Atractotomus* (MacPhee 1976). With the exception of *L. lineolaris*, most of the apple-feeding mirids are facultatively predacious and thus are considered natural enemies as well as pests.

4.20 Thrips (Thysanoptera: Thripidae)

Thrips are serious and widespread crop pests worldwide, but have few representatives in the apple pest complex. The most common species is *F. occidentalis* (Pergande). The adults are attracted to blooming plants and are often present in the orchard on blooming weeds. When apple blossoms open, they move to developing fruits. Their feeding activities (sucking mouth-parts) cause a condition called 'pansy spot' on sensitive cultivars, and they leave a small oviposition scar in the centre of the pansy. The damage is most apparent on light-coloured cultivars, often colouring over on deeply coloured sports. The pear thrips, *Taeniothrips inconsequens* (Uzel) is primarily a pest of pear and sugar-maple, but is an occasional pest of apple.

4.21 Sawflies (Hymenoptera: Tenthredinidae)

Sawfly *Hoplocampa testudinea* (Klug) is a widespread and serious pest of apple in Europe (Giraud et al. 1996). The adults appear during bloom and lay eggs in the flower, giving rise to larvae that burrow and feed in the fruit. The adults may be monitored with white sticky panels.

4.22 Mature-Fruit Feeders

4.22.1 Codling Moth

BIOLOGY Codling moth, *C. pomonella* (L.), is the main direct pest of apples worldwide and has been extensively studied. It is not reported as present in Japan Taiwan, Korea or eastern China, but is otherwise cosmopolitan. It is present in the urban areas of the Brazilian apple-growing area, but it has not yet invaded the orchards. There are typically between one and four generations per year, depending on the climate. The level of infestation on untreated apple trees can reach 100 % of fruit infested, with evidence of multiple 'stings' or larval attacks. The economic threshold for codling moth is low (*c.* 1 % damaged fruit), even for crops that are not exported. These factors have combined to make this pest one of the greatest scourges for apple growers. It is also one of the most researched and consequently best understood insect pests. The absence of the insect from Asian growing regions has led to stringent procedures, including fumigation of apples and other potential host fruit with methyl bromide (e.g. Maindonald et al. 1992).

Female moths oviposit single eggs on or near developing fruit. Larvae hatch out and locate apples on the basis of an apple fruit volatile, (*E*,*E*)-α-farnesene (Sutherland and Hutchins 1972). Larvae then begin to enter the fruit and make their way to the core to feed on the seeds, like other members of the genus *Cydia* (Witzgall et al. 1996b). The entrance hole is frequently plugged with frass. Mature larvae emerge from the fruit with a characteristic exit hole. Diapausing fifth-instar larvae overwinter in cocoons in suitably protected locations under the bark of the host tree or on the ground. Factors contributing to population regulation of codling moth have been the subject of considerable research. There appears to be general acceptance of the findings of Geier (1963) that limited supply of fruit and overwintering sites are the key factors limiting codling-moth populations on unmanaged trees. The main recorded hosts are apple, European pear, nashi (Asian pear), Chinese pear and quince. Walnut and plum are consistently attacked, while peach, nectarine and apricot are also recorded hosts, and damage can be significant in some situations. Differences in the host preference, development, diapause, phenology and population dynamics have been found for strains or races of the moth taken from apple, plum or walnut host plants (Barnes 1991). The removal of alternative or abandoned host trees can therefore make an important contribution to control by reducing migration of the pest into smaller orchards.

4.23 Detection and Inspection Methods

Pheromone traps have been used for detecting adult male codling moths since the initial pheromone identification (Roelofs et al. 1971). This is one of the best understood and most widely used pheromone monitoring systems. A number of different management approaches have been based on pheromone-trap detection of males, including forecasting female moth flight and oviposition from sustained male flight activity used with day-degree accumulation (Riedl 1976). Direct observation of damaged apples during the growing season is another obvious method of monitoring the pest population, although detection of a direct pest at harvest is usually too late for economic production where there is a single generation.

4.24 Chemical Control

For much of the twentieth century, chemical control was the most widespread method of pest control. However, after usage and selection for populations with genetic resistance to arsenic (Hough 1928), followed by the same pattern with dichlorodiphenyltrichloroethene (DDT) (Glass and Fiori 1955), orchardists have switched to other broad-spectrum insecticides. Development of resistance to other insecticides has occurred, although it has not always occurred in all countries or been documented adequately.

Organophosphates were the next chemical group used in many countries (azinphosmethyl, phosmet, diazinon and phosalone), but resistance is now widely recorded (Barnes and Moffitt 1963; Bush et al. 1993; Varela et al. 1993; Blomefield 1994; Knight et al. 1994).

In Europe, more selective insecticides have been increasingly used, including juvenoids (such as fenoxycarb (Charmillot 1989)), chitin synthesis inhibitors (diflubenzuron, triflumuron, chlorfluazuron and teflubenzuron) and ecdysone agonists (e.g. tebufenozide and methoxyfenozide) (Heller et al. 1992). However, there are also examples of resistance to these compounds (Moffitt et al. 1988; Sauphanor and Bouvier 1995). In addition, avermectin (a macrocyclic lactone fermentation product) has some efficacy (Cox et al. 1995), as does spinosad, another fermentation product. The advantage of more selective insecticides is the reduced impacts on natural enemies, permitting the maximum contribution of biological control against other pests. Petroleum oils have been used as ovicides (Webster and Carlson 1942), although earlier products often caused phytotoxicity. More recently, highly refined and purified products have been shown to have good efficacy (Riedl et al. 1995) and have reduced phytotoxicity problems. Particle films (Unruh et al. 2000) also have some efficacy against codling moth.

Mechanical control, using bands on tree trunks to collect diapausing larvae, has also been used, but these do not collect the proportion of the population that falls to the ground directly. They can be effective if used in conjunction with other tactics (e.g. Judd et al. 1997).

4.25 Biological Control

A number of biological control agents of codling moth are available which control the moths by predation (Knight et al. 1997) or parasitism (Hassan 1989) of eggs and neonate larvae (MacLellan 1972). The cryptic habit of the larval stages (including diapause) offers some protection against natural enemies. In some situations, bird predation of diapausing larvae can be significant (Wearing and McCarthy 1992). However, the high levels of damage typically observed in the absence of controls indicate that biological control alone is not sufficient to, to maintain the pest below the economic threshold level.

4.26 Mating Disruption

Sex pheromones which could prevent or disrupt mating in codling moth has been has been practiced worldwide, based on promising results with a range of formulations (Charmillot 1978; Moffitt and Westigard 1984; Gut et al. 1992; Minks and van Deventer 1992; Judd et al. 1997). The mechanisms by which disruption acts are not entirely clear (Minks and Cardé 1988) and it may be possible to use pheromone-related compounds to improve results (Witzgall et al. 1996a).

Mating disruption is inversely density dependent and therefore works best at low pest densities in sites without significant immigration. It is not as effective in situations where the pheromone cloud is difficult to maintain or in close proximity to unmanaged populations. Mating disruption of codling moth is now commercially accepted in several countries. This has occurred in part because of the failure of conventional insecticides, due to resistance, as well as the intrinsic environmental and worker safety of pheromone products. Although codling-moth mating disruption is not yet registered in all European countries, it has been widely used in some areas (e.g. northern Italy). The relatively higher cost of this technique slows its adoption, especially in warmer regions where two applications per season of the dispenser are necessary.

4.27 Mass Trapping and Attracticidal Control

Mass trapping is not very effective method against codling moth (Proverbs et al. 1975), because of the cost and practical difficulties of deploying sufficient stations. As with mating disruption, the tactic aims to prevent mating and therefore pest progeny. However, whereas in mating disruption males can survive to find a mate the next night, this is not possible where males have been removed from the system, which represents a potential strength of the approach. If droplets containing sex pheromone and a fast-acting insecticide are used instead of traps (Charmillot et al. 1996), then the costs can be somewhat reduced. It may also be possible to develop multiple-species attracticides (Suckling and Brockerhoff 1999).

4.28 Sterile-Insect Technique

Although it is technically feasible (e.g. Proverbs et al. 1982), sterile-insect release is expensive and has several important limitations. Most importantly, it requires mass rearing with specialized capital-intensive facilities, excellent quality control to maintain mating competitiveness with feral insects, geographical isolation, political support and ongoing investment in the event of movement of contaminated fruit. There are apparently few regional orchard industries that meet these criteria.

4.29 Microbial Control

The most promising microbial control against codling-moth neonate larvae is a granulosis virus (Tanada 1964), which has been tested extensively in Europe (Audemard et al. 1992), including the UK (Glen and Payne 1984), New Zealand (Wearing 1990) and the USA (Westigard and Hoyt 1990). In hot climates with high levels of solar radiation, the persistence of the virus in the field is poor (about 1 week), making frequent applications necessary. However, its effectiveness against high pest populations, in combination with mating disruption, offers organic apple growers an effective way of reducing pest populations to levels at which mating disruption can operate effectively. Commercial use of the virus has unfortunately been limited by the costs of production using live insects. Industrial-scale production offers reduced costs to growers (M. Guillon, personal communication), which should assist adoption in future.

4.30 Oriental Fruit Moth and Other Grapholita (= Cydia) Species

Grapholita molesta (Busck) and other members of the *Grapholita* genus, such as *G. lobarzewskii* (= *Cydia lobarzewskii*) and *G. janthinana* (*Cydia janthinana* (Dup.)) are sometimes recorded as pests of apple (Kalman et al. 1994). In several countries, *G. molesta* (or oriental fruit moth) is reported to be increasingly important as a pest of apples (e.g. Reis et al. 1988; Pollini and Bariselli 1993). These species typically feed on shoots early in the season, as well as fruits later in the season. The biology and options for control are similar to those for codling moth, but the pest status may not always warrant intervention. Within the past few years, oriental fruit moth has emerged as a major pest in several Midwestern and eastern US growing districts, surpassing codling moth in importance.

4.31 Tephritid Fruit Flies (Diptera: Tephritidae)

True fruit flies of the family Tephritidae (Aluja and Norrbom 2000) deposit eggs directly into the flesh of developing fruit, particularly fruit approaching readiness for harvest. The tiny puncture made through the skin of fruit during egg-laying is difficult to detect without magnification and may remain so even when underlying flesh has decayed substantially during larval feeding. Commonly, infested fruit are detected only after a few days of exposure to room temperature following purchase by an unwary consumer. Three species of tephritid flies have been recorded as key pests of apple which include, the apple maggot fly, *R. pomonella* (Walsh) (native to North America), the Mediterranean fruit fly, *Ceratitis capitata* (Wiedemann) (native to Africa) and The South American fruit fly, *Anastrepha fraterculus* (Wiedemann) (native to South America) (Sugayama et al. 1998) have spread to different apple growing areas causing considerable damage to commercial apple production.

Sometime during the past two centuries, all three species expanded their host range to include apples. In the process, they have escaped most of their natural enemies (particularly parasitoids), which provide some biological control of fruit-fly eggs or larvae in native host fruit. Currently, all three pest species are managed primarily by applications of organophosphate insecticides, although in some areas the preharvest interval dictates the use of pyrethroids. Applications are timed in accordance with the occurrence and abundance of captures of invading adults by monitoring traps placed on perimeter trees.

Predictive phenology models (Jones et al. 1989) have been useful in determining the timing of emergence. In some cases, confining insecticide application only to perimeter trees or baiting perimeter trees with odour–visual traps has provided effective control (Cohen and Yuval 2000; Prokopy et al. 2000). Even though there are no known cases of insecticide resistance in any tephritid fly, the need for continuous protection of apples by insecticide residue over the course of the 2–3-month period of susceptibility to fly oviposition is prompting some growers to seek alternative approaches to fly management.

4.32 Leaf-Rollers (Lepidoptera: Tortricidae)

4.32.1 Biology

Leaf-rollers have only an indirect physiological impact on the tree, since they feed on the fruit surface rather than the seeds. While the impact on the tree may be negligible, the impact of fruit feeding on grower returns is a direct one. Leaf-rollers emerge as a major concern in many orchards that apply selective controls for codling moth, as well as for exporters forced to meet quarantine tolerances with a nil threshold. Larvae typically web foliage together and many also feed directly on the fruit surface. This cryptic habit has often made insecticidal control difficult. Fruit damage is visible as scarring or corking or as rots associated with open wounds in storage, and larvae occasionally enter the apple calyx. Injury to fruits destined for fresh and especially export markets has the most significant economic impact, compared with that of processing grade apples.

Leaf-roller biology differs in several important ways from the internal feeding tortricid species (van der Geest and Evenhuis 1991). Many have much wider host ranges and feed on leaves as well as fruit (Chapman and Lienk 1971). Their external life habit is accompanied by larval dispersal through ballooning, typically followed by the establishment of a larval nest on shoots or the undersides of leaves. Larger larvae are able to relocate to fresh nests and use their silken thread for both nest construction and escape. Many species are multivoltine, with up to four generations per year. Unlike codling moth, few leaf-roller species are geographically wide-

Table 4.1 List of leaf-roller pests affecting apple in various regions (Source: Beers et al. 2003)

Species	Common name	Distribution
Adoxophyes orana (Fischer von Röslerstamm)	Summer-fruit tortrix	Europe, Asia
Archips argyrospila (Walker)	Fruit-tree leaf-roller	North America
Archips breviplicanus (Walsingham)	Asiatic leaf-roller	Asia
Archips podana (Scopoli)	Great brown-twist moth	Europe, Asia
Archips rosana (L.)	European leaf-roller	Europe, USA
Archips xylosteanus (L.)	Apple leaf-roller	Eastern Europe
Argyrotaenia velutinana (Walker)	Red-banded leaf-roller	Eastern USA
Choristoneura rosaceana (Harris)	Oblique-banded leaf-roller	North America
Epiphyas postvittana (Walker)	Light brown apple moth	Australia, New Zealand
Pandemis heparana (Denis and Schiffermüller) *Pandemis*	leaf-roller	Europe
Pandemis limitata (Robinson) *Pandemis*	leaf-roller	North America
Pandemis pyrusana Kearfott *Pandemis*	leaf-roller	Western USA
Platynota flavedana Clemens	Variegated leaf-roller	Eastern USA
Platynota idaeusalis (Walker)	Tufted apple-bud moth	Eastern USA

spread. Instead, apple-growing regions typically have a unique complex of leaf-roller species (Chambon 1986).

4.33 Host Range and Detection Methods

The leaf rollers attacking apple have a very wide host range and some of the hosts of light brown apple moth, *Epiphyas posvittana* include kiwifruit, chrysanthemum, hawthorns, cotoneaster, jasmine, privet, lychee, macadamia nut, lucerne pines, peach, apricot, pears, oaks, blackberry, raspberry, potato, clovers, broad bean, grapevine, currants, roses, citrus, poplars and blueberries. Detection and monitoring of leaf rollers have been done using pheromone traps (Suckling and Karg 2000). Modern diagnostic methods using DNA have been used for correct identification of species proving support to taxonomic studies (Sin et al. 1995; Gleeson et al. 2000) (Table 4.1).

4.34 Biological Control

The reduction in use of broad spectrum insecticides on apples resulted in increased biological control activity of leaf rollers. For example, there was a considerable spread in the distribution of the parasitoid wasp *Colpoclypeus florus* Walkerused for control of oblique-banded leaf-roller, *Choristoneura rosaceana* (Harris) in Washington (Suckling et al. 1998).

4.35 Cutworms and Fruit Worms (Lepidoptera: Noctuidae)

Although minor in importance in comparison with the tortricids, several species are capable of fruit feeding later in the season. The larvae excavate shallow round holes in the fruit, rendering them unmarketable. Dome of the examples include: The spotted cutworm (*Xestia c-nigrum* (L.)), Bertha army worm (*Mamestra configurata* Walker), variegated cutworm (*Periodroma saucia* (Hübner)), black cutworm (*Agrotis ipsilon* Hufnagel) and the western yellowstriped army worm (*Spodoptera praefica*). More recently, a new species, *Lacanobia subjuncta* (Grote & Robinson), was recorded from Washington State (Landolt 1998) and has become an important pest in some areas.

4.36 Fruit-Stinging Insects (Hemiptera)

Pests in this group are also orchard invaders and damage levels are often highest around the orchard borders. The surrounding habitat is a primary determinant of the intensity of attack. The most common example are the stinkbugs (Pentatomidae; *Euschistus conspersus* Uhler and *Acrosternum hilare* (Say)), but the western boxelder bug (*Leptocoris rubrolineatus* Barber; Hempitera: Rhopalidae) has similar pest status. Damage usually occurs in the latter part of the season and is characterized by a spongy, depressed area *c.* 1 cm in size surrounding the feeding puncture. Externally, damage can resemble physiological disorders such as bitter pit, but the tissue beneath the skin does not turn brown.

4.37 Miscellaneous Opportunists

A number of insects are attracted to ripening or overripe fruit and will either create a point of entry or enlarge damage due to other causes (splits, stem punctures, etc.). Vespid wasps are often found in orchards near harvest and, although they are primarily predacious, they chew holes in ripe fruit and pose a hazard to harvesters. Nitidulid beetles are also attracted to ripening fruit and can be found feeding under the surface. Earwigs are orchard residents that are usually predacious, but will also chew or enlarge holes in fruit. They can curl up in the stem cavity and make their way into the packing-house. The dock sawfly, *Ametastegia glabrata* (Fallén), tunnels into the fruit, especially those close to the ground, in order to find an overwintering shelter.

4.38 Foliage Feeders

Several species of insect pests attack and feed on foliage, with the primary damage being loss of photosynthetic capacity due to loss of chlorophyll and disrupted osmotic balance. From the perspective of plant productivity, specifically yield parameters, the effect of chlorophyll loss is controversial. No clear and uncontested relationships have been established, although it seems clear from the body of literature that there is not a directly proportional relationship between loss of chlorophyll and loss of photosynthetic capacity.

Trees are capable of sustaining a certain degree of foliar damage without any measurable loss in yield; thus, the critical question becomes: 'How much damage?' The studies performed attempting to establish such relationships quantitatively have been restricted in interpretation to the particular combination of cultivar, climate and growing regime in which they were conducted and, as a consequence, the results and interpretation have been quite variable.

The implication of some level of tolerable damage has a critical implication for IPM: the latitude for biological control. In many cases, some level of the pest population must survive in order for the natural enemy to survive (unless there is an alternative host). Unlike pests of quarantine importance or direct fruit feeders, there is a greater window of opportunity for non-pesticidal control measures, since the need for control is not so immediate or triggered at a low threshold. Given the societal emphasis on reduction in pesticide use, this characteristic should be more fully exploited in the future. Many of the foliage feeders are classed as secondary (in importance) or induced pests. The latter classification implies that they would not have achieved pest status without pesticide inputs directed at a primary (often direct) pest. Again, this points to the opportunity to regulate this group using non-pesticidal methods, or only on an occasional basis.

4.39 Mesophyll Stylet Feeders

This group feeds on cellular contents (including chlorophyll) by penetrating the leaf surface (often from the underside), killing only one or a group of cells at each feeding site. The damage appears as speckling (leafhoppers) or bronzing (tetranychid and eriophyid mites), depending on the size of the mouth-parts and the depth of penetration. Reduction in photosynthesis may follow extensive feeding, due possibly to a combination of chlorophyll loss and/or stomatal closure caused by water loss.

4.40 Tetranychid (Spider) Mites

European red mite (*P. ulmi* Koch) and two-spotted spider mite (*T. urticae* Koch) are serious pests of apple worldwide. Other species (*T. mcdanieli, Tetranychus viennensis, Eotetranychus carpini, Bryobia* spp.) cause a similar type of damage, but are regional in distribution. A few species of tenuipalpids (false spider mites) and tarsonemids (e.g. *Cenopalpus pulcher* Canestrini & Fanzago) are apple pests in some regions (Jeppson et al. 1975b).

The biological control of spider mites has been achieved with varying degrees of success. The predatory mites belonging to family Phytoseiidae (*Typhlodromus occidentalis, Typhlodromus pyri* Scheuten, *Amblyseius fallacis* (Garman), *Ambylseius andersoni* (Chant), *Neoseiulus californicus* (McGregor)) are the most successful biological-control agents (Jeppson et al. 1975a; Kostianinen and Hoy 1996). Different species are better adapted to different growing regions; for example, *T. occidentalis* is ideally suited to the arid climate of the western USA, whereas *T. pyri* requires a more humid, temperate climate (Beers et al. 1993). *T. pyri* is the most important mite predator in the temperate regions of Europe (excluding Scandinavia and the Mediterranean region). It is widely used for European red-mite control, often through the release of organophosphate (OP)- resistant strains (Blommers 1994). The number of years needed to achieve successful spider-mite control may vary between 1 (temperate conditions) and 3 (cooler conditions). *T. pyri* does not occur in the Mediterranean area, where summers are too hot and dry. *A. andersoni* is the most important predator in these areas, where its natural populations can very successfully control the pest populations (García-Marí et al. 1989).

Several predatory mite species have adapted well to orchard spray regimes, and this is, in large part, the reason why integrated control programmes have been possible (Hoyt 1969). In addition, several species are reared commercially and sold for release in orchards either as an inoculative measure or as a sort of 'living pesticide'; some strains have been selected for tolerance to pesticides (Hoy and Knop 1981; Roush and Hoy 1981). Other families also contain predatory species useful in apple orchards (e.g. Trombidiidae, Anystidae, Stigmaeidae and Tydeidae); however, these predators usually play a supporting role to the phytoseiids. In the mid-Atlantic area of the USA, a predatory coccinellid (*Stethorus punctum* (LeConte)) provides the greatest degree of biological control, whereas a related species in the western USA (*Stethorus picipes* Casey) plays only a minor role. Several groups of predatory bugs (especially mirids in the genera *Campylomma, Campyloneura, Blepharidopterus, Atractotomus*) will prey on mites and may play an important role in biological control.

4.41 Eriophyid Mites

There are two groups of eriophyids mites causing plant deformities (galls or blisters). The apple-rust mite, *Aculus schlechtendali* (Nalepa), is widely distributed and common, but rarely considered a serious pest. While high populations can cause leaf bronzing and premature terminal bud set (Hull et al. 1986), it is considered a quasi beneficial species in some areas in that it provides an alternative prey for predatory mites (Hoyt 1969). Sensitive cultivars (e.g. 'Golden Delicious') may be russeted by feeding in the calyx area, which occurs shortly after bloom. Examples of the gall formers attacking apple are few. Burts (1970) reported on two closely related species *Eriophyes* (*Phytoptus*) *pyri* (Pagenstecher) and *Eriophyes mali* (Burts), both of which may attack apple. They cause blisters on the leaves and fruit, leaving the latter scarred and deformed. The current spray programme has made these mites rare.

4.42 Leaf-Miners

Several species of leaf miners mine on apple leaves in their larval stages. They lay eggs on the surface of the leaves and newly hatched larvae penetrate the leaf directly from the egg. The shape of the mine is usually characteristic of the species or group: the gracillariid leaf-miners (*Phyllonorycter* (= *Lithocoletis*) *blancardella*, *Phyllonorycter elmaella*, *Phyllonorycter crategella*), or socalled 'tentiform' leaf-miners, produce a distinctive dome-shaped mine with white feeding specks visible from the upper leaf surface. Two species of lyonetid moths (*Leucoptera malifoliella* (= *scitella*) and *Lyonetia clerkella*) produce a blotch and sinuous mine, respectively (Alford 1984). Several species of coleophorid moth (case-bearers) also form mines, but these are usually minor pests. The cryptic habit of the larvae presents some challenges for chemical control. Either the adult must be targeted with applications sufficient to cover the entire flight period or the pesticide must penetrate the leaf surface in order to deliver the toxicant to where the larvae are feeding. True systemic insecticides are now rare and, because of residue problems, few are being developed. Insecticides with translaminar activity are sufficient and typically present few problems in the registration process. While several effective insecticides are registered for use against leaf-miners, biological control has been reasonably well studied and partially implemented. Parasitic wasps (e.g. *Pnigalio flavipes*, *Pnigalio marylandensis*, *Apanteles ornigis*) are regionally abundant and can provide substantial levels of control. However, hymenopterous parasitoids, as a group, tend to be less tolerant of broadspectrum insecticides, and biological control is easily disrupted.

4.43 Skeletonizers

There are several species of arthropods from various groups that skeletonize leaves, but none are specialists on apple and their significance is sporadic and local. Examples include the apple and thorn skeletonizer (*Eurtomula pariana*; Lepidoptera: Choreutidae) and the pear slug (*Caliroa cerasi*; Hymenoptera: Tenthridinidae).

4.44 Bulk Leaf Feeders

This is a varied group, comprised mostly of polyphagous Lepidoptera. Many are pests of deciduous forest trees, which can use apple as a host in the absence of pesticide residues. Examples include the notodontid moths *Datana ministra* and *Schizura concinna* and several species in the lasiocampid/ lymantriid group (*Orygia antiqua*; *Euproctis chrysoheoea*, brown-tail moth; *Euproctis similis*). The winter moth (*Operophthera brumata*) is occasionally an important pest of apple in Europe. The autumn webworm (*Hyphantria cunea*; Arctiidae) is an example of a gregarious nest maker, which forms a large web (up to 50 cm long) and devours all leaf material inside it. Other gregarious lepidopterans include the tent caterpillars (*Malacosoma americana, Malacosoma fragilis* and *Malacosoma disstria;* Lasiocampidae) and the ermine moths (Yponomeutidae, e.g. *Yponomeuta malinellus* (apple ermine moth)); and others in the genera *Swammerdamia* and *Paraswammerdamia* are capable of using apple as a host. Currently, these are primarily pests on unsprayed back-garden trees, but they represent a rich pool of potential insect species that may respond to our changing pest-control regimes.

4.45 Structural Feeders

The group is defined as those attacking plant parts other than fruits and foliage, that is, branches, trunk and root systems. The group is a varied one taxonomically, and several of the pests included cross the damage-classification boundaries as defined here. While some of these pests can cause sufficient damage to cause tree death, as a group they are generally considered less important than the fruit and foliage feeders.

4.46 Superficial Woody-Tissue and Shoot Feeders

Two groups of Homoptera (scales and mealybugs) are widespread and sometimes important pests of apple. San Jose scale (*Quadraspidiotus perniciosus* (Comstock)) is widely distributed and, left unchecked can cause reduced tree vigour or even mortality. Scales feed primarily through tree bark, forming large encrustations that devitalize the tree. Mealybugs (especially Pseudococcidae) also suck plant juices, but usually choose more tender tissues (shoots and leaf axils) as feeding sites. In the latter case, the primary damage is not from removal of plant product, but rather the production of honeydew (liquid drops of excrement rich in simple sugars). Honeydew dripping on fruit can cause fruit russeting on sensitive cultivars or can support the growth of sooty mould, a superficial but unsightly fungal growth.

Both scales and mealybugs are considered to be induced secondary pests, which would occur only at low levels if their natural enemy complex were not decimated by broad-spectrum pesticides. Currently, the pre-bloom use of horticultural spray oils appears to keep scales in check, although this activity is probably supplemented by in-season use of organophosphates. Mealybugs, on the other hand, can be extremely persistent once established (usually in large, older trees) and even an intense spray programme can only keep them in check, not eradicate them. Both species will infest the fruit towards the latter part of the season, especially when populations are high. A red ring appears around the scale that settles on fruit; mealybugs usually move to the calyx, where detection is difficult during packing operations. Feeding in the calyx end causes a softening and deterioration, which may be exacerbated by long-term storage. Quarantine measures and food contamination are issues with these two groups of pests.

The psyllids (Homoptera: Psyllidae) are key pests of pear, but one species, *Psylla mali* (apple sucker), is a corresponding pest of apple in some regions. Like pear psylla, this pest feeds on shoots and produces honeydew, with the attendant problems for fruit and vegetative growth. However, its importance on apples is minor in magnitude compared with the related species attacking pear. Another large and important group of homopterans (aphids) may also be classed as shoot feeders. This group has specialized in phloem feeding and is a large and successful group of pests on many crops. The aphids that feed on apple may use it as their only host or as the primary or overwintering host, with a different plant species as a summer host. The two life-history patterns have a definite influence on management.

Apple aphid (*Aphis pomi*) is a serious pest of apple in most of the apple growing areas of the world. It spends its entire life cycle on apple, reproducing by parthenogenesis for the greater part of the season. Winged (alate) forms are produced under high population levels to colonize new host plants, and in the autumn sexual forms are produced, which mate and lay overwintering eggs. Both leaves and shoots are attacked, and some level of growth reduction is assumed under heavy attack; however, most of the concern for this pest involves the production of honeydew and sooty mould and the resulting fruit contamination.

Several other common aphid species are heteroecious, although their damage may be quite distinct from that of apple aphid, e.g. the rosy apple aphid (*Dysaphis plantaginea* or *Dysaphis devecta*) which also feeds on shoots and leaves, but injects a salivary toxin, which severely deforms both organs. In addition, the toxin causes fruit deformity on sensitive cultivars. This pest colonizes a herbaceous weed host during the summer; thus control measures must occur fairly early in order to be effective. Woolly apple aphid uses elm as the alternative host in some areas, but is functionally monophagous in the northwestern USA and Europe.

Feeding by both the root and shoot colonies produces galls; typically, the aboveground galls (which occur in leaf axils) are pruned off and are of little significance. Several species of *Rhopalosipum* (*R. fitchii* and *R. insertum*) are occasional pests of apple, using one of the *Gramineae* as their summer host. Only very heavy infestations, which can infest developing fruitlets, are considered damaging.

Aphids have a rich and varied naturalenemy complex that prey on them, including lacewings (*Chrysopa* and *Hemerobius*), coccinellids (ladybirds), various parasitic wasps and a variety of predatory mirids (e.g. *Campylomma*, *Deraeocoris*, *Orius*). Despite this, aphids often escape from biological control. Many of their predators are generalists and will only be attracted to large aphid populations (i.e. after the point where control is needed or desired). A number of broad-spectrum pesticides used in orchards are toxic to one or more of these natural enemies and disruption of biological control early in the season may preclude stable regulation for the rest of the season. Woolly apple aphid was one of the earliest targets (1920s and 1930s) of a widespread introduction of a biological-control agent, the parasitic wasp *Aphelinus mali*. This wasp was introduced in many of the areas around the world where woolly apple aphid had also been introduced and was successfully established in most areas (Yothers 1953). It is still thought to provide the primary means of biological control today, although the generalist predators described above also play a role.

4.47 Wood-Boring Insects

Several families of Lepidoptera attack the cambium of the trunk and major scaffold limbs, and prolonged attack can girdle and kill these organs. The clearwing moths (Sesiidae) have several species that attack various fruit and ornamental trees and at least one species that infests apple (*Synathedon myopaeformis*; UK and continental Europe). One species of tortricid moth (cherry-bark tortrix, *Enarmonia formosana*) causes a similar type of damage.

While rarely a problem in sprayed orchards, these insects are difficult to control because it is not easy to kill larvae in their galleries with pesticides; thus pesticidal control measures must be directed at the adults. Typically they are univoltine, with a prolonged flight, making continuous coverage a necessity. The scolytid beetles comprise some of the more serious forest pests, and several species in the genera *Scolytus* and *Xyloborus* are pests of apple. The larvae form distinctive galleries in the wood, and adults often bore into shoots just below buds, causing weakening and breakage. In general, these insects are usually attracted to trees that are already weakened by some other pest or disease, although young trees can suffer damage when they are close to a source of emerging adults. Coleopteran wood-borers in the families Buprestidae and Cerambycidae may also attack apple, but are rare in sprayed commercial orchards.

4.48 Root-System Pests

This is a restricted group of pests and has received little attention as they cause only occasional damage or because of their cryptic life history. The larvae of scarab beetles (several species in the genera *Polyphylla* and *Pleocoma*) feed on roots and can be locally severe. Trees on sandy soils where the grubs thrive may suffer long-term damage; orchardists will often replant repeatedly, trying various combinations of fertilizer or watering to promote tree growth, when in fact the root system is being systematically destroyed.

Soil fumigation is currently the best remedy to allow trees sufficient time to establish before the beetles reinfest the orchard. Soilapplied pesticides are widely discouraged because of groundwater contamination issues. While biological-control agents are known, their management is little studied or applied. Entomophagous nematodes (injected into the soil) may alleviate the problem, but their effect is not well studied in tree fruits.

One very large species of cerambycid beetle (*Prionis* sp.) can attack apple; control measures are similarly difficult. Weevil (curculionid) larvae are known to attack the root system on occasion, but the extent of damage is not well defined. The adults of some species may also be problematic when they feed on fruits, fruit stems or foliage. Woolly apple aphid is the only truly ubiquitous root pest of apple (see above), although typically only the aerial colonies are treated.

In conclusion apple pest management is continually evolving in response to changing horticultural practices, the genetic structure of insect populations and the needs of the society. The emphasis is on the use of fewer and safer residues on food products, reduced environmental impact and the concept of sustainable agricultural production. There has been increased regulation of pesticide use worldwide and incentive programmes that promote reduced-impact pest management programmes. With the globalization of fruit marketing, it is likely that all countries of the world want to conform to production and pest-management practices in conformity to these concepts.

4.49 Insect Pests of Pear

A wide diversity of insects attack pear orchards. The main dangerous pests included: *Cydia pomonella* (L.), *Argyrotaenia ljungiana* (Thunberg), *Pandemis cerasana* (Hübner) and sometime *Grapholita molesta* (Busck). Oriental fruit moth *Cydia molesta* is a highly damaging invasive pest (Najar-Rodriguez et al. 2013). Its primary host is peach (*Prunus persica*), whereas some other fruits such as pear (*Pyrus communis*) are considered secondary hosts. *Cydia molesta* is an important pest of pear fruit late in the growing season.

Wallis et al. (2013) reported that Pear leafcurling midge (*Dasineura pyri*) is a gall midge in the family Cecidomyiidae a persistent pest in New Zealand. Whilst mature trees can withstand considerable damage, feeding by larvae can cause severe distortion (galling) of developing leaves on younger trees. Apart from obvious leaf damage, PLCM activity is difficult to detect so the recent development of the synthetic sex pheromone provides a useful monitoring tool for this pest.

Plum curculio, *Conotrachelus nenuphar* has been reported as a major pest of stone and pome fruits e.g., apples, pears, peaches and cherries (Shapiro-Ilan et al.

2013). Shapiro-Ilan et al. (2013) reported that entomopathogenic nematodes (*Steinernema spp*. and *Heterorhabditis spp*.) are virulent to ground-dwelling stages of *C. nenuphar* and may be incorporated for integrated management of weevils.

Sandhu and Sohi (1986) reported that local Pathar Nakh pear (*Pyrus pyrifolia*, Nakai) are attacked by 18 species of insects, 2 species of mites and 1 bird species. Bortolotti et al. (2012) studied the biology of Miridae bugs in pear orchards that cause fruit deformations.

Insecticides applied at specifics times according to the presence of the pest, effectively contain the damage.

Pear psylla *Cacopsylla bidens* (Šulc 1907) has been reported as a serious pest of pear in in Egypt Mohamed et al. (2013) and Israel (Shaltiel-Harpaz et al. (2014)). Fountain et al. (2013) reported two species of pear psyllids *Cacopsylla pyri*, (L.) and *C. pyricola* (Foerster) in Europe which have developed resistance to many insecticides and are difficult to control with conventional methods. Though insect predators can naturally regulate the populations of these psyllids, but their populations are often inadequate for effective biocontrol.

Pear psylla (*Cacopsylla pyri*) has also been recorded as most damaging pests of pear in Hungary (Sipos et al. 2013). Kaolin particle film technology was an effective control method for pear psylla, not only in regions of arid climate, but also under climatic conditions of Hungary. Bell (2013) reported that pear psylla is serious pests of pear throughout North America and Europe. Inhibition of nymphal feeding has been identified as a key component of resistance, being correlated with inhibition of oviposition, mortality, and delayed development. Singh and Sharma (2013) studied the biology of fruit flies, *B. dorsalis* and *B. zonata* and found that in fruit crops the number of egg punctures in peach and Kinnow were less to that of pear and guava. The number of days taken by maggots for coming out of fruits was minimum in guava followed by pear and peach.

4.50 IPM in Pears

Valente et al. (2012) stated that Voliam Targo is a new insecticide and acaricide, composed by chlorantraniliprole and abamectin, effective on several key pests of fruits and vegetables. Wang et al. (2007) reported that pear fruit bagging is one of the most important practices in producing high quality fruits. Kutinkova et al. (2005) highlighted the importance of integrated plant protection system (involving the use of pesticides, pheromone traps, resistant cultivars, soil cultivation, irrigation and other non-chemical methods) for the control of pear pests in Bulgaria.

Samietz et al. (2013) emphasized the importance of forecasting system in order to optimize timing of monitoring, management, and control measures of insect pests in fruit orchards on the basis of relationships between temperature and stage specific development rates for the relevant stages of the life cycles under controlled laboratory conditions for major pests of apple, pear, cherry and plum which included : Rosy apple aphid (*Dysaphis plantaginea*), apple sawfly (*Hoplocampa testudinea*), codling moth (*Cydia pomonella*), smaller fruit tortrix (*Grapholita lobarzewskii*), apple blossom weevil (*Anthonomus pomorum*), summer tortrix (*Adoxophyes orana*), pear psylla (*Cacopsylla pyri*), European cherry fly (*Rhagoletis cerasi*), red spider mite (*Panonychus ulmi*), and plum fruit tortrix (*Grapholita funebrana*).

Steenwyk et al. (2008) reported that Codling moth, *Cydia pomonella* can be controlled by mating disruption with one to two supplemental broad-spectrum insecticide applications. The broad-spectrum supplemental insecticide applications suppress beneficial insects that suppress the pear psylla, *Cacopsylla pyricola* Foerster, populations and other insect and mite pests of pears. Thus the development of a sustainable pear pest management system requires the development of environmentally benign and effective codling moth insecticides. According to the above authors, Spinetoram is a reduced risk second-generation spinosyn insecticide with an ecotoxicological profile similar to spinosad. Field results show that spinetoram provides equivalent or superior control of codling moth and pear psylla compared to the grower standard.

Bibliography

Abel CA, Wilson RL, Luhman RL (2003) Pollination efficacy of *Osmia corniforence* and *Osmi lifnaria* subsp. *lignaria* (Hymenoptera: Magachilidae) on three Bassicaceae grown under field cages. J Entomol Sci 38:545–552

Abrol DP (1989) Energy intake and expenditure in *Andrena ilerda* and *A. leaena* (Hymenoptera: Andrenidae). Korean J Apic 4:41–47

ACTA (2001) Index phytosanitaire. Association de Coordination Technique Agricole, Paris, 724 pp

Adams WM (1990) Green development: environment and sustainability in the third world. Routledge, London

Alford DV (1984) A colour atlas of fruit pests – their recognition, biology and control. Wolfe Publishing, London, 320 pp

Aluja M (1994) Bionomics and management of *Anastrepha*. Annu Rev Entomol 39:155–178

Aluja M, Norrbom AL (2000) Fruit flies: phylogeny and evolution of behavior. CRC Press, Boca Raton, 999 pp

Ambrose JT (1990) Apple pollination. In: N.C. Apple Production Manual. North Carolina Agricultural Extension Service, AG-415

Angeles DE (1991) Mangifera altissima. In: Coronel RE, Verheij EWM (eds) Edible fruits and nuts, Plant Res. S.E. Asia 2 (PROSEA). Pudoc, Wageningen, pp 206–207

Arneson PA, Losey JE (1997) Integrated pest mangement-lecture notes. Cornell University, Tirana, Albania

Audemard H, Burgerjon A, Baudry O, Bergere D, Breniaux D, Delay JC, Desvaux R, Formantin C, Gendrier JP, Tarbouriech MF (1992) Evaluation of 100 trials of carpovirusine, a granulosis virus preparation to control codling moth *Cydia pomonella* L. in apple orchards. Acta Phytopathol Entomol Hung 27:45–49

Avilla J, Bosch D, Sarasúa MJ, Costa-Comelles J (1993) Biological control of *Panonychus ulmi* in apple orchards in Lleida (NE of Spain). Acta Hort 347:267–272

Bailey J, Hoffman M, Olsen K (1988) Black light monitoring of two avocado insect pests. Calif Agric 42:26–27

Bajwa WI, Kogan M (1998) Compendium of IPM Definitions (CID): a collection of IPM definitions and their citations in worldwide IPM literature. Integrated Plant Protection Center (IPPC), Oregon State University, Oregon

Ball BV, Pye BJ, Carreck NL, Moore D, Bateman RP (1994) Laboratory testing of a mycopesticide on non-target organisms: the effects of an oil formulation of Metarhizium flavoviride applied to Apis mellifera. Biocontrol Sci Technol 4:289–296

Barnes BN (1990) FFTRI monitoring manual for orchard research institute. Fruit and Fruit Technology Research Institute, Stellenbosch, 21 pp

Barnes MM (1991) Codling moth occurrence, host race formation, and damage. In: van der Geest LPS, Evenhuis HH (eds) Tortricid pests: their biology, natural enemies and control. Elsevier, New York, pp 313–327

Barnes BN (1999) Fruit fly baiting in deciduous fruit doing the right things right. Deciduous Fruit Grow 49:5–7

Barnes BN, Blomefield TL (1997) Goading growers towards mating disruption: the South African experience with Graphotita molesta and Cydia pomonella (Lepidoptera: Tortricidae). IOBC WPRS Bull 20:45–46

Barnes BN, Eyles DK (2000) Feasibility of eradicating *Ceratitis* spp. fruit flies from the Western Cape of South Africa by the sterile insect technique. In: Tan KH (ed) Area wide control of fruit flies and other insect pests. Penerbit Universiti Sains, Pulau Pinang, pp 449–455

Barnes MM, Moffitt HR (1963) Resistance to DDT in the adult codling moth and reference curves for guthion and carbaryl. J Econ Entomol 56:722–725

Batra SWT (1982) The hornfaced bee for efficient pollination of small farm orchards. US Department of Agriculture, A.R.S. Miscellaneous Publication, Washington, DC, p 1422

Batra SWT (1989) Japanese hornfaced bees, gentle and efficient new pollinators. Pomona 22:3–5

Batra SWT (1994) *Anthophora pilipes villosula Sm.* (Hymenoptera: Anthophoridae), a manageable Japanese bee that visits blueberries and apples during cool, rainy, spring weather. Proc Entomol Soc Wash 96:98–119

Beaument P (1993) Pesticides, policy and people. Pesticide Trust, London

Beers EH, Brunner JF, Willet MJ, Warner GM (1993) Orchard pest management: a resource book for the Pacific Northwest. Good Fruit Grower, Yakima, Washington DC, 276 pp

Beers E, Hull L, Jones V (1994) Sampling pest and beneficial arthropods of apple. In: Buntin D, Pedigo L (eds) Handbook of sampling methods for arthropods in agriculture. CRC Press, Boca Raton, pp 383–415

Beers EH, Suckling DM, Prokopy RJ, Avilla J (2003) In: Ferree DC, Warrington IJ (eds) Apples botany, production and uses. CABI Publishing, Wallingford, pp 489–520

Bell RL (2013) Host resistance to pear psylla of breeding program selections and cultivars. American society for horticultural science, Alexandria, USA. HortScience 48(2):2143–145

Benedek P, Finta K (2005) Méhmegporzás a gyümölcstermelésben. Mezőgazdasági Tanácsok 14/3:20–23

Benedek P, Nyéki J (1995) Role of bee pollination in the fruit set and yield of self-fertile and self-sterile apple, sour cherry and plum cultivars. Int J Hortic Sci 27(3–4):34–37

Benedek P, Nyéki J (1996a) Fruit set of selected self-sterile and self-fertile fruit cultivars as affected by the duration of insect pollination. Acta Hortic 423:57–63

Benedek P, Nyéki J (1996b) Relationship between the duration of insect pollination and the yield of some apple cultivars. Int J Hortic Sci 2(3–4):93–96

Benedek P, Nyéki J (1997) Yield of selected apple cultivars as affected by the duration of bee pollination. Acta Hortic 437:207–211

Benedek P, Manninger S, Virányi S (1976) Megporzás mézelő méhekkel. Mezőgazdasági Kiadó, Budapest

Benedek P, Nyéki J, Lukács GY (1989) A méhmegporzás intenzitásának hatása az alma kötődésére és termésére. Kertgazdaság 21(3):8–26

Benedek P, Nyéki J, Soltész M, Erdős Z, Skola I, Szabó T, Amtmann I, Bakcsa F, Kocsisné Molnár G, Vadas Z, Szabó Z (2000) The effect of the limitation of insect pollination period on the fruit set and yield of temperate-zone fruit tree species. Int J Hortic Sci 6(1):91–95

Bentley JW, O'Neil RJ (1997) On the ethics of biological control of insect pests. Agr Hum Values 14(3):283–289

Bhagat KC, Masoodi MA, Koul VK (1988) A note on the occurrence of a coccinellid predator in Kashmir. Curr Res, Univ Agr Sci, Bangalore 17:49–51

Bhartiya SP, Thakur DR, Kar PL, Chadha TR (1983) Morphological variability of flowers in apple cultivars(Mx domestica Borkh.). Prog Hortic 15:200–203

Birch LC (1948) The intrinsic rate of natural increase in an insect population. J Anim Ecol 17:15–26

Blomefield T (1994) Codling moth resistance. Is it here, and how do we manage it? Deciduous Fruit Grow 44:130–132

Blomefield TL (1997) Managing resistance of codling moth, Cydia pomonella (L.) in South African pome fruit orchards, Combating resistance. IARC-Rothamsted, Harpenden, pp 14–16

Blommers LHM (1994) Integrated pest management in European apple orchards. Annu Rev Entomol 39:213–241

Boavida C, Neuenschwander P, Schulthess F (1992) Spatial distribution of *Rastrococcus invadens* Williams (Homoptera: Pseudococcidae) in mango trees. J Appl Entomol 114:381–391

Boivin G, Stewart RK (1982) Identification and evaluation of damage to McIntosh apples by phytophagous mirids (Hemiptera: Miridae) in southwestern Quebec. Can Entomol 114:1037–1045

Boller E, Avilla J, Gendrier JP, Jörg E, Malavolta C (1998) Integrated production in Europe. 20 years after the declaration of Ovrannaz. IOBC/WPRS Bull 21(1), Dijon, France, 41 pp

Bortolotti PP, Caruso S, Nannini R, Casoli L, Montermini A (2012) Evaluation of different insecticides for control of Miridae bugs, causal agents of pear fruit deformations

Bosch J (1994a) Improvement of field management of *Osmia cornuta* (Latrielle)(Hymenoptera: Megachilidae). Apidologie 25:71–83

Bosch J (1994b) Nesting behaviour of mason bee Osmia cornuta (Latr) with special reference to its pollinating potential(Hymenoptera:megachilidae). Apidologie 25:84–93

Bosch J, Kemp WP (1999) Exceptional cherry production in an orchard pollinated with blue orchard bees. Bee World 80(4):163–173

Bosch J, Kemp WP (2002) Developing and establishing bee species as crop pollinators: the example of *Osmia* spp. (Hymenoptera: Megachilidae) and fruit trees. Bull Entomol Res 92(1):3

Bosch J, Kemp WP, Trostle GE (2006) Cherry yields and nesting success in an orchard pollinated with Osmia lignaria (Hymenoptera: Megachilidae). J Econ Entomol 99:408–413

British Columbia Ministry of Agriculture Fisheries, and Food (1994)

Brittain WH (1933) Apple pollination studies in the Annapolis Valley, N.S., Canada 1928 1932. Bull Dept Agric Can New Ser 162:1–198

Brunner JF, Hoyt SC, Wright MA (1982) Codling moth control – a new tool for timing sprays, EB 1072. Washington State University Cooperative Extension Service, Pullman, Washington DC, 4 pp

Bull BC, Raupp MJ, Hardin MR, Sadof CS (1993) Suitability of five horticulturally important armored scale insects as hosts for an exotic predaceous lady beetle. J Environ Hort 11:28–30

Burnip GM, Suckling DM, Shaw PW, White V, Walker JTS (1995) Monitoring graphania mutans (Noctuidae) in apple orchards. In: Proceeding of the Forty-eighth New Zealand plant protection conference. New Zealand Weed and Pest Control Society, Palmerston North, pp. 125–129

Burrell AB, Parker RG (1932) Pollination of the McIntosh apple in the Champlain valley. Third Progress Report. Am Soc Hort Sci Proc (1931) 28: 78–84

Burts EC (1970) Biology of blister mites, *Eriophyes* spp., of pear and apple in the Pacific Northwest. Melanderia 4:42–53

Bush MR, Abdel Aal YAI, Rock GC (1993) Parathion resistance and esterase activity in codling moth (Lepidoptera: Tortricidae) from North Carolina. J Econ Entomol 86:660–666

Campbell CW, Knight AJ (1983) Production de litchis. Communication at XIII Congreso de N.O.R.C.O.F.E.L Puerto de la Cruz. Ministerio de Agricultura Espana (En Prensa)

Cane JH (2005a) An effective, manageable bee for pollination of Rubus bramble fruits, Osmia aglaia. In the 9th international Rubus and Ribes symposium, Pucon, p 459–463

Cane JH (2005b) Pollination potential of the bee Osmia aglaia for cultivated red raspberries and blackberries (Rubus : Rosaceae). Hortscience 40:1705–1708

Carter DJ (1984) Pest Lepidoptera of Europe with special reference to the British Isles. W. Junk, Dordrecht, 431 pp
Chadha KL, Choudhary ML (2007) Report of the working group on horticulture, plantation crops and organic farming for the XI Five Year Plan (2007–12), GOI, Planning Commission, Jan 2007
Chambon JP (1986) Les Tordeuses nuisibles en arboriculture fruitière. INRA, Paris, 118 pp
Chang KP (1970) A study of Anoplophora macularia, aserious insect pest of litchi. Taiwan Agr Q 6(3):133–143
Chapman PJ, Lienk SE (1971) Tortricid fauna of apple in New York (Lepidoptera: tortricidae); including an account of Apples' occurrence in the state, especially as a naturalized plant, Special publication, Cornell University. NY State Agricultural Experiment Station, Geneva, 427 pp
Charmillot PJ (1978) Réduction des captures de carpocapse (*Laspeyresia pomonella* L.) par inhibition des mâles due à la diffusion d'attractif sexuel synthétique en verger. Bulletin de la Société Entomologique Suisse 51:5–12
Charmillot PJ (1989) Insect growth regulators (IGR), mimics of juvenile hormone, as morphogenetical and ovicidal means of control against orchard tortricids. Entomol Exp Appl 51:59–70
Charmillot PJ (1990) Mating disruption technique to control codling moth in Western Switzerland. In: Ridgway RL, Silverstein RM, Inscoe MN (eds) Behavior-modifying chemicals for insect management. Marcel Dekker, New York, pp 165–182
Charmillot PJ, Pasquier D, Scalco A, Hofer D (1996) Essais de lutte contre le carpocapse *Cydia pomonella* L. par un procédé attracticide. Mitt Schweiz Entomol Ges 69:431–439
Cochran WG (1977) Sampling techniques. Wiley, New York, 3 rd ed
Cohen H, Yuval B (2000) Perimeter trapping strategy to reduce Mediterranean fruit fly damage on different host species in Israel. J Econ Entomol 93:721–725
Cox DL, Knight AL, Biddinger DJ, Lasota JA, Pikounis B, Hull LA, Dybas RA (1995) Toxicity and field efficacy of avermectins against codling moth (Lepidoptera: Tortricidae) on apples. J Econ Entomol 88:708–715
Crane JE (1876) The relation of bees to fruit culture. In 3rd Biennial report of the Vermont state board of agriculture, pp 253–270
Crane E, Walker P (1984) Pollination directory for world crops. International Bee Research Association, London, 184 pp
Croft BA (1975) Integrated control of apple mites, Extension bulletin E-825. Michigan State University, East Lansing, 12 pp
Cunningham IC (1984) Mango insect pests. Australian mango research workshop. CSIRO, Melbourne, pp 211–224
Currie RW, Winston ML, Slessor KN, Mayer DF (1992) Effect of synthetic queen mandibular pheromone sprays on pollination of fruit crops by honey bees (Hymenoptera: Apidae). J Econ Entomol 85:1293–1299
Daku L, Norton GW, Pfeiffer DG, Luther GC, Pitts CW, Taylor DB, Tedeschini J, Uka R (2000) Farmers' knowledge and attitudes towards pesticide use and olive pest management practices in Vlora, Albania: a baseline survey. IPM CRSP working paper 001–1, Virginia Polytechnic Institute and State University
Danka RG, Collison CH, Hull LA (1985) Honey bee (Hymenoptera: Apidae) foraging during bloom in dimethoate-treated apple orchards. J Econ Entomol 78:1042–1047
Dashad SS, Sharma JK (1994) Foraging rates of insect visitors on blooms of various cultivars of apple (Malus domestica Borkh). Ann Biol 10(1):75–79
Dashad SS, Sharma JK, Yadav PR (1994) Loose pollen grains carrying capacity of insect visitors apple bloom. Ann Biol 10(1):173–175
de Laroussilhe F (1980) Le Manguier. G.P. Masoneuve & Larose, Paris
De Liñán C (ed) (2001) Vademécum de Productos Fitosanitarios y Nutricionales. Ediciones Agrotécnicas SL, Madrid, 671 pp
DeBach P (1964) Biological control of insect pests and weeds. Chapman & Hall, London, 844 pp

Degrandi-Hoffman G, Hoopingarner R, Baker KK (1985) The influence of honey bee 'sideworking' behavior on cross-pollination and fruit set in apples. HortScience 20:397–399
DeGrandi-Hoffman GD, Mayer I, Terry DL (1995) Validation of PC-REDAPOL: fruit set prediction model for apples. J Econ Entomol 88(4):965–972
Delaplane KS, Mayer DF (2000) Crop pollination by bees. CABI Publishing, Oxon
Dent D (2000) Insect pest management, 2nd edn. CABI Publishing, Ascot, p 399
DPPQS, Directorate of Plant Protection, Quarantine and Storage (2001) Integrated pest management package, Government of India. (IPM package no: 28, 31, 32, 34, 37)
Drew RAI, Hancock DL (1994) The *Bactrocera dorsalis* complex of fruitflies (Diptera: Tephritidae: Dacinae) in Asia. Bull Entomol Res Suppl Ser 2:68
Driggers BF (1937) Five years' experiments with lead arsenate–summer oil in codling moth control. J Econ Entomol 30:407–413
Eyer JR (1937) Ten years' experiments with codling moth bait traps, light traps, and trap bands. Bulletin No 253. New Mexico Agriculture Experiment Station Bulletin, Las Cruces, New Mexico, 67 pp
Falcon LA, Huber J (1991) Biological control of codling moth. In: van der Geest LPS, Evenhuis HH (eds) Tortricid pests, their biology, natural enemies and control. World Crop Pests. 5: 355–369
Farahbakhch G, Moini M (1975) Olive pests in Iran, Report of plant pests and diseases research institute. National Agricultural Research Organization, Ministry of Agriculture and Natural Resources, Tehran, pp 1–14
Fell RD (1986) Foraging behaviors of Apis mellifera L. and Bombus spp. on oilseed sunflower (Helianthus annuus L.). J Kans Entomol Soc 59:72–81
Fell RD (1995) Pollination. In: Hogmire H Jr (ed) Mid-Atlantic orchard monitoring guide. Natural Resource, Agriculture, and Engineering Service, Ithaca, pp 255–265
Fisher DV, Upshall WH (eds) (1976) History of fruit growing and handling in United States of America and Canada – 1860–1972. Regatta City Press, Kelowna, 360 pp
Flint ML, van den Bosch R (1981) Introduction to pest management. Plenum Press, New York
Follett PA, Duan JJ (eds) (2000) Nontarget effect of biological control. Kluwer, Boston, 316 pp
Fountain MT, Nagy C, Harris A, Cross JV, Ioriatti C, Altindisli FÖ, Børve J, Escudero-Colomar LA, Lucchi A, Molinari F (2013) Importance of naturally occurring predators for pear sucker control. International Organization for Biological and Integrated Control of Noxious Animals and Plants (OIBC/OILB), West Palaearctic Regional Section (WPRS/SROP), Dijon, France. IOBC/WPRS Bull 91:117–125
Free JB (1962) The effect of distance from pollinizer cultivars on the fruit set on trees in plum and apple orchards. J Hort Sci 37:262–271
Free JB (1964) The behaviour of honeybees on sunflowers (*Helianthus annus* L.). J Appl Ecol 1:19–27
Free JB (1966) The foraging areas of honeybees in an orchard of standard apple trees. J Appl Ecol 3:261–268
Free JB (1970) Insect pollination of crops. Academic, London, 544 pp
Free JB (1993) Insect pollination of crops, 2nd edn. Academic, London, 684 pp
Galan-Sauco V (1990) Los frutales tropicales en los subtropicos. Ed. Mundi Prensa, Madrid, pp 83–84
García-Marí F, Costa-Comelles J, Ferragut F, Laborda R (1989) Lutte intégrée contre les acariens dans les vergers de pommiers de Lleida (Espagne). Annales de l'Association Nationale pour la Protection des Plantes 2(1):501–518
Gardner KE, Ascher JS (2006) Notes on the native bee pollinators in New York apple orchards. J N Y Entomol Soc 114:86–91
Gathmann A, Greiler H-J, Tscharntke T (1994) Trap-nesting bees and wasps colonizing set-aside fields: succession and body size, management by cutting and sowing. Oecologia 98:8–14
Geier PW (1963) The life history of the codling moth, *Cydia pomonella* (L.) (Lepidoptera: Tortricidae), in the Australian Capital Territory. Aust J Zool 11:323–367
Giornate Fitopatologiche (2012) Milano Marittima (RA), 13–16 marzo 2012, pp 139–142

Giraud M, Baudry O, Orts R, Gendrier JP (1996) Protection intégrée pommier–poirier. Centre Technique Interprofessionnel des Fruits et Légumes, Paris, 277 pp

Glass EH, Fiori B (1955) Codling moth resistance to DDT in New York. J Econ Entomol 48:598–599

Gleeson D, Holder P, Newcomb R, Howitt R, Dugdale J (2000) Molecular phylogenetics of leafrollers: application to DNA diagnostics. N Z Plant Prot 53:157–162

Glen DM, Payne CC (1984) Production and field evaluation of codling moth granulosis virus for control of *Cydia pomonella* in the United Kingdom. Ann Appl Biol 104:87–98

Glukhov MM (1955) Honey plants. Izd. Selkhoz Lit-ry. 512 pp

Gold CS, Nankinga C, Niere B, Godonou I (2003) IPM of banana weevil in Africa with emphasis on microbial control. In: Neuenschwander P, Borgemeister C, Langewald J (eds) Biological control in IPM systems in Africa. CABI Publishing, Wallingford, pp 243–257, 400 p

Golez HG (1991) Bionomics and control of the mango seed borer, *Noorda albizonalis* Hampson (Pyralidae: Lepidoptera). Acta Hort 291:418–424

Griggs WH (1970) The status of deciduous fruit pollination. In the indispensable pollination. Ark Ag Ext Serv Misc Publ 127:185–210

Guédot C, Horton DR, Landolt PJ (2009) Attraction of male winterform pear psylla to female-produced volatiles and to female extracts and evidence of male-male repellency. Entomol Exp Appl 130(2):191–197

Gupta JK, Goyal NP, Sharma JP, Gautam DR (1993) The effect of placement of varying numbers of *Apis mellifera* colonies on the fruit set in apple orchards having different proportions of pollinisers. In: Veeresh GK, Uma Shankar R, Ganeshaiah KN (eds) Proceedings of the international symposium on pollination in the tropics. International Union for Studies on Social Insects, Bangalore

Gut L, Brunner J, Knight A (1992) Mating disruption as a control for codling moth and leafrollers. Good Fruit Grow 43:56–60

Haji FNP, Barbosa FR, Lopes PRC, Moreira AN, de Alencar JA, Ferreira RCF (2004) Monitoring mango pests within an integrated production programme in Brazil. Acta Hort 645:163–165

Hameed SF, Sharma DD, Agarwal ML (1992) Integrated pest management in litchi. National seminar on recent developments in litchi production held at Rajindra Agriculture University, Bihar, Pusa, 30–31 May, Abstract of papers, pp 38

Hansen JD, Amstrong JW (1990) The failure of field sanitation to reduce infestation by mango nut weevil, cryto mangiferae in Hawaii. Trop Pest Manag 36:359–361

Hara H, Shinohara A (2013) A slug sawfly, *Caliroa matsumotonis* (Hymenoptera: Tenthredinidae), injurious to peach and pear trees in Japan and Korea. Appl Entomol Zool 48(3):379–386

Hare DJ (1994) Sampling arthropod pest in citrus. In: Pedigo L, Buntin D (eds) Handbook of sampling methods for arthropods in agriculture. CRC Press, Boca Raton, pp 417–432

Haseeb M (2007) Current status of insect pest problems in guava. Acta Hort 735:453–467

Hassan SA (1989) Selection of suitable *Trichogramma* strains to control the codling moth *Cydia pomonella* and the two summer fruit tortrix moths *Adoxophyes orana* and *Pandemis heparana* (Lepidoptera: Tortricidae). Entomophaga 34:19–28

Havenith C (2000) Pollination with solitary bees – an opportunity for fruit growers. Erwerbsobstbau 42:44–50

He DP (2001) An overview of integrated management of insect pests in litchi orchards of Guangdong. Acta Hort 558:401–405

Heller JJ, Mattioda H, Klein E, Sagenmuller A (1992) Field evaluation of RH 5992 on lepidopterous pests in Europe, Brighton crop protection conference. British Crop Protection Council, Brighton, pp 59–65

Hill D (1975) Agricultural insect pests of the tropics and their control. Cambridge University Press, New York

Hough WS (1928) Relative resistance to arsenical poisoning of two codling moth strains. J Econ Entomol 21:325–329

Howard PA, Michalenko EM, Sage GW, Jarvis WF, Meylan WM, Basu DK, Beauman JA, Gray DA (1991) Handbook of environmental fate and exposure data for organic chemicals. Lewis Publishers, Chelsea

Hoy MA, Knop NF (1981) Selection for and genetic analysis of permethrin resistance in *Metaseiulus occidentalis*: genetic improvement of a biological control agent. Entomol Exp Appl 30:10–18

Hoyt SC (1969) Integrated chemical control of insects and biological control of mites on apple in Washington. J Econ Entomol 62:74–86

Hoyt SC, Burts EC (1974) Integrated control of fruit pests. Ann Rev Ento 19:231–252

Huang MD, Li SX (1989) The damage and economic threshold of citrus leafminer, Phyllocnistis citrella Stainton to citrus, Studies on the IPM of citrus insect pests. Academic Book and Periodical Press, Guangzhou, Guangdong

Huda AKS, Luck J (2008) Early warning of pest/diseases for selected crops using climate information and crop simulation modeling approach under climate change scenarios. Proceeding of international symposium on agrometeorology and food security, Hyderabad, 18–21 Feb 2008. pp 11

Hull LA, Beers EH, Grimm JW (1986) Action thresholds for arthropod pests of apple. In: Frisbie RE, Adkisson PL (eds) Proceedings of a national symposium for IPM on major agricultural systems, October 8_11, 1985. Texas A&M Press, Washington, DC, pp 274–294

Humphry-Baker P (1975) Pollination and fruit set in tree fruits. British Columbia Department of Agriculture, Victoria

Hutchinson WZ (1886) The relation of bees to horticulture, benefits and injuries. Mich Hort Soc Proc 16:13–15

Hutson R (1926) The relation of honey bees to fruit pollination in New Jersey. NJ Agric Exp Sta Bull 434:1–32

Jaworska M (1992) Biological control of *Hoplocampa testudinea* Klug (Hymenoptera: Tenthredinidae). Acta Phytopathol Entomol Hung 27:311–315

Jeppson LR, Keifer HH, Baker EW (1975a) Biological enemies of mites, Mites injurious to economic plants. University of California Press, Berkeley, 614 pp

Jeppson LR, Keifer HH, Baker EW (1975b) Mites injurious to economic plants. University of California Press, Berkeley, 614 pp

Johansen CA, Mayer DF (1990) Pollinator protection. A bee and pesticide handbook. Wicwas Press, Cheshire, 212 pp

Jones VP, Davis DW, Smith SL, Allred DB (1989) Phenology of apple maggot (Diptera: Tephritidae) associated with cherry and hawthorn in Utah. J Econ Entomol 82:788–792

Jothi DB, Tandon PL, Verghese A (1994) Hot water immersion as a quarantine treatment for Indian mangoes infested with the oriental fruitfly, Bactrocera dorsalis (Hendel) (Diptera:Tephritidae). FAO Plant Protect Bull 42(3):158–159

Judd GJR, Gardiner MGT, Thomson DR (1997) Control of codling moth in organically-managed apple orchards by combining pheromone-mediated mating disruption, postharvest fruit removal and tree banding. Entomol Exp Appl 83:137–146

Kalman S, Eyorgy D, Tiborne G, Molna J (1994) Occurrence of the appleseed moth (*Grapholitha lobarzewskii* Now.) and hawthorn berry moth (*Grapholitha janthinana* Dup.) in Hungarian apple orchards. Novenyvedelem 30:327–332

Kaul V, Kesar YK (2003) Incidence and management of lepidopteran fruit borers of guava (Psidium guajava L.) in Jammu, India. J. Asia-Pacific Entomol 6(2):201–205

Kaul V, Malik GH, Shankar U, Monobrullah M (2007a) Incidence and management of olive psylla, *Euphyllura pakistanica*. Indian J Entomol 69(4):331–340

Kaul V, Tiku AK, Shankar U, Monobrullah M (2007b) Green stink bug (Hemiptera: Pentatomidae) recorded as a new pest of olive in India. J Asia-Pacific Entomol 10(1):81–83

Kemp WP, Bosch J (2002) Blue orchard bees – another six – legged secret to improved fruit tree pollination. Am Bee J 142:501–503

Keenan M (1997) In: Smith D, Beattie G, Broadley R (eds) Citrus pests and their natural enemies. Deptt. of Primary Industries, Q 197030, Brisbane, 272 pp

Kendall DA (1973) Viability and compatibility of pollen on insects visiting apple blossom. J Appl Ecol 10:847–853

Kendall DA, Solomon ME (1973) Quantities of pollen on bodies of insects visiting apple blossom. J Appl Ecol 10:627–634

Kevan PG (1988) Pollination: crops and bees. Ministry of Agriculture and Food, Ontario, p 72

Khangura JS, Kanwar JS, Singh H, Singh D (1992) Pest management in litchi (Litchi chinensis Sonn.). National seminar on recent developments in litchi production held at Rajindra Agriculture University, Bihar, 30–31 May, Abstract of papers, pp 40–41

Kitamura T, Maeta Y (1969) Studies on the pollination of apple by Osmia: (III) Preliminary report on the homing ability of Osmia cornifrons (Radoszkowski) and O. pedicornis Cockerell

Klein AM, Vaissiére B, Ca ne JH, Steffan-Dewenter I, Cunningham SA, Kremen C et al (2007) Importance of crop pollinators in changing landscapes for world crops. Proc R Soc Lond B Biol Sci 274:303–313

Knight AL, Brunner JF, Alston D (1994) Survey of azinphosmethyl resistance in codling moth (Lepidoptera: Tortricidae) in Washington and Utah. J Econ Entomol 87:285–292

Knight AL, Turner JE, Brauchla B (1997) Predation on eggs of codling moth (Lepidoptera: Torticidae) in mating disrupted and conventional orchards in Washington. J Entomol Soc British Columbia 94:67–74

Kostianinen TS, Hoy MA (1996) The phytoseiidae as biological control agents of pest mites and insects: a bibliography, Monograph 17. University of Florida Agricultural Experiment Station, Gainesville, 355 pp

Kozar F, Brown MW, Lightner G (1994) Spatial distribution of homopteran pests and beneficial insects in an orchard and its connection with ecological plant protection. J Appl Entomol 117:519–529

Krunić MD, Stanisavljević LZ (2006a) The biology of European orchard bee Osmia cornuta (Latr.) (Hymenoptera: Megachilidae). Faculty of Biology of University of Belgrade, Belgrade

Krunić MD, Stanisavljević LZ (2006b) Supercooling points and diapause termination in overwintering adults of orchard bees Osmia cornuta and O. rufa (Hymenoptera: Megachilidae). Bull Entomol Res 96:323–332

Kuhn ED, Ambrose T (1982) Foraging behavior of honey bees on 'Golden Delicious' and 'Delicious' apple. J Am Soc Hortic Sci 107:391–395

Kuhn ED, Ambrose JT (1984) Pollination of 'Delicious' apple by megachilid bees of the genus Osmia (Hymenoptera, Megachilidae). J Kans Entomol Soc 57:169–180

Kumar N, Suresh J, Anbu S (2007) Problems and prospects of guava cultivation under Tamil Nadu conditions. Acta Hort 735:329–334

Kutinkova H, Andreev R (2004) Integrated pest management in sweet cherry (*Prunus avium* L.) orchards in Bulgaria. J Fruit Ornam Plant Res 12:4147

Kutinkova H, Subchev M, Light S, Lingren B (2005) Interactive effects of ethyl (*2E,4Z*)-2,4-decadienoate and sex pheromone lures to codling moth: apple orchard investigations in Bulgaria. J Plant Prot Res 45:49–52

Landolt PJ (1998) *Lacanobia subjuncta* (Lepidoptera: Noctuidae) on tree fruits in the Pacific Northwest. Pan-Pacific Entomol 74:32–38

Latimer LP (1931) Pollination studies with the McIntosh apple in New Hampshire. Am Soc Hortic Sci Proc 1930:386–396

Levin MD (1986) Using honey bees to pollinate crops. US Department of Agriculture, Leaflet 549

Li M, Cong B, Li J, Yang G, Yang G, Yang C (2004) Mason bee and its application on fruit tree pollination. J Jilin Agric Univer 26:422–425

Lichtenberg E, Zilberman D, Archibald SO (1990) Economics and pesticides. Chapter 5 In: Zilberman D, Lichtenberg E (eds) Economic perspectives on pesticide use in California: a collection of papers. Berkley University, California, Nov 1990

Loken A (1958) Pollination studies in apple orchards of western Norway. In 10th International Congress Entension Proceedings 4: 961–965. Montreal, 17–25 Aug 1956

MacDaniels LH, Heinicke AJ (1929) Pollination and other factors affecting the set of fruit, with special reference to the apple. N.Y. (Cornell) Agric Expt Sta Bull 497: 47 pp

MacLellan CR (1972) Codling moth populations under natural, integrated and chemical control on apple in Nova Scotia (Lepidoptera: Olethreutidae). Can Entomol 104:1397–1404

MacPhee AW (1976) Predictions of destructive levels of the apple-stinging bugs *Atractotomus mali* and *Campylomma verbasci* (Hemiptera: Miridae). Can Entomol 108:423–426

Maeta Y, Kitamura T (1974) How to manage the Mame-ko-bee(Osmia cornifrons Radoszkowski) for pollination of fruit crops. Ask Co. Ltd

Maindonald JH, Waddell BC, Birtles DB (1992) Response to methyl bromide fumigation of codling moth (Lepidoptera: Tortricidae) eggs on cherries. J Econ Entomol 85:1222–1230

Malo SE, Campbell CW (1994) The guava. Horticultural sciences department fact sheet HS-4. Florida cooperative extension service, institute of food and agricultural sciences, University of Florida, Apr 1994

Mani M, Krishnamoorthy A (1990) Evaluation of the exotic predator *Cryptolaemus montrouzieri* Muls. (Coccinellidae, Coleoptera) in the suppression of green shield scale *Chloropulvinaria psidii* (Maskell) (Coccidae, Homoptera) on guava. Entomon 15:45–48

Mani M, Krishnamoorthy A, Singh SP (1989) The impact of the predator, *Cryptolaemus montrouzieri* Muls. on the pesticide resistant populations of the striped mealybug *Ferrisia virgata* (Ckll.) on guava in India. Insect Sci Appl 11:167–170

Masoodi MA (1985) Growth response of *Lymantria obfuscate* Walker in relation to tannin content in different host foliage. Indian J Entomol 47(4):422–426

Mathur RN (1975) Psyllidae of the Indian subcontinent. Indian Council of Agricultural Research, New Delhi, 429 pp

Mathur AC, Tondon PL (1974) Litchi mite can be controlled. Indian Hort 19:11–12

Matsumoto S, Abe A, Maejima T (2009) Foraging behavior of Osmia cornifrons in an apple orchard. Sci Hortic 121:73–79

Mayer DF (1994) Sequential introduction of honey bee colonies for pear pollination. Acta Hortic 367:267–269

Mayer DF, Johansen CA, Burgett DM (1986) Bee pollination of tree fruits. Pacific Northwest Extension Publication, PNW 0282

Mayer DF, Britt RL, Lunden JD (1989a) Evaluation of bee scent as a honey bee attractant. Am Bee J 129(1):41–42

Mayer DF, Johansen CA, Lunden JD (1989b) Honey bee (*Apis mellifera* L.) foraging behavior on ornamental crab apple pollenizers and commercial apple cultivars. HortScience 24:510–512

Mayer DF, Miliczky ER, Lunden JD (1990) Pollination of pears, Pear production in the Pacific Northwest. University of California Press, Davis

Mayer DF, Patten KD, Macfarlane RP (1994a) Pear pollination with managed bumble bee (Hymenoptera: Apidae) colonies. Melanderia 50:20–23

Mayer DF, Patten KD, Macfarlane RP, Shanks CH (1994b) Differences between susceptibility of four pollinator species (Hymenoptera: Apoidea) to field weathered insecticide residues. Melanderia 50:24–27

McGregor SE (1976) Insect pollination of cultivated crop plants. U.S. Department of Agriculture Handbook Number 496, Washington, D.C

Metcalf RL, Luckmann WH (1975) Introduction to insect pest management. Wiley, New York, 587 pp

Miliczky ER, Mayer DF, Lunden JD (1990) Notes on the nesting biology of *Andrena* (*Melandrena*) *nivalis* Smith (Hymenoptera: Andrenidae). J Kans Entomol Soc 63:166–174

Miller JC, Hanson PE, Dowell RV (1987) The potential of gypsy moth as a pest of fruit and nut crops. Calif Agric 41:10–12

Mills NJ, Carl KP (1991) Parasitoids and predators. In: van der Geest LPS, Evenhuis HH (eds) Tortricid pests. Their biology, natural enemies and control. Elsevier, Amsterdam, pp 235–252

Minks AK, Cardé RT (1988) Disruption of pheromone communication in moths: is the natural blend really most efficacious? Entomol Exp Appl 49:25–36

Minks AK, van Deventer P (1992) Mating disruption of codling moth and fruit tree leafrollers in Dutch apple orchards: testing of commercial products. Bull OILB SROP 15:72–75

Moffitt HR, Westigard PH (1984) Suppression of the codling moth (Lepidoptera: Tortricidae) population on pears in southern Oregon through mating disruption with sex pheromone. J Econ Entomol 77:1513–1519

Moffitt HR, Westigard PH, Mantey KD, Van de Baan HE (1988) Resistance to diflubenzuron in the codling moth (Lepidoptera: Tortricidae). J Econ Entomol 81:1511–1515
Mohamed GH, Ouvrard D, Fatma AM (2013) Pear psylla *Cacopsylla bidens* (Šulc, 1907): a new pest on pear trees in Egypt (Hemiptera: Psylloidea). Egypt Acad J Biol Sci: Entomol 6(3):49–52
Mohyuddin AI, Mahmood R (1993) Integrated control of mango pests in Pakistan. Acta Hort 341:467–483
Monzon VH, Bosch J, Retana J (2004) Foraging behavior and pollinating effectiveness of Osmia cornuta and Apis mellifera (Hymenoptera: Megachilidae, Apidae) on 'Comice' pear. Apidologie 35:575–585
Morin C (1967) Cultivo de Frutales Tropicales. Librerias ABC, Lima, Per'u, pp 225–229
Mullen JD (1995) Estimating environmental and human health benefits of reducing pesticide use through integrated pest management programs. Master Thesis, Virginia Polytechnic Institute and State University, 1995
Muller H (1883) In: Thompson DW (ed) The fertilisation of flowers. Macmillan, London, 669 pp
Murray RC (1991) Arthropod pests of mango in the Caribbean Islands. Review Crop Protection University, Kingston
Najar-Rodriguez A, Orschel B, Dorn S (2013) Season-long volatile emissions from peach and pear trees in situ, overlapping profiles, and olfactory attraction of an oligophagous fruit moth in the laboratory. J Chem Ecol 39(3):418–429
Naumann K, Currie RW, Isman MB (1994a) Evaluation of the repellent effects of a neem insecticide on foraging honey bees and other pollinators. Can Entomol 126:225–230
Naumann K, Winston ML, Slessor KN, Smirle MJ (1994b) Synthetic honey bee (Hymenoptera: Apidae) queen mandibular gland pheromone applications affect pear and sweet cherry pollination. J Econ Entomol 87:1595–1599
Navon A, Cohen M, Keren S, Levski S, Harel M, Dunkelblum E (1997) Trapping of the leopard moth, *Zeuzera pyrina* L. (Cossidae: Lepidoptera) with sex pheromone traps for monitoring the pest experiments in apple and pear orchards. Alon Hanotea 51(8):358–366
Nelson KC (1994) Participation, empowerment, and farmer evaluation: a comparative analysis of IPM technology generation in Nicaragua. Agr Hum Values 11:109–125, Spring-Summer 1994
Nijjar GS (1972) Litchi cultivation. Punjab Agriculture University, Ludhiana
Norton GW, Mullen J (1994) Economic evaluation of IPM programs: a literature review. Report 448–120. Cooperative Extension, Blacksburg
Norton GW, Rajotte EG, Gapud V (1999) Participatory research in integrated pest management: lessons from the IPM CRSP. Agr Hum Values 16:431–439
Oatman ER, Legner EF, Brooks RF (1964) An ecological study of arthropod populations on apple in northeastern Wisconsin: insect species present. J Econ Entomol 57:978–983
Overholser EL (1927) Apple pollination studies in California. Calif Agric Expt Sta Bull 426: 17 pp
Pandey V, Patel MG, Chaudhari GB, Patel JR, Bhatt BK, Vadodaria RP, Shekh AM (2003) Influence of weather parameters on the population dynamics of mango hopper. J Agrometeorol 5(1):51–59
Parker FD, Batra SWT, Tepedino VJ (1987) New pollinators for our crops. Agric Zool Rev 2:279–304
Pasqualini E, Boselli M, Scannavini M (2013) Spirotetramat (Movento®) in control of *Cacopsylla pyri*: field evaluation of efficacy, application timing and selectivity for *Anthocoris nemoralis*. Giornate Fitopatologiche 2012, Milano Marittima (RA), 13–16 marzo 2012, pp 125–132
Paulson GS, Hull LA, Biddinger DJ (2005) Effect of a plant growth regulator prohexadione-calcium on insect pests of apple and pear. J Econ Entomol 98(2):423–431
Peach ML, Alston DG, Tepedino VJ (1994) Bees and bran bait: is carbaryl bran bait lethal to alfalfa leafcutting bees (Hymenoptera: Megachilidae) adults or larvae? J Econ Entomol 87:311–317
Peach ML, Alston DG, Tepedino VJ (1995) Sublethal effects of carbaryl bran bait on nesting performance, parental investment, and offspring size and sex ratio of the alfalfa leafcutting bee (Hymenoptera: Megachilidae). Environ Entomol 24:34–39
Pedigo LP (2001) Entomology and pest management, 4th edn. Prentice Hall, New Jersey

Peña JE (2004) Integrated pest management and monitoring techniques for mango pests. Acta Hort 645:151–161

Pena J, Johnson F (2001) Insect management in guava. Document ENY- 412. Florida cooperative extension service, institute of food and agricultural sciences, University of Florida, Nov 2001

Peng RK, Christian K (2005) Integrated pest management in mango orchards in the Northern Territory Australia, using the weaver ant, *Oecophylla smaragdina*, (Hymenoptera: Formicidae) as a key element. Int J Pest Manag 51(2):149–155

PengFei L, LingQiao H, Wang ChenZhu L (2012) Identification and field evaluation of pear fruit volatiles attractive to the oriental fruit moth, *Cydia molesta*. J Chem Ecol 38(8):1003–1016

Peterson SS, Baird CR, Bitner RM (1992) Current status of the alfalfa leafcutting bee, *Megachile rotundata*, as a pollinator of alfalfa seed. BeeScience 2:135–142

Phillips EF (1933) Insects collected on apple blossoms in western New York. J Agric Res 46:851–862

Pilling ED, Jepson PC (1993) Synergism between EBI fungicides and a pyrethroid insecticide in the honeybee (*Apis mellifera*). Pesticide Sci 39:293–297

Pimental D, Acquay H, Biltonen M, Rice P, Silva M, Nelson J, Lipner V, Giordano S, Horowitz A, D'Amore M (1992) Environmental and economic costs of pesticide use. BioScience 42(10):750–760

Pollini A, Bariselli M (1993) *Cydia molesta*: pest on the increase and defence of pome fruits. Infor Agrar 49:19–21

Prasada Rao GSLHV, Beevi SP (2008) Forewarning tea mosquito bug helopeltis antonii sign. (miridae:hemiptera) in cashew. In: Proceeding of international symposium on agrometeorology and food security, Hyderabad, 18–21 Feb 2008, pp 203

Prokopy RJ, Wirth CB, Leskey TC (1999) Movement of plum curculio adults toward host trees and traps: flight versus walking. Entomol Exp Appl 91:385–392

Prokopy RJ, Wright SE, Black JL, Hu XP, McGuire MR (2000) Attracticidal spheres for controlling apple maggot flies: commercial-orchard trials. Entomol Exp Appl 97:293–299

Proverbs MD, Newton JR, Logan DM (1966) Orchard assessment of the sterile male technique for control of the codling moth, *Carpocapsa pomonella* (L.) (Lepidoptera: Olethreutidae). Can Entomol 98:90–95

Proverbs MD, Logan DM, Newton JR (1975) A study to suppress codling moth (Lepidoptera: Olethreutidae) with sex pheromone traps. Can Entomol 107:1265–1269

Proverbs MD, Newton JR, Campbell CJ (1982) Codling moth: a pilot program of control by sterile insect release in British Columbia. Can Entomol 114:363–376

Rabb RL (1972) Principles and concepts of pest management. In: Proceedings of the national extension pest-management workshop. Purdue University, Purdue, pp 6–29

Rai KM, Gupta BP (1983) Role of honey bees and other insects as pollinators in apple and pear. Indian Bee J 45(2/3):56–57

Rao VP, Ghani MA, Sankaran T, Mathur KC (1971) A review of the biological control of insects and other pests in South East Asia and the pacific region. CIBC Tech Commun 6:149

Reis W, Nora I, Melzer R (1988) Population dynamics of *Grapholita molesta*, Busck, 1916, and its adaptation on apple in south Brazil. Acta Hort 232:204–212

Reissig WH, Nyrop J, Straub R (1998) Oviposition model for timing insecticide sprays against plum curculio (Coleoptera: Curculionidae) in New York State. Environ Entomol 27:1053–1061

Riedl H (1976) Forecasting codling moth phenology based on pheromone trap catches and physiological- time models. Can Entomol 108:449–460

Riedl H, Halaj J, Kreowski WB, Hilton RJ, Westigard PH (1995) Laboratory evaluation of mineral oils for control of codling moth (Lepidoptera: Tortricidae). J Econ Entomol 88:140–147

Roelofs WL (ed) (1979) Establishing efficacy of sex attractants and disruptants for insect control. Entomological Society of America, Lanham, 97 pp

Roelofs WL, Comeau A, Hill AS, Milicevic G (1971) Sex attractant of the codling moth: characterization with electroantennogram technique. Science 174:297–299

Roling N, van de Fliert E (1994) Transforming extension for sustainable agriculture: the case of integrated pest management in rice in Indonesia. Agric Human Values Spring–Summer: 96–108

Roush RT, Hoy MA (1981) Genetic improvement of *Metaseiulus occidentalis*: selection with methomyl, dimethoate, and carbaryl and genetic analysis of carbaryl resistance. J Econ Entomol 74:138–141

Roversi A, Ughini V (1986) Ricerche sulla biologia fiorale del ciliego dolce. Ann Fac Agraria UCSC (Piacenza) 26:189–203

Roy RS, De BN (1950) Control of litchi mite. Indian J Hort 7(1):16

Samietz J, Höhn H, Razavi E, Schaub L, Graf B, Ioriatti C, Altindisli FÖ, Børve J, Escudero-Colomar LA, Lucchi A, Molinari F (2013) Decision support for sustainable management of the main orchard pests with the Swiss forecasting system SOPRA. IOBC/WPRS Bull 91:243–251

Sampson BJ, Cane JH, Kirker GT, Stringer SJ, Spiers JM (2009) Biology and management potential for three orchard bee species (Hymenoptera: Megachilidae): Osmia ribifloris Cockerell, O. lignaria (Say) and O. chalybea Smith, with emphasis on the former. Acta Hortic 810:549–556

Sandhu GS, Sohi AS (1986) Insects and other pests of sand pear and their control in the Punjab. Punjab Hortic J 26(1/4):46–55

Sauco GV, Menini UG (1989) Litchi cultivation. FAO plant production and protection paper no. 83, Food and Agricultural Organization of the United Nations, Rome, 205 pp

Sauphanor B, Bouvier JC (1995) Cross-resistance between benzoylureas and benzoylhydrazines in the codling moth, *Cydia pomonella* L. Pestic Sci 45:369–375

Scott-Dupree C, Winston M, Hergert G, Jay SC, Nelson D, Gates J, Termeer B, Otis G (eds) (1995) A guide to managing bees for crop pollination. Canadian Association of Professional Apiculturalists, Guelph

Sekita N (2000) Managing Osmia cornifrons to pollinate apples in Aomori prefecture, Japan. Acta Hortic 561:303–308, pp. 413–420

Shaltiel-Harpaz L, Soroker V, Kedoshim R, Hason R, Sokalsky T, Hatib K, Bar-Ya'akov I, Holland D (2014) Two pear accessions evaluated for susceptibility to pear psylla *Cacopsylla bidens* (Šulc) in Israel. Pest Manag Sci 70(2):234–239

Shankar U, Kaul V, Rai S (2007) New recode of anar fruit borer, Deudorix isocrates (Fabricius) on aonla. J Res, SKUAST-J 6(2):292–295

Shapiro-Ilan DI, Wright SE, Tuttle AF, Cooley DR, Leskey TC (2013) Using entomopathogenic nematodes for biological control of plum curculio, *Conotrachelus nenuphar*: effects of irrigation and species in apple orchards. Biol Control 67(2):123–129

Sharma PL (1961) The honeybee (Apis indica) population among insects visiting temperate-zone fruit flowers and their role in setting fruit. Bee World 42:6–7

Sharma DD (1985) Major pests of litchi in Bihar. Indian Farming 35(2):25–26

Sharma DR (2006) Management of insect pests of tropical and sub-tropical fruit crops. In: Proceedings of advance training course on emerging trends in economic entomology, Hisar, India, 21 Feb–13 Mar 2006, pp 123–135

Sharma DR, Singh S (2006) Management of insect pest of temperate fruits. In: Proceedings of advance training course on emerging trends in economic entomology, Hisar, India, 21 Feb–13 Mar 2006, pp 111–122

Sheffield CS (2014) Pollination, seed set and fruit quality in apple: studies with *Osmia lignaria* (Hymenoptera: Megachilidae) in the Annapolis Valley, Nova Scotia, Canada. J Pol Ecol 12(13):120–128

Sheffield CS, Westby SM, Kevan PG, Smith RF (2008) Winter management options for the orchard pollinator *Osmia lignaria* Say (Hymenoptera: Megachilidae) in Nova Scotia. J Entomol Soc Ont 139:3–18

Sin FYT, Suckling DM, Marshall J (1995) Differentiation of the endemic New Zealand greenheaded and brownheaded leafroller moths by restriction fragment length variation in the ribosomal gene complex. Mol Ecol 4:253–256

Singh SP (1989) Achievements of AICRP on biological control, Tech. Bull. No. 2. Biological Control Centre, NCIPM, Faridabad, 20 pp

Singh G (1991) Loss assessment, ecology and management of mango fruitfly, *Dacus* sp. Acta Hort 291:425–436

Singh S, Sharma DR (2013) Biology and morphometry of *Bactrocera dorsalis* and *Bactrocera zonata* on different fruit crops. Indian J Agr Sci 83(12):1423–1425

Singh A, Trivedi TP, Sardana HR, Dhandapani A, Sabir N (2001a) Resource manual on validation and promotion of IPM. National Centre for IPM, New Delhi, p 119

Singh SP, Rao NS, Ramani S, Poorani J (eds) (2001b) Annual report 2000–01, project directorate of biological control, Bangalore. PDBC, Bangalore, 218 pp

Sipos P, Óbert N, Marczali Z, Markó V (2013) Efficacy of kaolin particle films against pear psylla (*Cacopsylla pyri*) in Hungary. Növényvédelem 49(4):151–160

Smith D, Beattie G, Broadley R (1997) Citrus pests and their natural enemies: IPM in Australia. Queensland, Department of Primary Industries, Brisbane, 272 pp

Srivastava AS, Masoodi MA (1985) Influence of host plants on the development and survival of *Lymantria obfuscate*. FAO Plant Protect Bull 33(2):67–69

Steffan-Dewenter I, Schiele S (2008) Do resources or natural enemies drive bee population dynamics in fragmented habitats. Ecology 89:1375–1387

Steiner H, Altner G, Baggiolini M, Celli G, Schneider F (1977) Vers la production agricole intégrée par la lutte intégrée. IOBC/WPRS Bull 1977/4, Dijon, France, 153 pp

Stern VM, Smith RF, Van Den Bosch R, Hagen KS (1959) The integrated control concept. Hilgardia 29:81–101

Suckling DM, Brockerhoff EG (1999) Control of lightbrown apple moth, *Epiphyas postvittana* (Lepidoptera: Tortricidae) using an attracticide. J Econ Entomol 92:367–372

Suckling DM, Karg G (2000) Pheromones and semiochemicals. In: Rechcigl N, Rechcigl J (eds) Biological and biotechnical control of insect pests. CRC Press, Boca Raton, pp 63–99

Suckling DM, Burnip GM, Walker JTS, McLaren GF, Shaw PW, Howard CR, White V, Fraser J (1998) Abundance of leafrollers and their parasitoids on selected host plants in New Zealand. N Z J Crop Hort Sci 26:193–203

Sugayama RL, Kovaleski A, Liedo P, Malavasi A (1998) Colonization of a new fruit crop by *Anastrepha fraterculus* (Diptera: Tephritidae) in Brazil: a demographic analysis. Environ Entomol 27:642–648

Sutherland ORW, Hutchins RFN (1972) α-Farnesene, a natural attractant for codling moth larvae. Nature 239:170–171

Syed RA, Ghani MA, Murtaza M (1970) Studies on trypetids and their natural enemies in West Pakistan. III. *Dacus (Strumeta) zonatus* (Saunders). Tech Bull Commonw Inst Biol Control 13:1–16

Tanada Y (1964) A granulosis virus of the codling moth, *Carpocapsa pomonella* (L.) (Olethreutidae, Lepidoptera). J Insect Pathol 6:378–380

Tandon PL, Verghese A (1985) World list of insect, mite and other pests of mango. Technical document no 5. IIHR, Banglore, p 22

Thakur JN, Rawat US, Pawar AD (1989a) Investigations on the occurrence of natural enemies of San Jose scale, *Quadraspidiotus perniciosus* Comstock (Hemiptera: Coccidae) in Jammu & Kashmir and Himachal Pradesh. Entomon 14:143–146

Thakur JR, Sharma PC, Gupta PR (1989b) Occurrence of olive psylla, *euphyllura pakistanica* on olive in India: a new record. Trop Pest Manag 35(3):331

Thomson JD, Goodell K (2001) Pollen removal and deposition by honey bee and bumblebee visitors to apple and almond flowers. J Appl Ecol 38:1032–1044

Toepfer S, Gu H, Dorn S (1999) Spring colonisation of orchards by *Anthonomus pomorum* from adjacent forest borders. Entomol Exp Appl 93:131–139

Tomlin C (ed) (2000) The pesticide manual, 12th edn. Crop Protection Publications, Farnham, 873 pp

Torchio PF (1985) Field experiments with the pollinator species osmia lingnaria propinquaCresson in apple orchards: V. (1979–1980 studies) methods of introducing bees, nesting success, seed counts, fruit yields(Hymenoptera: Megachilidae). J Kans Entomol Soc 58:448–464

Torchio PF (1987) Use of non-honey bee species as pollinators of crops. Proc Entomol Soc Ont 118:111–124
Torchio P (1988) The blue orchard bee: an alternative pollinator of apples. Utah Sci 49:2–9
Torchio PF (1990a) Diversification of pollination strategies for U.S. crops. Environ Entomol 19(6):1649–1656
Torchio PF (1990b) *Osmia rubifloris*, a nature developed as a commercially managed pollinator of blueberry (Hymenoptera: Megachilidae). J Kans Entomol Soc 63:427–436
Torchio PF (1994) The present status and future prospects of non-social bees as crop pollinators. Bee World 75:49–53
Unruh TR, Knight AL, Upton J, Glenn DM, Puterka GJ (2000) Particle films for suppression of the codling moth (Lepidoptera: Tortricidae) in apple and pear orchards. J Econ Entomol 93:737–743
Vandame R, Meled M, Colin M-E, Belzunces LP (1995) Alteration of the homing-flight in the honey bee *Apis mellifera* L. exposed to sublethal dose of deltamethrin. Environ Toxicol Chem 14:855–860
Valente M, Liguori R, Baldessari M, Tolotti G, Allegri A, Manucci F, Pelliconi F, Cristiani C, Alvisi G, Ponti D, Pradolesi G, Boselli M, Scannavini M (2012) Voliam Targo, a new solution against lepidoptera, psylla and mites: efficacy and timing in pome fruits. Giornate Fitopatologiche 2012, Milano Marittima (RA), 13–16 marzo 2012, pp 109–118
van der Geest LPS, Evenhuis HH (eds) (1991) Tortricid pests. Their biology, natural enemies and control. Elsevier, Amsterdam, 808 pp
van Steenwyk RA, Dunley JE, Webster A, Oliveira CM (2008) Spinetoram – a second generation spinosyn insecticide for pear pest management. Acta Hort 800(2):867–874
Varela LG, Welter SC, Jones VP, Brunner JF, Riedl H (1993) Monitoring and characterization of insecticide resistance in codling moth (Lepidoptera: Tortricidae) in four western states. J Econ Entomol 86:1–10
Varela LG, Elkins RB, van Steenwyk RA, Sánchez EE, Sugar D, Webster AD (2011) Evolution of secondary pests in California pear orchards under mating disruption for codling moth control. Acta Hort 909:531–542
Veerish GK (1989) Pest problems in mango – world situation. Acta Hort 231:551–565
Vergnani S, Caruso S (2013) Evaluation of new insecticides for the control of pear psylla (*Cacopsilla pyri*, homoptera psyllidae). Università di Bologna, Bologna, Italy, Giornate Fitopatologiche 2012, Milano Marittima (RA), 13–16 marzo 2012, 2012, pp 119–123, 5 ref
Verma LR, Chauan P (1985) Distribution, abundance and diversity of insect pollinators in apple orchards of Shimla hills. Indian J Ecol 12:286–292
Verma LR, Dulta PC (1986) Foraging behaviour of honeybee *Apis cerana* and *A. mellifera* in pollinating apple flowers. J Apic Res 25:197–201
Verma LR, Rana RS (1994) Further studies on the behaviour of *Apis cerana* and *Apis mellifera* foraging on apple flowers. J Apic Res 33(3):175–179
Vicens N, Bosch J (2000a) Pollinating efficacy of *Osmia cornuta* and *Apis mellifera* (Hymenoptera Megachilidae, Apidae) on 'Red Delicious' Apple. Environ Entomol 29(2):235–240
Vicens N, Bosch J (2000b) Weather-dependent pollinator activity in an apple orchard, with special reference to *Osmia cornuta* and *Apis mellifera* (Hymenoptera Megachilidae and Apidae). Environ Entomol 29(3):413–420
Villeirs EADE, Matthee WA (1973) The occurrence of bark borers on litchi and other fruit tree. Citrus Grow Sub-Trop Fruit J 477:17–19
Vincent C, Chouinard G, Bostanian NJ, Morin Y (1997) Peripheral-zone treatments for plum curculio management: validation in commercial apple orchards. Entomol Exp Appl 84:1–8
Vincent C, Chouinard G, Hill SB (1999) Progress in plum curculio management: a review. Agr Ecosyst Environ 73:167–175
Waite MB (1895) The pollination of pear flowers. US Dept Agric Div Veg Path Bull 5: 86 pp
Waite MB (1899) Pollination of pomaceous fruits. Yearbook US Dept Agric 1898:167–180
Wallis DR, Shaw PW, Zydenbos SM (2013) The use of sex pheromone lures to compare pear and apple leafcurling midge phenology. N Z Plant Protect 66:270–273

Wang SM, JiangYong W, ZhiHan W, JuanXia Y, Huan G (2007) Summary and prospect of the control of pear disease and insect pests by bagging. J Northwest A & F Univ Nat Sci Ed 35:141–146

Wearing CH (1975) Integrated control of apple pests in New Zealand. 2. Field estimation of fifth-instar larval and pupal mortalities of codling moth by tagging with cobalt-58. N Z J Zool 2:135–149

Wearing CH (1990) Granulosis virus control of codling moth in Nelson. In: Proceedings of the Forty-third New Zealand weed and pest control conference. New Zealand Weed and Pest Control Society, Dunedin, pp 317–321

Wearing CH, Charles JG (1978) Integrated control of apple pests in New Zealand. 14. Sex pheromone traps to determine the applications of azinphos-methyl for codling moth control. In: Proceedings of the New Zealand weed and pest control conference. New Zealand Weed and Pest Control Society, New Plymouth, pp 229–235

Wearing CH, McCarthy K (1992) Predation of codling moth, *Cydia pomonella* L. by silvereyes, *Zosterops lateralis* (Latham). Biocontrol Sci Technol 2:285–295

Webster RL, Carlson C (1942) Ovicidal value of light mineral oils for the codling moth. J Econ Entomol 35:530–533

Westigard PH, Hoyt SC (1990) Codling moth (Lepidoptera: Olethreutidae): evaluation of the granulosis virus for control in Pacific Northwest apple and pear orchards. Melanderia 46:14–18

Whalon ME, Croft BA (1984) Apple IPM implementation in North America. Annu Rev Ent 29:435–70

Whitehead VB, Rust DJ (1972) Control of fruit piercing moth, *Sorrodes partita* (Fabr.) (Lepidoptera: Noctuidae). Phytophylactica 4:9–12

Williams IH (1996) Aspects of bee diversity and crop pollination in the European Union. In: Matheson A, Buchmann SL, O'Toole C, Westrich P, Williams IH (eds) The conservation of bees. Academic, New York, pp 63–80

Witzgall P, Backman AC, Svensson M, Bengtsson M, Unelius CR, Vrkoc J, Kirsch PA, Ioriatti C, Löfqvist J (1996a) Potential of a blend of E8, E10-12OH and E8, E10-12Ac for mating disruption of codling moth, *Cydia pomonella* L. (Lep., Tortricidae). J Appl Entomol 120:611–614

Witzgall P, Chambon J-P, Bengtsson M, Unelius RC, Appelgren M, Makranczy G, Muraleedharan N, Reed DW, Hellrigl K, Buser H-R, Hallberg E, Bergström G, Tóth M, Löfstedt C, Löfqvist J (1996b) Sex pheromones and attractants in the Eucosmini and Grapholitini (Lepidoptera, Tortricidae). Chemoecol 7:13–23

Yaşar B, Uysal O (2013) Evaluation of the efficacy of different traps in capturing apple blossom beetle (*Epicometis hirta* (Poda, 1761)) (Coleoptera: Scarabaeidae). Türkiye Entomoloji Dergisi 37(2):169–177

Yothers MA (1953) An annotated bibliography on Aphelinus mali (Hald.), a parasite of the woolly apple aphid, 1851–1950, Bulletin E-861. United States Department of Agriculture, Washington, DC, 61 pp

Zouiten N, Hadrami IE (2001) Olive tree psyllid: state of knowledge and perspective of control. Agricultures 4(10):225–232

Chapter 5
Stone Fruits

Apricots, cherries, peaches, plums and almonds are collectively called as stone fruits. They are susceptible to various pests and diseases. These pests must be managed to produce quality fruits and ensure survival of trees. Pests are not the only cause of poor flowering. Poor fruit set can also occur for a variety of other reasons such as damage to pistil and stamens before the flower blooms prevents pollination and fertilization and cool, wet weather during bloom discourages pollination.

5.1 Pollination Requirements and Pollinators of Different Stone Fruits

5.1.1 Peach **Prunus persica** *(L.)*

Batsch flowers have five petals surrounded by small sepals and 15–30 pollen laden anthers with single pistil which reaches a single ovary containing two ovules which ultimately develop into one seeded fruit after fertilization (Stewart et al. 1967). The flowers are highly attractive to honeybees and other nectar–pollen collecting insects. Stigma is receptive immediately after opening the flower and remains so for 4–7 days (Fig. 5.1).

5.1.1.1 Pollination Requirements

Pollination in peach and nectarine is relatively simple because only one ovule per ovary needs to be fertilized. Varieties range from completely self-sterile to completely self-fertile (McGregor 1976). However, even self-fertile varieties can benefit from insects transferring pollen from anthers to receptive stigmas. Khan and Rahman (1930) concluded that cross pollination is necessary to obtain good yields and that bees are the chief agent for cross pollination. The set was higher in unbagged

D.P. Abrol, *Pollination Biology, Vol. 1*, DOI 10.1007/978-3-319-21085-8_5

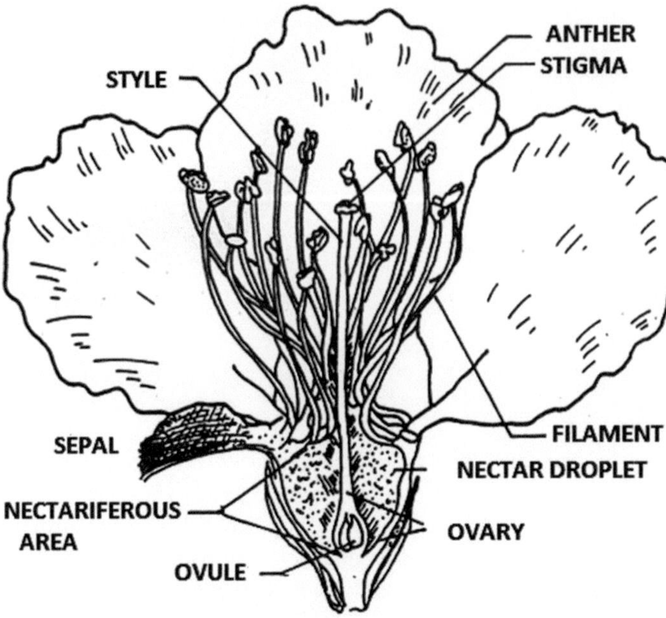

Fig. 5.1 Longitudinal section of peach flower (McGregor 1976)

than bagged flowers (Sharma 1961; Mann and Singh 1981a). Similar, results were obtained by Bulatovic and Konstantinovic (1962) who obtained better set on various species with exposed flowers than with selfed ones. Langridge et al. (1977) found that pollination of peaches by *A. mellifera* resulted in increased fruit set by 2.9 times and fruit weight by 2.6 times as compared to those from which bees were excluded. These references clearly indicate that, pollinating insects are of value even for the self-fertile cultivars of peaches. Several investigators have recommended different number of colonies for effective pollination of peaches. Banner 1963 recommended 0.2–0.3 colonies of honeybees per acre and 0.5–0.7 colonies of bees per ha. Scott-Dupree et al. (1995) recommended 1 colony per acre and 2.5 colonies per ha.

5.1.1.2 Pollinators

Most cultivars of peach (*Prunus persica*) are reported to be self fertile and only a few as self fertile (Kanato et al. 1967). Randhawa et al. (1963) studied the floral biology of peach. The flowers opened after 0600 h, most of them were open by 1000 h and remained open during night. The stigma was receptive for 3 days. Bhalla et al. (1983) reported that peach was visited by honeybee, *A.c. indica, Andrena reticulata, Ceratina hieroglyphica, Bombus sp., Syrphid flies, Wasps (Vespa sp.)* and some beetles (Plates 5.1 and 5.2).

Plate 5.1 (**a**) Fruit flies probing for oviposition on guava fruit (**b**) Magnified view of fruit fly oviposition on guava (**c**) Damaged litchi fruits due to fruit borer (**d**) Litchi fruit borer larva (**e**) Leaf curl symptoms developed due to attack of peach (**f**) Peach fruit fly probing for oviposition on (**g**) *Coccinela dimitida* grub feeding on peach aphids (**h**) Methyl eugenol traps hanging on peach tree for monitoring peach fruit fly

Plate 5.2 A view of *Apis mellifera* apiary in fruit orchard. A cherry tree in bloom, *Apis cerana* on apple flowers. *Apis dorsata* on peach flowers, *Apis mellifera* on peach flowers, A. dorsata on guava flowers, A view of *Apis mellifera* apiary in citrus orchard, *Bombus haemorrhodalis* on citrus flowers, *Apis cerana* and *Apis dorsata* on citrus flowers, A. view of apiary in litchi orchard, litchi bloom with insect pollinators, *Apis dorsata* on litchi flowers, *Apis florea* on ber flowers, *Apis dorsata* on guava, *A. dorsata* on litchi flowers. Apis cerana on strawberry flowers, *Apis florea*, flies and other pollinators on strawberry flowers, well formed and malformed fruits in strawberry and aphids attacking strawberry plants

Plate 5.2 (continued)

Kumar et al (1989) observed as many as 17 species of bees on peach and apricot. Kumar et al (1989) recorded *Xylocopa* as the only bee species on plum bloom. Abrol and Bhat (1989) recorded *Xylocopa valga* as important pollinator of peach, plum and pear flowers. Rana et al. (1994) reported that dipterans constituted 11.8 % of the insect visitors in case of plum. Abrol et al (2005) found that honey bees *Apis dorsata, A. mellifera, A cerana* and *A. florea* were the dominant flower visitors and comprised more than 73.79 % and 93.91 % of the total flower visiting insects in peach and plum flowers, respectively. The other important insects frequenting peach and plum flowers were *Xylocopa* sp., *Andrena* sp., *Megachile* sp., *Musca* sp. and *Syrphus sp.* Randhawa et al. (1963) reported honey bees as the most important pollinator of peaches. However, Yokozawa and Yasui (1957) reported that the Dipterans were the most common floral visitors when the weather was cloudy and rainy.

5.1.1.3 Pollination Recommendations and Practices

Numerous studies have indicated that bees are beneficial to peaches, and recommend to growers that action be taken to increase the number of insect pollinators in the orchard. It is fortunate that peach flowers are attractive and ample pollination is obtained free when conditions are favourable, with bees coming long distances. Jorgensen and Drage (1953) considered bees necessary. Kelly (1964) made a study relating to cost of peach growing and found that an average of only one hive per 16 acres was used. Benner (1963) recommended one strong colony of honey bees for each three to five acres of orchard just coming into bearing but stated that in older orchards one good colony of bees for each acre might be needed.

5.1.2 Plum

The flowers of plum are about 2.5 cm wide having five petals, several stamens, one style with two lobed stigma. The ovary is with one compartment and two ovules out of which usually one develops (Fig. 5.2). The stigma is receptive immediately after opening the flower and flower remains opened for 3–5 days. Anthers usually release pollen when the flower is fully open. If the pollination does not occur immediately, the flowers drop. Flowers occur in clusters of 1–3 along with new growth branches. The flowers are attractive to bees, the quantity and concentration of nectar varies among different cultivars.

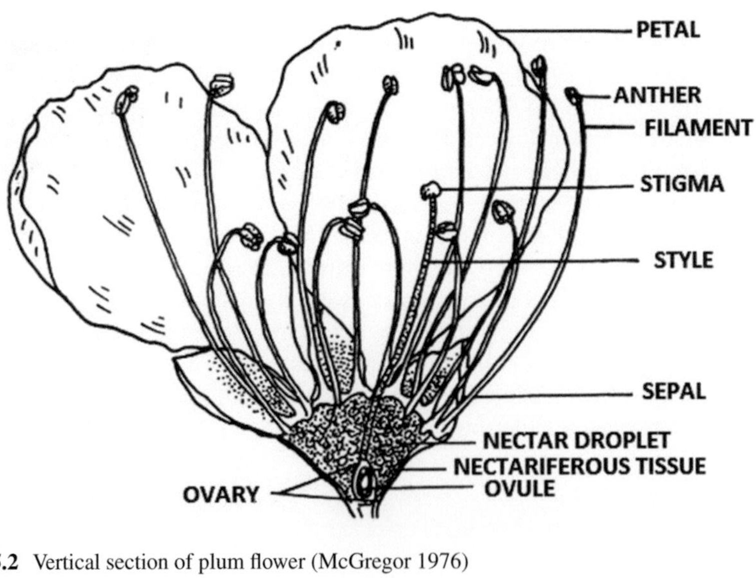

Fig. 5.2 Vertical section of plum flower (McGregor 1976)

5.1.2.1 Pollination Requirements

Several studies have established that plum cultivars vary from completely self-incompatible, in which they set no fruit with their own pollen, to complete self-compatibility, where a full crop is set from the plants' own pollen. (Griggs 1970; Griggs and Hesse 1963; Backhouse 1911, 1912; Hendrickson 1916, 1918, 1919a, b, 1922a, b, 1923, 1930; Luce and Morris 1928; Marshall 1920; MacDaniels 1942; Philp and Vansell 1932, 1944; Waugh 1898). Pollinating agents are necessary for transfer of pollen from anthers to the stigma on all cultivars (Alderman and Angelo 1933). Pollination requirements have been reported to vary among different cultivars. For example, Thompson and Liu (1972) reported that the Italian prune is fully self-fruitful and do not require bees for pollen transfer. Similarly, Dickson and Smith (1953) found that except for the Italian prune and Stanley, all European cultivars in Canada are self-unfruitful and require mixed plantings and cross-pollination. They further found that the 'Burbank' and the 'Shiro', the main Japanese cultivars are also self-unfruitful and require insect pollination. Free (1962) showed that fruit set on plum trees decreased with increasing distance from the pollenizer tree. Trees close to pollenizer trees had a greater set on the sides facing the pollenizers than those on the other sides, indicating that the pollen was not thoroughly distributed over the tree.

5.1.2.2 Pollinators

The honeybees have been recognised as the most important pollinators of plums (Buchanan 1903; Free 1962; Hendrickson 1930; Hooper 1936; Kinman 1938, 1943; MacDaniels 1942), although bumble bees, various species of wild bees, flies and blowflies have also been reported as important pollinators (Brown (1951). Honey bee attractants may be helpful under suboptimal pollination conditions such as poor weather, competing bloom, or when the bee-keeper is prevented from depositing hives in orchard interiors. Webster et al. (1985) showed in plum orchards that honey bee hives fitted with pollen traps fielded a higher proportion of pollen foragers compared to non-modified hives. The flight season of imported horned-face bees (*O. cornifrons*) coincides with plum bloom in Maryland (Batra 1982). However, populations of non-managed bees are generally too small to support commercial pollination needs in orchard crops (Scott-Dupree and Winston, 1987).

5.1.2.3 Pollination Recommendations and Practices

Stephen (1961) recommended one colony per acre and suggested that bees be moved in orchards at one-third bloom stage. Griggs and Hesse (1963) also recommended at least one strong colony of honey bees per acre with four or five frames of brood and enough bees to cover eight frames. They suggested that colonies be placed in groups of 5–10 in the orchards. Similarly, many other investigators have also recommended one (McGregor 1976; Standifer and McGregor 1977; Crane and Walker 1984; Kevan 1988; Scott-Dupree et al. 1995) to two colonies per acre

(Mayer et al. 1986) for successful pollination of plum flowers. Most growers take some steps to see that bee colonies are in or near their orchards.

5.1.3 Apricot

Apricot flowers are white in colour borne singly or double on very short stems. The flower contains 30 stamens with one pistil. The flowers are attractive to bees for both pollen and nectar (Fig. 5.3).

5.1.3.1 Pollination Requirements

Pollination in Apricot is equally beneficial as in case of other fruit crops. Even the self-sterile varieties when interplanted gave better yields (Hootman 1935). Jusubov (1957) reported that some cultivars were self-fertile and whereas others were completely self-sterile. Griggs (1970) identified two self-incompatible varieties namely "Perfection" and "Riland". Stanger et al. (1973) reported that some varieties of apricot were unable to produce large crop without cross-pollination. McGregor (1976) and Ahmad et al. (1986) reported that some varieties produced fruits only when cross-pollinated whereas in others there was significant increase in production when cross-pollinated. Guleryuz et al. (1999) observed that fruits sets were generally

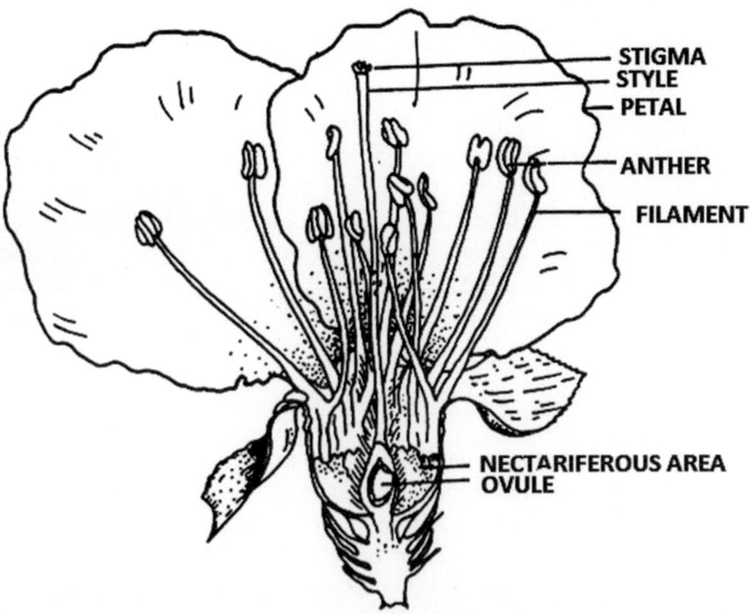

Fig. 5.3 Longitudinal section of apricot flower (McGregor 1976)

lower of self-pollinated trees than free-pollinated ones. Cross-pollination has been reported to improve fruit set and ripening period (Szabo et al. 1999). Albuquerque et al. (2000) reported that the chances for fertilization would not be increased by pollination in apricot. McLaren et al. (1996) reported that time of cross-pollination is shorter in apricot than apple as the flower in former has shorter life period than the later.

5.1.3.2 Pollinators

Murneek (1937) reported that insect were equally necessary for proper pollination and setting of fruit in self-sterile or self-fertile varieties. Stark (1944) observed that honeybees were the most important pollinators for commercial fruit production in apricot. Jorgensen and Drage (1953) stated because of the sticky pollen insects are needed for pollen transfer from stamens to stigma. Stanger et al. (1973) found that apricot generally produced larger crop when honeybee pollinated. McGregor (1976) reported that honeybees were the primary pollinating agents of apricot. Benedek and Nagy (1995) found that floral constancy of honeybees varied between different plant species. Benedek et al. (1995) mentioned that there were no significant differences among cultivars with regard to bee pollination in apricot.

Austin et al. (1996) reported honeybee *Apis mellifera* as the only significant flower visitors of apricot. Batra and Richards (1997) reported that in the great Himalayan range, fruit trees were pollinated primarily by halictine bees (*Lasioglossum matianense*) and bumblebees. Among other insects, *Apis laboriosa* was the only honeybee found in this high northern area. These were diverse bee species that pollinated the apricot (*Prunus armeniaca*) and other cultivated as well as wild fruit. Szabo et al. (2000) reported from Hungary that apricot flowers producing pollen and nectar were attracted sufficiently attractive to bees. Bee visits varied at different locations and with growing season. Most of the bees visited apricot flowers for pollen, however, nectar gatherers were also observed. Bee pollination was necessary not only for the self-incompatible varieties but also to enhance the yield of self-fertile varieties. This means that not only self-sterile but also self-fertile fruits clearly depend on insect (bees) pollination. They further reported that additional bee pollination was needed for all kinds of temperate-zone fruit tree species when bee visitation was insufficient (Benedek et al. 2000). These studies clearly indicate that insect pollination was required on self-sterile cultivars and was at least beneficial to the self-fertile cultivars.

5.1.3.3 Pollination Recommendations and Practices

Like peach and nectarine, apricot also depends upon pollinating insects for commercial crop production. It is necessary that bees may be distributed throughout the orchard. Corner et al. (1964) recommended one colony of honeybees per acre, but the colonies need to be distributed in small numbers throughout the orchard (McGregor 1976; McLaren et al. 1995). The addition of extra beehives increased the number of bees/tree but did not increase fruit set. They further suggested that an average of 4 bees/30 s/tree were sufficient to produce a satisfactory crop (McLaren et al. 1996). Szabo et al. (2000) recommended two to four bee colonies per hectare in commercial plantations for the whole flowering period.

5.1.4 Cherry **Prunus** *spp.*

Both Sweet cherry (*Prunus avium*) and sour cherry (*P. cerasus*) flowers are white in colour and occur in clusters of 2–5 on short lateral spurs. Each flower is about 2.54 cm in diameter, has five petals, one upright pistil with an ovary and two ovules, and about 30 stamens (Fig. 5.4). The flower opens for 3–5 days. The stigma is receptive when the flower opens. The anthers start opening shortly after opening of the flower and continue into the second day (Srivastava and Singh 1970). The sweet cherry nectar is much richer in sugar (55 % sugar) (Vansell 1942) which attracts wide variety of insects. Bees visit flowers for nectar and pollen throughout the day.

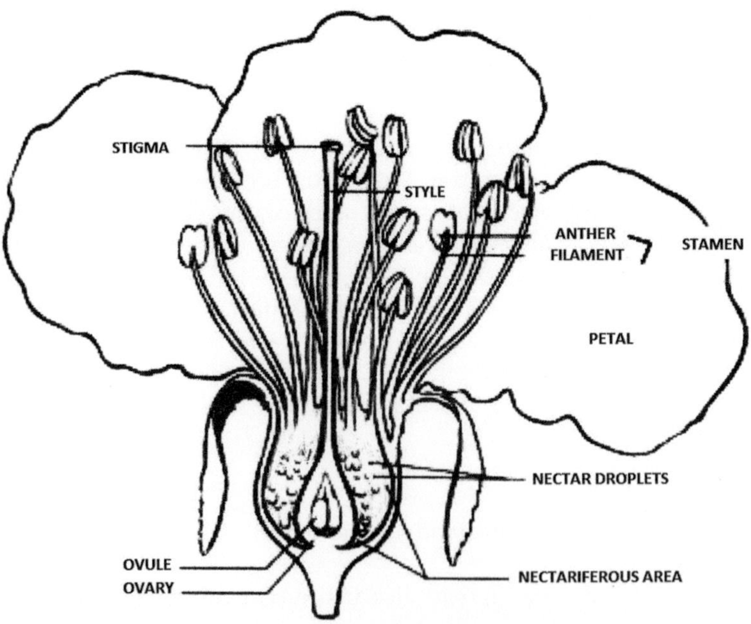

Fig. 5.4 Vertical section of cherry flower (McGregor 1976)

5.1.4.1 Pollination Requirements

The sweet cherry has been shown to be self-sterile or self unfruitful (Einset (1932). Murneek (1930) found that even self-fertile cultivars benefited from insect pollination in unfavourable seasons. Amount of fruit set due to insect pollination has been investigated by several investigators. Shoemaker (1928) reported an average increase of 35 % fruit set in sweet cherries. Griggs et al. (1952) reported an overall average increase of 21–32 % set. Free and Spencer-Booth (1964) reported decreasing production with increased distance from the pollenizer row of sweet cherries. In conclusion, all cherries are basically incapable of automatic self-pollination. Mayer et al. (1988b) reported that ovules of some cherry flowers begin to degenerate even before the flower opens; therefore, it is important that pollination occurs as soon as possible after the flower opens. They found that first 20 % of flowers that open, if pollinated, will set higher quality fruit than the remaining 80 %. Almost all sweet cherry varieties require cross-pollination with pollen from a suitable pollenizer cultivar. It is therefore, essential that pollenizer varieties should be interplanted in orchards in order to encourage good cross-pollination. Pollinators in sufficient numbers especially honeybees must be present to transfer pollen from pollenizer to main varieties.

Number of colonies required per ha has variously been reported by different workers foe effective pollination of sweet cherry. It varied between 2.5 (Schuster 1925; Tufts and Philp 1925; Luce and Morris 1928; Marshall et al. 1929), 2.5–5.0 (Levin 1986; Scott-Dupree et al. 1995), 3.7–6.2 (Kevan 1988) and 5 colonies per ha (Mayer et al. 1988b).

5.1.4.2 Pollinators

The possibility of pollination by wind in cherries has been ruled out by many investigators (Roberts 1922; Burtner 1923; Murneek 1930; Claypool et al. 1931; Brown 1968). Honeybees have been reported as the most important and efficient pollinators of cherry flowers comprising more than 90 % of the total insect visitors (Mayer et al. 1988a). Mayer et al. 1988b also reported that cherry nectar is rich in sugar 930–45 %), however, more than 70–95 % of honey bee forgaers collected pollen. Abrol (1989) reported *Xylocopa valga, Colletes eous, Lasioglossum sp., Halictus sp.., Andrena flavipes Syrphus sp., Apis cerana indica,* and *Apis mellifera* as important pollinators of cherry flowers. Mattu et al. (1994) observed *Syrphus, Eristalis,Fannia, Musca* and *Dolichopus* on cherry bloom. Abrol (2005a) found that more than 11 species of insects belonging to order Hymenoptera and Diptera frequented cherry flowers. Of all the flower visiting insects, honey bees *Apis mellifera* and *Apis cerana* were the most abundant and comprised more than 55 % of the total flower visiting insects (Plate 5.2). Percentage fruit set was significantly higher at 100 m (28.2 %) from the apiary than at 500 m (9.0 %). Similarly, number of colonies/ha also significantly influenced the fruit set fruit yield /plant in the cherry orchards. Gupta et al. (1990) reported that honeybees *A. c. indica* F. and

A. mellifera L. foraged both for nectar and pollen on cherry flowers. Abrol (1989), when studied the insect pollination of cherry (*Prunes avium* L.), it was found that flowers covered with muslin cloth did not set any fruit while those left for open pollination set 38–56 % fruits.

With either sweet or sour cherries, it is important that pollination occurs as soon as possible after anthers begin releasing pollen. Honey bee hives should be placed in the orchard in groups of 4–12 and no further apart than 100 yards (91.4 m). With sweet cherry, hives should be moved into the orchard on the day bloom begins or 1 day earlier. With sour cherry, it is better to move hives into the orchard at least 1 day after bloom begins because sour cherry nectar is less attractive to bees and a higher degree of orchard bloom is necessary to keep honey bees from leaving the orchard and seeking richer forages.

5.1.4.3 Pollination Recommendations and Practices

In general, several investigators have recommended one strong honeybee colony per acre for effective pollination of sweet cherry flowers (Schuster 1925; Tufts and Philp 1925; Luce and Morris 1928; Marshall et al. 1929; Murneek 1930; Philip 1930; Philp 1947; Auchter and Knapp 1937; Stephen 1961). Brown (1968) in Oregon recommended four to five colonies per acre. Eaton (1962) stated that strong colonies should be brought into the sweet cherry orchard on or before the day the first flowers open, because placement in the orchard even 1 day late could result in a reduced crop. Nevkryta (1957) recommended four to five colonies per hectare. Skrebtsova and Iakovlev (1959) showed that with 3.8 colonies per hectare, 15 % of all flowers set fruit but with 2.8 colonies only 13 % set.

5.1.5 Almond

Prunus amygdalus Batsch, family Rosaceae

The almond (*Prunus dulcis*) flower has one pistil with one ovary containing two ovules, 10–30 stamens, and five pinkish-white petals. Usually only one ovule develops into a fruit. Nectar is excreted into a floral cup at the base of the pistil (Fig. 5.5). Nectar foragers are active on almonds throughout the day if weather permits, but pollen foragers are most active during midday. The honey bee is the primary insect visitor to almond flowers.

5.1.5.1 Almond Pollination Requirements

Almond flowers are completely self incompatible and require cross-pollination with a pollen from different plant and variety for any fruit production. The pollination requirements of almond are extraordinarily high. While most other orchard crops need 5–10 % fruit-set to make a viable crop, almond requires 30–60 % (Traynor

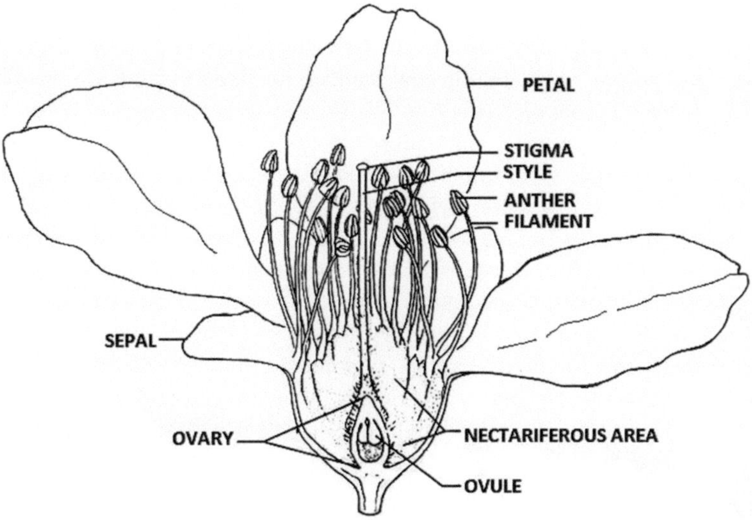

Fig. 5.5 Vertical section of almond flower (McGregor 1976)

1993). Not all pollinated flowers set fruit, and unlike apple and other orchard crops, almond is not thinned. This means that virtually 100 % of the flowers must be cross-pollinated to achieve an acceptable crop.

In planning an orchard, it is necessary to select compatible varieties whose bloom periods overlap. Varieties should be chosen so that they bloom within 3 days of each other. If possible, it is preferable to select pollenizers that bloom shortly before the main variety rather than after. Varieties also differ in frost-hardiness and in quantities of pollen produced. Agricultural extension services and crop consultants can help one select varieties for a particular location.

If possible, it is a good idea to plant blocks of mid- and late blooming varieties next to each other (with the usual 1:1 or 1:1:1 interplant of compatible varieties within each block). By placing bee hives between blocks, one can conceivably get twice as much work out of the rented hives; bees will pollinate the mid-blooming block 1 week and the late block the next.

McGregor (1976) recommended 5–7 colonies of honeybees per ha for almond pollination, whereas Thorp and Mussen (1979) suggested a minimum of 2.5–7 colonies. Levin 1986 recommended 7–10 colonies per ha. Traynor (1993) and Scott-Dupree et al. (1995) reported that five colonies of honeybees were adequate for successful pollination of almonds.

5.1.5.2 Honey Bees as Almond Pollinators

Almond (*Prunus amygdalus*) was found to flower at low temperature when honeybee colonies were in weakest condition; hence fruit set remained low. Therefore, need for honeybee pollination was greatly felt (Muttoo 1953). Mann and Singh (1981b) reported that honey bees were the commonest visitors to almond flowers. However, pollination requirements for this crop in Asia are not known. Abrol et al. (1990) observed mosquitoes and musca in low numbers in Jammu and Kashmir whereas Singh (1988) recoded four species of *Eristalis*, one species of *Episyrphus*, *Metasyrphus*, *Scaveva* and *Orthellia* as most important pollinators of almond in Himachal, India.

Honey bees are the most important pollinator of almond (Ferreres in Mexico 1947; Gagnard in Algeria 1954; Griggs in California 1970; Muttoo in India 1950; Purdie and Winn in Australia 1964, 1965). They readily visit almond flowers and effectively pollinate them. Because almond's pollination demands are so high, it is doubly important to use strong colonies. The goal is to have colonies at peak strength at the time of peak bloom (McGregor 1976). Abrol (1988) recoded *Apis cerana*, *Xylocopa valga* and *Lasioglossum spp*. as dominant flower visitors and important pollinator of Almond. He found that flower buds covered with muslin cloth did not set any fruit while 100 uncovered buds set 30 fruits indicating that almond exclusively depend on insect for pollination. The observations made on flowering phenology, insect visitors and environmental conditions showed that a period of 10–12 was the most crucial for almond pollination. He suggested that as almonds flower in March when the weather is unfavourable for foraging insects the induction of late flowering will improve fruit setting.

However, because the bloom season for almond is so early in the year when weather is cool and mostly inclement. Honey bees do not fly readily until temperatures exceed 12.8 °C. One way to minimize this problem is to put hives in sunny, wind-protected locations to encourage good flight. If cool temperatures persist, it may be necessary to increase hive density in order to compensate for reduced flight efficiency. With many other crops it is advisable to bring in bees after some bloom has already begun. This is not the case with almond because the earliest blooms are the most productive. Moreover, there is usually little else blooming to distract the bees, and honey bees eagerly visit almond as soon as it starts blooming. So, there is no reason to delay the arrival of bees and most growers bring in bees as soon as the crop blooms, or earlier. McGregor (1976) recommended placing colonies within an orchard in groups spaced at a distance of 160 m apart. But it seems that having strong colonies is more important than their distribution in the orchard. As with other crops, bee hives placed together in large groups perform better than hives distributed singly. Hives are usually deposited in groups of 16, 24, or 32. Webster et al. (1985) showed in almond orchards that honey bee hives fitted with pollen traps fielded a higher proportion of foragers collecting pollen compared to non-modified hives. This potential benefit, however, was partially offset by a slower rate of growth in those colonies with traps.

5.1.5.3 Mason Bees as Almond Pollinators

Two orchard mason bees species such as blue orchard bee (*O. lignaria propinqua*), native of North American and the orange orchard bee (*O. cornuta*) from Spain, when introduced into California almond orchards, both bees collect almond pollen, occupy man-made nests, and develop normally on a diet of almond pollen and nectar (Torchio 1981a, b; Torchio et al. 1987). These bees are good pollinators; in Spain, each *O. Cornuta* female visits 9500–23,600 almond flowers during a season, and only three females per tree are needed to maximize pollination (Bosch 1994b). However, the technology for their large-scale, practical use is not yet developed.

5.1.5.4 Pollination Recommendations and Practices

Pollination in almond is essential for any fruit production and depends exclusively on bees for the purpose. Vansell and Griggs (1952) recommended that there should be either one pollenizer row of trees for every three rows of the main variety, or two rows of pollenizer trees for each two of the main variety. They further recommended that two to three strong colonies of honey bees be used per acre. Woodrow (1932), Purdie and Winn (1964), and Sheesley and Poduska (1970a, b, c) showed that strong colonies of bees were much more effective than weak ones, particularly at lower temperatures, such as those likely to occur during almond blossom time. Griggs et al. (1952) counted 20–30 bees on each of two almond trees caged with a colony of honey bees. Griggs and Iwakiri (1960) counted 150–200 bees per tree in the open, which they considered fair to good bee activity. The studies indicate that at least two to three strong colonies per acre may be required for maximum production of almonds. The colonies should be distributed within the orchard in small groups one-tenth mile apart. The colonies should be in the orchard at the beginning of flowering and should remain until flowering on the main cultivar has ended.

5.2 Pollination Management

There has been considerable reduction in crop yields due to reduction in number of pollinating insects due to destruction of nesting habitats of bees and excessive use of pesticides (Bohart 1951; Bohart 1962; Stephen 1955; Williams 1982; Crane and Walker 1983; Kosior 1987; Torchio 1991). The blooming of plant species early in the year when weather is inclement prevents the pollination activity of bees thereby limiting fruit production (Gould 1939). Unfavourable weather can adversely affect the reproductive success of both pollinating insects and the plants depending on them for pollination (Eisikowitch and Galil 1971; Martinez del Rio and Burquez 1986; Bergman et al. 1996). Information on relationship between thresholds for foraging of bees and climatic conditions such as temperature, wind, humidity and solar radiation is necessary to evaluate the potential performance of a pollinator on different crops and in different climatic areas (Vicens and Bosch 2000b). For proper management of candidate pollinator it is essential to have basic knowledge of the life cycle and developmental biology to rear them

Weather not only affects the pollinator activity, but can also influence fruit or seed production as cold temperatures during early spring can result in low pollen production and reduced pollen viability (Faust 1989), as well as ovule sterility (Williams 1969). Post-flowering weather conditions also have an important effect on fruit set. Insufficient sunlight hours in the 10 day period immediately after flowering significantly reduces fruit set even after an optimal pollination period (Williams 1991).

Solitary bees (Parker et al. 1987; Torchio 1987, 1991; Crane 1991; Cane 1997; Heard 1999) inspite of their great diversity have not been fully utilized for pollination purposes and honeybees *A. mellifera* continues to be the main pollinator and honey producer in the world.

Lower yields in trees located far from pollinator nesting sites or from pollinizer rows may also be an indication of insufficient pollen transfer (Free 1962).

The promising pollinator species should show a preference for foraging on flowers of the target crop (Torchio 1976; Maeta 1978; Marquez et al. 1994) have high rates of stigma contact. *Apis mellifera* bees sometimes collect nectar from side that results in low rates of stigma contact, whereas pollen collectors are much more efficient pollinators of these crops (Roberts 1945; Free 1960; Robinson 1979; Bosch and Blas 1994; Monzon 1998; Vicens and Bosch 2000a). Rates of stigma contact are sometimes also related to the body size of the pollinator relative to the size of the crop flower. In general, small bee species are not the best pollinators of large-flowered plants (Inouye 1980).

The flowering period of the crop should coincide with the activity period of the candidate pollinators. For fruit trees, with short seasonal flowering periods, univoltine pollinators with relatively short nesting periods are desirable (Parker et al. 1987). Futhermore, basic knowledge on the nesting behaviour and nesting requirements and nesting substrates should be available to allow for large-scale management (Bosch and Kemp 2002). Cavity-nesting *Megachilidae* normally accept a variety of artificial or semi-artificial nesting materials, including holes drilled in wood or styrene, wood or styrene grooved boards, paper or cardboard tubes, and reed sections (Yamada et al. 1971; Maeta 1978; Torchio 1982, 1984; Bosch 1995). Adequate nesting materials need to be not only attractive to female bees, but also affordable and manageable. Nesting materials should also offer protection from light, excessive heat and humidity, parasites and predators. Progeny from less-preferred nesting materials tend to be excessively male-biased and/or express high developmental mortality (Torchio 1982a, b, 1984a, b; Bosch 1995).

5.3 Pest Management in Stone Fruits

Stone and nut fruits are grown in temperate and subtropical climates. Number of insect pests damaging peach, plum, apricot and cherry is around 80, 60, 30 and 40, respectively. About two dozen insect pests attack fruit trees. Many of the insect problems are common to fruits grown in temperate areas and are described on the basis of parts of the trees they attack.

5.4 Sucking Insects

5.4.1 Aphids

The main aphid species associated with stone fruits in India are the peach leaf curl aphid, *Brachycaudus helichrysi (*Kaltenbach), peach black aphid, *Pterochloroids persicae (Chol), green peach aphid, Myzus persicae* Sulzer, cherry aphid *Myzus cerasi* (F.), mealy plum aphid, hyalopterus pruni (Goeff.), and walnut aphids *Chromaphis juglandico*la (Kaltb) and Callaphis juglandis (Goeze). Among stone fruits, most serious is the peach leaf curl aphid and on walnut, *C.juglandicola* is of common concern. Some species of economic concern are briefly described below:

Peach leaf curl aphid, *Brachycaudus helichrysi* (Kaltenbach);

The sucking of sap from vegetative buds and unfolding leaves results in curling up and paling of leaves. Flowers appear sickly and dull. Severe infestation may interfere with fruit set, results in premature fruit drop and subnormal fruit formation. The badly curled leaf-whole whorls usually get dried and shed. It also acts as an important vector of non-persistent potato viruses.

Brachycaudus helichrysi is a heterocyclic aphid it infests stone fruits from autumn to beginning of summers (October to April–May) depending upon altitude. However, there had been difference of opinion about prevalence of sexual generation in India. While exclusive parthenogenetic reproduction has been reported from lower altitudes (<1525 m) of Uttarakhand and Shimla hills (2286 m), egg laying has been reported in plains as well as at higher altitudes by others. However, a detailed study carried out in Himachal Pradesh at 1100–2300 m altitude revealed alternation of generation on peach (Gupta and Thakur 1993). Autumn migrant gynoparae appear on peach trees and remain active throughout October and November; they deposit a progeny of oviparae by parthenogenetic viviparity. The mature apterous oviparae are fecundated by alate migrant males coming from secondary host(s) on peach foliage or buds. They usually deposit eggs singly on the current growth either in bud axils or under the scales during November–December. Eggs hatch in 20–28 days during December–mid January. The neonate fundatrix makes random movements to settle in bud axils. Development of fundatrix and laying of fundatrigene is mainly dependent on state of the host and temperature. In mid hills fundatrices commence larviposition on early varieties from January and onwards and give birth to about 50 young ones in life span of about a fortnight. Migrant-alates appear from March end to May and shift to secondary hosts. This species passes summer and rainy season on alternate/secondary hosts, mainly belonging to compositae (at least 20 plant species). *Erigeron canadensis*, *Trifolium pratense*, *Cineraria sp.* and *Ageratum conyzoides* are its *main* host in Himachal and Uttaranchal.

5.4.1.1 Peach Mealy Aphid, *Hyalopterus pruni* (Geoffroy)

It is another aphid pest reported from Kashmir, Punjab and Maharashtra on peach, plum and apricot. In Europe and Russia, its known secondary host is the reed, *Phragmites communis* on which many generations are passed from July to September before migrating to plums, hence also known as reed aphid. In India summer is passed on *Arundo donax* and *Phragmites karka.* Gynoparae and males return to winter host around October and afterwards. In areas where winters are harsh, sexual generation is completed and eggs are laid on peach shoots during December–January. It infests lower surface of leaves, succulent parts of shoots, fruits and peduncles; infested leaves get slightly twisted. The produce honey dew in abundance imparting a glisten to foliage.

5.4.1.2 Peach Brown/Black Aphid, *Pterochloroides persicae* (Cholodkovsky)

This aphid is known from the Middle East, the coast of the Caspian sea, India, Pakistan and Egypt infesting not only peach but also apricot, almond, plum, cherry and apple. Usually, adults and nymphs are found in August, the first few alates in September and a higher proportion of alates in late October and early November; the proportion of alates declines significantly in December and January. Parthenogenetic viviparous reproduction occurs in early January. Its large size brown/black colouration resembling the colour of peach shoots and congregation pattern make it conspicuous on stems and tender shoots. The aphid colonies are found mainly on large branches and also on the trunk of young peach trees, usually on the lower surface or on the shady side. Infestation is characterized by dark staining of the ground or grass beneath the trees, caused by sooty mould developing on the copious amount of honeydew excreted by the aphids.

5.4.1.3 Peach Green Aphid, *Myzus persicae* (Sulzer)

It is a worldwide distributed truly polyphagous and cosmopolitan species, peach being its primary host. Colonies of the aphid appear on leaves, floral buds and young fruits of stone and pome fruit trees. These are usually green but pale yellow to pinkish. It is a holocyclic in temperate regions with severe winter, but is exclusively parthenogenetic in subtropical areas. It is a well known vector.

5.4.1.4 Walnut Aphids, *Chromaphis juglandicola* (Kaltb.) *and Calaphis juglandis (Goeze)*

These aphids are usually active during April–September, with a peak population in spring. Sexual morphs appear at leaf drop and mated oviparous aphids deposit their hibernating eggs in basal cracks of host tree buds usually more eggs are found on small peripheral branches (1–2 cm thick). Walnut green aphid is of much common occurrence and its action threshold is 15 aphids/leaflet. Heavy infestation results in highly significant reduction of light coloured and totally edible kernels and significant increase in the proportion of mouldy and shrivelled kernels. An early, heavy but short-lived outbreak of the dusky veined aphid reduces nut size but not kernel quality, whereas later prolonged infestations have more effect on quality. A variety of natural enemies feed on walnut aphids and parasitiods *Trioxys pallidus and Praon himalayensis* are usually found parasitizing it. *For management of C. juglandicola, Trioxys pallidus* has widely been used as a biological control agent in Europe and America. It has better net reproductive rate (53.85 female progeny/female) and innate capacity to increase (0.385) and lesser doubling time (1.8 days) than its host. The annual gain to the California walnut industry was estimated to be in excess of s 0.5 million as a consequence of introduction of *T. pallidus* (Bosch et al. 1979). Now, a strain of this parasitiod, resistant to azinophos methyl has been developed, which can be used to suppress the walnut aphid.

Management of Aphids

Population of most of the aphids in the beginning of spring is low and at that time, the natural enemy activity is also least. It is appropriate to spray stone fruits at pre-bloom/pink bud stage, especially on peach, where healthy fruit can be harvested without insecticide application. Invarably, systemic insecticides, viz. Oxydemeton methyl 0.025 % or dimethoate 0.03 % or quinlphos 0.025 % are used at pink bud stage for its suppression, though contact insecticides are effective. Often there is no need of post bloom spray as residual population if any, is taken care of by their natural enemies, e.g. coccinellids like *Coccinella septumpunctuata* L. and *Hippodamia variegata* (Goeze), larvae of a variety of syrphid and chamaemyiid flies, anthocorid bugs (*Anthocoris min*ki Dhorn), chrysopids, hemerobids and parasitiods, the population of which build up in March–April. Above all, these natural enemies need protection from harmful insecticides. Regular pruning of current vegetative growth during December, which normally carries eggs, reduces infestation in coming season. Removal of alternate/secondary hosts in and around orchards also helps in reducing the source of infestation.

5.5 Scale Insects

Both armoured and soft scales infest stone and nut fruit trees. Among armoured scales (Diaspididae), the San Jose scale is of common occurrence on most of pome and stone fruits trees. Besides, *Childaspis asiatica* has been observed on apple and plum in Himachal Pradesh, *Margonella longispina* (Morgan) on olive and walnut, *Parlatoria oleae* (Colvee) on apple, apricot, pear, plum and sweet cherry, peach white scale *Pseudaulacaspis pentagona* Targioni on peach, plum, almond and cherry in Uttaranchal, Assam and West Bengal and *P. manni* (Green) on walnut in

Himachal Pradesh. Among unarmoured or soft scales, scales insects of Eulecanium genus are of concern on stone fruits in Jammu and Kashmir, Himachal Pradesh and Uttaranchal.

Eulecanium spp.: The species existing in Uttaranchal (*E. coryli*) and Himachal Pradesh (*E. tiliae?*) appear to be different. The former predominantly on peach and the latter on plum. Both species are univoltine, but mode of reproduction is sexual in *E.coryli* by production of males in March–April while in *E.tiliae*, it is exclusively by pathogensis.

Like other scale insects, the stages of development of *Eulecanium* include egg, two nymphal instars, and adult female. Seasonal history is more or less alike. Adult females appear in February–March and swell due to voracious sap feeding to become dome shaped. They secrete honeydew in abundance which attracts a variety of flies, ants and other insects. Eggs are laid under the body throughout April (200–1500 eggs/female). The female dies after egg laying and its abdomen shrinks against hardened dorsal integument. Such globular females are generally seen on 1 year old twigs. Eggs remain protected under the dead female and hatch during May-July. Crawlers leave the dome shaped body of the mother through the anal slit and usually settle on under side of the leaves along midrib and main veins. They moult to second instar in August and migrate to 1–2 year old twigs during September–October. The insect overwinters in this stage till march when it moults to become adult female. Natural enemies put effective check on the population growth and these are parasitioids *Coccophagus* sp. nr. *Ishii* (Aphelinidae) and *Blastothrix sericea* (Encrytidae) causing high parasitization (upto 73 %), *fungus Rhinocladiella* (parasitizing about 13.6 % second instar nymphs), and coccinellid predation. This scale insect can be effectively controlled by the pre-bloom application of dormant spray oil or by summer spray oil application (after the crawler emergence) if the population of the pest is high and warrants chemical intervention, insecticides suggested for the sanjose scale can be applied after the fruit harvest.

5.6 Fruit Borers and Blossom Pests

Fruit moth, *Cydia* sp. (Tortricidae:Lepidoptera): fruit moth Cydia sp infestation has been noticed on apricot and plum in Himachal Pradesh, India. A fruit may contain up to seven larvae depending upon its size and stage of development. The larva makes gallery of variable shape and size in mesocarp by feeding on the flesh and damage is conspicuous around the stone. Adults are on the wings at dusk during April end to May. They usually lay eggs singly on fruit, which hatch in 1–2 weeks and larvae enter the fruit. Pupation takes place in cocoon on the tree, near its base or any suitable site. Developing larvae of second generation are more destructive. They become full fed by August end, over winter as pre pupae in silken web and pupate in April. Collection and destruction of fallen fruits and debris harbouring larvae reduce the inoculum of the pest in the next year. In orchards, where dormant spray oil in combination with fenitrothion 0.05 % is sprayed, the incidence of the pest is negligible. In hot spots, two sprays of insecticides, like fenitrothion 0.05 %, malathion 0.05 %, quinlphos 0.05 % or carbaryl 0.1 %, first 1 month before the anticipated date of harvest and other about 2 weeks later, can be applied. Release of *Trichogramma embryophagum* or *T. cacoeciae pallidium* during egg laying period and spray of *Bacillus thuringiensis* subspecies *kurstaki* 7–10 days after release of the parasitioids, can also be effective treatments and need testing.

Fruit Flies, *Bactocera spp.* (Tephritidae: Diptera) The attack of fruit flies is generally localized in warm and humid fruit growing areas. Some of the important species include Oriental fruit fly, *Bactocera dorsalis* Hendel, peach fruit fly, *B. zonata* (Saunders) and Ethiopian melon fly, *B. ciliata* (Loew.). Localized infestation of fruit flies in specific pockets is an indicator of the fact that their self dispersal is restricted and their survival depends upon climatic conditions conducive for their overwintering. It is feasible to check their spread by imposing domestic quarantine. To contain the pest in endemic area, concerted efforts of inhabitants in form of campaign are required. Sanitary and some cultural methods provide relief from fruit fly pestilence. Collection of infested fruit and deep burying (about 15–30 cm below the soil surface) and prompt disposal of unmarketable fruit from the orchard so as to prevent the attraction of the flies to lay eggs in them are likely to reduce the infestation of in the next season. Digging of tree basin and area around it in winters to disturb and expose the overwintering pupae to natural enemies and climatic adversities destroys good proportion of the hibernating population of the pest. Since flies deposit eggs inside the developing fruit and maggots are internal feeders, insecticidal treatment of ripening fruit crops that too near harvest is uneconomical, unethical and undesirable, though some insecticides like deltamethrin 0.0025 %, fenitrothion 0.05 % and dimethoate 0.03 % are reported to be effective when sprayed 4 and 2 weeks prior to the harvest. Better would be the application of bait of the fruit flies, or attracting the adults to poisoned bait traps/stations (methyl eugenol) as attractant to *B. dorsalis* and *B. zonata,* cue lure for *(B. cucurbitae)* or yellow sticky traps/ yellow traps treated with pyrethroid. A bait containing the biologically produced spinosad, has recently been recommended and approved for agricultural use in the USA for suppression of olive fruit fly and is proving to be effective when applied frequently throughout the period of pit hardening to harvest. The best results can be obtained if all these If all these techniques are used in combination.

Walnut Weevil, *Neomecyslobus porrectirostris* (Marshall) (Curculionidae, Coleoptera) It is a specific and serious pest of walnut in the Himalayas and sub-mountainous regions wherever walnut is grown at the altitude range of 1800–3000 m. However, not much work has been carried out on this pest of significance, perhaps for the reason that the walnut is not cultivated in well organized orchards. In some areas, fruit loss may be above 50 %. There are two generations, one from April to July and other from June to September. It overwinters in adult stage. The weevils feed on leaves, petioles, delicate shoots, buds and fruit. Female makes a pit on the rind of the fruit and inserts eggs in it. The duration of egg, larval and pupal stage is 3–5, 13–22, and 9–17 days, respectively. First generation grubs do much damage as they bore and feed on kernel turning it into a black mass; infestation may result into fruit drop. Second generation grubs bore in the rind and find the shell hard to enter. One fruit harbours 1–3 grubs. Grubs are parasitized by a tachinid fly, Hamidegeera villenuvei Bar. There is no satisfactory method to control the pest. Timely collection of attacked and fallen fruit and their burning on campaign basis can provide a good relief from the pest.

FruitSuckingMoth,*Calpeophideroides*Guen.(Noctuidae,Lepidoptera) Although large numbers of moth-species have been reported sucking juices from various temperate fruits in different parts of the world, so far only the attack of C.ophideroides has been observed on peaches and nectarines in Kumaon hills, Uttranchal. The moths pierce fruits, suck juice from them; as a result they rot and fall to the ground. Since their larvae feed on a variety of wild growing plants, it is difficult to control it in larval stage. No satisfactory method is known to contain the pest and its damage. Poisoned bait based on fermenting fruit juices, may give some relief. Recently in china, control of Oraesia excavate (Calpe excavate) in stone fruit orchard was shown possible by black lamp traps and bagging of fruit.

5.7 Xylophagous Pests

Stem and shoot borers (Apriona cineria, Aeolesthes holosericea, and a.sarta; Zeuzera sp.) and root borers (Dorysthenes hugelii) and shot-hole borers damage pome, stone and nut fruit trees. These have already been covered under pests of pome fruit and their management. The pests which are of concern on stone and nut furits are flat headed stem borer, Sphenoptera lafertei, bark eating caterpillar, Indarbela quadrinotata and peach twig borer, Anarsia lineatella. The cherry trees suffer heavily with the attack of Zeuzera multistrigata in some areas.

Flat headed stem borer (Jewel Beetles), Sphenopotera lafertei Thom. (Buprestidae, Coleoptera): Incidence of flat headed stem borer is of common occurrence on stone fruits, mainly on cherry and peach. This beetle is considered as one of the causes fro dwindling of cherry plantation in Himachal Pradesh, where infestation on cherry is as high as 74.5 %. The decreasing order of severity of attack on stone fruits is chery, peach, plum and almond and there is no relation in altitude and degree of infestation (Sharma et al. 1990). Sharma and Gautam (1994) found the infestation of s. lafertei on both cultivars of almond, viz. Drake and Non Pariel planted on four root stocks (wild peach, bitter almond, Sloh hybrid and behmi) but the trees raised on bitter almond were comparatively less susceptible to its attack. While screening plum and almond varieties for infestation to S. lafertei, Khajuria and Sharma (1998) found no attack on Beauty, Laroda and Peach Plum, and in almond, Badamjor Spillo 3, Bruce, Hybrid 15, Ne Plus Ultra, Monaqua and Texas remained free from attack during the study period (1992–1994). Among prunes, Cacanska Rana and Secer were least susceptible. In Haryana, the pest was seen to damage 45 % peach trees. It attack trees of all sizes but worst affected are the older, weak, injured, unhealthy and sun burnt trees. Orchards located in southern aspect are severely damaged and potion of trees confined to southern sunny portion is attacked. Its damage is seen on stem and branches (2.5 cm to 58 cm thick). The cultivar Matchless was found most tolerant among screened peach cultivars in Haryana (Lakra et al. 1980). The damage is inflicted by larvae feeding on frsh bast and then boring into the sapwood. They make the feeding zone powdery and sometimes girdle branches and stem. Samll oval holes (1 cm long and 6 mm broad) are present at attacked portion. Adults are small (10–13 mm) coppery black beetles, commonly called jewel beetle, which feed on foliage.

In mid hills, the beetle completes two generations during April to October and larvae/prepupae of third generation overwinter. Beetles from overwintering generation emerge by mid March, feed on foliage and lay white spherical eggs singly in crevices, all over the tree trunk and main branches. The duration of egg, four larval instars and pupal stage is 16.1–18.5, 66.5–167, and 14.3–20.6 days. The neonate white larvae with a small yellow head and greatly enlarlgleld prothoracic segment bores into the bark and from the entry hle, a lump of resinous gum oozes out. As larva grows, it penetrates deeper into the sapwood. The full-fed grub (18–24 mm) excavates a small pupation chamber (4–6 × 2–4 mm and 1 cm deep) in the wood and pupates in it. In plains (Haryana), it completes 3–4 (including one in the winters) generations.

5.7.1 Management

Proper training and pruning of the trees such that the stem is not directly exposed to the sun provide some protection against this pest. Further, maintenance of tree vigour is essential. The dead or heavily infested branches with gum oozing out should be cut and destroyed. Sharma et al. (1999) reported that wrapping of 1 cm thick layer of dry grass or 1 layer of gunny burlap on exposed stem and branches right from February to November, provided complete protection of trees against this pest. If the incidence is high, spraying of infested branches with 0.1 % methyl parathion, fenitrothion or DDVP when egg laying begins, provide satisfactory control. Recently, Sharma et al. (2003) found that three applications of 0.04 % chlorpyrifos or lindane, synchronizing with the activity period of the pest, i.e. February, June and October, significantly reduced the incidence of the pest in Punjab.

Bark eating caterpillar, Indarbela quadrinotata Wlk. (Metarbelidae, Lepidoptera): It is a pest of subtropical fruits in India. Its incidence has been recorded on plum, peach, pear, pomegranate, Grewia, Bauhinia and to a lesser extent on apple and pear in mid hills up to 1500 m altitude in Himachal Pradesh. Among stone fruits, plum is the preferred host and a tree may bear about 13 holes confined to forking of branches or to depressions on the stem and branches that are more than 4 cm in diameter. Eggs are laid under loose bark or in cracks and crevices of bark in small clusters of 15–25 during April–June. Hatching occurs in 8–11 days and after 2–3 days, larvae bore into the trunk or main branches usually at forking places. During day time, larva remains in the undivided tunnel and comes out at night to feed on the bark under silken web interspersed with fine chips of wood and faecal pellets. The infested trees often suffer from canker. Full-fed larvae (46–52 mm) are with dark brown head and dirty brown body. They may feed up to April (larval duration of 9–10 months), when pupation occurs. Pupal period lasts for 3–4 weeks. Thus, it has only one generation in a year.

Old and weal trees in the neglected orchards are more susceptible to the attack. The control measures suggested for shoot borers would also be effective against this pest. Injecting dichlorvos or methyl parathion 0.05 % in the live borer holes or painting of the bark around the borer holels with monocroptophos 0.05 % has been found giving relief against this pest.

Peach Twig Borer, *Anarsia lineatella* Zeller (Gelechiidae, Lepidoptera) It is an important pest of peach, plum and apricot in different fruit growing areas. Tiny grayish moth is active at dusk, lays about 170 eggs singly or in small groups of 2–3 on shoots, twigs, fruit or leaves during life span of about a week. The egg, larval, prepupal and pupal period lasts for 6–11, 18–35, 3–9, and 6–14 days, respectively and completes the life cycle in 40–70 days. There are three generations in a year, the full-fed third instar larvae of third generation over-winter under flimsy silken webs or in silk-lined cracks on the barak. Laarvae resume feeding by end of Maarch and bore into the new growth (twigs or terminal buds) making small feedings galleries, which kill the new growth. The full-grown reddish-pink larva is 12–18 mm long. Pupation takes place on the bark in pupal cocoon. The second-generation larvae almost feed on fruit. The chemical control methods suggested for apple leaf folder would also take care of twig borer infestaltion. To initiate control measures, sex pheromone monitoring can be done.

5.8 Foliage Feeders

Insects feeding on the foliage of pome fruits also damage the foliage of stone and nut fruits and most of these belong to order Coleoptera, that too to Scarabaeidae family (e.g. cockchafers adoretus duvauceli Blanchard, A epipleuralis Arrow, A. horticola Arrow, Anomala dimidiate Hope, A. flavipes arrow, a. lineatopennis Blanchard, A. rufiventris Redtenbacher, Brahmina coriacea (Hope), Hoplosternus furicaudus (Ancy), Macronota quadrilineata Hope, Microtrichia cotesi Brenske, cherry green beetle Mimela passerine Arrow, etc.), Chrysomelidae (Altica spp., walnut blue beetle Monolepta erythrocephala Baly, almond chrysomelid Mimastra cyanura Hope, peach chrysomelid Hoplasoma sexmaculata Hope, etc.) and Curculionide (almond weevil Mylocerus laletivirens Marshal, grey weevil Mylocerus undecimpustulatus maculosus Desbrochers, plum weevil Amblyrrhinus poricollis Boheman, apricot weevil Emperorrhinus defoliator Marshall, etc.). Among lepidopteran defoliators, leaf roller Archips termias Meyrick (Tortricidae), tent hairy caterpillar Malacosoma indica (Walker) (Lasiocampidae), plum hairy caterpillar Euproctis fraternal (Moore), Indian gypsy moth Lymantria obfuscate Wlk., hairy caterpillar Dasychira horsfieldi Strand and D. mendosa Hubner (all belonging to Lymantriidae). In Himachal Radesh at high altitude (dry temperate zone) and in Kashmir, larvae of a lycaenid, the walnut blue Chaetoprocta odata Hewitson defoliate walnut trees. These defoliators feed on leaves and sometimes create complete defoliation. However, damage inflicted on the crop is not severe.

Some of these have been described under pome fruits and management strategy if needed remains the same. Besides foliage feeders, phytophagous mites also damage foliage in stone and nut fruits. The important species include two spotted spider mite Tetranychus urticae Koch and Oriental mite Eutetranychus orientalis (Klein).

Bibliography

Abrol DP (1988) Pollination studies in almond (*Prunus amygdalus* L.). Curr Sci 57(7):398–399

Abrol DP (1989) Insect pollinators of cherry (*Prunus avium* L.). J Anim Morphol Physiol 36(2):243–245

Abrol DP (1993) Intervarietal differences in pollinator attractiveness and fruit set in almond cultivars. In: Veeresh GK, Ganeshaiah KN, Umashankar (eds) Pollination in tropics. Proc Intl Symp Pollin Trop. IUSSI, Bangalore, pp 120–122

Abrol DP (1995) Energetics of nectar production in some almond cultivars as a predictor of floral choice by honeybees *Apis cerana indica* F. and *Apis mellifera* L. (Hymenoptera: Apidae). Proc Indian Nan Sci Acad B 61(4):285–290

Abrol DP (2005a) Foraging ecology of insect pollinators visiting cherry flowers and their impact on fruit production. Korean J Apicul 20(2):159–168

Abrol DP (2005b) Pollination energetics. J Asia Pacific Entomol 8(1):3–14

Abrol DP, Bhat AA (1989) New record of *Xylocopa valga* Gerstacker (Hymenoptera: Anthophoridae) from India. Curr Sci 58(1):41

Abrol DP, Bhat AK, Khan AR (1990) Factors affecting pollination activity of honeybee and fruit set in almond. In: Veeresh GK, Kumar ARV, Shiva-shankar T (eds) Social insects: an Indian perspective. IUSSI Indian Chapter/Oxford and IBH Publishing, Bangalore/New Delhi, pp 50–53

Abrol DP, Sharma D, Monobrullah M (2005) Abundance and diversity of pollinating insects visiting peach and plum flowers and their impact on fruit production SKUAST. J Res 4(1):38–45

Ahmad R, Muzaffar N, Makhdomi SMA, Khan S (1986) Honeybee research programme, Progress report 1977–85. Pak Agri Res Council, Islamabad, 172 pp

Alburquerque N, Burgos L, Egea J (2000) Consequences to fertilization of the developmental stage of apricot ovules at anthesis. J Hortic Sci Biotech 75(6):662–666

Alderman WH, Angelo E (1933) Self and cross sterility in plum hybrids. Am Soc Hortic Sci Proc 29:118–121

Allen FW (1929) Plum growing in California. Calif Agr Ext Sen Cir 34: 65 pp

Anderson DL (2000) Variation in the parasitic bee mite *Varroa jacobsoni* Dud. Apidologie 31:281–292

Anderson DL, Trueman JWH (2000) *Varroa jacobsoni* (Acari: Varroidae) is more than one species. Exp Appl Acarol 24:165–189

Anonymous (1970) Buzzzz. Almond Facts 35(1):27

Auchter EC, Knapp HB (1937) Orchard and small fruit culture, 3rd ed. Wiley, New York, 627 p

Austin PT, Hewett E, Noiton DA, Plummer JA (1996) Cross-pollination of sundrop apricot (*Prunus armeniaca* L.). NZ J Crop Horticul Sci 24(3):287–294

Backhouse W[O] (1911) Self-sterility in plums. Gard Chron 1296:299

Backhouse W[O] (1912) The pollination of fruit trees. Gard Chron 1352:381

Batra SWT (1982) Biological control in agroecosystems. Science 25:134–139

Batra SWT, Richards KW (1997) Fruit-pollinating bees of the Garhwal Himalaya, U.P., India. In: Proceedings of the 7th international symposium on pollination, Lethbridge, Alberta, 23–28 June 1996, pp 325–328

Benedek P, Nagy C (1995) Honeybee flower constancy on some fruit species. Hort Sci 27(3–4):38–42

Benedek P, Nyeki J, Szabo Z, Gulcan R (ed), Aksoy U (1995) Bee pollination of apricot: variety features affecting bee activity. In: Tenth international symposium on apricot culture, Izmir, 20–24 September 1993, Acta-Horticulturae No 384, pp 329–332

Benedek P, Nyeki J, Soltesz M, Erdos Z, Skola I, Szabo T, Amtmann I, Bakcsa F, Kocsis-Molnar G, Vadas Z, Szabo Z (2000) The effect of limitation of insect pollination period on the fruit and yield of temperate-zone fruit trees species. Int J Hor Sci 6(1):90–95
Benner B (1963). Fruit and vegetable facts and pointers: peaches. United Fresh Fruit and Vegetable Association, Washington, DC, 3d rev. and expanded ed. p 11 (total pagination not known)
Bergman P, Molau U, Holmgren B (1996) Micrometeorological impacts on insect activity and plant reproductive success in an Alpine environment, Swedish Lapland. Arctic Alpine Res 28:196–202
Bhalla OP, Verma AK, Dhaliwal HS (1983) Foraging activity of insect pollinators visiting stone fruits. J Entomo Res 7(2):91–94
Bohart GE (1951) Alfalfa seed growers of Utah should protect their wild bees. Utah Agric Expt Sta Farm House Sci 12:32–33
Bohart GE (1962) How to manage the alfalfa leaf-cutting bee (*Megacilile rotundata* Fabr.) for alfalfa pollination. Utah Agricultural Experiment Station, Circular 144
Boiler CA (1953) Pollination of stone fruits. Oreg State Hort Soc Proc 45:122–125
Bosch J (1992) Parasitism in wild and managed populations of the almond pollinator *Osmia cornuta* Latr (Hymenoptera: Megachilidae). J Apic Res 31:77–82
Bosch J (1994a) Improvement of field management of *Osmia cornuta* (Latreille) (Hymenoptera, Megachilidae). Apidologie 25:71–83
Bosch J (1994b) *Osmia cornuta* Latr. (Hymenoptera: Megachilidae) as a potential pollinator in almond orchards. Releasing methods and nest hole-length. J Appl Entomo 117:151–157
Bosch J (1994c) The nesting behavior of the mason bee *Osmia cornuta* (Latr) with special reference to its pollinating potential (Hymenoptera, Megachilidae). Apidologie 25:84–93
Bosch J (1995) Comparison of nesting materials for the orchard pollinator *Osmia cornuta* (Hymenoptera: Megachilidae). Entomo Gene 19:285–289
Bosch J, Blas M (1994) Foraging behaviour and pollinating efficiency of *Osmia cornuta* and *Apis mellifera*' on almond (Hymenoptera: Megachilidae and Apidae). Appl Entomo Zoo 29:1–9
Bosch J, Kemp WP (1999) Exceptional cherry production in an orchard pollinated with blue orchard bees. Bee World 80(4):163–173
Bosch J, Kemp WP (2000a) Development and emergence of the orchard pollinator, *Osmia lignaria* (Hymenoptera: Megachilidae). Environ Entomol 29:8–13
Bosch J, Kemp WP (2000b) Developmental biology and rearing methods for *Osmia* bees used as crop pollinators. In: Sommeijer MJ, de Ruijter A (eds) Insect pollination in greenhouses. CIP-DATA Koninklijke Bibliotheek, Den Haag, pp 119–126
Bosch J, Kemp MP (2001) How to manage the blue orchard bee, *Osmia lignaria*, as an orchard pollinator. Sustainable Agriculture Network, Washington, DC
Bosch J, Kemp WP (2002) Developing and establishing bee species as crop pollinators: the example of *Osmia* spp. (Hymenoptera: Megachilidae) and fruit trees. Bull Entomol Res 92(1):3–16
Bosch RV, Horn R, Mattteson P, Frazer BD, Messenger PS (1979) Biological control of the walnut aphid in California: impact of parasite *Trioxys pallidus*. Hilgardia 47:1–13
British Columbia Ministry of Agriculture, Fisheries and Food (1994) Tree fruit production guide for commercial growers in interior districts. Victoria
Brittain WH (1933) Field studies in the role insects in apple pollination. Bulletin of Agricultural Canada, Ottawa, No 182, pp 91–157
Brown AG (1951) Factors affecting fruit production in plums. Fruit Year Book 1950(4):12–18
Brown K (1968) Pollination of royal Ann (a-10) in the Willamette valley. Polk County (Oregon) Agr Ext Serv Agent, 4 pp. [Mimeographed.]
Buban T (1996) The effect of temperature and environmental pollution on fertilization. In: Nyeki J, Soltesz M (eds) Floral biology of temperate zone fruit trees and small fruits. Akademiai Kiado, Budapest, pp 170–172
Buchanan RE (1903) Contribution to our knowledge of the development of *Prunus americana*. Iowa Acad Sci Proc 11:77–93
Buchmann SL, Nabhan GP (1996) The forgotten pollinators. Island Press, Washington, DC, 292 pp

Bulatovic S, Konstantinovic B (1962) The role of bees in the pollination of the more important kinds of fruit in Serbia. In: Burtner JC (eds) 1st international symposium on pollination Proc., Copenhagen, Aug 1960. Commun 7, Swedish Seed Growers' Assn, pp 167–172. 1923. Latest cherry pollination studies. Better Fruit 182: 5–6, 23–24

Burtner JC (1923) Latest cherry pollination studies. Better Fruit 182(5–6):23–24

Burrill RM, Dietz A (1981) The response of honeybees to variations in solar radiation and temperature. Apidologie 12:319–328

Cane JH (1997) Ground-nesting bees: the neglected pollinator resource for agriculture. Acta Hor 437:309–324

Cane JH (2001) Exotic non-social bees (Hymenoptera: Apoidea) in North America: ecological implications. In: Strickler K, Cane JH (eds) For non-native crops, whence pollinators of the future? Thomas Say Publications in Entomology: Proceedings, Entomological Society of America, Lanham (in press)

Chandler WH (1957) Deciduous orchard. Henry, Kimpton

Claypool LL, Overley FL, Overholser EL (1931) Washington sweet cherry pollination studies in 1931. In: 27th Ann. Mtg. Wash. State Hort. Assoc. Proc., December 1, 2, and 3 at Yakima, Wash., pp 171–174

Claypool LL, Overley FL, Overholser EL (1932) Sweet cherry pollination in Washington for 1931. Amer Soc Hort Sci Proc 28:67–70

Coe FM (1934) Cherry pollination studies in Utah. Utah Agr Expt Sta Bul 245, 53 pp

Conners CH (1922a) Peach breeding a summary of results. Amer Soc Hort Sci 19th Ann Mtg Proc:108–115

Conners CH (1922b) Fruit setting of the J. H. Hale Peach. Amer Soc Hort Sci 19th Ann Mtg Proc:147–151

Conners CH (1926) Sterility in peaches. Hort Soc NY Mem 3:215–221

Cook AJ, Green EC (1894) Symposium on bees and fruit-fertilization, again. Gleanings Bee Cult 22:448–451

Coote G (1895) Fruits and vegetables. Oreg Agr Expt Sta Bul 34:17–32

Corbet SA, Williams IH, Osborne JL (1991) Bees and the pollination of crops and wild flowers in the European Community. Bee World 72:47–59

Corner J, Lapins KO, Arrand JC (1964) Orchard and honeybee management planned tree-fruit pollination. Min Agr, Victoria, British Columbia, Apiary Cir 14, 18 pp

Crane E (1991) *Apis species* of tropical Asia as pollinators, and some rearing methods for them. Acta Hort 288:29–48

Crane E, Walker P (1983) The impact of pest management on bees and pollination. Published by Tropical Development and Research Institute, London, 232 pp

Crane E, Walker P (1984) Pollination directory for world crops. London: Int Bee Res Assoc, 184 pp

Cullinan EP (1937) Improvement of stone fruits. US Dept Agr Year Book 1937:665–748

Detjen LR (1945) Fruitfulness in peaches and its relationship to morphology and physiology of pollen grains. Del Agr Expt Sta Bul 257 (Tech Bul 34), 24 pp

Dickson GH, Smith MV (1953). Fruit pollination. Ontario Agr Col Cir 172, 6 pp

Eaton GW (1959) A study of the megagametophyte in *Prunus avium* and its relation to fruit setting. Can J PI Sci 39:466–476

Eaton GW (1962) Further studies on sweet cherry embryo sacs in relation to fruit setting. Rep Hort Exp Stn Prod Lab, Vineland

Einset O (1932) Experiments in cherry pollination. NY Agr Expt Sta (Geneva) Bul 617, 13 pp

Eisikowitch D, Galil J (1971) Effect of wind on the pollination of *Pancratium maritimum* L. (Amaryllidaceae) by hawkmoths (Lepidoptera: Sphingidae). J Anim Ecol 40:673–678

Eves JD, Mayer DF, Johansen CA (1980) Parasites, predators and nest destroyers of the alfalfa leafcutting bee, *Megachile rotundata.* Washington State University, Agricultural Experiment Station, Western Regional Extension Publication 32, p 15

Faust M (1989) Physiology of temperate zone fruit trees. Wiley, New York

Ferreres LP (1947) [polinizacion del almendro por los insectos.] Rancho mex. 3(18):17–20, 75. [in spanish.]
Free JB (1960) The pollination of fruit trees. Bee World 41:141–169
Free JB (1962) The effect of distance from pollinizer varieties on the fruit set on trees in plum and apple orchards. J Hortic Sci 37:262–271
Free JB (1993) Insect pollination of crops, 2nd edn. Academic Press, London
Free JB, Spencer-Booth Y (1964) The effect of distance from pollenizer varieties on the fruit set of apple, pear and sweet-cherry trees. Jour Hort Sci 39:54–60
Gagnard JM (1954) Systematic characters and the sterility in almonds cultivated in Algeria.] Algeria Inst Agr Ann 8(2): 163. [In French] Abstract in Bee World 39:192
Gardner VR (1913) A preliminary report on the pollination of the sweet cherry. Oreg Agr Expt Sta Bul 116, 37 pp
Gould HP (1939) Why fruit trees fail to bear. US Dep Agric Leafl 172
Grieve P (1879) Bees as fertilizing agents. Gard Chron 11:204
Griggs WH (1970) The status of deciduous fruit pollination. In the indispensable pollination. Ark Ag Ext Serv Misc Pub 127:185–210
Griggs WH, Hesse CO (1963). Pollination requirements of Japanese plums. Calif Agr Expt Stat Ext Serv Leaflet 163, n.p
Griggs WH, Iwakiri BT (1960) Orchard tests of beehive pollen dispensers [traps] for cross-pollination of almonds, sweet cherries, and apples. Amer Soc Hortic Sci Proc 75:115–128
Griggs WH, Vansell GH, Iwakiri BT (1952) The use of beehive pollen dispensers in the pollination of almonds and sweet cherries. Proc Am Soc Hortic Sci 60:146–150
Guleryuz M, Bolat I, Karayiannis I (1999) Investigation on characteristics of apricot cultivars in Erzincan – Turkey. In: Proceedings of the XIth international symposium on apricot culture, Veria-Makedonia, Greece, 25–30 May 1997, Acta-Horticulturae. 1(488), pp 139–142
Gupta PR, Thakur JR (1993) Sexual generation and overwintering of the peach leaf curling aphid *Brachycaudus helichrysi* (Kalt.) in Himachal Pradesh, India. Ann Appl Bio 122(2):215–221
Gupta JK, Reddy MCM, Kumar J (1990) Pattern of nectar secretion in wild cherry, Prunus puddum Roxb, and the associated foraging behaviour of Apis cerana indica F and *Apis mellifera* L. Apidologie 21:11–16
Heard TA (1999) The role of stingless bees in crop pollination. Annu Rev Entomol 44:183–206
Hendrickson AH (1916) The common honeybee as an agent in prune pollination. Calif Agr Expt Sta Bul 274:127–132
Hendrickson AH (1918) The common honeybee as an agent in prune pollination. Calif Agr Expt Sta Bul 291:215–236
Hendrickson AH (1919a) Plum pollination. Calif Agr Expt Sta Bul 310, 28 pp
Hendrickson AH (1919b) Five years results in plum pollination. Amer Soc Hort Sci Proc 15:65–66
Hendrickson AH (1922a) Wonder work of bees. They make millions for the fruit growers. Bees that returned to the orchardist $100 per colony. Gleanings Bee Cult 50:226–229
Hendrickson AH (1922b) Further experiments in plum pollination. Calif Agr Expt Sta Bul 352:247–266
Hendrickson AH (1923) Prune growing in California. Calif Agr Expt Sta Bul 328, 38 pp
Hendrickson AH (1930) The essentials of plum pollination. Blue Anchor [Sacramento] 7(2):8–9, 31–32
Hooper CH (1930) The study of pollination in relation to cherry orchards. Gard Chron 88(2293):475–476
Hooper CH (1936) Plums; notes on their pollination, order of flowering of varieties and insect visitors to the blossoms. Jour Sol-East Agr Col [Wye, Kent] 38:131–140
Hootman HD (1935) Importance of pollination and the honey-bee in fruit yields. Midwest Fruitman 8(9):3–4, 9–10
Inouye DW (1980) The terminology of floral larceny. Ecology 61:1251–1253
Jorgensen C, Drage CM (1953) Pollination of Colorado fruit. Colo Agr Expt Sta and Ext Serv Bul 427-A, 13 pp

Jusubov AM (1957) Pollinators for new apricot varieties in the central belt. Sad i Ogorod 2:47–48. [In Russian.]
Kanato K, Yoshida M, Kurihara A, Makino Y (1967) Studies on pollen sterility of peach. Hiratsuka Hort Res Sta Bul Ser A, 6:91–104. [In Japanese, English tables and summary.]
Kearns CA, Inouye DW (1993) Techniques for pollination biologists. University Press of Colorado, Niwot
Kelly BW (1964) Factors relating to the cost of producing peaches in Pennsylvania, 1959–63. Pa Agr Ext Serv Farm Mangt Pub 19, 20 pp
Kevan PG (1988) Pollination: crops and bees. Ontario Ministry of Agriculture and Food, Publication 72
Kevan PG, Baker HG (1983) Insects as flower visitors and pollinators. Annu Rev Entomol 28:407–453
Khajuria DR, Sharma JP (1998) Field screening of different cultivars of plum, prunes and almonds against flat-headed peach tree borer, Sphenoptera lafertei Thomson. J Insect Sci 11(2):159–161
Khan KS, Rahman A (1930) Some observations on the pollination of peaches (*Prunus persica* Benth. and Hook.). Agric J India 25(6):492–494
Kinman CF (1924) Plum and prune growing in the pacific states. US Dept Agr Farmers' Bul 1372, 59 pp
Kinman CF (1938) Plum and prune growing in the pacific states. US Dept Agr Farmers' Bull 1372, 55 pp
Kinman CF (1943) Plum and prune growing in the pacific states. US Dept Agr Farmers' Bul 1372, rev, 55 pp
Kosior A (1987) Impact of economic activity upon bumblebee Bombus Latr population in the west Bieszczady Mts. Ochrona Przyrody 45:239–262
Kostina KF (1966) Stepen' samooplodnosti sortov i gibridov abrikosa različnyh ekologo-geografičeskih grupp. Sel'skohozyaystvenaya Bio 1:352–355
Krombein KV (1967) Trap-nesting wasps and bees: life histories, nests and associates. Smithsonian Press, Washington, DC
Kumar J, Rao KVK, Gupta PR, Dogra GS (1989) Temporal distribution of bees (Hymenoptera: Apoidea) on spring flowering crops in mid hills of Himachal Pradesh, India. Indian Bee J 51(2):55–58
Lakra RK, Kharub WS, Singh Z (1980) Comparative efficacy of some banding materials against mango mealybug, Drosicha mangiferae Green in Haryana. Indian J Ent 42(2):170–176
Langridge DF, Jenkins PT, Goodman RD (1977) A study on pollination of dessert peaches cv "Crawford". Aus J Agr Res Ani Hus 17:697–699
Larsson FK, Tengo J (1989) The effects of temperature and body size on the mating pattern of a gregariously nesting bee, *Colletes cunicularius* (Hymenoptera: Colletidae). Ecol Entomol 14:279–286
Levin MD (1986) Using honey bees to pollinate crops. US Department of Agriculture, Leaflet 549
Luce WA, Morris OM (1928) Pollination of deciduous fruits. Wash Agr Expt Sta Bul 223, 22 pp
Macdaniels LH (1942) Notes on the pollination of the Italian prune. Am Soc Hortic Sci Proc 40:84–86
Macdaniels LH, Heinicke AJ (1929) Pollination and other factors affecting the set of fruit with special reference to the apple. NY (Cornell) Agr Expt Sta Bul 497, 47 pp
Maeta Y (1978) Comparative studies on the biology of the bees of the genus Osmia in Japan, with special reference to their management for pollination of crops (Hymenoptera, Megachilidae). Bull Tohoku Nat Agric Exp Stn &num 57, 221 pp
Maeta Y, Kitamura T (1974) How to manage the Mame-Ko bee (Osmia cornifrons, Radoszkowski) for pollination of fruit crops. Ask Co Ltd, 16 pp
Mahanoglu G, Eti S, Gulcan R (ed), Aksoy U (1995) Effects of artificial pollination on the fruit set level and fruit quality in some early ripening apricot cultivars. In: Tenth international symposium on apricot culture, Izmir, Turky, 20–24 September 1993, Acta Horticulturae No. 384, pp 397–400
Mann GS, Singh G (1981a) A note on activity and abundance of flower visiting insects of peach (Prunus persica L.) at Ludhiana (Punjab). Prog Hort 13(3/4):25–27
Mann GS, Singh G (1981b) Activity and abundance of flower visiting insects of almond (Prunus amygdalus Matsch) at Ludhiana (Punjab). J Bombay Nat Hist Soc 78(3):617–6 18

Marquez J, Bosch J, Vicens N (1994) Pollens collected by wild and managed populations of the potential pollinator *Osmia cornuta* Latreille. J Appl Entomo 117:353–359

Marshall G (1920) Principles of economics, 8th edn. Mac Millan, London

Marshall RE, Johnston S, Hootman HD, Wells HM (1929) Pollination of orchard fruits in Michigan. Mich Agr Expt Sta Spec Bul 188, 38

Martinez del Rio C, Burquez A (1986) Nectar production and temperature dependent pollination in *Mirabilis jalapa* L. Biotropica 18:28–31

Masoodi AA, Bhagat KC, Koul VK, Bhat OK (1987) Seasonal incidence of dusky veined walnut aphid Callaphis juglandis Kaltenbach Aphididae Homoptera in Kashmir Valley. Environ Ecolo (Kalyani) 5(3):572–574

Masoodi MA, Koul VK, Bhagat KC, Bhat OK (1989) Bioecology of dusky veined walnut aphid callaphis juglandis on walnut in Kashmir, India. Indian J Plant Prot 17(1):139–142

Mattu VK, Chaudhary DK, Kumar L (1994) Foraging ecology of *Apis cerana* F.and *Apis mellifera* L. in pollinating stone fruit. PEST Manag Eco Zoo 2(1):35–39

Mayer DF (1980) Honey bees and peaches. Research Report of the West Orchard Pesticide Conference 54, 26

Mayer DF, Johansen CA (1988) How to reduce bee poisoning. Washington State cooperative extensions Leaflet No. 15. WREP, pp 12

Mayer DF, Lunden JD, Kious CW (1988a) Effects of dipping alfalfa leafcutting bee nesting materials on chalkbrood disease. Appl Agric Res 3:167–169

Mayer DF, Lunden JD, Rathbone L (1988b) New ideas in cherry pollination. Proc Wash State Hortic Assoc 83:228–229

Mayer DF, Johansen CA, Burgett M (1986). Bee pollination of tree fruits. Publication, Pacific Northwest Cooperative Extension, No PNW 0282

Mayer DF, Britt RL, Lunden JD (1989) Evaluation of BeeScent as a honey bee attractant. Am Bee J 129(1):41–42

McGregor SE (1976) Insect pollination of cultivated crop plants. USDA, Agri Handbook No 496, 411 pp

McLaren GF, Fraser JA, Grant JE, Gulcan R, Aksoy U (1995) Pollination compatibility of apricots grown in central, Otago, New Zealand. International Society for Horticultural Science, The Hague

McLaren GF, Fraser JA, Grant JE (1996) Some factors influencing fruit set in cv Sundrop apricot. NZ J Crop Hort Sci 24:55–63

Monzon VH (1998) *Biologia de Osmia cornuta* Latr. (Hymenoptera: Megachilidae) y su utilizacion como polinizador de peral (Pyrus communis). Ph.D. thesis, University of Barcelona, Barcelona

Morrill FL (1899) Bees and fruit. Gleanings Bee Cult 27:430–431

Murneek AE (1930) Fruit pollination. Mo Agr Expt Sta Bul 283, 12 pp

Murneek AE (1937) Pollination and fruit setting. Mo Agr Expt Sta Bul 379, 28 pp

Muttoo RN (1950) Honey bees and fruit crops. Ind J Hort 7(3/4):17–20

Muttoo RN (1953) Flora of Jeolikote. Indian Bee J 15:49–54

Naumann K, Currie RW, Isman MB (1994) Evaluation of the repellent effects of a neem insecticide on foraging honey bees and other pollinators. Can Entomol 126:225–230

Nevkryta AN (1957) [Distribution of apiaries for pollinating cherries.] PchelovodstVo 34(4):34–38. [In Russian.] AA-373158

Nyeki J, Szabo Z (1996) Fruit set of self and pollen pollinated peach flowers under Hungarian ecological conditions. In: Acta Horticulturae – Proceedings of the third international peach symposium. ISHS Secretariat, Leuven, pp 177–180

Parker FD, Batra SWT, Tepedino VJ (1987) New pollinators for our crops. Agric Zool Rev 2:279–304

Philip GL (1930) Cherry culture in California. Calif Agric Ext Serv Cir 46, 43 pp

Philp GL (1947) Cherry culture in California. Rev Calif Agr Ext Serv Cir 46

Philp GL, Vansell GH (1932) Pollination of deciduous fruits by bees. Calif Agr Ext Serv Cir 62, 26 pp

Philp GL Vansell GH (1944) Pollination of deciduous fruits by bees. Calif Agr Ext Serv Cir 62, rev, 26 pp
Purdie JD, Winn RA (1964) Almond pollination-honey bee activity. So Austral Agr 68(5):152–157
Purdie JD, Winn RA (1965) Almond pollination: honey bee activity. Australian Bee J 46:17–19
Rana RS, Verma LR, Mattu VK (1994) Foraging activity and abundance of insect pollinators on plum bloom. In: Pollination biology-environmental factors and pollination. Rajendra Scientific Publishers, Hisar, Haryana, .pp 20–31
Randhawa GS, Yadav IS, Nath N (1963) Studies on flowering, pollination and fruit development in peach grown under subtropical conditions. Indian J Agr Sci 33(2):129–138
Roberts RH (1922) Better cherry yields. Wis Agr Expt Sta Bul 344, 30 pp
Roberts RH (1945) Blossom structure and setting of 'Delicious' and other apple varieties. Proc Am Soc Hortic Sci 46:87–90
Roberts D, Congdon NB (1955) The relationship of nectar secretion (volume) and sugar concentration to insect pollination of plums (*Prunus* spp.). NZ J Sci Tech Sect A 37(3):196–206
Robinson WS (1979) Effect of apple cultivar on foraging behaviour and pollen transfer by honey bees. J Am Soc Hort Sci 104:596–598
Schultz JH (1948) Self-incompatibility in apricots. Am Soc Hort Sci Proc 51:171–174
Schuster CE (1925) Pollination and growing of the cherry. Oreg Agr Col Expt Sta
Scott-Dupree CD, Winston ML (1987) Wild bee pollinator diversity and abundance in orchard and uncultivated habitats in the Okanagan Valley, British Columbia. Can Entomol 119:735–745
Scott-Dupree C, Winston M, Hergert G, Jay SC, Nelson D, Gates J, Termeer B, Otis G (eds) (1995) A guide to managing bees for crop pollination. Canadian Association of Professional Apiculturalists
Sharma PL (1961) The honeybee [*Apis indica]* population among insects visiting temperate-zone fruit flowers and their role in setting fruit. Bee World 42:6–7
Sharma HK, Gautam DR (1994) Evaluation of some almond root stocks for tolerance to flat headed peach tree borer *Sphenoptera laferti* Thompson. Pest Manag Econ Zoo 2:153–155
Sharma HK, Chander R, Thakur JR (1990) Incidence of flat headed peach tree borer *Sphenoptera laferti* Thompson(Coleoptera: Buprestidae) on stone fruits in himachal Pradesh. Indian J Agric Sci 60:785–787
Sharma HC, Mukuru SZ, Mayasa E, Were J (1999) Breakdown of resistance to Sorghum midge *Stenodiplosis* sorghicola. Euphytica 109:131–140
Sharma HC, Taneja SL, Kameshwara Nao N, Prasada Rao KE (2003) Evaluation of sorghum germplasm for resístance to insects. Information Bulletin No. 63. International Crops Research Institute for the Semi-Arid Tropics (ICRISAT), Patancheru 502 324, Andhra Pradesh, India, 184pp.
Sheesley B, Poduska B (1970a) Grading bee colony strength. Almond Facts 35(5):22–24
Sheesley B, Poduska B (1970b) Strong honeybee colonies prove value in almond pollination. Calif Agr 24(8):4–6
Sheesley B, Poduska B (1970c) Relative values of honeybees colonies for almond pollination. Gleanings Bee Cult 98(8):486–491
Shoemaker JS (1928).Cherry pollination. Ohio Agr Expt Sta Bul 422, 34 pp
Singh S (1988) Insect pollination studies on Almond. Ph.D. thesis, Dr. Y.S. Parmar University of Horticulture and Forestry, Solan
Skrebtsova ND, Iakovlev AS (1959) [Effectiveness of saturated pollination of cherries by bees.] Pchelovodstvo 36(5):25–26. [In Russian.] AA-154/61
Slate GL (1970) Apricots, nectarines and almonds. Horticulture 48(5):42, 47–48
Soltesz M (1996) Requirements for successful fruit set in orchards. In: Nyeki J, Soltesz M (eds) Floral biology of temperate zone fruit trees and small fruits. Akademiai Kiado, Budapest, pp 257–286
Srivastava RP, Singh I (1970) Floral biology, fruit-set, fruit-drop, and physico-chemical characters of sweet-cherry (Prunus avium L.). Indian J Agr Sci 40:400–420
Standifer L, McGregor SE (1977) Using honey bees to pollinate crops. US Department of Agriculture, Leaflet 5
Stanger W, Thorp R, Foote L (1973) Honeybee pollination in California. Div Agri Sci, Uni Calif Leaflet No 75-LE, 9 pp

Stark AL (1944) Fruit pollination – a problem in Utah. Farm and Home Sci 5(4):5–6
Stephen WP (1955) Alfalfa pollination in Manitoba. J Econ Entomol 48:543–548
Stephen WP (1961) Bees and pollination of stone fruits. Oreg State Hort Soc Ann Rpt 53:78–79
Stephenson AG (1981) Flower and fruit abortion: proximate causes and ultimate functions. Ann Revof Ecol Sys 12:253–279
Stewart, N., Luckwill, L. C., Mealy, A. G., and others. 1967. The pollination of fruit crops. Sci Hort 14 and 15:1–68
Szabo Z, Nyeki J, Andrasfalvy A, Szalay L, Pedryc A, Karayiannis I (1999) Evaluation of some Romanian apricot varieties in Hungary. In: Proceedings of the XIth international symposium on apricot culture, Veria-Makedonia, Greece, 25–30 May 1997, Acta-Horticulturae, 1(488), p 211–214
Szabo Z, Nyeki J, Benedek P, Erdos ZM, Soltesz M (2000) Bee pollination and association of apricot varieties. Int J Hort Sci 6(4):20–24
Tepedino VJ, Torchio PF (1989) The influence of nest-hole selection on sex ratio and progeny size in *Osmia lignaria propinqua* (Hymenoptera: Megachilidae). Ann Entomol Soc Am 82:355–360
Thompson MM, Liu LJ (1972) Pollination and erratic bearing in 'italian prunes' Amer. Soc Hort Sci Proc 97:489–491
Thompson M, Brenner D, Anwer R, Ahmed Z, Bhatti S (1988) Biogeographic survey and collection of temperate fruit and nut genetic resources in Northern Pakistan. IUCN, Gilgit, 105 pp
Thorp RW, Mussen E (1979) Honey bees in almond pollination. University of California Cooperative Extension Service, Leaflet 2465, University of California, Oakland
Torchio PF (1976) Use of *Osmia lignaria* Say (Hymenoptera: Apoidea: Megachilidae) as a pollinator in an apple and prune orchard. J Kan Entomol Soc 49:475–482
Torchio PF (1981a) Field experiments with the pollinator species, *Osmia lignaria propinqua* Cresson in almond orchards: I, 1975 studies (Hymenoptera: Megachilidae). J Kan Entomol Soc 54:815–823
Torchio PF (1981b) Field experiments with the pollinator species, *Osmia lignaria propinqua* Cresson in almond orchards: II, 1976 studies (Hymenoptera: Megachilidae). J Kan Entomol Soc 54:824–836
Torchio PF (1982a) Field experiments with the pollinator species, *Osmia lignaria propinqua* Cresson in apple orchards: I, 1975 studies (Hymenoptera: Megachilidae). J Kansas Entomol Soc 55:136–144
Torchio PF (1982b) Field experiments with the pollinator species, *Osmia lignaria propinqua* Cresson in apple orchards: II, 1976 studies (Hymenoptera: Megachilidae). J Kansas Entomol Soc 55:759–778
Torchio PF (1984a) Field experiments with the pollinator species, *Osmia lignaria propinqua* Cresson (Hymenoptera: Megachilidae) in apple orchards: III, 1977 studies. J Kansas Entomol Soc 57:517–521
Torchio PF (1984b) Field experiments with the pollinator species, *Osmia lignaria propinqua* Cresson (Hymenoptera: Megachilidae) in apple orchards: IV, 1978 studies. J Kansas Entomol Soc 57:689–694
Torchio PF (1987) Use of non-honey bee species as pollinators of crops. Proc Entomol Soc Ontario 118:111–124
Torchio PF (1989) In-nest biologies and development of immature stages of three Osmia species (Hymenoptera, Megachilidae). Ann Entomol Soc Am 82:599–615
Torchio PF (1990) Diversification of pollination strategies for U.S. crops. Environ Entomol 19:1649–1656
Torchio PF (1991) Bees as crop pollinators and the role of solitary species in changing environments. Sixth Int Symp on Pollination. Acta Hortic 288, pp 49–61
Torchio PF, Asensio E (1985) The introduction of the European bee, *Osmia cornuta* Latr, into the U.S. as a potential pollinator of orchard crops, and a comparison of its manageability with *Osmia lignaria propinqua* Cresson (Hymenoptera: Megachilidae). J Kan Entomol Soc 58:42–52
Torchio PF, Asensio E, Thorp RW (1987) Introduction of the European bee, *Osmia cornuta,* into California almond orchards (Hymenoptera: Megachilidae). Environ Entomol 16:664–667

Traynor J (1993) Almond pollination handbook. Kovak Books, Bakersfield
Tufts WP, Philp GL (1925) Pollination of the sweet cherry. Calif Agr Expt Sta Bul 385, 28 pp
van den Bosch R, Hom R, Matteson P, Frazer BD, Messenger PS, Davis CS (1979) Biological control of the walnut aphid in California: impact of the parasite Trioxys pallidus. Hilgardia 47:1–13
Vansell GH (1942) Factors affecting the usefulness of honeybees in pollination. Circ US Dept Agric 650:1–31
Vansell G, Griggs WH (1952) Honey bees as agents of distribution. USDA Yearbook:88–107
Vermeulen L, Pelerents C (1965) Effect of the honeybee on fruit setting. Fruitrev [Belgium]:1–4. [In Dutch.] AA-792/71
Vicens N (1997) Potencial pol. linitzador en pomeres i inversio parental en l'abella solitaria *Osmia cornuta* (Latreille) (Hymenoptera, Megachildae). Ph.D. dissertation, Universitat de Barcelona, Barcelona
Vicens N, Bosch J (2000a) Pollinating efficacy of *Osmia cornuta* and *Apis mellifera* (Hymenoptera: Megachilidae, Apidae) on 'Red Delicious' apple. Environ Entomol 29:235–240
Vicens N, Bosch J (2000b) Weather-dependent pollinator activity in an apple orchard, with special reference to *Osmia cornuta* and *Apis mellifera* (Hymenoptera: Megachilidae and Apidae). Environ Entomol 29:413–420
Vicens N, Bosch J, BIas M (1993) Analisis de Ios nidos de algunas *Osmia* (Hymenoptera: Megachilidae) nidificantes en cavidades preestablecidas. Orsis 8:41–52
Vij V, Harish K, Mehrotta N, Aulakh P, Thind S, Arora P (1999) Evaluation of almond varieties under arid-irrigated conditions of Punjab. Indian J Hort 56:233–237
Waugh FA (1898) Pollination of plums. Vt Agr Expt Sta 11th Ann Rpt 1897–98: pp 238–262
Webster TC, Thorp RW, Briggs D, Skinner J, Parisian T (1985) Effects of pollen traps on honey bee (Hymenoptera: Apidae) foraging and brood rearing during almond and prune pollination. Environ Entomol 14:683–686
Weiss K (1957) The dependence of the cherry harvest on the number of colonies present. Deut Bienenw 8(7):124–126, German
Wellington Westwood MN (1993) Temperate-zone pomology, physiology and culture. Tllnber Press, Portland
Williams RR (1969) Factors affecting pollination in fruit trees. In: Luckvill LC, Cutting LV (eds) Physiology of tree crops. Academic Press, London, pp 193–207
Williams PH (1982) The distribution and decline of British bumble bees (*Bombus* Latr.). J Apicul Res 21:236–245
Williams JT (1991) Seasonal light requirements in a fruit orchard. Fruit Notes 56:20–21
Williams IH (1994) The dependence of crop production within the European Union on pollination by honey bees. Agri Zoo Rev 6:229–257
Winston ML (1987) The biology of the honey bee. Harvard University Press, Cambridge, MA
Woodrow AW (1932) The comparative value of different colonies of bees in pollinati0n. Jour Econ Ent 25:331–336
Yakovleva LP (1975) Utilization of bees for pollination of entomophilous farm crops in the USSR. In: Proceedings of III international symposium pollination, Prague, 2, pp 199–208
Yamada M, Oyama N, Sekita N, Shirasaki S, Tsugawa C (1971) The ecology of the megachilid bee *Osmia cornifrons* and its utilization for apple pollination. Bull Aom Apple Exp Stan 15:1–80 (in Japanese)
Yokozawa Y, Yasui A (1957) Studies on the pollination of peach.] 1. Insect visitors to the flowers of peach. Hort Assoc Jap Jour 26(3):185–191. [In Japanese, English title and summary.]
Zimmerman M, Pyke GH (1988) Reproduction in *Polemonium*: assessing the factors limiting seed set. Am Nat 131:723–738

Chapter 6
Nuts

6.1 Nut Species

Fruits with hard shell with an edible kernel are such as chestnut, persimmon, walnut, olive, pecan, hazel nuts and pistachio are called nuts. They have imperfect flowers and produce very small dry pollen grains which float on air. Their stigmatic surface is large and they produce very large number of pollen grains per ovule. The inadequate knowledge about pollination mechanism has been a limitation for its genetic improvement. Therefore, there is need to understand the reproductive mechanisms in these nuts to improve them through breeding techniques. They include

6.2 Chestnut *Castanea* spp., Family Fagaceae

Chestnut is a deciduous tree having pistillate and staminate flowers borne on the same tree at different locations. Staminate flowers appear on cylindrical catkins with 10–20 stamens and 6-parted calyx. Staminate catkins appear first near the terminal portion of the shoot and the pistillate flowers arise at the base of one or more bisexual catkins near the upper portion of the shoot (Westwood 1978). Three pistillate flowers surrounded by multibracted involucres make up an involucre, each floret capable of producing three nuts. The staminate flowers shed pollen before the pistillate flowers open. The female flowers are usually ready to receive the pollen when the male flowers of the bisexual catkins shed the pollen. However, distinct flowering pattern is observed in different cultivars which remains constant in each season (Zimmerman 1972).

D.P. Abrol, *Pollination Biology, Vol. 1*, DOI 10.1007/978-3-319-21085-8_6

6.2.1 Pollination Requirements

Chestnut is considered to be self sterile (Reed 1941). McKay (1939) reported male sterility in *C. crenata* tree and found that flowers of *C. mollissima* set 1.3 %, 34.9 % and 68.1 % fruit when self, cross and open pollinated, respectively. Indicating the need for pollen transfer between trees. Crane et al. (1937) also found that chestnuts are more or less self sterile but set better when interplanted with other cultivars.

6.2.2 Pollinators

In chestnut large quantities is pollen is produced which is carried by wind (Crane et al. 1937; Clapper 1954). However, several investigators have suggested that bees visit the staminate flowers for both nectar and pollen (Hazslinszky 1955; McKay 1939). McKay (1972) stated that honeybees are highly beneficial in the transfer of pollen and visit the staminate flowers in large numbers. McGregor (1976) observed half the population of insects on chestnut flowers as honeybees due to the presence of sweet chestnut woods all around the plantation. In general honeybees tend to disregard chestnut flowers if more attractive flowers are available in the vicinity. The relative abundance of other insects such as butterflies, syrphid flies and beetles occurred in a similar way as observed in other wind pollinated plants (Porsch 1950).

6.2.3 Pollination Recommendations and Practices

Sokolov and Chernyshov (1980)in USSR stated that honeybee pollination improved nut yield and they recommended 1.5 colonies per hectare. Watanabe et al. (1964) reported that fruit set was greatest in rows adjacent to pollenizer rows than on the decreasing 3d to 10th rows. They recommended that pollenizer cultivars be set in the ratio of 1:1 or 1:2 of the main cultivar. It is this evident thus chestnut is both wind and insect pollinated.

6.2.4 Pests of Chestnut

Chestnuts have relatively few pests of minor importance as compared to many other perennial trees. The most common pests include weevils, beetles, mites, leaf hoppers, leaf miners, gall forming wasps etc. Two species of weevils have been reported attacking chestnuts. They are mottled brownish coloured with nettle like beak. The large weevil is ¾ in. long whereas small weevil is half of that length. The adults do not cause any significant damage. The females use their long beaks to drill holes into the bur reaching the developing nuts and deposit eggs there. The developing larvae (grubs) feed on meat of the nuts rendering unfit for human consumption. After attaining maturity in October–November, the full grown grubs leave the nuts through a exit hole and enter the soil for pupation. The best way to reduce infestation is to pick up the dropped fruits and place them in containers so that emerging grubs do not escape. Another most important pest of chestnut is Japanese beetle which causes considerable defoliation by its feeding activities during July–August. The best manage these beetles is to pick up them early in the morning before commencement of their flight, however, this method has the limitation when dealing with large trees. Besides, a number of other insect pests attack chestnut in different parts of the world. Their nature of damage and distribution is given in Table 6.1

6.3 Persimmon

The persimmon, *Diospyros kaki* L. F., (Family Ebenaceae) trees are dioecious, with single pistillate flowers and usually three staminate flowers in a group. Generally, a male tree must pollinate a female tree before fruit can form. However, sometimes inconsistencies occur. Male flowers may arise on female trees or perfect flowers containing male and female parts may self-pollinate (Fig. 6.1). All flowers are greenish-yellow, with male flowers grouped in threes and female flowers appearing singly. If female flowers are not pollinated, they may wither or form small fruits that do not mature and that drop prematurely from trees.

Persimmon flowers are generally wind pollinated but benefit from insect pollination. Generally insects play an important role in pollen transfer from staminate to the pistillate flowers but Fletcher (1942) reported that wind may also contribute to some extent. Persimmon flowers are a good source of nectar and pollen and visited by bees throughout the day (Pellett 1947). Gould (1940) reported that pollination problems exist with persimmons as some cultivars set fruits without pollination whereas in others either there is no fruit setting or fruit drop prematurely.

6.3.1 Pollinators

Persimmon flowers do not require pollinating insects for commercial fruit production but give significantly enhanced yield when visited by insects (Sharma 1961). Fukae et al. (1987) reported that for satisfactory fruit production, a flower required at least 20 honeybee visits. Results on fruit production were better on trees open to honeybee pollination for 7 days or those trees whose flowers were artificially pollinated; but 1–2 days pollination by honeybees was unsatisfactory. It was calculated that one colony was sufficient for 4 ha of the crop. The use of hive inserts was found to be very effective. Hand pollination was effective when weather conditions were poor. Important pollinators for collection of nectar and pollen include bumblebees and honeybees which help in transfer of pollen.

Table 6.1 Insect pests of chestnut (Source: Hill 1983)

Name of pest	Common name	Family	Distribution	Nature of damage
Glossonotuus acuminatus (F.)	Chestnut tree hopper	Membracidae	USA	Adults and nymphs infest foliage; suck sap; sooty moulds often present
Cinalathura folial (Theo.)	Big aphid	Aphididae	China	
Cinara pinea Mord.	Black pine aphid	Aphididae	China	
Longistigma caryae (Harris)	Giant bark aphid	Aphididae	USA	
Pterochlorus tropicalis V.d.G	Chestnut aphid	Aphididae	China	
Eulecanium tiliae (L.)	Hazel scale	Coccidae	S. Europe	Adults and nymphs infest twigs; suck sap
Parthnolecanium corni (Bche.)	Plum scale	Coccidae	Europe, Asia	
Melanaspis obscura (Cmst.)	Obscure scale	Diaspididae	USA	
Zeuzera pyrina L.	Leopard moth	Cossidae	Europe, USA	Larvae bore branches
Tischeria spp.	Leaf miner	Tischeriidae	Europe	Larvae mine leaves
Coleophora spp.	Casebearers	Coleophoridae	S. Europe	Defoliating larvae in tiny cases
Synanthedon vespiformis (L.)	Yellow legged clear wing	Yellow legged clear wing	Europe	Larvae bore wood under bark
Phyllonorcycter messaniella (Zeller)	Blister moth	Gracillaridae	S. Europe	Larvae make leaf blister mines
Cydia splendana (Hbn.)	Tortricid	Tortricidae	Europe	Larvae bore fruits
Datana ministra (Drury)	Yellow necked caterpillar	Notodontidae	USA	Larvae eat leaves, may defoliate
Dryocosmus kuriphilus Yas.	Chestnut gall wasp	Cynipidae	China	Larvae gall flowers
Rhopalomyia castaneae Felt	Chestnut gall wasp	Cecidomyiidae	USA	Larvae bore trunk
Melittomma sericeum (Harris)	Chestnut bud gall midge	Lymexylidae	USA	
Sp. indet.	Chestnut tomber worm	Cerambycidae	Yugoslavia	Larvae bore into trunk and branches, under the bark
Agrilus billneatus Weber	Two lined chest nut borer	Buprestidae	USA	
Agrilus spp.	Jewel beetles	Buprestidae	Europe, Asia, N. America	
Chrysobothris spp.	Flat headed borers	Buprestidae	Europe, Asia, N. America	

(continued)

Table 6.1 (continued)

Name of pest	Common name	Family	Distribution	Nature of damage
Curculio elephus Gyll.	Elephant weevil	Curculionidae	Europe, Asia	Larvae bore nuts
Curculio spp.	Chestnut weevils	Curculionidae	USA	
Cyrtepistomus castaneus (Roe.)	Chestnut Asiatic oak weevil	Curculionidae	Asia, USA	Adults eat leaves; larvae bore nuts
Phyllobius spp.	Common leaf weevil	Curculionidae	Europe	Adults feed on leaves whereas larvae in soil on roots
Xyleborus dispar (F.)	Twig borers	Scolytidae	Europe	Adults bore twigs
Oligonychus bicolor (Banks)	Oak red mite	Tetranychidae	USA	Adults and nymphs scarify (Scorch) foliage by feeding
Eutetranychus hicoriae (McG.)	Hickory scorch mite	Tetranychidae	USA	
Panonychus ulmi (Koch)	Red spider mite	Tetranychidae	Europe, Asia, N. America	

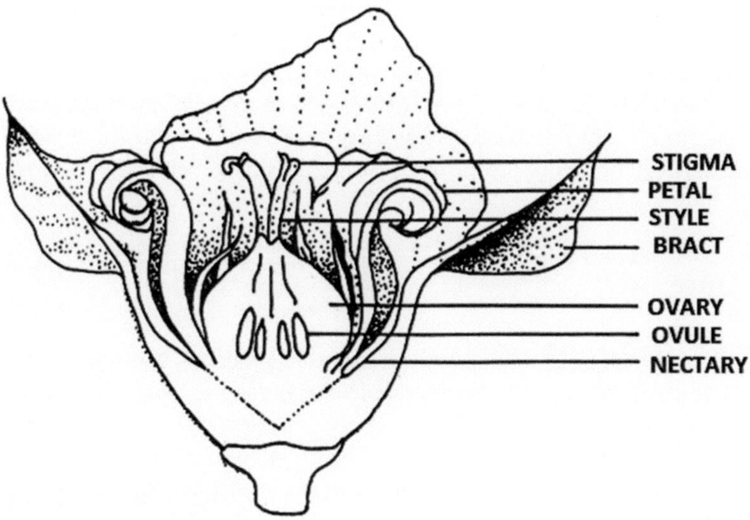

Fig. 6.1 Longitudinal section of persimmon flower (Free 1993)

To set fruit, female persimmon trees need pollinizers to supply pollen and pollinators to deliver it. Although wind generally pollinates nut trees, bees are primary pollinators for fruit trees. Establishing bee colonies in backyard orchards improves pollination, which increases fruit set. Planting one male tree for every eight female trees ensures adequate supply of pollen. 3–4 bee hives should be provided /ha for better pollination and good yield (Fukae et al. 1987).

6.3.2 Persimmon Pests

Persimmon has very few pests except mealybug (*Pseudococcus longispinus*) which are found under the calyx of fruit. The other pests include scale insects (*Parthenoleucanium sp.* and *Hemiberlesia sp.*), fruit flies (*Bactrocera spp.*), twig girdlers and borers (*Chrysobothris sp.*), persimmon psylla (*Psylla sp.*), leaf rollers (*Hyplocala rostrata*) and mite (*Orthotydeus sp.*). The other pests which are occasionally observed include white flies, thrips and mites. Some of the vertebrate pests attacking root and fruits are squirrels, deer, coyotes, rats, opossums and birds. In addition to above stink bugs and scale insects also attack the persimmon trees. The scale insects attacking persimmon of major concern include the grape-myrtle scale *Eriococcus lagerstroemiae*, horned wax scale *Ceroplastes pseudoceriferus*, and elongated cottony scale *Phenacoccus aceris*. The stink bugs of major concern rendering the fruit unmarketable are: brown marmorated stink bug *Halyomorpha halys*, brown-winged green bug *Plautia stali*, and bean bug *Riptortus clavatus*. Machine oil spray during winter is necessary against the scales. A Japanese gall-forming thrips *Ponticulothrips diospyrosi* first discovered in Korea is spreading in other areas (Son et al. 2009). Steven and Sale (1985) recorded three species of armoured scale insects, greedy scale (*Hemiberlesia rapax* (Comstock)), latania scale (*H. lataniae* (Signoret)) and oleander scale (*Aspidiotus nerii Bouche'* attacking persimmon leaves and fruit. Control measures against armoured scale insects should start when trees are dormant or before flowering to ensure the first generation is controlled.

6.4 Olive

Olive, *Olea europaea* L., (family Oleaceae) tree produces enormous number of flowers. However, the proportion of complete or perfect flowers which determines the final outcome of the crop. Generally trees producing moderate intensity of flowers tend to have the higher fruit set. The pollination process in olives is very complicated because of the flowers and scanty pollen production in some cultivars resulting in inadequate pollination.

The olive flowers are cream-colored to white fragrant that develop in the axil of the leaf. A typical olive flower has four lobed gamosepalous calyx four corolla, yellowish white in colour, alternating with sepals, and two stamens that produce pollen copiously and little, if any, nectar (Free 1993) (Fig. 6.2). The flower opens before the pollen is released thereby allowing cross-pollination to occur before selfing could take place. It has been found that a mature olive tree produces about 500,000 flowers, of which, only 1–2 % set fruit that reaches maturity (Lavee et al.

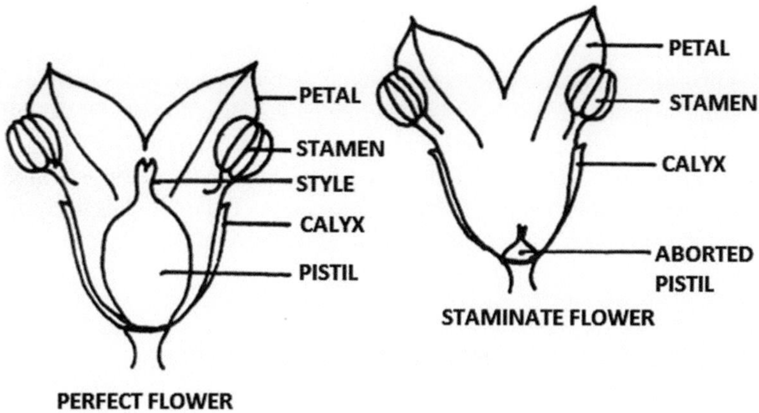

Fig. 6.2 Typical outlines of olive flowers showing both normal aborted pistils (**a**) Perfect flower, (**b**) Staminate flower

1996). Pollination requirements vary greatly among different cultivars as most cultivars are self-fertile, some self-sterile, and others intermediate (Crider 1922; Mort 1952; Pierce 1896 Free 1993; Crossa-Raynaud 1984; Lavee 1996). Honey bees visit olive flowers for nectar and pollen collection (Silvestri et al. 1947).

6.4.1 Pollination Requirements

The concept of crop pollination in olives was given by Muller (1883) who found that most of the cultivars yielded better crops when two or more were interplanted. The pollination requirements of different cultivars of olives have been reported to vary considerably (Crider 1922; Bradley et al. 1961). Even self-fertile cultivars benefit from cross-pollination (Bradley et al. 1961; Hartmann and Opitz 1966; Griggs et al. 1975). In a study conducted in Italy, Hartmann and Opitz (1966) reported that most varieties were self-sterile, a few self-fertile, and some partially self-fertile. They further reported that both in Portugal and in California satisfactory crops were produced when two cultivars are interplanted.

6.4.2 Pollinators

The olive flowers are generally pollinated by wind. Honey bees though visit olive flowers for pollen but are not essential for pollination (Morettini 1941; Griggs et al. 1975). Activities of honeybees supplement pollen transfer by wind pollination. Honeybees do not collect pollen from olive flowers as in other plants. In order to increase the visits of honeybees to olive flowers there is a need to increase the concentration of honeybee colonies in or near the olive trees. In the absence of wind pollination honeybee pollination leads better crop yields. Canale and Loni (2010) reported that more than 380 insects s belonging to two orders, three families, eight genera and ten species of insects were collected on the olive flowers. Of the different flower visitors, honeybee *Apis mellif*era L. was the most abundant followed by three species of Halictidae such as Halictus maculates Smith, *Lasioglossum nitidulum* (F.)and *Lasioglossum villosulum* (Kirby). Six species of syrphid flies of Syrphidae, both Syrphinae and Milesiinae, were seen on olive flowers.

6.4.3 Pollination Recommendations and Practices

There is no information on the concentration that might be desired. Studies in this area would be productive.

6.4.4 Olive Pests

The olive tree is attacked by several pests and diseases. The most serious insect pests attacking olive include attacks include olive black scale *Saissetia oleae*, olive white scale *Mataceronema japonica*, olive psylla *Euphyllura pakistanica*, olive lace bug *Eteoneus sp*., *Bactrocera oleae* (*Dacus oleae*), *Prays oleae*, *Saissetia oleae*, *Hemiberlesia lataniae*, *Frankliniella occidentalis*, *Spilocaea oleaginea*, *Verticillium dahliae*, *Mycocentrospora cladosporioides*, *Pseudomonas syringae* pv. *savastanoi* and *Armillaria mellea*. Control of pests utilizes various methods, e.g. cultivation (El-Hakim and Kisk 1988), biological (El-Khawas 2000) or IPM (Elmore *et al.* 2001). Kaul et al. (2007) reported green stinkbug, *Nezara virudula* (L.) as new pest of olive in India. The other pests included olive psylla, olive black scale and scarab beetles. Singh et al. (1988) and Kapoor (1993)) recorded olive fruit fly *Bactrocera oleae* (Gmelin) from Jammu and Kashmir, And Ludhiana, India respectively.

6.5 *Bactrocera oleae* (Gmelin) (Olive Fruit Fly, Diptera: Tephritidae)

The olive fruit fly, *Bactrocera oleae* (Gmelin) is a very damaging pest of olive fruit. The adults are rarely seen. They lay eggs under the skin of the olive fruit and the emerging larvae feed on the olive flesh, leaving brown tracks and tunnels. Its larva causes damage by drilling tunnels into the fruit resulting in 30 % loss of the crop. The adult *Dacus* feeds on nectar, honeydew and other liquid food sources. *Dacus oleae* was first detected in California in 1998.

6.5.1 Morphology

Length of adult females is about 5 mm with a wing span of 10 mm. The wings are transparent having brown markings with spot at each tip. Both thorax and abdomen are black in colour. The thorax has three parallel lines whereas the abdomen is covered with grey pubescence. The posterior portion of the scutellum is yellow and odours are emitted by the adults (Economopoulos *et al.* 1971). The ovipositor is reddish in colour with black sheath.

6.5.2 Life Cycle

The insect has two to five generations per year in Mediterranean countries. It survives in the pupal form and overwinters several centimetres below the soil surface, the first adults appearing from March to May, depending on temperature and latitude. During summer a period of 6–10 days elapses before mating, and this period is longer with lower temperatures. Starting in June, the female of *Dacus* lays ten to 12 eggs daily, one per olive fruit, with total number of eggs 200–250 eggs during entire life span. The females deposit their eggs beneath the fruit skin with their ovipositor. From each egg a legless larva derives, which feeds on fruit pulp; this results in fruit drop. The egg, larval and pupal period has been reported in the range of 2–4, 10–14 and 10 days, respectively. Mating occurs at dusk and the females produce a multi-component pheromone to attract males. The mating occurs several times during their lifetime.

6.5.3 Control

Various ways of controlling *Dacus* have been used (Collier and Van Steenway 2003), including x-ray male sterilization (Economopoulos 1972), mass trapping (Haniotakis et al. 1986) and use of chemical repellents of females (Lo Scalzo et al. 1994). The most common method is the use of insecticide in bait or air sprays. Other means, more acceptable environmentally, include the use of radiation to sterilize males, pheromones and biological methods. The required radiation for sterilization of both sexes is 8–12 Krad. Furthermore, synthetic analogues of pheromone have been produced, such as 1,5,7-trioxaspirol (5.5) undecane, which is equally as attractive but does not last in the traps for longer durations.

In order to use synthetic pheromones, small pieces of paper or plywood are dipped in an aqueous (0.1 %) solution of deltamethrin for 15 min, then sex pheromone or ammonium bicarbonate added – both *Dacus* attractants. This method provides a low-cost means of control. Other methods include the use of traps for trapping adult insects, biological control and cultivation (early fruit harvesting).

Pheromones of fruit fly, *Bactrocera oleae* has not been studied extensively. The major component of the pheromone is racemic 1, 7-dioxaspiro (5.5) undecane, with low levels (3 %) of hydroxyl derivatives. The adult populations of fruit flies can be monitored using yellow sticky traps baited with sex pheromone and/or ammonium bicarbonate. The former attracts the male and the later females only.

Females need protein and are attracted by ammonia, which is a product of protein decomposition. The yellow colour of the trap attracts both male and female insects. The population of insects trapped is a function of insect population, temperature and humidity of the atmosphere. Another use of this trap is to determine the time of insecticide application, by monitoring the population of *Dacus* trapped.

For control by insecticide, two methods are available: (i) full coverage sprays; and (ii) bait sprays, both methods using OPs, such as dimethoate or fenthion. The number of sprays depends on temperature and humidity, or rainfall during the summer, and involves one to two cover sprays and/or six to seven bait sprays. The number of sprays is smaller (two to three) in hot and dry areas. However, the number of sprays depends on the insect population caught in the traps. It also depends upon the purpose for which the fruit is to be used whether for oil production or for table purposes. Organophosphates are replaced by the insecticide Spinosad in bait sprays, since this is a microbial product, very effective against *Dacus*, but with low toxicity to humans and animals. Spinosad is mixed with the new fruit fly bait known as GF-120, and applied weekly from mid-June up to just before harvesting time.

Another insecticide of low toxicity to vertebrates is a pyrethroid compound (deltamethrin) applied on yellow plywood or paper strips, plus sex pheromone and/or ammonium bicarbonate. *Dacus* is attracted and killed by the lethal dose of pyrethroid. The biological control of the olive fruit fly has been practiced in several areas. *Dacus oleae* is parasitized by the insect *Psyttalia* (or *Opius*) *concolor*. This was introduced to Italy, France, Greece and then California. However, *P. concolor* is ineffective against *Dacus*, since there is no synchronization between the biological stages of the two species, i.e. when female *P. concolor* appears in the spring, no fruit fly larvae are available.

6.6 *Saissetia oleae* (Olive Black Scale, Hemiptera: Coccidae)

Saissetia oleae is a serious pest in different parts of the world (Katsoyannos 1992; Van Steenwyk et al. 2002; Kovanci and Kumral 2004). It is native to South Africa and is widespread throughout Mediterranean countries and California. The same scale can also be found in other fruiting trees such as fig, citrus, apple, etc.

6.6.1 Morphology

The colour of the young adult is brown, becoming black later. The shape of mature females is hemispherical, and their size is 2–5 mm in length and 1–4 mm in width. The characteristic feature of this pest is the appearance of an 'H' shape on its back.

The insect lays eggs under the female scale; their length is 0.3 mm and their colour becomes pink within 2–3 days initially, and reddish orange before hatching. From the eggs after hatching nymphs appear, small in length (0.4 mm) and yellow to brown in colour. These nymphs start to crawl for some days until they find a suitable site for feeding. They are called crawlers and, after a period of feeding lasting 3–8 weeks, their size doubles to 1.0–1.3 mm, with a recognisable dorsal 'H' (second instar nymphs). The next stage is the immature adult stage, which is 2 mm in length and brown; lastly, the pest enters the adult stage. The laying capacity of one female black scale is from 1200 to 4000 eggs.

6.6.2 Life Cycle

This species overwinters in the form of a nymph and early in spring becomes adult, starting to lay eggs in May. After hatching crawlers appear in July, feeding on leaves, shoots or sometimes on fruit. The insect has one or two generations depending on weather conditions, pruning and pest control.

6.6.3 Economic Damage

Since the insect feeds on tree sugars, these trees become weak, leaf abscission occurs and flower bud differentiation is significantly reduced. Furthermore, young scales produce honey-like substances. This leads to a secondary infection by the fungus *Capnia*, resulting in deterioration of fruit quality and a decrease in photosynthesis.

6.6.4 Control

The insect can be controlled by cultivation practices, insecticidal sprays or by natural enemies of the black scale. In trees left unpruned or lightly pruned and a relatively dense canopy, damage and infestation are severe. Regular pruning exposes the black scale to hot and dry conditions and increases its percentage mortality. However, if the weather is cool and humid this favours black scale, especially when the tree canopy is dense. In unpruned olive trees the shady and moist environment inside the tree canopy protects the scale, which can then survive even very hot summers. Furthermore, irrigation may create favourable conditions for the black scale population to increase. Due to the existence of the scale's covering, pesticides are effective only at the crawler stage during the summer. When the population of black scale is low to medium oil treatments during the dormant period are effective. The most effective pesticides are OPs. There are many natural enemies, such as *Metaphycus helvolus*, *M. bartletic*, *Scutellista cyanea*, *Chrysoperla* spp., *Hippodamia convergens* and *Hyperaspis* sp. From these enemies only *Hyperaspis* spp. can significantly reduce the population of black scale.

6.7 *Prays oleae* (Bernard) (Olive Moth, Lepidoptera: Yponomeutidae)

This insect is known also as olive kernel borer. It is widespread in the Mediterranean area and in other olive-growing countries, causing severe damage, though a little less harmful than the effects caused by *Dacus*. The moth feeds on olive trees, cultivated and wild, and all olive cultivars are prone to damage if no control measures are taken. Other species where this moth can be found include other genera of the Oleaceae family such as *Ligustrum*, *Jasminum* and *Phillyrea. Prays oleae* has three generations: the first on floral buds, the second in the fruit endocarp and the third on the leaves where the insect overwinters.

6.7.1 Life Cycle

The insect overwinters as an adult and appears in spring; the females deposit their eggs in the calyx of flowers. The larva, appearing after 9–12 days, feeds on the floral parts – initially on anthers and subsequently on stigma, style and ovary. The larvae later connect the entire inflorescence with a silky thread. The second generation attacks the young fruits close to the stalk and the larva enters the endocarp, where it stays for 80–135 days. After this period the larva leaves the endocarp and exits via the same opening, resulting in fruit drop. The adult insects appear after 8–14 days. The third generation lays eggs on the upper leaf surface. The hatched larvae (7–14 days) enter the leaf by drilling tunnels and digest foliar material. At the end of winter the adults appear and a new cycle begins.

6.7.2 Economic Damage

The damage caused by this moth ranks second only to that of *Dacus oleae*, and is serious with high insect populations or low flowering conditions. Damage affects flowers, fruits and leaves and results in fruit drop.

6.8 Chemical and Biological Control

6.8.1 Cultivation Practices

Cultivation practices: Unfavourable climatic conditions greatly limit population of this insect. Pruning and maintenance of open canopies reduce RH, and with RH below 50 % or with temperatures above 30 °C egg death results. That is why this insect does not appear in dry areas.

6.8.2 Biological Control

Biological control: The insecticidal bacterium *Bacillus thuringiensis* Berliner (subp. Kurstaki) when sprayed is very effective in reducing larval populations. Furthermore, the insect *Chrysoperla carnea* Stephens is a natural enemy of *P. oleae*, since it is predatory to the insect's eggs, larvae and pupae.

6.8.3 Chemical Control

Chemical control: The use of insecticides such as dimethoate controls population levels. **Pheromones:** The use of female sex pheromones is useful as a control method (Campion et al. 1979; Kavallieratos et al. 2005).

6.9 *Palpita unionalis* (Hübner) (Jasmine Moth, Lepidoptera: Pyralidae)

6.9.1 Life Cycle

This pest is active at night. The newly hatched larvae feed on young and older ones on the parenchyma of older leaves. The first generation is completed in 45–48 days and the mean total life of the first generation is second generation in 38 days (Alexopoulou-Vassilaina and Santorini 1973; El-Kifl et al. 1974; Amin and Saleh 1975; Badawi et al. 1976; Shehata et al. 2003). The larvae drill tunnels having a crescent-like shape, and can also infect inflorescences and young fruits causing fruit drop, since they feed on the seed embryo.

6.10 Control

6.10.1 Biological Control

Biological control: *Bacillus thuringiensis* has toxic effects on moth larvae when they feed on it. Use of the commercial product marketed under the name Dipel 2X at a concentration 0.4 g/l achieves 100 % eradication of the larvae within 48 h.

6.10.2 Cultivation Practices

Cultivation practices: Sucker eradication reduces the problem, since the insect prefers to lay its eggs on the suckers. The larval stage lasts 16 days at 17–23 °C and 15 days at 21.6–25.5 °C for the first and second generations, respectively.

6.10.3 Chemical Control

Chemical control: Oil spraying is effective when the infestation is low. It is preferable to spray at night or early in the morning if the temperature is greater than 32 °C during the day. When the infestation is severe, the insecticide Carbaryl (Sevin 80S) is included in the spray solution. This spray is, however, very destructive to most natural enemies of *Palpita*. Another insecticide that might be incorporated is Methidathion (Supracide 25WP).

6.11 *Frankliniella occidentalis* (Pergande)

This insect is also known as western flower thrips. Many plant species are hosts of this insect.

6.11.1 Morphology and Life Cycle

The adults of *F. occidentalis* are very small insects (1 mm). The thorax is orange in colour and the abdomen is brown. Adults overwinter in weeds and in other places in the orchard. During spring they lay eggs and deposit them in buds, shoots and flowers. After hatching, nymphs feed on shoots, fruits or leaves. The nymphs, after attaining full size and development, drop to the ground and pupate in protected places. The maximum adult population occurs in June and, as the weeds dry out, the adults move to various crops such as olive trees. This insect has five to six generations per year. The damage is due to the fact that thrips feed on leaves and shoots but, because they absorb sap from the fruits, the affected fruit shows scars and is rendered unsuitable for processing. This problem is serious with green olives intended for processing.

6.11.2 Control

Various insecticides are recommended for thrips, and cultivation techniques are very effective. One such technique is to disc the area close to the olive grove that is covered with weeds, to avoid an increase in thrips population and subsequent migration to olive orchards.

6.12 *Parlatoria oleae* (Colvée) (Olive Scale, Homoptera: Diaspididae)

Parlatoria oleae is also known as olive scale. It has a wide spread distribution. This is a serious pest and requires systematic pesticide treatment every year to reduce economic losses. *Parlatoria* can be found in about 200 plant species.

6.12.1 Morphology and Life Cycle

The adult is 2 mm in length with a greyish oval covering. The young female's body under the scale cover is reddish to deep purple; the covering of the male is elongate, white and flat. The insect has two generations per year. The emergence of the first-generation crawlers occurs in late June to early July, when the young fruits are attacked and deformed. The second generation in August causes semi-circular dark purple–black scales on the green fruits; this damage is severe in table olives. Infestation may also occur on branches and leaves, which results in a significant decrease in productivity.

6.12.2 Economic Damage and Control

This insect can cause leaf abscission, drying of twigs and reduction in photosynthesis; however, the most serious damage is that on the fruits (Foda 1973). These become disfigured, with purple spotting on green olives, which then become unsuitable for processing Biological control of *Parlatoria* can be done by two parasites namely Aphytis *maculicornis* MASI and *Coccophagoides utilis* Doutt. If chemical treatments are necessary two sprays are required, the first in late May–June for the first generation and second in July–August for the second generation. Applications are effective if carried out at the stasge when crawlers appear.

6.13 *Hemiberlesia lataniae* (Signoret) (Latania Scale, Hemiptera: Diaspididae)

6.13.1 Morphology

Latania scale is a scale similar in size to the adult olive scale. The difference is in the waxy shell covering, which is more conical, with a small black spot to one side of the centre. Also, the body of the female is yellow while it is reddish purple on the olive scale. The insect produces many generations per year.

6.13.2 Life Cycle

The insect overwinters in the form of nymph (second instar). Early in spring it matures and the female insect lays eggs in batches of 15–20, located beneath the scale covering. After hatching the crawlers start to feed, appearing in May, July and September. Latania, except on olive trees, feeds on avocado, acacia, euonymus, kiwi fruit, rose and other species.

6.13.3 Economic Damage

The scales infest leaves, bark and fruits, producing a dark purple spot of very clear outline. Infested fruits, especially for table olive processing, become worthless.

6.13.4 Chemical and Biological Control

For biological control the species *Aphytis melinus* and other parasites are used. However, where chemical treatment is necessary oil sprays or oil sprays plus Sevin 80S or Supracide 25WP are used. Spraying is performed in late May and in late July to August. The best results are obtained when scale crawlers start moving onto the fruit. Treatment between these two sprays is not necessary. Furthermore, a postharvest treatment (October/November) is also effective.

6.14 *Aspidiotus nerii* (Oleander Scale (Bouche) or Ivy Scale, Hemiptera: Diaspididae)

Many plants are hosts to this insect such as camellia, cherry, grapefruit, lemon, magnolia, orange, rose and olive, where infestations may damage the fruit.

6.14.1 Morphology and Life Cycle

It resembles greedy scale, the only difference being that the covering of its scale is less convex. It overwinters in the form of adult females, which start to lay eggs in April. The insect has two generations, in April and in July/August.

6.14.2 Economic Damage and Control

During heavy infestations this scale infests olive fruit, producing green spots on purple fruit and certain deformations. The infested fruit becomes worthless. Biological control agents may decrease the scale population.

6.15 *Phloeotribus scarabaeoides* (Bernard) (Olive Beetle, Coleoptera: Scolytidae)

This is known also as olive beetle and is present in all Mediterranean areas. The adult is 2.0–2.5 mm long.

6.15.1 Life Cycle

This insect overwinters in the form of either larva, nymph or adult in shoot tunnels. The female drills galleries in the branches and trunk. In these channels mating takes place and the female lays eggs. Within 1–2 weeks larvae appear, which open small channels under the bark. In another 3–4 weeks the larva is transformed into an adult, which leaves the branch. The insect has four generations, the adults appearing in March/April, April–July,

August/September and August–November. The infested trees have low vigour and low productivity.

6.15.2 Control

- Eradication of the pruning material by burying, before the insect lays its eggs.
- Burning the suckers, which can host a large number of larvae.
- Keeping the trees healthy by appropriate fertilization, irrigation and pruning.
- Application of organophosphate insecticides is recommended when the infection is severe.

6.16 *Oxyenus maxwelli* (K.) (Olive Mite)

This mite is widespread in Mediterranean and other countries.

6.16.1 Morphology and Life Cycle

It is very small in size, with four legs, with body colour yellowish to orange. It overwinters in the form of adult in the tree bark. The females lay eggs in early spring through to summer. During periods of high temperature and low humidity its population is reduced.

6.16.2 Economic Damage and Control

It infests mainly young olive leaves, these leaves developing a silver colour and longitudinal curl. The mite also infests inflorescences, causing pistil abortion. Some of the symptoms due to infestation include inflorescence abscission and necrosis and drop of buds. In cases of severe infestation chemical control is recommended.

6.17 *Zeuzera pyrina* L. (Leopard Moth, Lepidoptera: Cossidae)

Damage due to this moth has been reported in species from over 30 botanical families. In olives the insect is present mostly in Mediterranean countries, and also Syria. The insect can damage all olive varieties: resistant varieties have not yet been reported.

6.17.1 Life Cycle

Adults lay their eggs during summer and, after hatching, the larvae attack lateral buds and drill sub-cortical channels. The larvae exit the branch by creating a hole and subsequently enter a new branch. Their presence is obvious by the sawdust released from the opening (frass). The biological cycle lasts 1 year, with larvae being fully developed in late winter and overwintering in the form of chrysalids. The adults occur in May to June.

6.17.2 Economic Damage and Control

The tunnels opened cause serious damage and young trees may dry out. The secondary branches also die and tree vigour is reduced. For control we immerse a piece of cotton wool in a liquid producing toxic vapours and with that we plug the opening of the tunnel. Furthermore, in young trees spraying with systemic insecticides may solve the problem.

6.18 Walnut

Walnuts are monoecious and generally self-fertile. The staminate flowers are borne on catkins that develop from lateral buds (Tsurkan and Pyntya 1987). Each flower on catkin has 13–18 anthers and each anther has 900 pollen grains. It is estimated that about 1.8 million pollen grains are produced per catkin (Impiumi and Ramina 1967). The pollen is very small and easily carried by wind. In some varieties time of pollen shedding does not overlap with the receptivity of female flowers and require another variety for pollen. The over pollination results in withering of flower and consequently no fruit set.

Pollination requirements in walnut orchards are inadequately understood. Pollination relationships in this species are more complex than is typical of other orchard crops. The species is self compatible, however because it is also dichogamous, the period of overlap between pollen shed and pistillate flower receptivity may be inadequate for maximum productivity. It has been reported that pollen loads in the range of 30–50 times the number of ovules is necessary for set.

6.18.1 Walnut Pests

Walnut trees are attacked by large number of insects. Several investigators have recorded variable number of pests in different locations. Kearby (1975) recorded 62 species of insects associated with black walnut plantations in Missouri,

USA. Nixon and McPherson (1977) reported 300 species from walnut plantations in southern Illinois, but he stated only few of them are of significance. Marshall (1989) reported 65 species of insects attacking the genus Juglans and black walnut. The pests of concern include: walnut shoot moth (*Acrobasis demotella* Grote), ambrosia beetle (*Xylosandrus germanus*) (Blandford), cicada (Magicicada spp.), the walnut shoot moth, walnut caterpillar (*Datana integerrima* Grote & Robinson), fall webworm (*Hyphantria cunea*) (Drury), oystershell scale (*Lepidosaphes ulmi* L.) and ambrosia beetle and the black walnut curculio, *Conotrachelus retentus* (Say). The most serious pests that feed on walnut are nut feeder,defoliator, shoot borer and sucking insects.

Walnut trees are susceptible to the attack of several pests such as walnut weevil (*Alcides porrectirostris* Marsha), walnut blue beetle (*Monolepta erythrecephale*), Sanjosescale (*Quadraspidiotus perniciosus* Comst), and walnut green aphid (*Chromaphis juglandicola* Kalt). Among the different pests prevalent in the walnut-producing areas, walnut weevil is considered serious in some places. Defoliation due to *Chaetopractus odata* has occasionally been observed in some areas. Haq (1962) reported *Lachnosterna* (*Holotrichia*) *longipennis* beetles as important defoliators of apple, walnut, cherry and strawberry in the hilly districts of Uttar Pradesh. This species is widely distributed in Uttrakhand from Jeolikote (4000 ft) to Chaubattia (7000 ft). Singh (1964) described as many as 35 species causing damage to many fruit and forest trees. Important species include B. coriacea, Brahmina crinicollis Burmeister, L. (H.) *longipennis*, *M. furcicauda*, *A. polita*, *A. rugosa*, *A. lineatopennis*, *A. rufiventris, A. dimidiata, A. bimarginatus, A. versutus, Adoretus duvauceli* Blanchard, *Cephaloserica thomsoni* (Brenske), *Hilyotrogus*, *holosericeus Redtenbacher* and *Mimela fulgidivittata* Blanchard which feed and cause damage to leaves of apple, apricot, cherry and walnut. Stem borer (*Aeolesthes sarta*) and bark beetles (*Scolytus* sp.), the pests of both fruit as well as forest trees, have emerged as a serious problem for apple and walnut. Codling moth (*Cydia pomonella*) is a pest of international importance. This pest attacks many fruit trees e.g. apricot, pear, walnut, etc. besides all cultivars and varieties of apple. Indian gypsy moth (*Lymantria obfuscata*), is a voracious polyphagous pest attacking many fruit and tree plantations including apple and walnut. Chaffer beetles (*Adoretus simplex, Brahmina coriacea*, *Holotrichia longipennis*) damage fruit crops especially High-density plantation of apple and walnut. Among 70 species of leaf feeding beetles recorded in temperate regions of India, scarabaeids, cetoniids, curculionids and chrysomelids in that order cause significant damage to pome and stone fruits. Walnut weevil (*Alcides porrectirostris*) prefers thin shelled varieties to the less valuable thick shelled walnut types. The kernel of the nut is damaged and the attacked fruit usually drops or is unmarketable. The dropping of the fruit starts from 2nd week of May till end of August. Another important pest associated with walnut is grey weevil (*Myllocerus* spp.). It feeds on foliage by biting holes in the lamina or by peripheral feeding on margins. The maximum damage is caused from mid May to mid June. Among leaf rollers, folders *Archips pomivora* is the most important. It causes maximum damage in June. It attacks foliage but injury on fruits and their rotting cause economic losses. *Chaetoprocta odata* (Hewitson) is considered as one of the most serious damage causing lepidopteron insect pests infesting walnut trees with particular reference to Kashmir (Masoodi and Trali 1987; Mir and Wani 2005; Khan et al. 2013) Table 6.2.

6.19 Management of Walnut Pests

6.19.1 Cultural Controls

Good cultural practices can minimize walnut pests: Maintaining trees from water stress can minimize spider mite infestations that usually don't occur in well-irrigated orchards. Shaking of trees to remove old mummy nuts and then destroying them can help reduce over-wintering stages of this insect. Pruning can help the plants to gain vigour and minimize attack of diseases and pests as densely shaded trees are more. Prompt harvesting can minimize damage by infestation from Walnut husk fly and kernel molds. Hairy caterpillar is another most serious pest resulting in stripping off foliage of trees. Burlapping of tree trunks with gunny bags and collection and destruction of gypsy moth larvae in early summer and egg masses in late summer is an efficient management method for the control of the pest. Tortix moth caterpillars after hatching from eggs in month of May bore into nutlets and feed on green hull inside. Walnut Tortrix moth infested fruits should be collected and buried deep in the soil so as to have reduced population in next year

6.19.2 Biological Control

Maintaining bird species richness can help to reduce codling moths . Barn owls, *Tyto alba* serve as important predators of pocket gopher, *Thomomys sp.* Bats also play a large role in insect control. Research shows they consume a considerable quantity and diversity of insects (Long 1998). The parasitic wasp Trichogramma platneri has been observed attacking codling moth eggs in walnuts (Mills et al. 1995). Three parasitoid species can help control codling moth which include: *M ridibundus*, *L. caudatus* (ichneumonids), and *M. rufipes* (a braconid). The two ichneumonid species are cocoon parasitoids and the braconid wasp is a larval parasitoid that attacks the mid-stage codling moth larvae inside fruit (Mills 1997). The parasitic wasp, *T. pallidus*, has eliminated walnut aphid as a pest in most orchards. *Bacillus thuringiensis* var. kurstaki has a selective larvacidal activity against all lepidopteran species including codling moth, navel orangeworm, and red-humped caterpillar.

Table 6.2 Insect pests attacking walnut in Kashmir

S. No.	Name of the pest	Scientific name	Order	Family
1	Walnut weevil	*Alcides porrect irostris* (Marshall)	Coleoptera	Curculionidae
2	Stem borer	*Aeolesthes sarta* (Solsky)	Coleoptera	Cerambycidae
3	Pin hole borer or shot hole borer	*Scolytoplatipus* spp.	Coleoptera	Scolytidae
4	Chaffer beetle	*Protecta neglecta Adoretus simplex*	Coleoptera	scarabaidae
5	Indian Gypsy moth or hairy caterpillar	*Lymantria obfuscata* (Walker)	Lepidoptera	Lymantriidae
6	Walnut leaf defoliater	*Chaetoprocta odata* (Hewitson)	Lepidoptera	Lycaenidae
7	Leopard moth	Zeuzera multistrigata Moore	Lepidoptera	Cossidae
8	European red mite	Panonychus ulmi (Coach)	Trombidiformes	Tetranychidae
9	Two spotted mite	Tetranychus urticae Coach	Trombidiformes	Tetranychidae
10	Tent hairy caterpillar	*Malacosoma indica* Walker	Lepidoptera	Lasiocampidae
11	Flat headed borer	*Sphenoptera lafertei* Thomson	Coleoptera	Buprestidae
12	Codling Moth	*Laspeyresia pomonella*	Lepidoptera	Totricidae
13	Red-Humped Caterpillar	*Schizura concinna*	Lepidoptera	Notodontidae
14	Walnut Husk Fly	*Rhagoletis completa*	Diptera	Tephritidae
15	Walnut Aphid,	*Chromaphis juglandicola*	Homoptera,	Callaphididae
16	Dusky Veined Aphid	*Callaphis juglandis*	Hemiptera:	Aphididae
17	San Jose Scale	*Quadraspidiotus perniciosus*	Hemiptera	Diaspididae
18	Walnut Scale,	*Quadraspidotus juglansregiae*	Hemiptera	Diaspididae
19	European Fruit Lecanium Scale	*Lecanium corni*	Hemiptera	Coccidae
20	Walnut blue butterfly	*Chaetoprocta odata*	Lepidoptera	Lycaenidae

6.19.3 Chemical Controls

Insect Growth Regulator (IGR) such as Tebufenozide (Confirm) binds to the ecdysone receptor protein causing the molting process of codling moth larvae higly accelated thereby providing good control is obtained. Development of Codlemone, a synthetic sex attractant pheromone is very effective as mating disruption in codling moth. Spinosad with a bait attractant has a novel mode of action that affects the insect nervous system at the nicotinic acetylcholine receptors. The damage by pests can be reduced to large extent by following proper management practices. The fallen infested fruits should be collected and buried deep into the soil to reduce infestation of walnut weevils. Spraying of chlorpyriphos in the month of July can check weevil infestation. The attack of stem borer and bark beetle can be minimized by plugging the infested plant with cotton impregnated with 0.1 % of dichlorvos and while washing the trees with time. Second instar larvae of Walnut blue butterfly (*Chaetoprocta*) are most damaging causing severe damage to the tree by defoliating its leaves. Spraying with imidacloprid or dimethoate during first week of May when larval population attains peak can reduce infestation. Similarly for management of aphids imidacloprid or dimethoate are very effective. The plants infested by blister beetles in nursery should be discarded than attempting to control the problem with pesticides but in case of severe damage spraying the plants with dicofol can provide effective control.

6.20 Pecan

Poor pollination in pecan orchards is a major factor reducing both yield and quality of nuts. Pecan is monoecious and male and female flowers are located separately on the same plant. The catkins produce large quantities of pollen which is carried by wind to the female flowers where fertilization takes place. To prevent self pollination, the tree matures its flowers at different times so that when pollen shedding takes place no receptive flowers are available. The best option to get quality and quantity of produce from pecan, it is important to plant different cultivars whose flowers mature at different times. There are two types of pecan trees, protandrous and protogynous in which male and female flowers mature first, respectively. Honeybees and other flower visiting insects have relatively little significance.

6.20.1 Insect Pests of Pecan

Many insects feed on the leaves, nuts, branches and buds of the pecan tree, reducing the production potential of trees. Some insects pests directly affect the nuts whereas others cause indirect damage, by feeding thereby reducing the productivity. Aphids and true bugs are the major pests of pecan causing damage to leaves and nuts, respectively. A list of important pests of pecan is given below in Table 6.3

6.21 Hazel Nut

Hazelnuts (*Corylus avellana* L.), are monoecious, and wind-pollinated, mostly dichogamous, and self-incompatible. More than 90 % of the plants are protandrous. It is therefore recommended that sufficient number of varieties be interplanted to provide pollen at the time female flowers are receptive.

Table 6.3 Insect pests attacking pecan trees

Yellow pecan aphid (*Monelliopsis pecanis* (Bissell))	They feed on leaves, reducing photosynthetic area and reduce water from leaves which impairs the growth of leaves, shoots and nuts
Black pecan aphid (*Melanocallis caryaefoliae* (Davis))	Black pecan aphid is more serious than Yellow pecan aphid. While feeding they inject a toxin into the leaf, the area turns yellow causing the leaf to drop affecting quality of yield
Stink bugs	Feed on nut kernels before the shell hardens. This feeding will cause black or brown spots on kernels
Pecan Phylloxera	The attack of this insect produces galls on new growing leaves, twigs and nuts may also be affected, overwinters as eggs in bark crevices. Use of dormant oil is recommended for control
Shot-hole borer	Shoot hole larvae feed on the wood and emerge as adults by cutting small holes in the bark. Removal of dead decaying wood and its burning is recommended for control
Shuckworm	Shuck worms damage the nuts causing them to drop. They overwinter as full grown larvae in shucks on the ground or on the trees. The dropped nuts should be gathered and destroyed
Pecan Weevil (PW)	Adult feeding causes younger nuts to drop, more mature nuts however, remain on the tree, but their quality is lost. Adults can be controlled by application of carbaryl. Sprays must be re-applied every 10–14 days (Sevin) or as long as weevils continue emerging, or until the shucks split, a period of 6–8 weeks
Pecan Leaf Scorch Mite (PLSM)	The pecan leaf scorch mite is the most important spider mite attacking pecans Defoliating outbreaks can occur. Bifenazate (Acramite) may be used for control. Predatory mites have shown promise for bio-control
Spittlebug	They are very common in pecan orchards. Overwinter as eggs in the slit of stems. Nymphs appear soon after nut set and move onto the new growth. They can be controlled by spraying chlorpyrifos
Foliage-Feeding Caterpillars	Many species of caterpillars feed on pecan including walnut caterpillar, Datana integerima; Fall webworm, Hyphantria cunea; Pecan cigar casebearer, Coleophora laticornella; Pecan catocala, Catocala obtuse or C. viduata; and Pecan budmoth, Gretchina bolliana. Spinosad (Spintor) is very effective and somewhat less disruptive of beneficial

6.21.1 *Pest Management*

Hazelnuts are I attacked by only a few pests which include the fruit tree borer (*Cryptophasa melanostigma*) which weakens the tunnels by boring into the wood. The hazel aphid (*Myzocallis coryli*) is another pest most problematic during growing season. The Big bud mite (*Phytoptus avellanae*) damages terminal buds which are enlarged several times compared to normal.

6.22 Pistachio *Pistacia vera* L. Anacardiaceae

6.22.1 Pollination in Pistachio

Pistachio is a deciduous and dioecious tree with male and female flowers born on different trees thereby necessitating cross pollination (Acar et al. 2001; Ak et al. 1996; Iisfendiyarogluglugu et al. 2001). Pistachio flowers has no petals to attract insect pollinators, therefore, pollination is carried by wind (Ak et al. 1996; Herrera 1997). It is there necessary to have enough male trees to insure adequate pollination and to get maximum nut production. Male and female trees are usually interplanted in the orchard with one male to eight or eleven females depending on the orchard (Herrera 1997; Acar et al. 2001; Ak et al. 1996).

6.22.2 Insect Pests of Pistachio

Mourikis et al. (1998) reported that pistachio is attacked by various species of insects which include *Capnodis tenebrionis* attacking roots of non irrigated trees., the coleoptera *Acrantus (=Chaetoptelius) vestitus*, *Esteneborus perisii* and *Sinoxylon sexdentatum* on branches and twigs of weakened trees, *Archips rosanus* and *Teleiodes decorella* on leaves. *Pseudocoeliodes rubricus* attacks the inflorescence of male trees only. The psyllid *Agonoscena* sp. causes premature defoliation. *Idiocerus stali*, when in large populations, cause withering of leaves and blight of young panicles. The most serious pests of pistachios are the microlepidoptera *Palumbina guerinii* an d the seed chalcid *Eurytoma plotnikovi* attacking fruits every year up to 90 % infestation. The scale insects such as *Anapulvinaria pistaciae*, *Saissetia oleae*, *Ceroplastes rusci*, *Melanaspis inopinata* and *Lepidosaphes pistaciae* have been observed in small populations. *Apomyelois ceratoniae* infests split fruits remaining on the trees after harvest leading to infestation in stored nuts. *Ephestia* sp. also infests stored nuts. Proper sanitation is the primary means of controlling thse pests under stored conditions. Sprays should be made only in case of heavy infestations.

Mehrnejad (2001) stated that injurious pests of pistachiocan be divided into three groups. The first group include those pests which attack on leaves, fruits and twigs such as (i) the common pistachio psylla, *Agonoscena pistaciae*; (ii) the pistachio twig borer, *Kermania pistaciella*; and (iii) the pistachio bugs, *Acrosternum heegeri*, *Acrosternum millieri*, *Apodiphus amygdali*, *Brachynema germari*, *Brachynema segetum*, *Lygaeus pandurus*. The second group included mites and other phytophagous insects such as the pistachio fruit moth, *Recurvaria pistaciicola*; the pistachio leaf hopper, *Idiocerus stali*; the pistachio scale insects including: the pistachio twig and fruit scale, *Pistaciaspis pistaciae*, the pistachio trunk and branch scale, *Melanaspis inopinata*; the pistachio bark beetle, *Hylesinus vestitus*; the pistachio leaf borer, *Ocneria terebinthina*; the pistachio common mites, *Tenuipalpus granati* and the third group includes minor pests which under certain conditions become injurious the pistachio leaf-rolling psyllid, *Megagonoscena viridis*; the pistachio root beetle, *Capnodis cariosa hauseri*; the pistachio weevil, *Polydrosus davatchii*; the pistachio eriophyid mites, *Aceria (=Eriophyes) pistaciae* and *Aceria (= Eriophyes) stephanii*; the pistachio fruit hull borer, *Arimania komaroffi*; the carob moth, *Apomyelois (= Ectomyelois) ceratoniae*; pistachio leaf miner, *Stigmella promissa*; pistachio soft scale, *Anapulvinaria pistaciae*; pistachio spherical scale, *Eulecanium rugulosum*; pistachio Noghi scale, *Salicicola* pistaciae; pistachio seed wasp, Eurytoma plotnikovi and pistachio seed chalcid, Megastigmus pistaciae. Since pistachio does not attract insects for pollination, the pest management dtrategies on affect on pollinating insects.

Bibliography

Abbott CE (1926) The kaki and the loquat. Fla State Hort Soc Proc 39:228–233

Acar I, AK BE, Kuzdere H (2001) An investigation on artificial facilities in pistachios by using an atomizer. pp 145–148

Ak BE, Acar I, Kaska N (1996) An investigation on the male determination for some female varities throughout five years (1992–1996) grown at Ceylanpiner State farm in Sanliurfa conditions. Ciheam Options Mediterrannes 33:99–104

Amin AH, Saleh MRA (1975) Seasonal activity of the olive leaf moths, *Palpita unionalis* (Hbuner) (Lepidoptera: Pyralidae), in Kharga-Oasis, New- Vally, Egypt. Ann Agr Sci, Fac Agr Ain Shams Univ 20(1):35–41

Annecke DP, Moran VC (1982) Insects and mites of cultivated plants in South Africa. Butterworths, Durban, 383 pp

Apenteng NV, Wallade SM (2005) Olive pests and their natural enemies in the central Eastern Cape Province of South Africa. Proceedings of 15th of entomological congress, Grahamstown, 10–13 Jul 2005. Available at http://www.uckac.edu/ppq

Badawi A, Awadallah AM, Foda SM (1976) On the biology of the olive leaf moth *Palpita unionalis* Hbn. (Lep., Pyralidae). Z Ang Entomol 80:103–1

Baumgartner J (2001) Birds, spiders naturally control codling moths. Tree fruit magazine, Apr 2001. pp 5–7

Bradley MV, Griggs WH, Hartmann HT (1961) Studies on self- and cross-pollination of olives under varying temperature conditions. Calif Agric 15(3):4–5

Broumas T, Haniotakis G, Liaropoulos C, Tomazou T, Ragoussis N (2002) The efficacy of an improved of the olive fruit fly, *Bactrocera oleae* (Gmelin) (Dipt. Tephritidae): pilot-scale feasibility studies. J Appl Entomol 126:217–223

Bryant GE (1922) Some new injurious Phytophaga from Africa. Bull Entomol Res 10:473–475

Bryant GE (1944) New species of African chrysomelidae (Halticinae: Coleoptera). Ann Mag Nat Hist 84:817–821

Campion DC, McVeigh LO, Polyrakis J, Michelakis S, Stavrakis G, Beevor PS, Hall DR, Nesbitt BF (1979) Laboratory and field studies of the female sex pheromone of the olive moth *Prays oleae*. Experientia 35:1146–1147

Canale A, Loni A (2010) Insects visiting olive flowers (*Olea europaea* L.) in a Tuscan olive grove. REDIA XCIII

Clapper RB (1954) Chestnut breeding, techniques and results. ii. Inheritance of characters, breeding for vigor and mutations. J Hered 45:201–208

Collier TR, Van Steenway KRA (2003) Prospects for integrated control of olive fruit fly are promising in California. Calif Agric 57(1):28–31
Condit IJ (1947) Olive culture in California. Calif Agric Ext Serv Cir 135, 36 pp
Costa C (1998) Olive production in South Africa. A handbook for olive growers. Agricultural Research Council, 124 pp
Crane HL, Reed CA, Wood MN (1937) Nut breeding. US Dept Agric Yearb 1937:827–889
Crider FJ (1922) The olive in Arizona. Ariz Agric Expt Sta Bull 94:491–528
Crossa-Raynaud P (1984) Quelques productions fruitières dépendant d'une pollinisation anémogame: noyer, noisetier, olivier, palmier-dattier, pistachier. In: Pesson P, Louveaux J (eds) Pollinisation et productions végétales. INRA, Paris
Crovetti A (1996) Plant protection: development of methodologies and the protection of production and the environment 225-250pp. In: Plaza and Janés Editores, South Africa, Enric Granados. 1996. World Olive Encyclopaedia. International Olive Oil Council. Barcelona, 479pp
Daku L, Norton GW, PfeifferDG, Luther GC, Pitts CW, Taylor DB, TedeschiniJ, Uka R (2000) Farmers' knowledge and attitudes towards pesticide use and olive pest management practices in Vlora, Albania: a baseline survey. IPM CRSP Working Paper 001–1, Virginia Polytechnic Institute and State University
de Oliveira D, Gomes A, Ilharco FA, Manteigas AM, Pinto J, Ramalho J (2001) Importance of insect pollinators for the production in the chestnut, *Castanea sativa*. Acta Hort (ISHS) 561:269–273. http://www.actahort.org/books/561/561_40.htm
Desneux N, Decourtye A, Delpuech JM (2007) The sublethal effects of pesticides on beneficial arthropods. Annu Rev Entomol 52:81–106
Dimos I, Koutsikopoulos C, Economopoulos A, Lykakis J (2003) The distribution of olive fruit fly captures with McPhail traps within an olive orchard. Phytoparasitica 31:124–131
Economopoulos AP (1972) Sexual competitiveness of y-ray sterilized males of Dacus oleae. Mating frequency of artificially reared and wild females. Env ento 1:490–497
Economopoulos AP (1979) Attraction of *Dacus oleae* (Gmelin) (Diptera, Tephritidae) to odor and color traps. Z Angew Entomol 88:90–7
Economopoulos AP, Giannakakis A, Tzanakakis ME, Voyadjoglou A (1971) Reproductive behavior and physiology of the olive fruit fly. 1. Anatomy of the adult rectum and odors emitted by adults. Ann Entomol Soc Am 64:1112–1116
Eglington L (2004) Winning the war against the olive beetle. Farmer's Weekly. 6 Feb. pp 40–41
El-Hakim AM, El-Sayed S (1982–1983) Studies on the infestation of olive fruits with the olive fruit fly, Dacus olea Gmel. in Egypt (Tephrytidae: diptera). Bulletin de la Societe Entomologique d'Egypte 64:221–226
El-Hakim AM, Kisk SA (1988) Cultural methods for the control of olive pests. Bull Fac Agric CAIRO Univ 39(1):345–351
El-Khawas MA (2000) Integrated control of insect pests on olive trees in Egypt with emphasis on biological control. Ph.D. thesis. Faculty of science, Cairo University Egypt. 247 pp
El-Kifl AH, Abedsalam AL, Rahhal AM (1974) Biological studies on the olive leaf moth *Palpita unionalis* Hbn. Bull Soc Entomol 58:337
Elmore CL, Cudney DW, Donaldson DR (2001) UC IPM pest management guidelines. Olive UC ANR publication, 3452. University of California, Berkeley
Farahbakhch G, Moini M (1975) Olive pests in Iran, Report of plant pests and diseases research institute. National Agricultural Research Organization, Ministry of Agriculture and Natural Resources, Tehran, pp 1–14
Farmer A (1952) Pests and diseases of persimmons in New Zealand and their control. Orchardist N Z 25:4–9
Fletcher WF (1942) The native persimmon. U S Dept Agr Farmers' Bul 685, 22 pp
Foda SMA (1973) Studies on Margaronia (Glyphodes) unionalis and its control. M.Sc thesis, Ain Shama University Cairo
Free JB (1970) Insect pollination of crops. Academic, London, p 544
Free JB (1993) Insect pollination of crops, 2nd edn. Academic Press, London

Fukae Y, Hamachi F, Tsujikawa Y (1987) A detailed analysis of the effects of honeybee introduction on the pollination of persimmons. Honeybee Sci 8:167–171
Gallaway J et al. (1999) Who loves gophers? Nut Grower Magazine. Sept 1999
Gould HP (1940) Oriental persimmons. US Dept Agr Leaflet 194, 8 pp
Greathead DJ (1976) Report on a survey for natural enemies of olive pests in Kenya and Ethiopia, September-December 1975. Commonwealth Institute of Biological Control, Delémont, 23 pp
Griggs WH, Hartmann HT, Bradley MV, Iwakiri BT, Whisler JE (1975) Olive pollination in California. Calif Agric Expt Sta Bull 869
Haniotakis GE, Francke W, Mori K, Redlich H, Schurig V (1986) Sex specific activity of (R)-(−)- and (S)-(+)- 1,7dioxaspiro[5.5]undecane, the major pheromone of *Dacus oleae*. J Chem Ecol 12:1559–1568
Haniotakis G, Kozyrakis M, Fitsakis T, Antonidaki A (1991) An effective mass trapping for the control of *Dacus oleae* (Diptera: Tephritidae). J Econ Entomol 84:564–569
Haq A (1962) Notes on the Bionomics of Lachnosterna longzpennis B1. (Melolo?thinae : Coleoptera). Indzan J Ent 24:220–221
Hartmann HT, Opitz KW (1966) Olive production in California. Calif Agr Expt Sta and Ext Serv Cir 540, 63 pp
Hawkesa NJ, Janesb RW, Hemingwaya J, John Vontasa J (2005) Detection of resistance-associated point mutations of organophosphate-insensitive acetylcholinesterase in the olive fruit fly, Bactrocera oleae (Gmelin). Pestic Biochem Physiol 81:154–163
Hazslinszky B (1955) The importance of the chestnut tree for beekeeping. Meheszet 3(6):109–110, [in Hungarian.] Aa-171/57
Herrera E (1997) Growing pistachio in New Mexico. Coop Ext Serv Circ 532
Hill DS (1983) Agricultural insect pests of the trpoics and their control, 2nd edn. Cambridge University Press, Cambridge, 746 pp
Iisfendiyarogluglugu M, Ozeker E, Misrili A, Saglam H (2001) Determination of pollinator characteristics of different Pistachio sp. In Manisa-Yunt mountain area, pp 267–270
Impiumi G, Ramina A (1967) Floral biology and fruiting of the walnut (*Juglans regia*) floral morphology and pollen transportation. Riv Orthoflorofruttic Ital 51:538–543
Kapoor VC (1993) Indian fruit flies (Insecta: Diptera: Tephritidae). Oxford & IBH Publications, New Delhi, p 228
Karamanlidou G, Lambropoulos AF, Koliais SI, Manousis T, Ellar DJ, Kastritsis C (1991) Toxicity of Bacillus thuringiensis to laboratory populations of the olive fruit fly (Dacus oleae). Appl Environ Microbiol 57:2227–2282
Kardoush MA, S Al-Bakeer (2007) Effect of pollen grain types on growth and development of cultivar of Pistacia vera L. (Ashouri), In: International conferrence on the technology of Hort. Productivity for Sustainable Development and Biodiversity, Aleppo-Syria, 2–4 Dec 2007. pp 9
Katsoyannos P (1992) Olive pests and their control in the Near East, FAO plant production and protection paper 115. FAO, Rome
Katsoyannos BI, Kouloussis NA (2003) Captures of the olive fruit fly *Bactrocera oleae* on spheres of different colours. Entomol Exp Appl 100:165–172
Kaul V, Tiku AK, Shankar U, Monobrullah M (2007) Green stink bug (Hemiptera: Pentatomidae) recorded as a new pest of olive in India. J Asia-Pacific Entomol 10(1):81–83
Kavallieratos NG, Tomanović Ž, Athanassiou CG, Starý P, Žikić V, Sarlis GP, Fasseas C (2005) Aphid parasitoids infesting cotton, citrus, tobacco, and cereal crops in southeastern Europe: aphid- plant associations and keys. Can Entomol 137:516–531
Kawagoe H (1955) Studies on the period during which the capacity for fertilization of the chestnut persists. okayama nogaku shikenjo rinji Hokoku/spec. Bull Okayama Agric Expt Stat 53:141–154, [in japanese.] abs. in plant breed. 28(4): 826. 1958
Kearby WH (1975) Insects that affect the growth and form of black walnut in a multicrop system. 66th Ann. Rept. Of the Northern Nut Growers Assn. pp 119–127
Khan SA, Bhatia S, Tripathi N (2013) Entomological studies of chaetoprocta odata, an important pest on walnut trees (Juglans Regia L.) in Kashmir valley. J Acad Ind Res (JAIR) 2(6):378–381

Kovanci B, Kumral NA (2004) V. Uluslararası Zeytin Yetiştiriciliği Sempozyumu. İzmir, 27 Eylül-2 Ekim, 68 pp
Lavee S (1996) Biologia e fisiologia dell'olivo. In: Enciclopedia Mondiale dell'Olivo, Consiglio OleicoloInternazionale, Madrid, pp 61–110
Lavee S, Allol R, Rapoporth F, Troncoso A (1996) The floral biology of the olive: effect of floral number, typeand distribution of fruit set. Sci Hortic 66:149–158
Lo Scalzo R, Scarpati ML, Verzengnassi B, Vita G (1994) Olea europaea chemical repellent to Dacus oleae females. J Chem Ecol 20:1813–1823
Long RF (1998) Bats feed on crop pests in the sacramento valley. Calif Agric 52(1):8–10
Lounsbury CP (1918) Report with appendices for the year ended 31st March 1918. Department of Agriculture. Union of South Africa: Blue Book. UG 37–39. Cape Times Limited. Government Printers, Cape Town
Mally C (1924) A means of catching the fruit nibbler (*Odontionopa serica* Gyll.). J Dept Agric United South Africa IX(4), Pretoria
Manousis T, Moore NF (1987) Mini-review. Control of *Dacus oleae* a major olive pest of olives. Insect Sci Appl 8(1):1–9
Marshall PT (1989) Insects that affect growth and quality of black walnut. In: The continuing quest for quality: proceedings, fourth Black Walnut symposium, 30 July–2 August 1989, Southern Illinois University at Carbondale
Masoodi MA, Trali AR (1978) Record of chaetoprocta odata hewitson (Lycaenidae: Lepidoptera) on walnut trees in Kashmir. Ind J Plant Protect 15:213, 10
Mathur RN (1975) Psyllidae of the Indian subcontinent. Indian Council of Agricultural Research, New Delhi, 429 pp
Mazomenos BE (1983) Biosynthese Van Een Sexferomoon Van de Olijfvlieg Dacus Olea Gmel Gent : Mazomenos, 137 pp
McGregor SE (1976) Insect pollination of cultivated crop plants. US Dept Agric Agric Handb 496:1–411
Mckay JW (1939) Male sterility in castanea. Amer Soc Hortic Sci Proc 37:509–510
Mckay JW (1942) Self-sterility in the chinese chestnut (castanea mellissima). Amer Soc Hortic Sci Proc 41:156–160
Mckay JW (1972) Pollination of chestnut by honey bees. North Nut Grow Assoc Ann Rpt 63:83–86
Mehrnejad MR (2001) The current status of pistachio pests in Iran. In: Ak BE (ed) XI GREMPA seminar on pistachios and almonds. Zaragoza: CIHEAM, pp 315–322 (Cah iers Option s Méditerran éen n es; n. 56)
Miller JW (1895) Experiments on pollination of olives. California Agricultural Experiment Station. Report 1894–1895
Mills NJ (1997) Integrating Parasitism into Codling Moth Control. Nut Grower, Mar 1997, pp 37–43
Mills NJ, Dixon PL, Pickel C, Buchner R, Olson W, Krueger W, Kelley K, Thompson W, Hasey J (1995) *Trichogramma* augumentation as a component of of the management of codling moths in walnuts, Walnut research reports 1994. Walnut Marketing Board, Sacramento, pp 134–144
Mills N, Pickel C, Mansfield S, McDougall S, Buchner R, Caprile J, Edstrom J, Elkins R, Hasey J, Kelley K, Krueger B, Olson B, Stocker R (2000) Mass releases of *Trichogramma* wasps can reduce damage from codling moth. Calif Agric 54(6):22–25
Mir GM, Wani MA (2005) Severity of infestation and damage to walnut plantation by important insect pests in Kashmir. Ind J Plant Protect 33(2):188–193
Montiel-Bueno A, Jones O (2002) Alternative methods for controlling the olive fly, *Bactrocera oleae*, involving semiochemical. Use of pheromones and other semiochemicals in integrated production. Int Org Biol Integr Control–West Palearctic Reg Sect (IOBC/WPRS) Bull 25:147–156
Morettini A (1941) L'incremento produttivo negli olivi Moraiolo e Frantoio con l'impiego di adatte varietà impollinatrici. L'Italic Agricola 9:631–639
Mort CH (1952) Fruitfulness in olives. Agr Gaz N S Wales 63:371–372

Mourikis PA, Tsourgianni A, Chitzanidis A (1998) Pistachio nut insect pests and means of control in greece. Acta Hort (ISHS) 470:604–611. http://www.actahort.org/books/470/470_85.htm
Muller H (1883) The fertilization of flowers. Macmillan, London
Myburgh AC (1952) Olive beetle in the Western Cape province. Farming in South Africa (April), Western province fruit station, stellenbosch popular series no. 115 (reprint no. 17). The Government Printer, Pretoria, pp 1–4
Nel A, Krause M, Khelawanlall N (2002) A guide for the control of plant pests, 39th edn, Directorate: agricultural productions inputs. National Department of Agriculture, Republic of South Africa, Pretoria
Neuenschwander P (1982) Beneficial insects caught by yellow traps used in mass trapping of the olive fly, Dacus oleae. Entomol Exp Appl 32:286–296
Nixon PL, McPherson JE (1977) An annotated list of phytophagous insects collected on immature black walnut trees in southern Illinois. Great Lakes Entomol 10(4):211–222
Nkonkobe Municipality (2004/2005) Integrated Development Plan (IDP). Final Draft
Noyes J (2003) Universal calcidoidea database. Available online at http://www.nhm.ac.uk/jdsml/researchcuration/projects/chalcidoids/hosts.dsml?index=HostsandsearchPageURL=indexHosts.dsmlandVALGENUS=sycophilaandVALSPECIES=%25
Ohno M, Sato S, Sawabe H (1958) The study of chestnut pollination. 1. The fruit set of chestnuts which were pollinized by the wetting pollen.l. Chiba Univ Faculty Hortic Tech Bull 6:129–135 [In Japanese, English summary.]
Ozeker E, Iisfendiyarogluglugu M, Misrili A (2005) Variation of fruit characteristics of Pistacia spp. Pollinated by different hybridization Pistachio types in Manisa-Yunt mountain area in Turkey. Ege Univ Ziraat Fak Derg 42(1):13–24
Pellett FC (1947) American honey plants. Orange Judd, NewYork
Pierce NB (1896) Olive culture in the united states. US Dept Agr Yearbook 1896:371–390, wPinhey 1946
Porsch O (1950) Geschichtliche Lebenswertung der Kastanienblüte. Österr Botsch Zagreb 97:269–321
Prestidge RA, Holland PT, Clarke AD, Malcolm CP (1989) Pesticides for use close to and during harvest of persimmons. Proceedings of the 42nd NewZealand Weed and Pest Control Conference, pp 195–199
Rapsi A (1982) Preliminary considerations on the capture of useful entomofauna during the use of chemiochromotrophic traps in the olive-grove. Frustula Entomologia 4:103–109
Reed CA (1941) The present status of chestnut growing in the United States. Amer Soc Hortic Sci Proc 39:147–152
Rice RE (2000) Bionomics of the olive fruit fly *Bactrocera* (*Dacus*) *oleae*. Univ Calif Plant Protect Q 10(issue 3):2–5
Sharma PL (1961) The honeybee population among insects visiting temperate zone fruit flowers and their role in setting fruit. Bee World 42(1):6–8
Sharp DS, Eskenazi B, Harrison R, Callas P, Smith AH (1986) Delayed health hazards of pesticide exposure. Annu Rev Public Health 7:441–71
Shehata WA, Abou-Elkhair SS, Stefanos S, Youssef AA, Nasr FN (2003) Biological studies on the olive leaf moth, *Palpita unionalis* Hübner (Lepid., Pyralidae) and the olive moth, *Prays oleae* Bernard (Lepid., Yponomeutidae). J Pestic Sci 76:155–158
Silvestri F, Morettini A, Zappi-Recordati A (1947) The bee as a pollinator of the olive. Olivicoltura 2(9):12–15. [in italian.]
Singh (1964) Temperate fruit pests in India. Entomol Soc 5:261–300
Singh K, Sharma VK, Shant PS (1988) Efficacy of banding materials and insecticidal foliar sprays against mango mealy bug, *Drosicha mangiferae* Green (Margarodidae: Homoptera). Pesticides 22(10):26–27
Skouras PJ, Margaritopoulos JT, Seraphides NA, Ioannides IM, Kakani EG, Mathiopoulos KD, Tsitsipis JA (2006) Organophosphate resistance in olive fruit fly, *Bactrocera oleae*, populations in Greece and Cyprus. Pest Manag Sci 63: 42–48. http://greek.ari.gov.cy/content/diakratiki/paper.pdf

Small S, DeStaebler J, Geupel GR, King A (1999) Landbird response to riparian restoration on the Sacramento river system, Preliminary results of the 1997 and 1998 field season. Point Reyes Bird Observatory, Stinson Beach

Sokolov VB, Chernyshov MP (1980) Chestnut (*Castanea sativa*) of the black sea area of the caucasus. Pchelovodstvo 1:22–23

Son JK, Ji Eun Yun, Park CG (2009) Insect pest problems of sweet persimmon in Korea. Acta Hort (ISHS) 833:325–330. http://www.actahort.org/books/833/833_52.htm

Stasinakis P, Katsares V, Mavragani-Tsipidou P (2001) Organophosphate resistance and allelic frequencies of esterases in the olive fruit fly *Bactrocera oleae* (Diptera: Tephritidae). J Agric Urban Entomol 18:157–168

Steven D, Sale PR (1985) Insect control trials on persimmons. In the Proceedings of 38th N.Z. Weed and Pest Control Conf.: 203–206. Organised By © 1985 New Zealand Plant Protection Society (Inc.) www.nzpps.org Refer to http://www.nzpps.org/terms_of_use.html

Stonedahl GM, Dolling WR, duHeaume GJ (1992) Identification guide to common tingid pests of the world (Heteroptera: Tingidae). Trop Pest Manag 38:438–449

Taylor EA (1955) Cantaloup production increased with honey bees. J Econ Entomol 48:327

Thakur JN, Rawat US, Pawar AD (1989a) Investigations on the occurrence of natural enemies of San Jose scale, *Quadraspidiotus perniciosus* Comstock (Hemiptera: Coccidae) in Jammu and Kashmir and Himachal Pradesh. Entomon 14:143–146

Thakur JR, Sharma PC, Gupta PR (1989b) Occurrence of olive psylla, *euphyllura pakistanica* on olive in India: a new record. Trop Pest Manag 35(3):331

Tsurkan IP, Pyntya MA (1987) Characteristics of pistllate flower bud formation in walnuts. Sadovodstvo i Vinogradrstvo Moldavii 1:27–30

Van Steenwyk RA, Ferguson L, Zalom FG (2002) Olive fruit fly. University of California Pest Management Guidelines. http://www.ipm.ucdavis.edu/PMG/r583301311.html. Accessed 6 Jun 2003

Vassilaina-Alexopoulou P, Santorini AP (1973) Some data on the biology of P*alpita unionalis* (Hübner) (Lepidoptera: Pyralidae), under laboratory conditions. Ann Inst Phytopathol Benaki 10(4):320–326

Vossen P, Varela L (2003) Olive fruit fly. University of California Cooperative Extension – Sonoma County. http://cesonoma.ucdavis.edu/HORTIC/olive_fly/olive_fruit_fly.pdf. Accessed 6 June 2003

Watanabe Y, Adachi M, Hiyama H (1964) Studies on the pollination in chestnut trees. 1. Influence of distance from the pollinizer upon bur-set in variety ginyose. Ibaraki Hortic Expt Stat Bull 1:7–12. [In Japanese, English summary.]

Wells L (2007) Pollination. Southeastern Pecan Grow Handb 81

Westwood MN (1978) Temperate zone pomology. Freeman, San Francisco

Wharton RA (1989) Classical biological control of fruit-infesting tephritidae. In: Robinson AS, Hooper G (eds) World crop pests, vol 3A. Elsevier, Amsterdam, pp 303–313

Whitehead VB, Rust DJ (1997) Control of fruit piercing moth, Sorrodes partita (Fabr.) Wood, B.W. 1997. Source of pollen, distance from pollenizer, and time of pollination affect yields in block-type pecan orchards. HortSci 32:1182–1185

Wood BW (2000) Pollination characteristics of pecan trees and orchards. HortTechnol 10:120–126

Zimmerman RH (1972) Juvenility and flowering in woody plants: a review. Hortscience 7:447–455

Zouiten N, Hadrami IE (2001) Olive tree psyllid: state of knowledge and perspective of control. Agriculture 4(10):225–232

Chapter 7
Berry Fruits

Berry fruit is a thin-skinned one-celled fleshy fruit with seeds scattered through the flesh. The important berries include: Raspberries, blueberries, strawberries, blackberries and cranberries. Most of the berry fruits depend upon or benefit from bee pollination, but pollinators are adversely affected due to insecticides used for controlling pests. It is therefore important to devise a pest management system to provide safety to pollinating insects for quality and quantity of berry production. Several studies have been conducted on the role of pollinating insects and management of insect pests on berry fruit production. Information on some of the important berry fruits is discussed below:

7.1 Cranberry

Vaccinium macrocarpon Ait., family Ericaceae

The cranberry plant is a semi evergreen creeping perennial that roots freely along the runners to form a mat. The cranberry flower resembles very closely to the neck and head of a crane (Marucci 1967a). The individual flower of cranberry is small about half inch across. The petals of newly opened flowers are white or only slightly pink. The petals begin to open in the morning and later curl back which results in the exposure of the stamens and the style. The flowers turn rosy pink, if the pollination is delayed. Stamens 5–8 are brown in colour and are placed tightly forming a tube like structure (Cross 1953; Darrow et al. 1924; Franklin 1940). The pollen is released at tube tip after the maturation of anthers. The pollen is relatively heavy and is not wind blown, nor is it likely to come in contact with its own stigma. The pollen fails to come in contact with its own stigma as pistal length is more than the length of the stamens, and being heavy it is also not easily blown away by wind. The ovary is four carpelled and contains 24–36 ovules.

The production of nectar and pollen in cranberries depends upon the prevailing weather conditions and location. Cranberries produce little nectar but are good source of pollen for bees (Caswell 1962) Marucci (1967a) found that cranberries are

D.P. Abrol, *Pollination Biology, Vol. 1*, DOI 10.1007/978-3-319-21085-8_7

poor producers of nectar and pollen and honeybees are not much attracted. Stricker (1953) reported that honeybees visit cranberries mainly for pollen. This reveals that the plant is more attractive to honeybees for pollen than the nectar. Shimanuki et al. (1967) showed that some colonies consistently collect more pollen from cranberries than other leading to the development of specially selected bees for cranberry pollination. Cranberry breeders can benefit by selecting plant strains that produce more nectar to attract pollinating insects.

7.1.1 Pollination Requirements

Pollination by insects is essential for quality and quantity of fruit production in cranberries. Placement of honeybee colonies proved very beneficial (Gates 1911; Franklin 1911) and the areas from which the bees were excluded bore at least half crop of berries. Several investigators have concluded the usefulness of bees in pollination of cranberries (Darrow 1924; Roberts and Struckmeyer 1942; Filmer and Doehlert 1955; Hutson 1924, 1925, 1926, 1927; Farrar and Bain 1946, 194, Bain 1946). Filmer and Doehlert (1959) recommended one colony of honeybees per 2–3 acres for optimum fruit production, though wind also helps in pollen transfer but its utility has been discounted by many investigators. Cranberries require 2–3 honey bee colonies per acre for adequate pollination. Despite the claim that wind is of minor importance to cranberry pollination (Filmer 1949), Papke et al. (1980) demonstrated that there is enough cranberry pollen being carried in the wind to provide a significant contribution to pollination. Moreover, there is some evidence that manually agitated cranberry plants could set fruit in the absence of bees (Roberts and Struckmeyer 1942). In contrast, several studies found contrary evidence and concluded that wind or the manual agitation of plants does not contribute to cranberry pollination (Filmer 1949; MacKenzie 1994).

7.1.2 Pollinators

Crane berries have limited self-compatibility which requires outcrossing by insect pollinators for optimum fruit yields (Sarracino and Vorsa 1991). Among, honeybees, Apis mellifera is most extensively used to provide pollination services for cranberry (Evans and Spivak 2006; Ratti et al. 2008). Cranberry flowers having poricidal anthers are also suited for buzz pollination (Buchmann 1983; Free 1993). Since, honeybees do not buzz-pollinate, they are considered to be less effective pollinators (Cane et al. 1996; Cane and Schiffhauer 2001; King and Buchmann 2003; MacKenzie 1994), although some pollen transfer may occur due to their foraging activities for nectar and pollen. The alternative pollinators include managed bumble bees (Bombus spp.; Cane and Schiffhauer 2003; Evans and Spivak 2006; Stubbs and Drummond 1997), leafcutter bees (Megachile spp; Cane et al. 1996; Cane and

Schiffhauer 2003; MacKenzie and Javorek 1997; Stubbs and Drummond 1997), and mason bees (Osmia spp.; Cane and Schiffhauer 2003; Stubbs and Drummond 1997).

7.1.3 *Pollination Recommendations and Practices*

Most of the investigators have recommended one colony for area ranging between 1 and 5 acres (1 colony for 2–3 acres (Filmer and Doehlert 1959), 1 colony for 5 acres (Doehlert 1940), 1 colony for 2 acres (Cross 1953, 1966; Firmer 1953; Farrar and Bain 1946), 1 colony per acre (Swenson 1958; Stewart 1970; Stewart and Marucci 1970). 0.5–1.0 honeybee hives per hectare (Broussard et al. 2011), On an average, one colony per acre is sufficient for optimum yields in crane berry.

7.1.4 *Insect Pests in Craneberries*

There are many pests in cranberry but *Sparganothis* fruitworm *(Sparganothis sulfureana)*, Blackheaded fireworm *(Rhopobota naevana)* and Cranberry tipworm *(Dasineura oxycoccana)* are the major ones. Sparganothis fruitworm (SFW) and Blackheaded fireworm (BHFW) are slightly less damaging to fruit as they initially prefer to feed on the new tissue of the growing tips of uprights and runners. They make "tents" (feeding shelters) of the tips. Second and third generations of these two species also feed on foliage but the *Sparganothis* fruitworm prefers to feed on and within the cranberries. Cranberry tipworm *(Dasineura oxycoccana)* is increasing in most cranberry-growing regions. This tiny midge lays its eggs in upright tips where the instars feed and eventually kill the growing tips. Cranberry tipworm, *Dasyneura oxycoccana* (Johnson), feeds on the youngest leaves of cranberry, *Vaccinium macrocarpon,* at the apex of growing shoots. Larvae rasp the leaves while feeding, and the cranberry plant responds by growing distorted, cupped leaves. Feeding eventually kills the apical meristem of the shoot (Smith 1890). Secondary laterals will grow if damage is early enough, but later larval generations will kill next year's fruit buds. Monitoring in early stages shall help to reduce crop loss due to these and other pests. The properly timed pest controls help to prevent defoliation and loss of flower buds in spring; to safeguard pollinators during bloom; to prevent damage to fruit in summer; to prevent damage to roots throughout the growing season; and to minimize pesticide residues in the environment. Pest control in cranberries is combination of chemical, cultural, biological and behavioural approaches. The details of each pest are as given below:

7.1.4.1 Blackheaded Fireworm

Blackheaded Fireworm (*Rhopobota naevana* (Hlon.). They overwinter as flat, yellowish eggs on the underside of leaves. The colour of the larvae is green or pale yellow with shiny black heads. Generation 1 larvae hatch in late April. The small brown moths hatch in late May through June and lay eggs. Second and third generation larvae may enter and feed on berries. Larvae tend to be in the same locations year after year and are greatest around the edges. Use one of the following insecticides: such as Diazinon 500E, Diazinon 50 W or Malathion 85EC with proper care as these products are toxic to bees.

7.1.4.2 Cranberry Tipworm

Cranberry Tipworm (*Daineura oxycoccana* (Johnson). Adult is a tiny (fly) midge that lays single eggs in upright tips. Injured growing tips from feeding die and turn dark brown. Leaves become cupped and slightly puckered. Use Diazinon 500E when 30 % or more of the samples contain tipworm eggs or larvae.

7.1.4.3 Sparganothis Fruitworm

Sparganothis Fruitworm (*Sparganothis sulfureana*). Yellow-headed, greenish yellow caterpillar (larvae) that feed on and in cranberry fruit. Young larvae have dark heads and can be confused with Blackheaded fireworm. Adults are distinctive yellow moths about 1 cm with a cross pattern on their folded wings. Second and third generation larvae prefer to feed on and within cranberries. Alternate hosts include highbush blueberry and yellow loosestrife, *Lysimachia terrestris*. Preserve the natural enemies, if absolutely necessary apply chemical control using Diazinon 500E.

7.1.4.4 Cranberry Fruitworm

Cranberry fruit worm (*Acrobasis vaccinii* (Riley): It is a potentially serious pest as it feeds only on the berries. The strong flying moths are dark brown with noticeable white band on the forewing. Larvae remain inside berry, eating and filling the berry with frass. Larvae may eat several berries.

7.2 Strawberry (*Fragaria* x *Ananassa*)

7.2.1 Flowering

The strawberry has many flowers in a cluster and a definite order of flowering (Valleau 1918). Cluster is a series of double branching parts bearing a flower in the crotch of each branch. The flower is in the first crotch of each branch. The first flower in the first crotch is termed as the *primary* flower; the two in the next two crotches as *secondary* flowers, the next four are tertiary, the next eight quarternary flowers, and further sixteen if they develop, are the quinary flowers. Primary flowers produce the largest fruit. The flowers are perfect in the most important commercial varieties, but some varieties have exclusively female flowers, exclusively male flowers, flowers with only a few stamens, or flowers with nonfunctioning stamens. Each perfect flower has five petals, many pistils and styles, and 24–36 stamens (Fig. 7.1). Stigmas become receptive before the anthers dehisce which encourages cross-pollination by insects. Pollination most likely occurs during the first 4 days after the flower opens (McGregor 1976), but some flowers start to dry by the second day (Connor 1970). Fertilized ovules stimulate the surrounding tissue to begin growing. Non-fertilized ovules do not grow and if there are many of them the berry will be misshapen, if it develops at all. Strawberry flowers produce nectar and pollen, but they are not always attractive to honey bees.

7.2.2 Strawberry Pollination Requirements

Most commercial varieties are self-fertile. Varieties differ considerably in their responsiveness to different modes of pollination. Certain morphological characteristics of the flowers are associated with high levels of self-pollination, namely, relatively long pollen grains and long stamens (Connor and Martin 1973). Poor distribution of pollen on to stigmas results in small, misshapen, devalued fruit called *nubbins*. Insect pollination is essential for enhancing crop productivity (Free 1970; Kapil 1970; Abrol 1992).

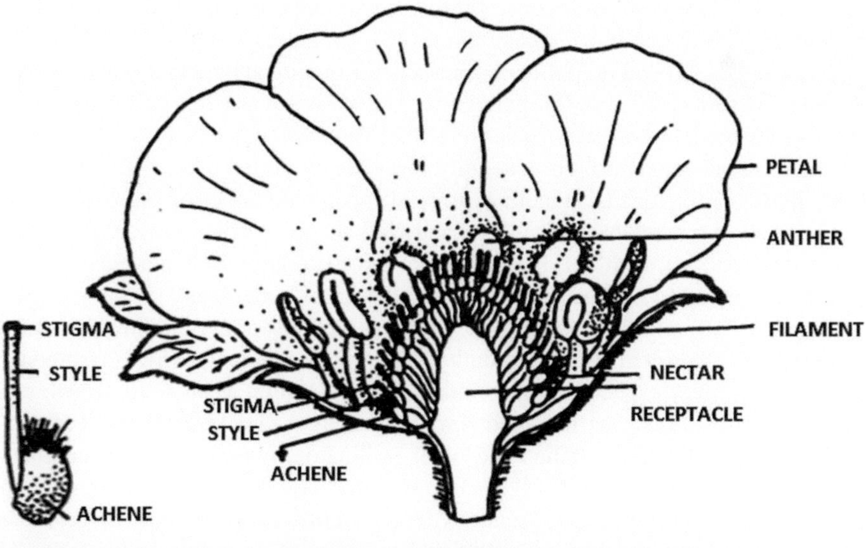

Fig. 7.1 Longitudinal section of strawberry with an achene and style (Free 1993)

7.2.3 Insect Pollinators

Strawberry is pollinated by self, wind, and bees. The relative contribution of each varies by variety, weather, and bee population size. On average, self-pollination accounts for 53 % of fruit development; wind adds an additional 14 % (=67 %), and bees add another 24 % (=91 %) (Connor and Martin 1973). Self-pollination is more difficult in varieties with short stamens, and for these the pollinating action of bees is more important. Honey bee visitation improves fruit yield and quality in strawberry in spite of the plant's self-pollinating habit (Antonelli et al. 1988). Free (1993) reported that honeybee pollination increased fruit set by 25 %, fruit yield by 18–100 % and malformed fruits decreased by 9–41 %. Almost every honey bee visiting a strawberry flower contacts the stigmas and anthers (Free 1968b); thus bees help distribute pollen to all pistils, which promotes a well-shaped fruit.

It takes many bee visits to spread pollen evenly and to optimize fruit-set and shape. In Russia, 16–19 honey bee visits per flower are considered adequate and fruit-set is optimized at 20–25 visits (Skrebtsova 1957). With 'Veestar' in Québec, Canada, most pollination is accomplished during the first 40 cumulative seconds of honey bee visitation which translates to the first four bee visits. Nearly 100 % of the pistils are pollinated after six bee visits. Bee activity is especially important in the large primary flowers which have many stigmas and produce the best fruit. There is a higher rate of pollination by wind and gravity in the secondary and tertiary flowers that have fewer stigmas (Chagnon et al. 1989). Mcgregor 1976 recommended 12.4–25 colonies of honeybees per hectare of strawberries for optimum fruit production. Williams 1994 recommended 2.5 and Scott-Dupree et al. (1995) 1.2 colonies of honeybee per hectare.

Weather conditions are often poor for bee activity during earlyspring strawberry bloom. The best way to compensate for this is with imported honey bee hives. This is especially important in large plantations (de Oliveira et al. 1991).

Strawberries depend on pollinating insects for quality and quantity of fruit production. Darrow (1927, 1937) reported that fruit set is not possible without pollination in a few varieties of strawberries. Free (1970) reported that insect pollination is essential for enhancing crop productivity. Bees are by far the most effective strawberry flower pollinators (Antonelli et al. (1988) compared to other insects or wind for increase in marketable, large sized and well shaped berries. Aras et al. (1996) noted that seed set, fruit set, berry weight, and maturation rate significantly and positively correlated with an increase in the density of honeybees.

Singh (1979) in India reported honey bees as the most important flower visitors of strawberry. Woo et al (1986) in South Korea reported that, hoverflies (syrphidae) and solitary bees, especially Andrenidae, were common visitors to strawberry flowers. Abrol (1989a, b) recorded *Apis cerana* as a predominant pollinator of strawberries in Kashmir, India. The other pollinators observed in few numbers included *Lasioglossum* species, *Xylocopa* spp. ants and flies. Pollinators used in a closed environment increased productivity and improved quality and saved labor requirement (Sakai and Matsuka 1988). Honeybees (*A. mellifera* and *A. cerana*) are most effective pollinators of strawberry (McGregor 1976). About 30 % of the world production of strawberry comes from a single cultivation in open field in the USA and Europe where honeybees are used to increase fruit set ratios (Bagnara and Vincent 1988; Free 1993); increase fruit weight (Chagnon et al. 1989; Moore 1969); decrease malformed fruit (Nye and Anderson 1974; Pion et al. 1980); increase receptacle development (Connor and Martin 1973); and, increase yield (de Oliveira et al. 1991). In Japan about 10 % of total area for greenhouse cultivation has been shared with strawberry since the end of 1960s (Katayama 1987a, b; Kitagawa 1985; Tsujikawa 1981). In Korea use of honeybees started at the end of 1970 and at the end of 1980. Then research on the use of honeybee as pollinator was initiated (Ahn et al. 1989, 1994). Abrol (1992) found that different cultivars of strawberry varied in their attractiveness to honeybee *Apis cerana* depending upon their nectar production characteristics. Levels of major sugars such as, glucose, fructose, sucrose and amino acids in the pollen and nectar of strawberry flowers were quite variable in different varieties but the ratio between sucrose and the two monosaccharides remained stable. Honeybees are being utilized for fruit production of strawberries in the plastic green houses in South Korea and Japan (Katayama 1987a, b; Sakai and Matsuka 1988; Ahn et al. 1989). As a result, the numbers of deformed fruits were lower and crop yields were higher than the green houses when honey bees were not used. Matsuka and Sakai (1989) reported that bee pollination significantly increased production in Japan. Several investigators (Free 1968; Katayama 1987a, b; Sakai and Matsuka 1988; Shimotori 1981) reported that the cultivated varieties of strawberry became bigger and sweeter, ripened earlier with less number of misshapen fruits when bee pollination was allowed. They have also reported the usefulness of honey bees in quality and quantity of fruit production in strawberries. Some other pollinators such as *Eristalis cerealis* (Diptera) and *Osmia cornifrons* (Kabayashi 1987; Meeta 1978) have also been recorded frequenting flowers of strawberry.

Connor and Martin (1973) and Connor (1975) reported that cultivars of strawberry in cloth cages which prevented insect pollination and eliminated or greatly reduced wind action, nylon insect cages but allowed wind pollination and open plots which allowed both insect as well as wind pollination, resulted in varied levels of fruit setting as 81 %, 93 % and 98 % respectively.

Anderson (1974) recorded more than 108 species of insects belonging to 35 families visiting strawberry flowers in Utah and USA. He categorized one species of honeybees, two species of Osmia, one species of Halictus and two species of *Eristalis* as the most important pollinators. Free (1968b) reported that very few bumble bees were visiting strawberry flowers and relatively cold weather deferred honey bees from doing so. Hooper (1932) reported that pollination is carried mainly by insects other than bees and especially by Diptera when it is cold.

Mommers (1961) reported that the cages with and without bees produced 86.9 kg and 74.0 kg of well formed fruit, and 3.3 kg and 4.1 kg of malformed fruit, respectively. Priore and Sannino (1979) in Italy reported that the pollination under greenhouse condition in the presence of honey bees resulted in 25 % increase in fruit set (68.47 %) and 70 % increase in fruit yield although 13 % of the fruit was deformed compared to 48 % when bees were absent. He also reported that the percentage of small, medium and large sized fruit was 11, 75 and 14 with bees and 19, 71 and 11 without bees. Similar results were obtained in Italy (Bonfante 1970; Pinzauti 1987; Burd 1970) in greenhouses and plastic greenhouses. Skrebtsova (1958)) reported that for the commercial growers who desire the maximum perfect berries in number provision of saturation pollination is necessary.

Jaycox (1970b) found that 11–15 visits are necessary to fertilize most of the ovules in each flower. Antonelli et al. (1988) reported that exposure of strawberry crop for 15 days or more was necessary to obtain acceptable fruit yield and berry weight. Several scientists have reported that honey bee population often increases the yield and is especially important in reducing the proportion of malformed fruit.

Several investigators have reported increased percentage of fruit set due to pollination to the extent of 25 % (Sannino and Priore 1979), 21 % (Moeller and Koval 1973) 20 % (Nye and Anderson 1974), 66 % (Lackett and Burkhardt 1979), 18–100 % (Blasse 1981). Moore (1969), Nye and Anderson (1974), Sannino and Priore (1979); Blasse and Haufe (1989); Goodman and Oldroyd (1988) reported decreased percentage of malformed fruit 32 %, 13 %, 9 %, 30 % and 41 % respectively. Lackett and Burkhardt (1979) reported decrease in percentage of culled fruits. Moeller and Koval (1973); Goodman and Oldroyd (1988) reported the increase in percentage to the extent of 7 % and 16 % respectively.

Skrebtsova (1957) reported that bees readily moved from one cultivar to another as they preferred some cultivars to others. She was also able to correlate this reference with the amount of pollen the different varieties produced and the pollen on the bodies of the bees working on them. Connor (1975) also reported the large differences in the attractiveness of different cultivars. Skowronek et al. (1985) reported that a strawberry fruit produced a mean of 0.6–1.7 mg pollen. Free (1968b) that on nearly every flower visit honeybees nectar-gatherers searched both stigmas and anthers. He further reported that bees collecting nectar had pollen loads also and so were collecting pollen incidentally. Antonelli et al. (1988) in the USA reported that 80 % of bees visiting strawberry flowers carried pollen loads while as Singh (1979) in India reported 60 % of *Apis cerana* carrying pollen loads. Percival (1955) reported that free anthesis does not take place below 14° centigrade and honeybee activity increased with temperature and reached at peak during midday. Connor (1970) reported that up to 79 % of honeybees collected pollen before noon in some areas. Kronenberg et al. (1959) reported that weather unsuitable for bees increased the percentage of malformed fruits. Kendall et al. (1971) in UK, reported that the fruit yield was increased in cages with *Eristalis tenax* (917 g/100 flowers) as compared to those which were caged without *E. tenax* (481 g/flowers).

7.2.4 *Insect Pests of Strawberry*

Anumber of insect pests have been reported attacking strawberry crop. They include:

7.2.5 Sucking Insect-Pests

7.2.5.1 Strawberry Aphids

Aphids infesting soft berry fruits (Blackman and Eastop 2000) are rarely abundant enough to cause economic damage. More often, their primary economic impact is as virus vectors. Aphids are not a major problem on field strawberries (Mossler and Nesheim 2003) however, in greenhouse the cotton aphid, *Aphis gossypii* Glover, can be a serious problem (Leclant and Deguine 1994). They feed on the underside of leaves sucking out plant sap. Their high population can reduce the vigor of the plant, making it susceptible to other pests. The honeydew that aphids excrete reduces fruit quality because of the development of a black sooty mold on the substrate. Moreover, this sooty mold reduces photosynthetic production and otherwise reduces the quality of the plant causing considerable economic injury. Three types of aphids have been found infesting the strawberry at various crop growth stages. They are as follows-

7.2.5.2 Strawberry Aphid, *Chaetosiphon fragaefolii*

It is pale green to yellowish in color. **Melon aphid, *Aphis gossypii:*** They are small, globular, and vary in color from yellowish green to greenish black.

7.2.5.3 Green Peach Aphid, *Myzus persicae*

The green peach aphid is green to greenish yellow in color and is more streamlined than the rounded melon aphid. Complexes of parasitoids and predators have been found suppressing the aphids infestation on strawberry plants (Kaplan and Eubanks 2002). Natural enemies found effective against the cotton aphid include lady beetles, Coccinella septempunctata and Hippodamia convergens, green lacewing, Chrysoperla carnea and hymenopteran wasps, Lysiphlebus testaceipes and Aphidius colemani (Howard et al. 1985; van driesche and Bellows 1996; Cross et al. 2001; Kaplan and Eubanks 2002).

7.2.5.4 Strawberry Spider Mites

Spider mites are the most important pest of field and greenhouse strawberries throughout the world (Oatman et al. 1985; Stonneveld et al. 1996; Walsh et al. 2002; Sato et al. 2004; Cloyd et al. 2006). They suck the sap from the plant and cause loss of plant vigour. Among mites, the two spotted spidermite *Tetranychus urticae* Koch is a major pest of strawberries having a high rate of fecundity and a short developmental time at high temperatures of 32 °C (Howard et al. 1985; Price and Kring 1991). High populations of TSSM can reduce foliar and floral development thereby decreasing the quality and quantity of mature fruit (Rhodes et al. 2006). Two spotted spider mite populations have become resistant to most acaricides due their short life cycle and high fecundity (Huffaker et al. 1969; Williams 2000; Cross et al. 2001; Stumpf and Nauen 2001; Sato et al. 2004). Outbreaks of TSSM have become more frequent over the last few decades due to increased use of pesticides in modern cultural practices. As a result, more growers are utilizing biological control as an alternative to chemical management (Huffaker et al. 1969; Escudero and Ferragut 2005; Rhodes and Liburd 2005; Rhodes et al. 2006).

Predatory phytoseiid mites, *Phytoseiulus persimilis* have become important elements of integrated pest management (IPM) in Florida strawberry production (Decou 1994) and have been found to be highly effective predators in controlling two spotted spidermite (Zhi-Qiang and Sanderson 1995). Two of the most commonly used phytoseiids are *Phytoseiulus persimi* and *Neoseiulus californicus* (McMurtry and Croft 1997; Cloyd et al. 2006) in greenhouse.

7.2.5.5 The Greenhouse Whitefly (GHWF) and Lygus Bug

The greenhouse whitefly (GHWF) (*Trialeurodes vaporariom*) and the lygus bug (*Lygus hespersus*) are major insect pests of strawberries. Both the GHWF and lygus bug can reduce quantity and quality of marketable strawberries and are shown to develop resistance to different group of insecticides (Kagabu 1999; Yamamoto 1999; Ware 2000). Imidacloprid was the first registered neonicotinoid followed by thiamexthoxam and acetamiprid, the latter two considered second-generation neonicotinoids (Yamamoto et al. 1998; Palumbo et al. 2001) for the control of sucking pests

7.2.5.6 Strawberry Thrips

Frankliniella occidentalis Pergande is a major strawberry pests in southern Brazil which causes russeting and wither in flowers and fruits reducing commercial value (Nondillo et al. 2008).

7.2.6 Lepidopteran Insect-Pests

7.2.6.1 Greasy Cut Worms

The greasy cut worm, *Agroits ispsilon* (Hufn.) is the most important pest on strawberry. Larvae of all instars of the cutworm are found to cause severe injury to strawberry plants. They are just below the surface of the ground beside the cut plants and eating of plants just above at or a short distance below the surface of the soil. These caterpillars often cut off the petioles of the leaves of strawberry plants or the stems of the young fruits, also attack the green and full grown fruits from below, often hollowing it out leaving the shell of the berry (Taha 1984).

7.2.6.2 Leaf Defoliators

Spodoptera litura are serious insect pests which completely defoliate and skeletonize the leaf area and reduce the growth and vigour of runners during summer months under partial to complete shed (Sharma et al. 2013).

7.2.7 Strawberry Leaf Folder

Entomopathogenic nematodes in the families Steinernematidae and Heterorhabditidae act as potential biological control agents (Kaya and Gaugler 1993). They can kill their hosts with their symbiotic bacteria (*Xenorhabdus* for steinernematids and *Photorhabdus* for heterorhabditids) in 24–48 h (Adams and Nguyen 2002 and Poinar 1990).

Fetoh et al. (2009) studied the impact of the beneficial nematodes and two biopesticides when used alone or in combination against the greasy cut worm, *Agrotis ipsilon* (Hufn.) in infested strawberry field as a new approach of integrated pest management. They reported that larvae and pupae of *A. ipsilon* were highly susceptible to the two nematode species, *Steinernema carpocapsal* (Sc) and *Heterorahabditis bacteriophora* (Hb) when used separately and the percentage mortality increase with increase the dose of nematodes. The concentration of 100 infective juveniles (IJs) was more effective than 25 IJs for both of the two species of nematodes used..

Spinosad is a fermentation product derived from naturally occurring actinomycetes soil bacterium *Saccharopolyspora spinosa* (Boek et al. 1994). Spinosad combines the best features of synthetic chemical insecticides and naturally derived biological insect control products. It has the efficacy, broad-spectrum activity against lepidopterous pests e.g. *Ostrinia nubilalis*, *Helicoverpa zea*, *Spodoptera spp.*, *Agrotis ipsilon*, *Trichoplusia ni*, *Plutella xylostella*, *Heliothis spp* and *Pieris rapae*. It is effective as bait for fruit flies *Ceratitis* spp. and *Bactrocera spp.* and some sucking lice and ants (Thompson et al. 2000). Spinosad also has a very favorable safety profile towards mammals (Sparks et al. 1995), reduced environmental risk and phytotoxicity (Harris and Maclean 1999) and safety to beneficial insects (Schoonover and Larson 1995 and Sparks et al. 1995). Moreover, the low toxicity of spinosad toward beneficial allows it to be incorporated into most (IPM) programs that heavily rely on predators and parasites (Bret et al. 1997).

7.2.8 Soil Arthropods

Soil arthropods such as ants, white grubs, wireworms, root weevils, mole crickets, termites and nematodes feed on the roots or underground part of the strawberry plants. Soil dwelling insect-pests can be controlled by making soil application at least 2 weeks before transplanting the runners in the main field. Crop rotation should be followed in nematodes infested field and planting materials should be raised in soil solarized beds. Crop rotation is the best method to control nematodes.

7.2.9 Termites

Termites are the greatest menace in subtropical condition where water table is found low. Besides, it is a big problem even in case of straw mulch used for strawberry production. Fruits are completely covered with the soil mass and entire fruits have been damaged by termites.

7.2.10 Snails and Slugs

Population of snails and slugs problems are faced in areas where weather is damps and high moisture condition prevails. They are most active during night and during cool and feed on ripening fruit creating holes, thus making fruit unmarketable. Control snails and slugs with Metaldehyde bait may be used for the control of snails and slugs.

7.2.11 Frugivorous Bird Pests

Many species of bird's especially insectivorous birds eat agricultural insect pests. However, some species of birds damage grape and fruit plantings. These birds cause damage by eating the entire berry or puncturing it and/or eating a portion of the fruit leaving it unmarketable and prone to disease. Birds also knock berries off bushes as they forage. Noise devices, bird scarer, reflecting ribbons, bird nets and many other home remedies have been tried, with varied success (Ames and Kuepper 2004). Bird netting was indicated to be the control method that worked best for the greatest number of growers. Auditory frightening devices which include bird distress calls, bottle rockets, cannons, sirens etc. were mentioned as being effective, but a few growers said that cannons specifically were not helpful. Research has shown that birds quickly habituate to these noises and to be successful growers should make sure that sound is at least 130 dB; sounds should cover a wide frequency range and be introduced at random intervals. The source of the sound should be moved frequently and combining visual deterrents and real danger (i.e. shooting) helped reinforce the frightening aspects of the noise. According to growers, the distress recordings seemed to provide the best control in the evening and early morning, which might be the most important times for bird control.

7.2.12 Integrated Pest Management for Strawberry Production

Effective control of insect pests in commercial strawberry crops can be done through the judicious use of insecticides/pesticides combined with sanitation and sound management agronomic practices. Among other strategies, site selection is an important activity wherein, avoid planting strawberry crop close to woody areas or plantation crops. For planting runners, select high quality runners which are free from pests and disease for transplanting. Try to follow the clean cultivation and remove the crop residues, debris and weeds from bunds and edges as they may shelter insect pests for their further multiplication. Beneficial biocontrol agents may be released to achieve the pest suppression.

Try to follow the proper cultural practices and purchase or borrow the disease free planting materials/runners or recommended resistant varieties for regional specific area. Removal of weeds in and around the vicinity of strawberry fields as they are the sources to harbour insect pests.

Biocontrol strategies hold a great promise for organic production of strawberries for better returns and consumer acceptability. Several biological control agents (BCAs) have been described as effective against insects such as thrips (Steiner et al. 2006), aphids (Easterbrook et al. 2006) and root weevil (Mahar et al. 2004); mites (Rhodes et al. 2006); nematodes and mollusk pest (Cross et al. 2001). More than dozens of BCA-based products are already commercialized and recorded throughout the world. Protected cultivation of strawberry provides more favourable conditions for exploitation of bio-control including introduction of insect predators and parasites. For example, Paranjpe and Cantliffe (2004) stated that two-spotted spider mites (*Tetranichus urticae*) can be controlled by predatory mites (*Neosiulus californicus* and aphids (*Aphis gossypii*) ladybug larvae (Coleomegilla *maculate)* when released in the strawberry fields. Certain entomopathogenic nematodes can be used as biological control agents to suppress a various number of economically important insect pests (Grewal et al. 2005; Shapiro-Ilan 2004).

7.3 Blueberry

Blueberries are native to North-America. They belong to genus *Vaccinium* and family Ericaceae. The most important blueberry species include: highbush (*Vaccinium australe*, *V. corymbosum*), lowbush (*V. angustifolium*, *V. myrtilloides*), and rabbiteye

(*V. ashei*). Lowbush blueberry develop from a fertilized seed spreading on the ground. Highbush and rabbiteye blueberries are upright with individual plants well suited to intense orchard management. On the other hand, highbush varieties are grown throughout much of Europe and North America; rabbiteye is predominantly grown in the southern US.

7.3.1 Flowering

The blueberry inflorescence is a raceme on the last several inches of a branch, except in mountain blueberry, where the flowers are borne in leaf axis either singly or in pairs. Fruit set is unlikely to occur if the flower is not pollinated within the first 3 days of opening. After successful fertilization, the ovary ripens within 2–3 months to produce a berry containing 65 seeds. The flower is well suited to buzz-pollinating bees. When the flower is vibrated by a buzz-pollinator, such as a bumble bee, the pollen falls out of the pollen pores and on to the insect which is usually probing the flower for nectar at the same time. The nectar and pollen are attractive to bees, but the flower morphology makes it difficult for some bees, including honey bees, to legitimately pollinate the flower. Some pollen is released even at slight pressures, and in this manner even non-buzz-pollinators can receive and transport pollen as long as they visit the flower at its opening (McGregor 1976; Delaplane 1995). Lyrene (1994) has suggested that short and wide corollae, large corolla apertures, and a short distance between the stigma and anthers are desirable characteristics that could make the blueberry flower more amenable to honey bee pollination. Highbush varieties can possess some of these characteristics, but the flower morphology of rabbiteye (*V. ashei*) is especially problematic for honey bees because the corollae tend to be comparatively long, the corolla apertures narrow, and the distance between anthers and stigma large (Ritzinger and Lyrene 1999). On average, the percentage of *Vaccinium* spp. pollen in pollen loads of foraging bees is higher with legitimate flower visitors (77.7 %) than with robbers (47 %) (Delaplane 1995).

7.3.2 Blueberry Pollination Requirements

Lowbush to highbush varieties are generally self-sterile to self-fertile. Rabbit eye varieties are self-sterile and require cross-pollination. All types of blueberries are benefitted by pollination by insects. Thus, bee activity is important for transporting pollen to receptive stigmas even in self-fertile varieties. Eck (1988) estimates that 60–80 % of blueberry flowers must set fruit in order to achieve a commercially viable yield.

7.3.3 Lowbush Pollination Requirements

Lowbush blueberry clones range from self-sterile to moderately selffertile (Hall and Aalders 1961). Cross pollination generally improves fruit-set in *V. angustifolium* (Wood 1968). The two main lowbush species *V. angustifolium* and *V. myrtilloides* often grow together naturally in cleared lands. The two occur in about equal proportions in cleared forest, whereas *V. angustifolium* predominates in abandoned farmland. The berries abort when *V. angustifolium* is pollinated with pollen from *V. myrtilloides*. Therefore, one can expect a maximum fruit-set of about 50 % in cleared forest where the two species occur. Fruit-set can be much higher in patches where *V. angustifolium* predominates. There would be probably fewer pollination problems in lowbush blueberry if plantings could be restricted to two or more good pollen-producing clones of the same species (Hall and Aalders 1961).

7.3.4 Highbush Pollination Requirements

Highbush varieties are largely self-fertile (El-Agamy et al. 1979), but cross-pollination has been reported to be beneficial resulting in increased number of seeds per berry, fruit-set, fruit size, and speed of ripening. Honey-bee-mediated cross-pollination of the southern highbush variety 'Sharpblue' increased yield of early-ripening 'Sharpblue' fruits by 140 %, increased heavy ($\geq$0.75 g) fruits by 13 %, and decreased small fruits by 66 %, all of which translated to a 43 % increase in early-market crop value (Lang and Danka 1991). Gupton and Spiers (1994) showed that varieties perform differently as pollen donors, have different effects on development of the berry. In crossing experiments with seven southern highbush varieties, selfing generally did not affect fruit-set, but it reduced number of seeds per berry, berry weight, and speed of ripening, thus confirming earlier studies. Pollinating southern highbush varieties with pollen from rabbiteye varieties does not affect fruit-set and speed of ripening, but it reduces number of seeds per berry and berry weight (Gupton and Spiers 1994).

7.3.5 Rabbiteye Pollination Requirements

Rabbiteye varieties are mostly self-sterile and require cross-pollination (El-Agamy et al. 1979). It has been reported that cross-pollination with other rabbiteye varieties improves fruit-set, size, and earliness of ripening (Gupton and Spiers 1994). It is, therefore, important to select varieties for interplanting that have similar chill hour requirements because these varieties have the most bloom overlap (Krewer et al. 1986).

7.3.6 Blueberry Pollinators

The blueberry flower is well suited for buzz-pollinators, that is, bees that are able to vibrate the flower rapidly to release pollen through the pores in the anthers. Buzzing, or sonicating, greatly increases the amount of pollen released by the flower and falling on to the bee, and the relative efficiency of single bee visits depends largely on whether that species can buzz-pollinate. But even non-buzz-pollinating species, such as honey bees, can pollinate blueberry when they are present in sufficient numbers and visit flowers legitimately.

7.3.7 Lowbush Blueberry Pollinators

Honey bees and many species of non-managed bees visit lowbush blueberry. Some wild bees are locally numerous and probably important pollinators (Morrissette et al. 1985). Populations of wild bees are generally too small or unpredictable to support commercial pollination needs in lowbush blueberry (Morrissette et al. 1985). Wild pollinators can be supplemented with honey bees. Yield in Newfoundland lowbush blueberry fields stocked with honey bees at 0.7 colonies acre^{-1} (1.7 ha^{-1}) was 54 % higher than in fields without supplemental honey bee hives (Lomond and Larson 1983). In Québec, 500 honey bee hives were introduced during bloom at one end of a large lowbush blueberry farm, and researchers measured bee densities and fruit characteristics at regular intervals in the field for up to 3.1 miles (5 km) from the hives (Aras et al. 1996). Honey bee densities decreased as distance from the hives increased and there was corresponding drop in seed set, fruit-set, berry weight, and speed of fruit ripening. Densities of wild non-honey bees were uniform across the field; thus, these results demonstrate the contribution of honey bees to lowbush blueberry pollination. Fruit-set was increased significantly in most cases in plots nearer to the bee shelters. Alfalfa leafcutting bees increased fruit-set as much as 30 % over background pollination provided by honey bees and wild bees. Rate of reproduction, however, was less than 0.2 loose cells per loose cell incubated. If these bees are to be used for commercial pollination of lowbush blueberry, it is probable that cells would have to be purchased annually.

7.3.8 Highbush Blueberry Pollinators

Highbush varieties range from self-fertile to highly self-fertile but cross-pollination between varieties has been reported to improve the quality of fruit set (Lang and Danka 1991). Large bee populations help transport large quantities of pollen and cross-pollinate highbush blueberry. Danka et al. (1993) reported that honey bee visitation to 'Gulfcoast' highbush variety increased speed of fruit ripening by 5 days and berry weight by 28 %. (Currie et al. 1992a). Solitary bees may be helpful highbush blueberry pollinators. The orchard bee *O. ribifloris* is an efficient pollinator in California highbush blueberry. As few as 300 nesting *O. ribifloris* females can pollinate 1 acre of highbush blueberry (741 ha^{-1}) (Torchio 1990). The solitary soil nester *Colletes validus* visits blueberry (many species, including highbush) in Maryland, works the blossoms legitimately, and nests in synchrony with crop bloom. It is a prime candidate for conservation efforts (Batra 1980).

7.3.9 Rabbiteye Blueberry Pollinators

Southeastern blueberry bees and bumble bees are the most efficient pollinators of rabbiteye blueberry, based on single-bee flower visits (Cane and Payne 1990). The south-eastern blueberry bee (*Habropoda laboriosa*) emerges, mates, and nests in February–April in close synchrony with blueberry bloom. Bumble bees are also important pollinators but their populations in early spring when rabbiteye is blooming are limited to nesting queens or to very small, young colonies with only a few workers. Thus, the species' maximum forager potential is not realized in time for rabbiteye bloom in that area. The carpenter bees rob nectar by cutting slits in the side of flowers without coming in contact with anthers or stigma. Honey bees are the most numerous bee visitor in blooming rabbiteye blueberry followed in descending order by bumble bee queens, bumble bee workers and carpenter bees. Percentage of bees collecting pollen is highest with bumble bee workers (76.3 %), followed by southeastern blueberry bees (60 %), bumble bee queens (38 %), honey bees (3.2 %), and carpenter bees (1 %). Bumble bee queens and workers carry the highest percentage (70.2 %) of *Vaccinium* spp. pollen on their bodies, followed by honey bees (67.7 %), south eastern blueberry bees (58.1 %), and carpenter bees (29.5 %). Thus, bumble bees and south-eastern blueberry bees are more diligent pollen collectors in rabbiteye blueberry, but even honey bees and carpenter bees can carry *Vaccinium* spp. pollen on their bodies (Delaplane 1995).

Recommended bee densities for blueberry pollination has been reported to vary considerably. (Delaplane and Mayer 2000). McGregor (1976) recommended 2.5, 12 and 25 colonies per hectare for optimum production of fruits. McCutcheon (1983) reported 5–12 colonies per ha. Scott-Dupree et al. (1995) found 2.5–10 colonies per hectare for optimum yields.

7.4 Blackberry

The blackberry (*Rubus fruticosus*) flowers are perfect, white in colour measuring about is 2.5 cm in diameter. They have four petals and 50–100 stamens clustered around 50–100 pistils. Each pistil after fertilization develops into a drupelet. The flowers produce both nectar and pollen.

7.4.1 Blackberry Pollination Requirements

There are over 400 species of cultivated and wild blackberries in North America. Some of the black berry varieties are self-sterile and need cross-pollination, and others partially self-fertile (McGregor 1976). Considering the cross-pollination benefits of wild blackberry, it may be advisable to preserve wild stands in order to provide pollination for nearby commercial plantings. However, wild stands can also harbour pests and diseases. Regardless of whether a blackberry plant is self-fertile or -sterile, bee pollinators provide the benefit of maximizing the distribution of pollen to all receptive stigmas. This increases the number of developing druplets and size of the fruit, thereby improving its shape. Crane and Walker (1984) recommended 7.5–10 colonies per ha for optimum yield in blackberries.

7.5 Blackberry Pollinators

Bees readily visit blackberry, and by pollinating the flower they improve the size and shape of this aggregate fruit. Honey bees sometimes increase yield and speed ripening even in self-fertile varieties, as shown for 'Thornless Evergreen' (van Praagh 1988). Gyan and Woodell (1987) bagged blackberry flowers, removed the bags, noted the first insect visitor, and counted the pollen grains deposited after one insect visit. On average, hover flies deposited 2.1 pollen grains; honey bees deposited 6.4 grains, and bumble bees deposited 8.3 grains.

7.5.1 Insect Pests of Highbush and Rabbiteye Blueberries

Several species of insect pests attack blueberries. The details are as given below:

7.5.2 Insect Pests of Flowers and Fruit

7.5.2.1 Flower Thrips

Flower thrips (*Frankliniella bispinosa* (Morgan) and Frankliniella tritici (Fitch)) are particular problems in highbush and rabbiteye blueberries. These species are most active during bloom, feeding on ovaries, pollen and corollas of the flowers and resulting in reduced pollination and seed set. Their damage on fruit appears as small round necrotic areas. Chilli thrips (*Scirtothrips dorsalis* Hood) are another pests attacking highbush and rabbiteye blueberries. They feed on young green tissues, leaves and fruits. Thrips belonging to the genera *Frankliniella* and *Scirtothrips* have a wide range of hosts and they travel in wind currents. They can be monitored by tapping flowers over white boards or using white sticky traps (Liburd et al. 2009). A number of biologically based pesticides are available for their control (Arevalo et al. 2009).

7.5.2.2 Blueberry Maggot

Blueberry maggot causes serious damage in blueberries. Eggs are generally laid under the skin of ripening or ripe berries, although some females oviposit into ‘full green’ fruit. Only one egg is laid per fruit and the maggot hatches 2–7 days later. The maggots are initially colourless and become whitish. The fruit with maggots are very soft and often have a small hole where the eggs were inserted. After feeding in the fruit for about 20 days, the larvae fall on to the ground to pupate. The blueberry maggot overwinters in the soil under bushes for 1–2 years, depending how much chilling the pupae receive over the winter. The adult is a little smaller than a housefly and has a black and dark grey body; there are distinctive black bands on its wings. It also has whitish markings on its thorax and thin bands of white on its abdomen.

The blueberry maggot is commonly monitored using yellow sticky traps baited with ammonium acetate. When flies are detected, insecticides are generally applied every 7–10 days throughout the season to prevent egg laying. Integrated pest management programmes have also been developed where pesticide application ceases after two applications, if no additional flies are detected (Burrack and Littlejohn 2011).

7.5.2.3 Cherry Fruit Worm

Cherry fruit worm (*Grapholitha packardi* Zeller) is a serious pest of blueberries in the mid- Atlantic and mid-western USA. Eggs are laid on to developing berries and leaves at about petal drop. The larvae enter the berries at the calyx cup and feed within them. They may move between berries but do not web them together. The larvae are initially white with black heads, and become pink with brown heads. After larval development, the larvae leave the berries and hibernate in burrows on weed stems or pruned blueberry stubs. They pupate in the early spring and emerge during bloom. The adults are dark-grey moth with a wingspan of about 9 5 mm. (3/8 in.); the wings have chocolate-brown bands. Traps can be used to track emergence and the abundance of adults. Insecticide sprays are commonly applied in areas where infestations are common

7.5.2.4 Cranberry Fruit Worm

Cranberry fruit worm (*A. vaccinii* Riley) is a widespread problem in the eastern half of North America. Eggs are deposited in the calyx cup of berries and, upon hatching, the larvae bore into the fruit, usually near the stem. The larvae feed on multiple berries, webbing the berries together and leaving their frass behind. The presence of the frass and webbing can be used to separate cranberry fruit worm from cherry fruit worm damage. The insect overwinters in a cocoon made of silk and soil particles. Adults are small with dark greyish-brown wings with two distinctive white patches on each forewing. Traps are used to monitor for this pest, two per acre. Pesticides are applied when the moths begin flight, and a degree-day model for timing these sprays.

7.5.2.5 Cranberry Weevil or Blueberry Blossom Weevil

Cranberry weevil or blueberry blossom weevil (*Anthonomus musculus* Say) is most serious pest of blackberry. The weevils sometimes feed on developing buds, but they are most active when inflorescences begin to open. They sometimes clip the flower pedicel, which dangles and eventually drops off the bush. Their damage also appears as tiny holes drilled into flower buds and corollas; infected flowers do not open and turn purple before falling to the ground. Leaf buds can also be attacked, completely destroying the buds or leaving small round holes in the earliest developing leaves.

The grub is small, legless, white and C-shaped with a brown head. The adult weevil is small (1.5–2 5 mm; 1/16–3/32 in.) and brown with whitish markings. Weevils are monitored by the beating tray method or counting the number of individual punctures in flower clusters. Control methods are required if there are more than five adults per bush or more than one puncture is found per five flower clusters (Liburd and Arevalo 2006).

7.5.2.6 Plum Curculio

Plum curculio (C. nenuphar Herbst) is an important pest of fruit crops that occasionally causes economic damage in blueberry, primarily in the mid- Atlantic and southern USA. The oviposition wound on the fruit is a diagnostic crescent-shaped scar that remains visible throughout the season. The larva is a white grub, 6 mm (1/4 in.) long, with no legs and a brown head. It feeds on a single fruit, which may fall to the ground. The larvae leave the fruit after feeding and pupate in the ground. Early varieties are most likely to be harvested while larvae are still in the fruit; however, fruit infested with plum curculio is usually so badly damaged that the berries drop prematurely rather than getting harvested along with sound fruit. Adults emerge in the mid-summer to autumn, and overwinter under debris. The adult is rarely seen, but can be identified as a small weevil 6 mm (1/4 in.) long with a long snout. The surface of the insect is predominantly brown and wrinkled, with grey, white and black specks. The adults 'play dead' when disturbed. Traps are available to monitor plum curculio and help time insecticide sprays.

Japanese beetle (*P. japonica* Newman): Japanese beetle is a major pest of blueberries. It is a shiny deep-green beetle, with dark-brown wing covers and an abdomen with white tufts along its sides. Damage appears as skeletonized leaves and scarred fruits. Leaf feeding is generally not a significant concern unless population numbers are extremely high; however, fruit feeding can significantly reduce quality and serve as an entry point for disease. These beetles also hang tightly on to berries and can ultimately contaminate packaged fruit.

The larvae prefer to feed on roots of grasses, and as a result are more common in sodden fields. The grubs are C-shaped, cream-coloured with brown heads and have three pairs of legs. Adults emerge in the summer as berries are beginning to ripen and are active for 6–12 weeks. Adults can be monitored beginning in mid-May with traps baited with pheromone, although placement of traps in crop fields can actually attract more beetles to a field. The grubs are the most susceptible stage for control using insecticides. Removal of grassy areas in and around fields during July and August can significantly reduce populations (Szendrei et al. 2005). Biological control agents are also available to help suppress populations such as the nematode, *Heterorhabditis bacteriophora*, and the bacteria, *Bacillus thuringiensis* (Bt) and *Bacillus papillae* (milky spore). There may be some difference among cultivars in susceptibility to Japanese beetle damage.

7.5.3 Insect Pests of Buds

7.5.3.1 Blueberry Gall Midge or Cranberry Tip Worm

Blueberry gall midge or cranberry tip worm (D. oxycoccana Johnson) is a widespread pest of highbush and rabbiteye blueberries in the south-eastern USA (Lyrene and Payne 1995). The adult is a tiny fly, with long legs, transparent wings and globular antennae. Larvae go through several colour changes from transparent to orange. Females lay eggs in floral and vegetative buds. Flower buds dry up and fall apart soon after infestation. Developing vegetative shoots are killed, resulting in a tip-burn that can be confused with frost damage. Mature larvae fall to the ground to pupate. There can be several generations produced each year. This pest can be monitored by examining shoots for the percentage of buds infested or by placing shoot tips into zip-locked bags and monitoring for larval emergence (Sarzynski and Liburd 2003; Yang 2005). For control, insecticides can be spayed during early bud development.

7.5.3.2 Blueberry Bud Mite

Blueberry bud mites (*A. vaccinii* Keifer) are a particular problem in the southeastern USA on rabbiteye and southern highbush, but are found in most blueberry production regions. The blueberry bud mite is too tiny to be seen by the naked eye. Heavily infested buds are reddish in colour and have rough bumps on their outer scales. As the buds open, the flowers desiccate and become distorted with distinctive red blisters (Weibelzahl and Liburd 2010). The resulting flowers often do not set fruit and the fruit that do develop have rough skins. Bud mites do not cause the vegetative tip-burn associated with gall midge damage. Epizootics by the acarine fungal parasite, *Hirsutella thompsonii* (Fisher), are at least partially responsible for the decline of blueberry bud mite during the summer and early autumn in the south-eastern USA (Weibelzahl and Liburd 2009). Timely pruning of old canes helps control this disease along with horticultural oils and miticides. Among four highbush cultivars in the field, Isaacs and Gajek (2003) found 'Burlington' and 'Rubel' to be the most highly infested, while 'Bluecrop' and 'Jersey' were the least susceptible.

7.5.4 Insect Pests of Foliage

7.5.4.1 Blueberry Aphid

Blueberry aphid (*I. pepperi* Mac. G.) is bright green and usually found on new succulent leaves and stem tips. It is of greatest importance due to its role as a vector of BBSSV. The largest individuals can be 4 mm (1/8 in.) in length. The adults give birth to live young without mating, and several generations of live-bearing females are produced each year, leading to very high densities by mid-season. Aphids overwinter as eggs on bushes. The feeding activity of the aphids produces honeydew which supports the growth of a black sooty mould. The primary economic damage of the blueberry aphid is as a vector for BBSSV. Aphids can be monitored by searching the succulent lower shoots on bushes weekly after bloom. Natural enemies usually keep aphid populations suppressed, but if fields are infected with virus or are composed of susceptible varieties, both broad-spectrum and selective insecticides are available for their control. Several species of parasitic wasps (Praon and Aphidius species) and predatory insects attack aphids and their eggs (Isaacs et al. 2008), so insecticides should be used that have lower toxicity to beneficial insects.

7.5.4.2 Leaf Rollers

Three species of leaf roller are common in the USA – red-banded (*Argyrotaenia velutinana* Walker), fruit-tree (Archips argyrospilus Walker) and oblique-banded (Choristoneura rosaceana Harris) – while the orange tortrix (Argyrotaenia citrana) is most common in the Pacific Northwest. The fruit-tree leaf roller adult is metallic brown with dark-brown spots on its wings. The obliquebanded leaf roller is tan with chocolate-coloured bands on its wings. The redbanded leaf roller has a complex pattern of colours on its wings including patches of brown, orange, tan and silver. The orange tortrix moth has wings that are pale yellowish-brown to grey in colour with darker mottling. Leaf rollers construct a shelter by rolling leaves with silk and pupate within their shelters. They sometimes tie flowers and green fruit together with silk. The larvae feed on flowers and the surface of berries, although their major importance is as a contaminant of harvested blueberries. They are easily dislodged from their shelters. Natural predators normally keep leaf roller numbers in check, although chemical insecticides are an option if numbers are too high. Pheromone traps are available to determine adult emergence, and growing-degree models have been developed to make predictions regarding egg hatch, larval development and optimal timing for control (Isaacs 2010).

7.5.4.3 Sharp-Nosed Leaf Hoppers

Sharp-nosed leaf hoppers (*S. magdalensis* Prov., *Scaphytopius ac*utus and *Scaphytopius frontalis*) are widespread pests that do not cause direct injury to blueberry bushes, but they vector the protoplasma that causes blueberry stunt disease. The sharp-nosed leaf hopper overwinters in blueberry leaves on the ground as an egg. Eggs hatch in the spring and the insect goes through five sedentary nymphal instars before becoming an adult in mid-summer. The dark brownish-black adults can travel great distances. Insecticides are available for leaf hopper control and their activity periods can be tracked using yellow sticky boards. The sharp-nosed leaf hopper attacks Vaccinium ashei and Vaccinium elliottii, but not in wild or cultivated V corymbosum (Meyer and Ballington 1990). The rabbiteye cultivars 'Premier' and 'Tifb lue' are resistant to the vector (Ballington et al. 1993).

7.5.4.4 White-Marked Tussock Moth

White-marked tussock moth (*Orgyia leucostigma* (J.E. Smith) is most common as a pest of blueberries. Mature larvae are large (30 mm; 1 1/4 in.) with distinctive coloration and hairs. They have a bright-red head with a yellowish body. The hairs can irritate the skin of the harvesting crew. Female moths lay large, hairy masses of eggs on blueberry branches. Frequent pruning and good weed management reduce the numbers of these moths, but if populations reach damaging levels, monitoring and control guidelines are now available (Isaacs and Van Timmeren 2009). Other locally important insect pests of highbush and rabbiteye blueberry

7.5.4.5 Blueberry Leaf Beetle

Blueberry leaf beetle (Colaspis pseudolavosa Riley) is most commonly found in poorly managed fields. The adults are shiny black and 4 mm (3/16 in.) long. Adults feed on leaves and skeletonize them, but they do most of their damage to younger leaves. High infestation during a cropping season can interfere with next year's yield. After several seasons of high infestations, bushes can be killed. Insecticides can help control this insect, but well-maintained fields rarely have significant infestations.

7.5.4.6 Blueberry Span Worm

Blueberry spanworm (*Itame argillacearia* Packard) is a minor pest, most commonly found in the northern USA on lowbush blueberries. Lowbush blueberries can be monitored for this insect by sweeping the foliage with nets; insecticides are sprayed when population numbers are high.

7.5.4.7 Citrus Thrips

Citrus thrips (*Scirtothrips citri* (Moulton)) have become a problem in California where blueberries are planted next to citrus. Citrus thrips feed on the new flush growth of blueberry plants, which causes stunting and likely affects yield and fruit quality. They are found in the blueberry canopy from May to early October. Regular insecticide sprays are being developed to control this insect

7.5.4.8 Leaf-Footed Bugs

Leaf-footed bugs (*Leptoglossus spp.*) are common where little pesticide is being sprayed. They are usually controlled by natural enemies. These bugs are brown, about 2 cm (4/5 in.) long, and their hind legs are shaped like a leaf. They damage fruit by poking holes into them. If population numbers become problematic, insecticides can be used to control them.

7.5.4.9 Scale

Scale (several species) is generally found in older fields on old wood, and can reduce bush vigour. They feed on the phloem and produce honeydew that supports sooty moulds. The scales are small waxy dots, 2–3 mm (1/12–1/8 in.) wide, on stems that cover a yellow insect. Population sizes are generally held in check by good pruning practices and several natural enemies.

7.5.4.10 Spotted-Wing Drosophila

Spotted-wing drosophila is a new pest of blueberries. Its spotted wings are characteristic of the species. They oviposit into intact fruit prior to harvest, and within a few days the fruit flesh starts to break down, leading to collapse of the fruit. The adult Spotted-wing drosophila lives for about 2 weeks and can lay more than 100 eggs in a day. Key factors in its control are monitoring and the timely application of insecticides with knockdown activity (Isaacs et al. 2010).

7.5.4.11 White Grub

White grubs (*Cyclocephala longula*) have recently become a major problem in southern highbush blueberries in California (Haviland and Hernandez 2012). They feed on plant roots, stunt plants and sometimes kill young newly planted bushes. The grubs pupate in May, and fly at dark from mid-June to mid-July; egg hatch occurs in mid-July. The nematode *H. bacteriophora* and the insecticide imidacloprid are effective in controlling the grub.

7.5.4.12 Nematode Pests of Highbush and Rabbiteye Blueberries

The most common nematodes found on blueberry plants are the root-lesion (*Pratyenchus spp.*), dagger (*Xiphinema ssp.*) and stubby-root (*Paratrichodorus ssp.*). The dagger nematode vectors the disease necrotic ringspot. All the nematodes are unsegmented roundworms that are almost invisible without magnification. They range in size from 2 5 mm (1/10 in. – dagger nematodes) to 1 mm (1/25 in. – stubby-root nematodes). Specialized laboratory procedures are necessary for their isolation and identification. To test for nematodes, soil samples should be taken during June and July (Pritts and Hancock 1992). The impact of nematodes in established plantings is not generally thought to be great; however, high nematode populations likely slow the growth of new plantings. It is not known what nematode levels cause economic damage, but pre-plant fumigation is recommended in the late summer or early autumn the year before planting.

7.5.5 Most Common Weeds of Highbush and Rabbiteye

7.5.5.1 Blueberry Fields

There are numerous weeds regularly found in blueberry fields (e.g. nutsedge, quackgrass, dandelion and dewberry). Sedges are perennial and while they look superficially like grasses, they belong to the family Cyperaceae rather than Poaceae, and have stems with three vertical rows. Horsetails are not flowering plants. They belong to the ancient family Equisetum and reproduce asexually via spores.

Management of weeds in blueberries is important for a number of reasons: (i) they compete with plants for water, nutrients and light; (ii) some serve as alternative hosts for insects and diseases; (iii) weeds growing close to blueberries reduce air flow which can favour fungal growth and harbor insects; (v) they provide habitat for vertebrate pests such as voles; (vi) they can compete for pollinators during bloom; and (vii) some produce fruit that can contaminate harvested blueberries.

Probably the most important weed management strategy is to eliminate all perennial weeds before planting. Once perennial weeds become established in a blueberry field, they become extremely difficult to remove. The key to perennial weed control is to eliminate them as much as possible during the year prior to planting, using a combination of cultivation and herbicide application (before they go to seed!).

Weeds can be controlled by cultivation, mulching and herbicide sprays (Majek 2006; Wise et al. 2010). It is critical that established fields are scouted regularly during the season to remove perennial weeds before they become established. It is also important to avoid using single herbicides repeatedly, as it can lead to an increase in resistant weeds. Specific herbicide recommendations can be found on local extension web sites.

7.6 Raspberries

The most important raspberries grown around the world is the European raspberry (*Rubus idaeus*). Most flowers are perfect, and each is about 1in (2.54 cm) wide with 5–12 petals, ~90 stamens, and ~90 pistils each of which has an ovary and a slender style (Redalen 1980; Jennings 1988). Thus, raspberry is an aggregate fruit, like strawberry and blackberry, with many pistils developing together as a single mass. When the flower opens, the immature anthers are bent over the immature styles, but shortly thereafter the outer stamens bend away from the styles and release pollen. Anthers mature and release pollen from the edge of the flower inward, during which time the styles grow and receptive stigmas emerge from their tips (Fig. 7.2). Only the innermost anthers have a chance to touch the outermost stigmas; this accomplishes selfing unless the stigmas had previously been cross-pollinated. Thus, both crossing and selfing are possible on the same flower. The petals begin to fall after 1 day of opening. A plant may be in various stages of flowering for 1–3 weeks. Raspberry is a rich source of nectar and pollen for bees.

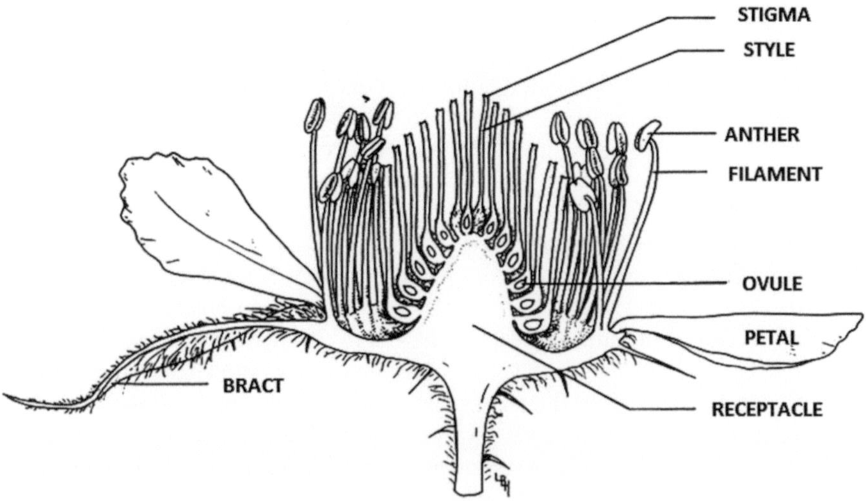

Fig. 7.2 Flower structure of raspberries

7.6.1 Raspberry Pollination Requirements

Raspberry is moderately self-fertile, but its flowering behavior gives ample opportunity for cross-pollination. Depending on the variety, cross-pollination with pollen from other varieties improves development of maternal tissues with an accompanying increase in fruit weight (Colbert and de Oliveira 1990). Bees help deliver pollen evenly among all receptive stigmas, thus optimizing fruit weight and shape even in self-fertile varieties (Kühn 1987). Because raspberry is an aggregate fruit, it requires multiple bee visits to adequately pollinate all the ovaries.

7.6.2 Raspberry Pollinators

Raspberry is very attractive to bees and other insects, and studies have demonstrated the yield-enhancing effects of insect pollination (Johnston 1929; Couston 1963; McCutcheon 1978). Five or six honey bee visits, or ~150 accumulated bee seconds per flower, are necessary in order to optimize the number of fertilized ovaries and fruit weight. This is easily accomplished in 1 day of good bee activity. Honey bees are most efficient if they are collecting pollen as well as nectar, rather than nectar only (Chagnon et al. 1991).

McGregor (1976) and Scott-Dupree et al. (1995) recommended 2.5 colonies per ha for quality and quantity of fruit production in raspberries.

The distribution of foraging honey bees in raspberry fields is uniform during warm and still weather. However, bees will concentrate on plants nearest their hives during overcast weather or wind (Murrell and McCutcheon 1977). Bumble bees are also excellent pollinators compared to honey bees as they forage for more hours in the day, tolerate more inclement weather, visit more raspberry flowers per minute, carry more raspberry pollen on their bodies (Willmer et al. 1994). These advantages over honey bees are explained by bumble bees' greater body size and superior abilities for cool-weather foraging. However, their natural populations are unpredictable domesticated bees are too expensive. Natural populations of other non-honey bees are generally too small to support commercial pollination needs in raspberry (Winston and Graf 1982). Neira et al. (1997) observed the entomofauna associated with flowers of raspberry (*Rubus idaeus* L., cv Meeker). Chagnon et al. (1991) studied honeybee foraging behaviour on raspberry (*Rubus idaeus* L.) pollination and found that number of drupelets and weight of berries increased with the number and length of visits.

7.6.3 Insects

Certain insects can be vectors of diseases, particularly viruses. In general, growers should protect the beneficial insects, such as pollinators (bees) and predators of other pests, and protect plants from the harmful insects that attack the roots, canes and fruits. Mites are not insects, but are found on the underside of the leaves and can cause economic damage, particularly under hot, dry conditions and under cover.

7.6.4 Insect Pests of Root

The raspberry crown borer (*Pennisetia marginata* Harris) is a serious pest of all raspberry fruits. It looks like a yellow jacket (family Vespidae), and is a clear winged moth with black and yellow bands on the abdomen. It takes 2 years to complete its life cycle. Adults appear in mid- to late summer (late July and August). Females can be seen on the foliage where they lay reddish-brown eggs on the underside of the leaf; the eggs take 30–60 days to hatch (Funt et al. 2004). The adults die in about 1 week. After hatching the larvae move to the base of the plant to enter the crown and roots the next spring, girdling the new cane before it goes to the root. In the second winter, it is in the root and by summer, the crown can be damaged. The larvae transform into a pupa and the adult emerges in mid- to late summer. The first symptoms are wilting, dying cane foliage and half-grown fruit (Bushway et al. 2008). Growers should destroy infected canes. All wild bramble plants in and near the planting should also be destroyed. If insecticides are available, a heavy application of insecticide (drenching) in the early spring should kill most larvae and spraying the soil with an insecticide in early fall can kill adults.

The raspberry cane borer (*Oberea bimaculata* Olivier) is a slender black beetle, 1.2 cm (. in) in length, with prominent antennae and usually two black dots on a yellow prothorax. It has a 2-year life cycle; adults appear in early to late summer, feeding on the tender green epidermis of cane tips and leaving brown patches or scars (Jennings 1988). The female creates two puncture rings around the cane, about 1.2 cm (. in) apart and about 15 cm (6 in.) from the cane tip, and deposits an egg between the rings (Bushway et al. 2008). Control can be achieved by removing and destroying the infected portion of the stem a few inches below the wilted part, immediately after the fi rst site of damage. Damaged canes should be destroyed during dormant pruning. Insecticides should be applied to control the adults in a late pre-bloom

Strawberry bud weevil (*Anthonomus signatus* Say) is a small 0.25-cm (0.1-in) long adult, reddish-brown in color with rows of pits or punctures along its back and two white spots with dark centers. The adult overwinters in fence rows, mulch or wooded areas and emerges as temperatures rise above 16 °C (60 °F). They move into raspberry fields and feed on immature pollen by puncturing the blossom bud with their snouts (Bushway et al. 2008).

In some areas this pest can cause economic damage and pesticide sprays may need to be applied at the pre-bloom stage (Funt et al. 2004). The red-necked cane borer (*Agrilus rificollis* Fabricus) or flat-headed cane borer appears in eastern North America, is seldom a serious pest and is a slender metallic black beetle about 0.6 cm long with a reddish or coppery thorax and short antennae; the larvae are white, legless, 1.9 cm long and flat-headed (Funt et al. 2004). Galls are formed in symmetrical swellings about 1.2–2.2 cm (1–3 in.) across, about 30 cm to 1 m (1–4 ft) above the soil line; canes often break at the swelling and larvae reach full size by fall.

If less than 5 % of the canes are aff ected, infected canes can be pruned out and burned, buried or destroyed (Funt et al. 2004). An insecticide can be applied in late pre-bloom (same as for the raspberry cane borer) if more than 5 % of the canes are affected, with another spray at or before petal fall until no more adults are found.

The raspberry cane maggot (*Pegomya rubivora* Coq.) is smaller than a housefly and its larvae (maggots) tunnel down the cane and girdle the cane from the inside, similar to the raspberry cane borer but with this insect there are no girdling marks. This damage may occur each year and the cane should be cut off below the infection when the symptoms appear in the summer; infected canes can also be cut and removed during dormant pruning.

Tree crickets (*Oecanthus sp.* Order *Orthoptera*) are small green-white insects with a slender body and dark antennae, which can be longer than its body. Both nymphs and adults can be seen on canes in summer. In late summer, females lay 30–80 eggs in the canes about 0.5 m (1.5 ft) from the tip; this weakens canes and they may be broken (Bushway et al. 2008). Diseases can then enter the injured part of the plant. Damaged (particularly spent) canes should be removed and destroyed after harvest. If damage is severe, an insecticide may be applied in late summer (Funt et al. 2004).

The raspberry cane midge (*Resseliella theobaldi* Barnes) is another serious pest and plays a role in midge blight. The first generation of adults emerge from the soil in spring (May and June) and lays their eggs in splits caused by internal growth stress on primocanes on certain cultivars; larvae are found 2 weeks later. They are translucent at first and turn pink to orange as they mature. The pupae spend 2–3 weeks in cocoons; the second generation appears in mid-summer; a third generation in late summer (August to September) is rare. No recommendations have been given on control (Jennings 1988).

7.6.5 Insects Affecting Leaves and Fruit

Different species of aphids such as the larger raspberry aphid (*Amphorophora agathonica*), the smaller raspberry aphid (*Aphis rubicola*) and small (leaf curling) raspberry aphids (*Amphorophora idaei*) are found attacking raspberries. They are major vectors of four major viruses of raspberries. These aphids feed on the leaves and cause the leaves to curl downward. They inject the virus into healthy plants resulting in mosaic, leaf curl and/or stunting thereby reducing yield by 50 % (Funt et al. 2004). The nursery plants are most susceptible to virus infections. The larger raspberry aphid transmits the raspberry mosaic virus complex resulting in the raspberry leaf curl virus. The larger raspberry aphid is about 0.3 cm (1/8 in.) long and is either yellow-green or pale bluish green; the smaller raspberry aphid is 0.16 cm (1/16 in.) long and pale yellowish green.

Aphids can be controlled by biocontrol agents like predators and parasites which attack aphids. Under natural conditionds heavy rains can reduce aphid populations. Apply insecticides only if aphid population is two or more aphids per cane tip (Funt et al. 2004).

A. idaei are small sedentary aphids that occur in Europe, form dense colonies on fruiting canes and primo canes causing a curl of the leaf and stunting and twisting of the shoot tips which become sticky with honeydew, and produce a large number of winged forms in early summer that migrate to young canes. Several related species, *A. rubifolii* (Thomas) and *A. spiraecola* (Hatch), also occur on raspberries in North America; *A. ruborum* (Borner) occurs on raspberries in Europe and South America. None of these is known to transmit viruses (Jennings 1988). Japanese beetles (*Popillia japonica* Newman), rose chafer (*Macrodactylus subspinosus* Fabricius), and green June beetle (*Cotinis nitida* Linnaeus) are insects (scarab beetles) that feed on the leaves, fl ower buds, and/or berry fruit. All three of these insects have one generation per year.

Chemical sprays may be needed at the time that these beetles first appear or at late pre-bloom before the blossoms open (Bushway et al. 2008) or during harvest, where it can be forecast that the insects will destroy approximately 20 % of the leaves. Insecticides may need to be applied more than once in a few days. Preharvest restrictions must be obeyed.

Another scarab beetle, the New Zealand grass grub (*Costelytra zealandica* (White)), also feeds on the foliage of raspberries. The most serious damage is done by the larvae on the roots, where a severe infestation will eat all the fi ner roots and the root epidermis of the plant up to ground level, killing the plants. Control methods for grass grub include use of insecticides, including diazinon and lannate. Another effective control measure is through the use of biocontrol agents including *Serratia entomophila*, which prevents the insect larvae from surviving.

Picnic beetle or sap beetles (*Glischrochilus quadrisignatus* Say and *Glischrochilus fasciatus* Olivier) are about 0.5–0.6 cm long, with four orange-yellow spots on the wing covers. They have one generation per year. The larva is white with a brown head and about 0.6–0.9 cm. They will overwinter in many different plant covers and as temperatures reach 16–18 °C (60–65 °F) in the spring, feed on fungi, pollen or sap of plants. Adults feed on ripe and over-ripe raspberries and other fruit and/or any other fermenting material (Bushway et al. 2008). One method of control is to place a bucket of over-ripe fruit and allow this to attract the beetles and trap them. Sanitation is the best method of control, with the berries picked frequently or at close harvest intervals so that over-ripe fruit does not occur. Keep ripe berries off the ground and bury culled fruit near packing plants (Funt et al. 2004).

The raspberry sawfly (*Monophadnoides geniculatus* Hartig) larvae are spiny pale-green worms. They are about 1.2–1.9 cm in length. Young larvae chew on the edges of leaves whereas older larvae chew everything except larger veins, causing a skeletonized appearance. The damage can result in considerable loss of yield. The adult (about 0.6 cm/. in) is a black fourwinged fl y with a yellow band on the abdomen and red markings (Funt et al. 2004). The female lays eggs singly on the top and bottom of the leaf and has one generation per year. It is common in North America, is not common and may occur in low numbers in Midwestern USA, but is not found in Europe (Jennings 1988). However, a second species (*Priophorus morio*) known as the small raspberry sawfl y, which has an appearance similar to *M. geniculatus*, but has two or more generations per year, has been reported to be a problem in greenhouse-grown raspberries in northeastern USA (Bushway et al. 2008). Chemical control is suggested at the early pre-bloom and late pre-bloom stage; this application will also control the fruitworm.

The raspberry are small, light brown beetles about 0.3 cm in length. As an adult, it emerges in early spring (late April in the USA and May in the UK), feeding fi rst on the growing point of the primocane as the leaves begin to open and then on the fl ower buds and young fruit, keeping the drupelet from developing and condemning any sample for fresh or processing markets (Jennings 1988). After the larvae hatch, they enter the blossom or young fruit. The larvae, which feed for about 30 days, are fully grown by early summer (July in the USA) (Funt et al. 2004). Early ripening cultivars may be more susceptible to the eastern raspberry fruit worm than late ripening ones. Because the larvae fall to the ground in early summer, autumn-fruitingraspberries often escape injury (Bushway et al. 2008). Cultivation of the ground in late summer can control the larvae, but cultivation that injures roots can cause more disease. Chemical control is mostly used at early pre-bloom and late pre-bloom. These sprays should control sawflies as well. Biological control may be available in some locations.

Tarnished plant bug (*Lygus lineolaris* Palisot de Beauvois) adults are about 0.6 cm in length, oval, somewhat fl at, greenish- (coppery) brown with reddish-brown markings on the wings and a distinguishing small, yellow-tipped triangle on the back. Nymphs resemble aphids but are more active (Bushway et al. 2008). They are pale green and about 0.025 cm (1/16 in.) when they fi rst hatch. Adults and nymphs have piercing/sucking mouth parts and are present on many plants, such as apple and peach trees and strawberries, until a frost in autumn. After eggs hatch in early spring, control weeds and do not mow forage crops, such as alfalfa, when brambles are flowering, because mowing encourages tarnished plant bugs to move (in the USA). They are found on the crop and weeds and then feed on flowers and developing fruit. Apply insecticides just before blossoms open and again before the fruit begins to color if there are 0.5 insect per cluster.

The two-spotted spider mite (*Tetranychus urticae* Koch) is a serious pest in North America and Australia, but is less common in Europe. Severe infestation can reduce yield and fruit quality, reducing consumer appeal because they look like brown dust on the fruit (Funt et al. 2004). The number of mites may be reduced under heavy rain or soaking sprays under high pressure. Natural enemies (predators), including predatory mites, lady beetle and lacewings, can be purchased from suppliers in the USA. Miticide sprays are applied as the population increases. Reducing certain types of insecticide spray can reduce the loss of some predators and increase the control of mites. Using a 10× magnifi er, monitor the underside of leaves; if there are 10–15 mites per leaf, chemical sprays may be justifi ed. Red spider mites

(*Panonychus caglei* Mellott and *P. ulmi* Koch) are found in North America and Europe, respectively, and are frequently found as greenhouse pests but are not generally found in the field (Jennings 1988). Mite damage in the greenhouse may look like mosaic or other disease. Potato leafhopper (*Empoasca fabae* Harris) occurs in North America. It is bright green and about 0.3 cm (1/8 in.) long. Young nymphs are smaller and light green; adults are identified when they move from side to side. Eggs are hatched within the leaves and stems. Nymphs on the underside of leaves and adults are very mobile and attack over 140 species of plants (Bushway et al. 2008). Potato leafhopper injury can be mistaken for herbicide injury, nutrient deficiency or symptoms of viral infection. Margins of affected leaves develop a light yellow color; new growth can be curled downward and stunted. Generally, damage can be found after the mowing of adjacent alfalfa fi elds in summer as the insect moves into raspberry plantings. Yellow jackets (*Paravespula*, *Vespa* and *Vespula* species) feed on ripe and/or injured fruit, particularly when weather conditions are dry. Because of their activity, picking can be difficult, particularly for pick-your-own customers. They generally build their nests underground or in old logs. They can be discouraged by sanitation, picking all ripe berries and/or removing over-ripe fruit.

7.6.6 Weed Control in Raspberries

Once plants are set, it is almost impossible to reduce perennial weeds in the row without harm to the plants. The control of certain weeds around and near the planting using a systemic herbicide can reduce many diseases from infected plants that can be harmful to an uninfected raspberry planting. Mulches and/or compost that are weed free may be used the first year but are not suggested in the second or consecutive years, particularly if the soil is heavy (clay) because it could cause more root rot. It is imperative in chemical weed control to apply the correct chemical at the best stage of weed growth and at the correct concentration, and finally, make sure there is adequate cover of the weed or soil area.

In conclusion, raspberries are subjected to many insects and diseases. Control of weeds and nematodes before planting can reduce disease pressure and improve yield and quality of the fruit. Starting with disease-free and essentially virus-free, high-quality plants can be one of the most important decisions a raspberry grower can make. Setting the plants into a well-drained soil that is weed free and maintaining good weed control throughout the life of the planting is vital for early returns on the investment. Annual insect, disease and weed control will be necessary as well as drip irrigation before, during and after harvest in areas defi cient in rainfall. The harvesting of ripe fruit at the proper time and maintaining short harvest intervals can be one of the best ways of reducing insects and diseases within the planting. Maintaining good sanitation practices, as mentioned in this chapter, and utilizing refrigeration will be beneficial for providing quality fruit to the customer.

7.7 Grapes

Vitis spp., family Vitaceae

The grape flowers borne in clusters usually bloom 6–10 weeks after the initiation os shoot depending on climatic conditions. There may be several hundred of flowers on the rachis, the main axis of the inflorescence. Branches arise from the rachis at regular intervals and divide to form the pedicles which bear individual flowers. Cultivated *Vitis vinifera* has two types of hermaphrodite flowers one with upright stamens and fertile pollen (Fig. 7.3) and the other has reflexed stamens and mostly sterile pollen. However, exclusively male or staminate and female or pistillate flowers are also present depending upon the variety. The bisexual flower consists of a single slender necked pistil and a two chambered enlarged ovary, each with two ovules. There are usually five stamens, however, the number may range from two to seven (Randhawa et al. 1960) even in the same cluster. The sepals reduced to a ring at the base of the flower. Five petals fused at the top form the calyptras or cap which gets detached at the base and drops off at the time of flowering. Grape flowers produce both nectar and pollen which attract large variety of insects (Pammel and King 1930). There are three type of grape flowers – perfect, staminate, and pistillate. Grape flowers are more important as pollen source than the nectar for flower visiting insects (Pellett 1947; Sharples et al. 1965; Davydova 1969).

7.7.1 Pollination Requirements

Grapes are generally self pollinated, some are self sterile and require cross-pollination (Beach 1892a, b, 1894, 1898; Booth 1911; Hedrick 1924) which may be achieved through wind or bees. It was interesting to find that flowers with recurved stamens could not self whereas those with upright ones selfed Beach (1892a, b).

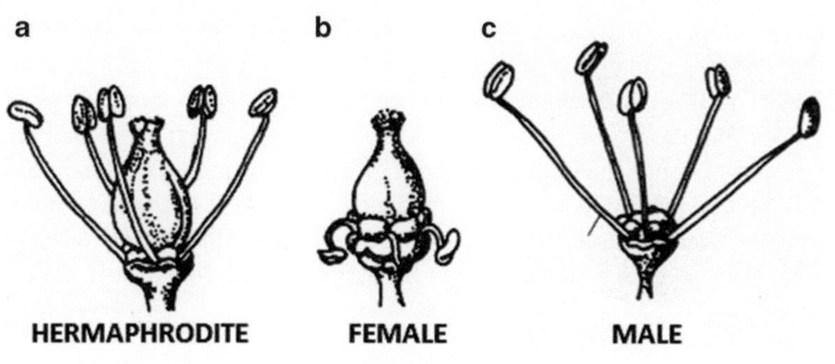

Fig. 7.3 Grape flower. *Upper* –bloom sequence of grape flower, (**a**) calyptras attached (**b**) calyptras separating, and (**c**) open flower (Babo and Mach 1990). *Below*: flower types- (**a**) hermaphrodite (**b**) female, and (**c**) male (Babo and Mach 1990)

Dorsey (1914) found that 11 self-sterile or partly sterile cultivars out of 95 with upright stamens whereas only two of 37 cultivars with reflexed stamens were partly fertile.

Sharples et al. (1965) on *V. vinifera* and Snyder and Harmon (1951) on the 'Cardinal' cv. compared the impact of bee pollination with open pollination and the plants caged to exclude pollinating insects. They found that berries in the bee cages and open plots had an average of 1.79 and 1.84 seeds each, compared to 1.65 seeds per berry in the no-bee cage. Gladwin (1937) reported that wind was the main factor for pollination and bees played only a minor role. Golodriga (1953) found that different cultivars of *V. vinifera* reacted differently to pollinating agents. Different cultivars of *V. vinifera* have been reported to benefit from insect pollination (Boem 1960; Dunne 1942; Hale and Jones 1956; Magoon and Snyder 1943; Marriott 1950). Gladwin (1937) reported that cross-pollination is not only essential in self-sterile cultivars but also beneficial in self-fertile cultivars. Golodriga (1953) emphasized on the importance of selecting the proper pollenizer cultivars for those cultivars that shed or produce inferior berries

7.7.2 Pollinators

Majority of the varieties under the species *vinifera* are self-pollinated, however, some self-sterile varieties require insects like beetles, halicitids, syrphids and honeybees for pollination (Armstrong 1945; Safran 1954; Crane and Walker 1984; Brantjes 1978). But there has been a diverse opionion on the role of pollinating agents. For example, Einset (1930) stated that insects can not be depended upon for their role in pollination, Similarly, Gladwin (1937) reported that bees have a minor importance in grapes pollination and wind may be the major agent affecting pollination.. Knuth (1908) and Munson (1899) gave credit both to insects as well as wind, although the stigma is not adapted for wind pollination as the amount of pollen produced is very small. Most of the investigators have given credit to honeybees as the most important pollinators of grapes (Husmann and Dearing 1913, 1916). Reimer and Detjen (1910) and Olmo (1943) reported honey bees and flies as the most important pollinators. Steshenko (1958) and Barskii (1956) reported that honey bees increased the weight of grape clusters by 23–54 %. Some other investigators (Dearing 1938) considered the *Halictus* bee as an excellent pollinator of grape flowers. Grapes have been reported to benefit from bee pollination. Laiok (1953) compared the production in six cultivars in cages and with bee visitation. He found that production was higher up to 5–15 % in cages with bees. He further found that production decreased as the distance from apiary increased.

7.7.3 Pollination Recommendations and Practices

There has been no agreement on pollination recommendations in grapes. As most of the breeders believe that physical structure of flower is such that it is capable of transferring its pollen from the anthers to the stigma. Dearing (1938) recommended the placement of colonies of honey bees in larger vineyards for maximum production. Davydova (1969) recommended one colony of honeybees per hectare for increased yield and quality. Barskii (1956) also recommended that one colony per 2–5 ha might be sufficient.

7.7.4 Insect Pests in Grapes

Insect pests serve as vectors for some pathogens, e.g. Pierce's disease bacteria, which is transmitted by members of the Cicadellidae (leafhoppers) known as sharpshooters, e.g. the glassy-winged sharpshooter (*Homalodisca coagulata*), and grape fanleaf virus, which is transmitted by nematodes. Some insect pests cause damage directly, e.g. grape root borer (*Vitacea polistiformis*) and *phylloxera*. In all grape-growing regions, roots, wood, shoots, leaves and fruits are subject to potential insect attack. Although the specific species of insect causing the damage may change in different grape regions, there is considerable similarity in their form. Some major insect pests for grapevines include phylloxera, leafhoppers, borers, mealybugs, leaf folders, cutworms and aphids. Grapes are also susceptible to infestation by many species of mites, which are not insects, but are included in the spider (arachnid) family.

7.7.5 *Phylloxera (***Daktulosphaira vitifoliae***)*

Phylloxera is an aphid-like insect of worldwide distribution. It is only known to attack grapes. When there is a point origin for phylloxera entering a vineyard, only isolated vines may be affected at first, with the area of dead and declining vines enlarging in a lens shape, sometimes quite rapidly (King and Buchanan 1986) in subsequent years as the insect spreads. In the case of vines infested as planting stock, the entire vineyard may show decline after a few years of growth, though soil type was found to have an influence on phylloxera populations (Chitkowski and Fisher 2005). High-sand content soils support smaller populations of phylloxera (Nougaret and Lapham 1928; Galet 1982; Pearson and Goheen 1988; Reisenzein et al. 2007), which may explain why some vineyards in infested areas have not succumbed to its effects.

The foliar form of phylloxera causes spherical galls on the underside of the leaf, as well as distortions and an exit hole on its upper side (Granett et al. 2001). The foliar form is known only on some cultivars in humid areas, such as in the north-eastern USA, where it originally evolved. The most serious damage occurs by the feeding of insect on grape roots, the materials injected into the tissue causing the formation of nodosities (galls on younger roots) and tuberosities (galls on older roots). Feeding by large numbers of phylloxera will kill a vine through the reduced capacity of its root system and potential infection of root wounds by pathogens (Omer et al. 1995).

Phylloxera spread is primarily through the action of man. The crawler form of phylloxera can be carried on vineyard machinery, although it can also be carried some distance by wind (King and Buchanan 1986). The winged form of the insect is rarely found, but is capable of movement over much greater distances (Granett et al. 2001).

Control of phylloxera in infested vines is not yet feasible, though some soil drenches have been tried. One problem with this approach is that the insect can live on roots far down into the soil profile and out of reach of applied chemicals, which are also difficult to apply in the clay soils that *phylloxera prefer* (Granett et al. 2001). Affected vines can be managed, through additional water and fertilizer inputs, to continue producing for some additional years but, in most cases, they will succumb to the pest and need to be replaced. Some biocontrol agents are being investigated, such as an insect pathogenic fungus (Kirchmair et al. 2004) and nematodes (English-Loeb et al. 1999), but it is unlikely that these methods will work in more than very specific situations. Successful management is therefore accomplished by avoiding the pest in the first place or grafting susceptible vines to resistant rootstocks, which have both been discussed earlier.

7.7.6 Leafhoppers

Leafhoppers are pests of many cultivated (and non-cultivated) crops and, since they have many alternate hosts, eradication of these from the vineyard is difficult. Leafhoppers feed through piercing the epidermis and taking up the cell sap. During this process there is exchange of material between the bug and the plant, which contributes to these insects acting as vectors for some disease organisms, most notably Pierce's disease. Symptoms of leafhopper feeding include fine chlorotic spots on leaves, leading to more general chlorosis or virus-like symptoms such as leaf curling and angular reddening on leaves (Winkler et al. 1974; Pearson and Goheen 1988). If the honeydew excreted from the insects falls on fruit, it can cause an unsightly sooty mould appearance, which reduces the quality of table grapes (Settle et al. 1986). As for most pests and diseases, a zero tolerance to their presence is not feasible, and one study has indicated that, at least from the standpoint of impact on vine health and productivity (crop loads and vine photosynthesis), a small population of leafhoppers can be tolerated (Candolfi et al. 1993).

Basal leaf removal is a good first line of defence against leafhopper infestation in vineyards, as this is the location where eggs are laid. Control of leafhoppers can also be through the use of insecticides, with pyrethroids being an option that has been in use for many years (Perring et al. 1999), as well as *Bacillus thuringiensis*, which is a bacterium that produces a chemical that disrupts cells in the insect gut. Synthetic insecticides such as organophosphates and carbaryl have been used, but seem likely to be phased out in the longer term due to toxicity concerns (Jenkins and Isaacs 2007a, b). There has also been research into successful biocontrol agents, such as wasps (Kido et al. 1984; Williams and Martinson 2000), which may form a component of integrated pest management systems.

7.7.7 Borer Insects

There are various types of borer insects that attack grapevines. In general, the adult forms are either beetles or moths that lay eggs in the soil near the vine or on the vine itself. Root and cane borer larvae hatch from the eggs and find roots or shoots, respectively, to infest. The larvae eat through the centre of the roots/canes/shoots, damaging the vascular tissue and causing water stresstype symptoms, or killing the vines outright. Control is best accomplished by removing alternate plant hosts near the vineyard and sanitation of infested material. Mating of the insects can be disrupted with pheromone release into affected areas (Johnson et al. 1991), which can be effective enough to negate the requirement for insecticides. At least two fungal pathogens of the larvae have also been reported (Dutcher and All 1978), which brings with it the promise of biological control.

7.7.8 Mealybugs

Mealybugs (species in the genus *Pseudococcus*) are a problem in vineyards because they vector diseases but, also, through their feeding, increase botrytis infection and physically infest the clusters, causing the appearance of sooty mould, which is unacceptable for table grapes. In many areas economic control can be accomplished through the use of natural predators, such as parasitic wasps (Berlinger 1977; Trjapitzin and Trjapitzin 1999), though there are many other types of potential control organisms (Walton and Pringle 2004).

7.7.9 The Grape Berry Moth

The grape berry moth (*Endopiza viteana*) is a serious pest vineyard located near the wooded areas (Botero-Garcés and Isaacs 2004). Although cultural practices that alternately bury and expose the overwintering stage of the insect are sometimes effective, chemical treatment is usually required. The sex pheromone of the grape berry moth has been successfully used for control (Martinson 1995; Witzgall et al. 2000), as has *B. thuringiensis* (Ifoulis and Savopoulou-Soultani 2004). In larger vineyards with less boundary per area, chemical control can be limited to the outside rows. In other vineyards damage is rare and no chemical control is required. The European grape berry moth is a different species (*Lobesia botrana*), causing similar grape damage and requiring equivalent methods of control, though some aspects may differ (Gabel and Renczés 1982; Witzgall et al. 2000).

7.7.10 Thrips

Thrips are very small insects that cause damage when the eggs are laid in fruit (sometimes called halo-spotting), or when they feed by rasping on the berry surface, which causes scarring. Damage is primarily visual, which is important in the table grape industry, but berry scarring can be an origin of berry splitting (Yokoyama 1979). Enhancing the plant diversity in vineyards can enhance natural control of thrips (Nicholls et al. 2000).

7.7.11 Beetles

Beetles can also cause damage, either through the larvae feeding on vine roots or from adults feeding on the leaves. The bark of the grapevine trunk is often host to a variety of insects, some of whom only come out to feed after dark.

7.7.12 Mites

Mites are not insects, but cause similar leaf symptoms. Spider mites, grape erineum mites, grape rust mites and false spider mites are found in most grapegrowing areas. Mites cause leaf damage by feeding, the leaves may brown or drop and reduce vine photosynthesis or buds can also become infested. Many problem mites are controlled naturally by predatory mites (Karban et al. 1995), so chemical applications to control mites must be carefully selected to prevent decimating the beneficial mites (English-Loeb et al. 1986). For example, sulphur applications for disease control can disrupt the life cycles of both damaging and beneficial predatory mites (Hanna et al. 1997).

7.7.13 *Chemical Control*

In general, chemical controls should not be applied unless the insect is present and in a form that is susceptible to the chemical. Insect life cycles vary for different insects, and the appearance of the vulnerable form may depend on macro or microclimate. Scouting, or monitoring, can be used to determine the presence and population of relevant insects. Vines should be examined in detail for the number and type of insects and, if the number seen is known to result in economic damage (through experience or published information), a chemical or other control should be applied to reduce the risk.

7.7.14 *Animal Pests*

7.7.14.1 Nematodes

Nematodes are generalized feeders present in most soils and many are serious pests of grapes. Root-knot nematodes (*Meloidogyne* spp.) cause general decline in grapevines, not usually killing them but rendering them more susceptible to other diseases or environmental stresses. Dagger nematodes (*Xiphinema* spp.) also cause vine decline but are also a vector of grapevine fanleaf virus (GFLV). Lesion nematodes (*Pratylenchus* spp.) also cause vine decline by root death (Nicol et al. 1999). Nematode populations tend to increase with time in vineyards (Ferris and McKenry 1974), and this has been associated with replant disorders. If a vineyard was to be planted into a soil with high nematode populations, this was traditionally corrected through soil fumigation (McKenry 1999). With recent and strict regulations regarding the use of methyl bromide, a commonly used pre-planting soil fumigant, alternatives such as the planting of crops that discourage nematode infestations can be tried (Nicol et al. 1999). The use of soil drenches to control nematode populations suffers the same problems as for phylloxera _ the organism can live quite deep into the soil profile and persist in the soil long after vines have been pulled out (Raski et al. 1965). Residual roots in the soil can carry the pests, so killing roots with systemic herbicide and removal of as many roots as possible is helpful.

7.7.14.2 Vine-Grazing Pests

Vine-grazing pests include many species of deer and rabbits. They do not usually consume significant amounts of plant in a healthy vineyard, but are particularly damaging in newly planted vineyards. They can defoliate vines in a new planting, eating the tender leaf blades while leaving the petioles still attached to the shoots. This can greatly delay the establishment of young vines through the loss of photosynthetic capacity. Rabbits, turkeys and other animals can also eat the bark from the young trunks, damaging or even killing vines.

7.7.14.3 Berry-Eating Pests

Berry eating pests include many mammals such as raccoons, skunks, opossums, rodents and others. Birds also are voracious berry consumers and can cause severe economic damage to grapes. Birds can be residents nesting in the vines, flocking (such as starlings) or migratory. Migratory birds travelling in large flocks are particularly dangerous, as they can consume a large amount of fruit in a short amount of time. There can be large differences in the extent of damage, depending on whether the vineyard is on a migration route or not. Although there are many possibilities for managing animal damage in vineyards, exclusion methods are most dependable for both mammalian and avian predation. Electric fences or barrier fences are both effective for mammals, though costly. Netting is effective for birds, although stationary noise and visual scare methods may suffice temporarily for moderate bird pressure (Aubin 1990). Many commercial operators in high pressure areas use a combination of nets, propane cannons and frequent movement/noise in the vineyard (e.g. motorbikes), with or without actual shooting of birds. Some growers find that shooting of birds is necessary for devices such as propane cannons to be effective.

Bird netting can be applied directly on the vines and tied under to prevent entry, or applied on supports over the vines. It must be applied before birds become habituated to feeding from the grapevines, and removed just before or soon after harvest. The latter is to prolong the life of the nets, which can last up to 10 years with care in application and storage. Though effective, nets require significant labour to apply and recover. New research on learning more about the behaviour of birds in vineyards may yield more effective and holistic management methods (Saxton et al. 2004a, b). Fences are expensive to erect, but can last a long time with continuing maintenance. Fences for berry-eating mammals must be close enough to the ground so the animal cannot crawl under, and must extend high enough to discourage those animals that can jump. Fences to keep out deer-like animals should be at least 3 m high if they are erected on uneven terrain (VerCauteren et al. 2006). An issue worthy of serious consideration when it comes to disease and pest control is the potential effects on neighbouring properties. Spray drift and noise used to manage these problems can often be objectionable to others, so it is best to communicate well with others in your area to ensure minimal disturbance. In some cases there are local regulations covering these types of activities, which should be taken into account when developing a vineyard.

7.7.14.4 Weeds

Weeds in vineyards use both nutrients and water needed for vine growth. Excessive weed growth can block light and hinder airflow as well. In some floor management systems no plants (weeds) are allowed to grow under the vines or between the rows. More machine travel in the vineyards has increased the need for a planted middle, which supports such activity, so vegetation is allowed to grow (at least for some of the year). These volunteer crops can also be called a cover crop, and may be suppressed during the highest water use season for the vines. Rather than use the volunteer plants as a cover crop, sowing a specific grass or other plant seed that has attributes better suited to vineyards is also practised.

Weeds can be prevented from growing, or be destroyed at some growth stage, by either mechanical or chemical methods. Chemical herbicides are used to kill established weeds, suppress weed growth in row middles and inhibit germination of weed seeds. It is possible completely to control weeds with chemical applications. Herbicides can be sprayed, wiped or broadcast. Some herbicides (contact) kill only exposed green tissue and can be used near the base of growing vines, although because only the part of the weed that was in contact with the herbicide is killed, it usually recovers. Other herbicides are translocated (systemic) by green tissues to roots and kill the whole plant. They must be applied when the vines are dormant or physically protected from contact, as they can also be taken up by the vine. Germination inhibitors are applied in conjunction with other herbicides to prevent weeds from growing rapidly in the space created by dying weeds. Weed control strategies must be developed that control the weeds without damaging the grapevines.

7.7.14.5 Pesticide Resistance

Disease-causing organisms, weeds and insects frequently become resistant to specific pesticides. Resistance to traditional insecticides in insect control has been shown with grape berry moth (Nagarkatti et al. 2002), suggesting the need for alternative methods. There are several weeds already resistant to glyphosate (Heap 1997; Powels et al. 1998). Many fungicides have become useless because of resistance development by their target pest organisms. Resistance develops because of the biology of the pesticide _ usually its specificity to the pest target and how frequently it is used. The greater the specificity, generally the greater the chance that a mutation will develop that overcomes it. It is very important not to overuse new and effective pesticides, which will encourage resistance development and possibly render them ineffective. Several strategies are involved, including limiting the number of applications per season, alternating use with pesticides of different modes of action or use in combination with a second pesticide of a different mode of action (Staub 1991; Clarke et al. 1997). Despite these actions, disease management using chemicals will be dependent on new pesticides being developed as either resistance develops or regulations force their withdrawal from the market.

7.8 Kiwifruit

Pollination appears to be a prime factor in kiwifruit production since inadequate pollination leads to small unmarketable fruits (Pyke and Alspach 1986). Successful pollination and fruit set depend upon receptivity of flowers during the few days following anthesis, so it is crucial to identify the main factor limiting the effective pollination period (González and Coque 1995). Both honeybees and wind play an important role in transfer of pollen to the female flowers, but the process is uneven and a proportion of female flowers often receive inadequate pollination (Bomben et al. 1999).

Kiwifruit (*Actinidia deliciosa*) vines are dioecious which means that vines have either male flowers or female flowers. Male vines produce more flowers than do females. The flowers are 1!s–2 in. (3.8–5 cm) diameter, have 5–6 white petals, and occur singly or in groups of three (Fig. 7.4). Petals change colour from white to yellow as they age. The female flower has 165–200 stamens that release sterile pollen, up to 41 functioning stigmas, and a swollen ovary containing up to 1500 ovules. The male flower has a non-functioning ovary and 134–182 stamens that release viable pollen (Hopping and Jerram 1979). Both male and female flowers release pollen but neither produces nectar. Female vines bloom for 2–6 weeks. Stigmas are receptive as soon as the flower opens and stay receptive for the next 7–10 days (Ford 1970; Sale 1983). Receptivity drops sharply after the fourth day, at least in the cultivar 'Hayward' in Spain (González et al. 1995).

Female anthers release pollen gradually through a longitudinal split that opens over a 5-day period (Goodwin 1986a). Male vines bloom for 2–4 weeks and their flowers also release pollen through a longitudinal split, but only for about 3 days.

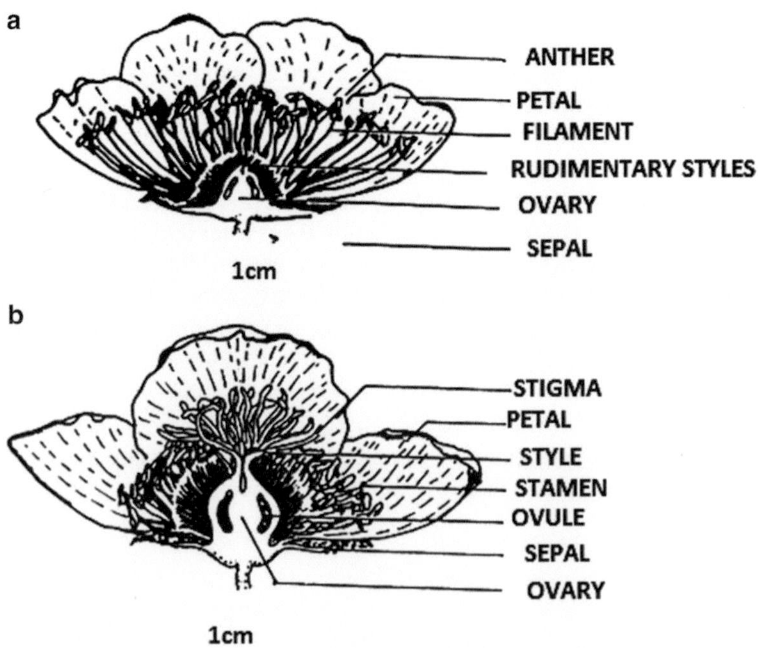

Fig. 7.4 Longitudinal section of (**a**) Staminate flower (**b**) Pistilate flower of Kiwi (McGregor 1976)

Male pollen is most abundant in an orchard in morning and early afternoon (Goodwin 1995). Female pollen is sticky and remains exposed at the split until it is picked up by an insect. Male pollen, on the other hand, is dry and readily falls from the flower when it is shaken.

7.8.1 Kiwifruit Pollination Requirements

Male and female plants must be interplanted in orchards to provide viable male pollen for female stigmas. There must be a synchrony between blooming time of male pollenizer cultivars well as the female cultivar. In New Zealand the recommended ratio of male vines to female and their configuration in the orchard are not firmly fixed. Orchards are planted in a configuration of one male to eight female plants with male vines as every third plant in every third row. Other options are to plant in a one male to five female configuration with males as every third plant in every second row or as every second plant in every third row (Free 1993). Male ratios as high as 1: 3 have been recommended (Sale 1984). A comparison of orchards with male densities ranging from 1: 3 to 1: 8 showed no differences in average fruit weight or seed number among orchards (Goodwin et al. 1999).

Some growers use overhead male vines which are trained to run at right angles across the tops of female vines (Sale and Lyford 1990). It is possible to train male vines with wire or wood supports so that they extend into all-female rows; this increases the chance of bees visiting male and female flowers on the same foraging trip. It has been suggested that male vines should probably not exceed 10 % of the total orchard area (Free 1993).

Kiwifruit vines do not flower prolifically, and 90 % of the flowers must set in order to achieve a commercially acceptable crop (Sale 1983). In New Zealand, 700–1400 seeds must set to produce fruit suitable for export (Hopping and Hacking 1983). Each female pistil needs about 3000 pollen grains to set 700 seeds (Hopping 1982).

7.8.2 Kiwifruit Pollinators

Honeybees are the most important pollinators of kiwi fruit but wind also plays a role in pollen transfer. Viable male pollen is easily shaken loose from flowers and borne by wind, so kiwi appears well suited to wind pollination. Sterile female pollen is sticky fruit and well suited to insect transport even though it has low nutritive value for bees (Schmid 1978; Jay and Jay 1993). Regardless of these characteristic differences, bees visit both male and female flowers to collect pollen. A successful pollinator must be able to carry viable male pollen to female flowers.

Costa et al. (1993) compared the relative performance of wind and honey bees in a 2-year experiment in Italy with specially-designed cages that excluded or included insects but did not significantly affect wind speed. During 1 year, vines in wind-only cages set fruit as well as those in bees + wind cages, as well as open-pollinated vines, and as well as hand-pollinated vines. The addition of bees to wind-pollinated cages increased fruit-set 1 year, yield both years, and average fruit weight 1 year. In both years, yield and average fruit weight were highest in hand-pollinated vines, but since it takes about 40 man-hours to hand-pollinate 1 acre of kiwifruit (100 h ha21) hand-pollination is not considered a practical alternative. Although wind can account for up to 29 % of commercial pollination (Clinch and Heath 1985), honey bees are the most abundant and practical pollinator of kiwifruit (Palmer-Jones and Clinch 1974; Clinch 1984). Honey bees do indeed pollinate kiwifruit. Average fruit weight and number of seeds increased in flowers that received one honey bee visit compared to fruit from flowers netted to exclude bees (Donovan and Read 1991). Fruit-set in France is 5.9 % in flowers pollinated by wind only but increases to 39.7 % in flowers pollinated by wind + one honey bee visit (Vaissière and Froissart 1996).

Honey bees often prefer flowers of one sex and return to those flowers on successive trips (Goodwin 1985; Goodwin and Steven 1993). Bees doing so presumably transfer little male pollen to female stigmas. This tendency is greatest when bee populations are low (Goodwin 1986b). Thus, by increasing bee density one can increase competition for pollen, increase bee movement between flowers, and increase the chances for pollination. In spite of their bias for one flower type, about 87 % of honey bees foraging on kiwifruit do carry pollen from both sexes, compared to 56 % of bumble bees (Macfarlane and Ferguson 1984). Bees favouring one type of flower can still pick up the opposite pollen from previous flower visitors or by contact with nestmates in the hive.

Honey bees concentrate on the anthers and rarely contact the stigmas uniformly while working a flower. For example, honey bees contact 25 % of the stigmas during one flower visit, compared to 48 % contact by bumble bee workers and 68 % contact by bumble bee queens (Macfarlane and Ferguson 1984). These data highlight the importance of multiple bee visits in order to achieve satisfactory pollination of a kiwifruit flower. Donovan and Read (1991) estimate that each flower needs about four honey bee visits in order to produce fruit of export size.

Feeding honey bee colonies sugar syrup may improve their efficiency as kiwifruit pollinators. Colonies fed 1 l (~1 qt) of sugar syrup every 1–2 days collect significantly more kiwifruit pollen compared to colonies not receiving syrup. Morning feedings seem to yield the best results, and daily smaller feedings are better than larger, less frequent feedings (Goodwin and Houten 1991). Feeding colonies syrup probably stimulates house bees to process syrup; thus there are fewer of them ready to relieve incoming nectar foragers of their nectar loads. This discourages nectar foragers from collecting more nectar and causes them to shift to collecting pollen which they can pack into cells without help from house bees. This speculation is only partly supported by field data in kiwifruit, but it remains the best explanation for the stimulating effects of syrup. More importantly, it is still uncertain whether increased pollen collection translates into increased kiwifruit pollination and yields.

Honey bee colonies often decline while they are in kiwifruit orchards probably because of competition from high colony densities and the poor nutrient quality of female kiwifruit pollen. Feeding neither colonies supplemental pollen substitute patties (Herbert and Shimanuki 1980) while they are in kiwifruit orchards does not affect the amount of kiwifruit pollen they collect nor the amount of honey they produce (Goodwin et al. 1994). In other words, feeding pollen substitute patties does not impair pollination (and probably improves overall colony condition), but does not increase subsequent honey production.

Large bee populations are important because honey bees are easily distracted away to more attractive crops, and kiwifruit requires a high percentage of fruit-set. The standard recommendation is 3.2 colonies acre21 (8 ha21) (Palmer-Jones and Clinch 1974). Even more colonies (>50 % above recommended rate) are needed when flowering is compressed, as is done with hydrogen cyanamide (Goodwin 1989; Goodwin et al. 1990). Placing hives in groups of 3–4 at the end of each block (1.2–2 acres, 0.5–0.75 ha) of vines disperses foragers evenly throughout the orchard, making it unnecessary to distribute hives singly along rows (Jay and Jay 1984). Associations of pollinating bee-keepers in New Zealand use published hive strength standards and contract independent auditors to sample members' hives for compliance. The goal is to ensure that beekeepers provide colonies of a consistent, acceptable standard for kiwifruit pollination. Alfalfa leaf-cutting bees are not promising pollinators of kiwifruit in New Zealand. In three field sites and one cage test, only four male leafcutting bees were seen on kiwi-fruit flowers in normal field conditions; pollen retrieved from bees was entirely non-kiwifruit pollen, and kiwifruit petals and leaves were unsuitable for nesting material (Donovan and Read 1988).

7.8.3 Pests of Kiwi Fruit

Mckenna et al (2009) determined the insect pests of *Actinidia arguta* kiwifruit in New Zealand. They found leafroller, Armoured scale insects and mealybugs. A number of other insects included several species of mites, fungal feeding beetles and psocids which were not of . primary concern. *Scolypopa australis* (passionvine hopper) and *Amphipsalta zelandica* (cicada) were present in high numbers. González (1986)) recorded the tortricids *Proeulia chrysopteris* and *P. auraria* and the diaspidids *Hemiberlesia rapax, Quadraspidiotus perniciosus, Aspidiotus nerii* and *H. lataniae* in the field and aleyroid *Trialeurodes vaporariorum* and the mite *Tetranychus urticae* in the nurseries. *T. urticae*, *Brevipalpus chilensis* and Eotetranychus sp. are also found in the plantations, but cause only local damage. The occasional pests observed include the coccid Saissetia oleae, the thysanoptera Heliothrips haemorrhoidalis, noctuid larvae, the curculionid Naupactus xanthographus and the snail Helix aspersa.

The key pests of kiwi fruits are two species of leaf rollers and three species of armoured scales (Berry 1989; Steven 1990). A number of secondary pests can also cause sporadic problems on kiwi fruit but generally pest control practices aim to protect the crop from leaf roller feeding damage and to ensure the fruit is free from insects at harvest. Calendar spraying was a simple and successful means of achieving this, but researchers recognized early on that such pest control practices were unsustainable, disruptive to the predator–parasite complex, had potential to lead to insecticide resistance, and carried unacceptable environmental risks.

In response to these threats, a research program specific to kiwifruit pests was set up in the early 1980s with the aim of reducing the reliance on organophosphate sprays and broadening the range of pest management tools available. Initial studies focused on developing an understanding of the biology of the key pest species. It was found that the fruit damage occurred immediately after fruit set up to 8 weeks and was due to a single endemic species, *Ctenopseustis obliquana* (Stevens et al. 1995). After this time, the risk of leafroller damage was shown to be low, but late-season infestations of a second endemic species, *Cnephasia jactatana*, were recognized as possible in some orchards. Concurrently, the number, timing and distribution of armoured scale generations in kiwifruit were determined (Greaves et al. 1994; Blank et al. 1996, 1997). This information allowed researchers to identify the periods when control measures would be critical, and conversely, the periods when insecticides could potentially be omitted from the spray schedule (Stevens et al. 1993; McKenna 1998a). However, pest pressure can vary markedly among kiwifruit blocks, and so it was considered important that growers were provided with a means of determining whether omission of a spray would be likely to put the crop at risk of pest damage. Using knowledge accumulated from the pest biology studies, systems for monitoring scale and leafroller populations in kiwifruit were developed and action thresholds identified (Steven et al. 1991; Blank et al. 1994; Stevens et al. 1997; McKenna 1998b). These monitoring systems are now an integral part of the KiwiGreen program and form the basis of the decision-making process on whether or not a spray is required in a kiwifruit block.

A second key component of the kiwifruit research program was targeted at identifying alternative, environmentally benign insecticides that would enable the production of residue-free fruit. Field studies showed products containing Bt could provide excellent control of leafrollers (McKenna et al. 1995; McKenna 1998a; Stevens and McKenna 1999), but an alternative was also needed for armoured scale control. Mineral oil was one such option, but trials in the 1970s had shown it to be phytotoxic to the crop (Sale 1972). But these problems were solved by the identification of factors that influence the occurrence of mineral oil damage to kiwifruit (McKenna and Steven 1993; McKenna et al. 1997) (Table 7.1).

Table 7.1 Common pests of kiwi fruit and their status as a production or quarantine problem (Source: Steven 1990)

Insect species	Scientific name	Production pest	Quarantine pest
Primary pests			
Leaf rollers	Ctenopseustis obliquana	√	√
	Cnephasia jactatana	√	√
Armoured scales	Hemiberlesia rapax	x	√
	Hemiberlesia lataniae	x	√
Secondary pests			√
Passionvine hopper	Scolypopa australis	√	√
Fuller's rose weevil	Asynonychus cervinus	x	√
Collembola	Xenylla maritima	x	√
Thrips	Heliothrips haemorrhoidalis	x	√
	Thrips obscuratus	x	√
	Nesothrips propinquus	x	√
Orbatid mites	Irgella bullager and others	x	√
	Tetranychus urticae	x	√
Fungal feeding beetle	Aridius sp., Orthoperus sp.	x	√

Bibliography

Abrol DP (1989a) Studies on abundance, diversity, behaviour and importance of native pollinators for crop production. Korean J Apic 4(2):25–40

Abrol DP (1989b) Studies on ecology and behaviour of insect pollinators frequenting strawberry blossoms and their impact on yield and fruit quality. Trop Ecol 30(1):96–100

Abrol DP (1992) Energetics of nectar production in some strawberry cultivars as a predictor of floral choice by honey bees. J Biosciences 17:41–44

Abrol DP, Andotra RS (2001a) Field toxicity of pesticides to honey bee Apis mellifera L. foragers. Korean J Apicul 16(1):19–26

Abrol DP, Andotra RS (2001b) Impact of pesticides on the foraging activity of honey bees Apis mellifera L.on Brassica compestris var, toria. Korean J Apicul 16(2):89–94

Abrol DP, Andotra RS (2003) Relative toxicity of some insecticides to Apis mellifera L. J Asian Pacific Entomol 6(2):235–237

Abrol DP, Kapil RP (1986) Factors affecting pollination activity of Megachile Lanata. Proc Indian Acad Sci, (Anim Sci) 95:757–769

Adams BJ, Nguyen KB (2002) Taxonomy and systematics. In: Gaugler R (ed) Entomopathogenic nematology. CABI Publishing, Wallingford, pp 1–34

Ahn SB, Kim IS, Cho WS, Choi KM (1989) Survey on the use of honeybee pollination of strawberries grown in plastic green house. Korean J Apicul 4(1):1–8

Ahn SB, Lee SH, Choi KM, Cho WS, Woo KS (1994) Optimum population size of honeybee for strawberry pollination in the vinyl house (Korean). Korean J Apicul 9:113–116

Ak K, Güçlü Ş, Tuncer C (2010) A new kiwi fruit pest, *Lymantor coryli* (Perris, 1853) (Colepotera: Scolytidae) in Turkey. Türkiye Entomoloji Dergisi 34(3):391–397

Allen WW, Grande SE (1963) Strawberry pollination. J Econ Entomol 56:823–825

Allison PA, Tomkins AR, Wilson DJ, Thomson C (1998) HortResearch client report 98/187, Intensive monitoring of passionvine hopper. HortResearch, Palmerston North, New Zealand, 16 p

Ames G, Born H, (2000) 'Strawberries: organic and IPM options' in ATTRA horticulture production guide. http://attra.ncat.org/attra-pub/PDF/strawberry.pdf. Accessed 14 Jun 2004

Ames GK, Kuepper G (2004) Overview of organic fruit production (Appropriate technology transfer for rural areas, Fayetteville, AK) http://www.attra.ncat.org/attra-pub/fruitover.html

Anderson LD (1964) The effect of sevin on honey bees. Gleanings in Bee Cult 92(6):358–364

Anonymous (2004) FAO, year book. Food and Agricultural Organization, Rome

Anonymous (2006) Directorate of horticulture. Jammu (J & K Government). Department of Horticulture, Jammu

Antonelli AL, Mayer DF, Burgett DM, Sjulin T (1988) Pollinating insects and strawberry yields in the Pacific Northwest. Am Bee J 128(9):618–620

Aras P, De Oliveira D, Savole L (1996) Effect of a honeybee (Hymenoptera:Apidae) gradient on the pollination and yield of lowbush blueberry. J Econ Entomol 89(5):1080–1083

Arevalo HA, Frau lo AB, Liburd OE (2009) Management of flower thrips in blueberries in Florida. Florida Entomol 92:14–17

Armstrong WD (1945) New varieties and pollination of muscadine grapes. Proc Am Soc Hortic Sci 43:450–452

Armstrong WD, Pickett TA, Murphy MM, Jr (1934) Muscadine grapes. Ga Agr Expt Sta Bul 185: 29 pp

Aubin T (1990) Synthetic bird calls and their application to scaring methods. Ibis 132:290–299

Bagnara D, Vincent C (1988) The role of insect pollination and plant genotype in strawberry fruit set and fertility. J Hortic Sci 63(1):69–75

Bain HF (1946) Blooming and fruiting habits of the cranberry in wisconsin. Cranberries 10(9):11–14

Ballington JR, Rooks SD, Milholland RD, Cline WO, Meyers JR (1993) Breeding blueberries for pest resistance in North Carolina. Acta Hortic 346:87–94

Barskii YAS (1956) Training bees to pollinate grapevines. Sad I Ogorod 4:64. [in Russian.]

Batra SWT (1980) Nests of the solitary bee, Anthophoraantiope in Punjab. India J Kans Entomol Soc 53:112–114

Beach SA (1892a) Notes on self-pollination of the grape. N.y. (geneva) agr. Expt Sta Ann Rpt 11:597–606

Beach SA (1892b) The self-pollination of the grape. Gard Forest 5:451–452

Beach SA (1894) The fertilization of flowers in orchards and vineyards, especially in its relation to the production of fruit. NY (Geneva) Agr Expt Sta Ann Rpt 13:633–648

Beach SA (1898) Self-fertility of the grape. N Y (Geneva) Agr Expt Sta Bul 157:397–441

Bergman HF (1954) Flowering and fruiting characteristics of the cranberry in New Jersey. Amer Cranberry Grow Assoc Proc 84:17, 22–20, 27

Berlinger MJ (1977) The Mediterranean vine mealy bug and its natural enemies in southern Israel. Phytoparasitica 5:3–14

Berry JA (1989) Ancistrocerus gazella (Vespoidea: Eumenidae); a first record for New Zealand. N Z Entomol 12:63–65

Berry JA, Morales CF, Hill MG, Lofroth BJ, Allan D J (1989) The incidence of three diaspid scales on kiwifruit in New Zealand. Proceedings of the 42nd New Zealand weed and pest control conference, pp 182–186

Blackman RL, Eastop VF (2000) Aphids on the World's crops: an identification and information guide. Wiley, Chichester, 466 pp

Blank RH, Gill GSC, Olsen MH (1994) Relationship between armoured scale infestations on leaves and fruit. In: Proceedings of the 47th New Zealand plant protection conference, 304–309 New Zealand Weed and Pest Control Society, Palmerston North, New Zealand

Blank RH, Gill GSC, Upsdell MP (1996) Greedy scale, *Hemiberlesia rapax* (Hemiptera: Diaspididae), phenology on kiwifruit leaves and wood. N Z J Crop Hortic Sci 24:239–248

Blank RH, Gill GSC, Dow BW (1997) Determining armoured scale distribution within kiwifruit blocks. Proceedings of the 50th New Zealand plant protection conference, 293–297. www.hortnet.co.nz/publications/nzpps/

Blasse W (1981) Erdbeerproduktion und Bieneneinsatz. Gartenbau 28(11):337–338

Blasse W, Haufe M (1989) Beeinflussung von Ertrag und Fruchtqualität durch Bieneneinsatz bei Erdbeere. Archiv für Gartenbau 37(4):235–245

Boek LD, Hang C, Eaton TE, Godfrey OW, Michel KH, Nakatsukasa WM, Yao RC (1994) Process for producing A 83543 compounds. US Patent No. 5:362–624
Boem EW (1960) Should you pollen-spray ohanez grapes? So Austral Dept Agr J 64(5):202–203
Bomben C, Malossini C, Cipriani G, Testolin R (1999) Long term storage of kiwifruit pollen. Acta Hortic 498:105–108
Bonfante S (1970) La impollinazione della fragola. Atti 4° Convegno Nazionale Fragola, Cesena, Italy, pp 291–296
Booth NO (1911) A suggestion in regard to the history of grape growing in America. Amer Soc Hort Sci Proc, 18th ann Mtg: 105ñ 112
Botero-Garces N, Isaacs R (2004) Movement by the grape berry moth, Endopiza viteana: displacement distance and direction. Physiol Entomol 29:443–452
Brantjes NBM (1978) Pollinator attraction of Vitis vinifera subsp. silvestris. Vitis 17:229–233
Bret BL, Larson LL, Schoonover JR, Sparks TC, Thompson GD (1997) Biological properties of Spinosad. Down to Earth 52:6–13
Broussard M, Rao S, Stephen WP (2011) Native bees, honeybees, and pollination in Oregon cranberries. Hortscience 46(6):885–888
Buchmann SL (1983) Buzz pollination in angiosperms. In: Jones CE, Little JR (eds) Handbook of experimental pollination biology. Van Nostrand Reinhold, New York, pp 73–113
Burd DW (1970) Tunnel strawberry pollination study: a step towards better fruit shape. Grower 73:1498–1499
Burrack HJ, Littlejohn K (2011) Rethinking blueberry maggot (Rhagoletis mendax) distribution and abundance in North Carolina: when area wide management is unintentional. Int J Fruit Sci (in press)
Bushway L, Pritts M, Handley D (eds) (2008) Raspberry and blackberry production guide (NRAES-35), Natural Resource, Agriculture, and Engineering Service (NRAES). Ithaca, New York
Candolfi M, Jermini M, Carrera E, Candolfi-Vasconcelos M (1993) Grapevine leaf gas exchange, plant growth, yield, fruit quality and carbohydrate reserves influenced by the grape leafhopper, *Empoasca vitis*. Entomol Exp Appl 69(3):289–296
Cane JH, Payne JA (1990) Native bee pollinates rabbiteye blueberry: highlights in agricultural research. Alabama Agric Res Sta 37(4):4
Cane JH, Schiffhauer D (2001) Pollinator genetics and pollination: do honey bee colonies selected for pollen-hoarding field better pollinators of cranberry vacciniurn macrocarpon? Ecol Entomol 26:117–123
Cane JH, Schiffhauer D (2003) Dose–response relationships between pollination and fruiting refine pollinator comparisons for cranberry [Vaccinium macrocarpon (Ericaceae)]. Amer J Bot 90:1425–1432
Cane JH, Schiffhauer D, Kervin LJ (1996) Pollination, foraging, and nesting ecology of the leaf-cutting bee Megachile (Delomegachile) addenda (Hymenoptera: Megachilidae) on cranberry beds. Ann Entomol Soc Am 89:361–367
Caswell JH (1962) Caswell Bee company – cape cod cranberry pollinators. Amer Bee J 102:222–223
Chagnon M, Gingras J, de Oliveira D (1991) Honey bee (Hymenoptera: Apidae) foraging behavior and raspberry pollination. J Econ Entomol 84:457–460
Chagnon M, Gingras J, de Oliverira D (1989) Effect of honey bee (Hymenoptera: Apidae) visits on the pollination rate of strawberries. J Econ Entomol 82(5):1350–1353
Chang YD, Lee MY, Young-II M (2001) Pollination on strawberry in the vinyl house by *Apis mellifera* L. and *A. cerana* Fab. Proceedings of the 8th pollination symposium. In: Benedek P, Richards K.W (eds). Acta Hort. 561, ISHS 2001
Chitkowski RL, Fisher JR (2005) Effect of soil type on the establishment of grape phylloxera colonies in the Pacific Northwest. Am J Enol Vitic 56:207–211
Cirnu LG, Fota, Grosu E (1978) Resultate pirivind plenizarea cu ajutoral albine lor a culturilor de capsuni (Fragarea sp.) in solaria. Apic Romania 53:12–14
Clarke EEK, Levy LS, Spurgeon A, Calvert IA (1997) The problems associated with pesticide use by irrigation workers in Ghana. Occup Mod 47:301–308

Clay OM (1886) A grape-grower defending the bees. Am Bee J 22:469

Clinch PG (1984) Kiwifruit pollination by honey bees I. Tauranga observations, (1978–81). N Z J Exp Agric 12:29–38

Clinch PG, Heath A (1985) Wind and bee pollination research. New Zealand Kiwifruit, 15 Nov

Cloyd RA, Galle CL, Keith SR (2006) Compatibility of three miticides with predatory mites *Neoseiulus callfornicus* McGregor and Phytoseiulus persimilis Athias-Henriot (Acari: Phytoseiidae). Hort Sci 41:707–710

Colbert S, de Oliveira D (1990) Influence of pollen variety on raspberry (Rubusidaeus L.) development. J Hered 81:434–437

Connor LJ (1970) Studies of strawberry pollination in Michigan. In the indispensable pollinators. Ark Agr Ext Serv Misc Pub 127:157–162

Connor LJ (1975) The role of cultivar in insect pollination of strawberries. In: Proceedings of the 3rd international symposium on pollination, Prague, pp 149–154

Connor LJ, Martin EC (1973) Components of pollination of commercial strawberries in Michigan. HortScience 8:304–306

Costa G, Testolin R, Vizzotto G (1993) Kiwifruit pollination: an unbiased estimate of wind and bee contribution. N Z J Crop Hortic Sci 21:189–195

Couston R (1937) Strawberry improvement U.S. Dept Agr Yearbook 1937: 445–495

Couston R (1963) The influence of insect pollination on raspberries. Scott Beekeeper 40:196–197

Crane E, Walker P (1984) Pollination directory for world crops. International Bee Research Association, London, 184 pp

Cross CE (1953) Cranberry flowers and the set of fruit. Cranberries 17(12):7–9

Cross CE (1966) Cranberry flowers and pollination. In research into action, mass. Univ Col Agr Coop Ext Serv Pub 435:27–29

Cross JV, Easterbrook MA, Crook AM, Crook D, Fitzgerald JD, Innocenzi PJ, Jay CN, Solomon MG (2001) Natural enemies and biocontrol of pests of strawberry in northern and central Europe. Bio Sci Technol 11:165–216

Currie RW, Winston ML, Slessor KN (1992a) Effect of synthetic queen mandibular pheromone sprays on honey bee (Hymenoptera: Apidae) pollination of berry crops. J Econ Entomol 85:1300–1306

Currie RW, Winston ML, Slessor KN, Mayer DF (1992b) Effect of synthetic queen mandibular pheromone sprays on pollination of fruit crops by honey bees (Hymenoptera: Apidae). J Econ Entomol 85:1293–1299

Danka RG, Lang GA, Gupton CL (1993) Honey bee (Hymenoptera: Apidae) visits and pollen source effects on fruiting of 'Gulfcoast' southern highbush blueberry. J Econ Entomol 86:131–136

Darrow GM (1927) Sterility and fertility in the strawberry. J Agr Res 34:393–411

Darrow GM (1937) Strawberry improvement. US Dept Agr Yearbook 1937:445–495

Darrow GM (1966) The strawberry. 447 pp. Holt. Rinehart and Winston, New York/Chicago/San Francisco

Darrow GM, Franklin HJ, Malde OG (1924) Establishing cranberry fields. U.S Dept Agr Farmers' Bul 1400, 37 pp

Davydova NS (1969) Possibilities of employing bees to pollinate Vineyards. In: 22d internationa. Apiculture Cong Proc, Munich: 176–180

de Oliveira D, Savoie L, Vincent C (1991) Pollinators of cultivated strawberry in Québec. Acta Hortic 288:420–424

Dearing C (1917a). The production of self-fertile muscadine grapes. Amer Soc Hort Sci Proc, 14th Ann Mtg: 30–34

Dearing C (1917b) Muscadine grape breeding. J Hered 8:409–424

Dearing C (1938) Muscadine grapes. U.S Dept Agr Farmers' Bul, 1785, 36 pp

Decou GC (1994) Biological control of the two-spotted spider mite (Acarina: Tetranychidae) on commercial strawberries in Florida with *Phytoseiulus persimilis* (Acarina: Phytopseiidae). Fla Entomol 77:33–41

Delaplane KS (1995) Bee foragers and their pollen loads in South Georgia rabbiteye blueberry. Am Bee J 135:825–826

Delaplane KS, Mayer DF (2000) Crop pollination by bees. CAB International, Wallingford, p 360

Dickey RD, Loucks KW (1938) Grape growing in florida. Fla Agr Expt Sta Bul 324, 36 pp

DoehlertCA (1940) The use of honeybees in cranberry bogs. Amer Cranberry Growers' Assoc Proc 70th Ann Mtg: 32–36

Donovan BJ, Read PEC (1988) The alfalfa leafcutting bee, *Megachile rotundata* (Megachilidae), does not pollinate kiwifruit, *Actinidia deliciosa var. deliciosa* (Actinidiaceae). J Apic Res 27:9–12

Donovan BJ, Read PEC (1991) Efficacy of honey bees as pollinators of kiwifruit. Acta Hortic 288:220–224

Dorsey MJ (L914) Pollen development in vitis with special reference to sterility. Minn Agr Expt Sta Bul 144: 60 pp

Dunne TC (1942) Pollen containing sprays for the cross-pollination of ohanez grapes. West Austral Dept Agr J 19((ser. 2) (3):210–213

Dutcher JD, All JN (1978) Survivorship of the grape root borer in commercial grape vineyards with contrasting cultural practices. J Econ Entomol 71:751–754

Easterbrook MA (1997) A field assessment of the effects of insecticides on the beneficial fauna of strawberry. Crop Prot 16:147–152

Eastwood B (1866) Complete manual for the cultivation of the cranberry. Orange-judd & co., New York, 120 pp

Easterbrook MA, Fitzgerald JD, Solomon MG (2006) Suppression of aphids on strawberry by augmentative releases of larvae of the lacewing *Chrysoperla carnea* (Stephens). Biocontrol Sci Tech 16(9):893–900

Eck P (1988) Blueberry science. Rutgers University Press, New Brunswick

Einset O (1930) Open pollination vs. hand pollination of pollen-sterile grapes. N.Y. (Cornell) Agr Expt Sta Tech Bul 162, 14 pp

El-Agamy SZA, Lyrene PM, Sherman WB (1979) Effect of mating system on time of ripening and fruit weight in blueberry. Proc Fla State Hortic Soc 92:258–259

English-Loeb GM, Flaherty DL, Wilson LT, Barnett WW, Leavitt GM, Settle WH (1986) Pest management affects spider mites in vineyards. Calif Agric 40:28–30

English-Loeb GM, Villani M, Martinson T, Forsline A, Conslie N (1999) Use of entomopathogenic nematodes for control of grape *Phylloxera* (Homoptera:Phylloxeridae): a laboratory evaluation. Environ Entomol 28(5):890–894

Escudero LA, Farragut F (2005) Life-history of predatory mites *neoseiulus californicus* and *phytoseiulus persimilis* (acari: phytoseiidae) on four spider mites species as prey, with special reference to *tetranychus evansi* (acari: tetranychidae). Biol Control 32:378–384

Evans EC, Spivak M (2006) Effects of honey bee (hymenoptera: apidae) and bumble bee (hymenoptera: apidae) presence on cranberry (ericales: ericaceae) pollination. J Econ Entomol 99:614–620

Farrar CL, Bain HF (1946) Honeybees as pollinators of the cranberry. Amer Bee J 86:503–504

Farrar CL, Bain HF (1947) Honeybees as pollinators of cranberries. Cranberries 11(9):6, 22–7, 23

Ferris H, McKenry MV (1974) Seasonal fluctuations in the spatial distribution of nematode populations in a California vineyard. J Nematol 6:203–210

Fetoh BES, Khaled AS, El-Nagar TFK (2009) Combined effect of entomopathogenic nematodes and biopesticides to control the greasy cut worm, Agrotis ipsilon (Hufn.) in the strawberry fields. Egypt Acad J Biol Sci 2(1):227–236

Filmer RS (1949) Cranberry pollination studies. In: Amer Cranberry Growers' Assoc Proc 80th Ann Conv: 14–20. American Cranberry Growers Association, New Jersey, USA

Filmer RS (1953) Cranberry pollination studies. In: Amer Cranberry Growers' Assoc Proc 84th Ann Conv: 28–36. American Cranberry Growers Association, New Jersey, USA

Filmer RS, Doehlert CA (1952) Use of honeybees in cranberry bogs. NJ Agr Expt Sta Bul 764, 4 pp

Filmer RS, Doehlert CA (1955) Use of honeybees in cranberry bogs. Gleanings Bee Cult 83:265–268

Filmer RS, Doehlert CA (1959) Use of honeybees in cranberry bogs. N. J Agr Expt Sta Cir 588, 4 pp

Filmer RS, Marucci PE, Moulter H (1958) Seed counts and size of cranberries. Amer Cranberry Grow Assoc Proc 88:22, 26–30

Ford I (1970) Investigations into the pollination of Chinese gooseberries. New Zealand Ministry of Agricultural and Fisheries Mimeograph, pp 1–12

Franklin HJ (1911) State bog report. Cape Cod Cranberry Grow Assoc Ann Rpt 24:16–28

Franklin HJ (1912) Report of cranberry substations for 1912. Mass Agr Expt Sta Rpt Part 1(209):234

Franklin HJ (1914) Reports on experimental work in connections with cranberries. Mass Agr Expt Sta Bul 150:37–62

Franklin HJ (1940) Cranberry growing in massachusetts. Mass Agr Expt Sta Bul 371: 44 pp

Free JB (1968a) The pollination of strawberries by honeybees. J Hortic Sci 43:107–111

Free JB (1968b) The foraging behaviour of honeybees (Apis mellifera) and bumblebees (Bombus spp.) on blackcurrant (Ribes nigrum), raspberry (Rubus idaeus) and strawberry fragaria x ananassa) flowers. J Appl Ecol 5:157–168

Free JB (1970) Insect pollination of crops. Academic Press, New York

Free JB (1993) Insect pollination of crops, 2nd edn. Academic, New York, p 684

Fry BO (1968). Cowart, a new self-fertile muscadine. Ga Agr Expt Sta Res Rpt 32, 6 pp

Funt RC, Ellis MA, Welty C (2004) Midwest small fruit pest management handbook, Bulletin 861. The Ohio State University Extension, Columbus

Gabel B, Renczés V (1982) Effects of design and siting of pheromone traps in monitoring the grape vine moth, *Lobesia botrana* (Lepidoptera, Tortricidae). Acta Entomol Bohemoslov 79:260–266

Galet P (1982) Les Maladies et le Parasites de la Vigne, Tome II. Les Parasites Animaux. Paysan du Midi, Montpellier

Gates BN (1911) The honey bee and cranberry growing. Cape Cod Cranberry Grow Assoc Ann Rpt 24:28

Gladwin FP (1931) Pollination with particular reference to the grape. Amer Fruit Grow 57(4):16–35

Gladwin FF (1937) Pollination with particular reference to the *grape*. Am Fruit Grow Mag 57:9.24–25,35: 3:16.35

Golodriga PI (1953) On the selection of Pollinizer varieties of grape. Agrobiologiya 5:105–110. [in Russian.]

González RH (1986) Pests of kiwi fruit in Chile. Revista Frutícola 1986 7(1):13–27

González MV, Coque M (1995) Stigmatic receptivity limits the effective pollination period in kiwifruit. J Amer Soc Hortic Sci 120(2):199–202

González MV, Coque M, Herrero M (1995) Stigmatic receptivity limits the effective pollination period in kiwifruit. J Am Soc Hortic Sci 120:199–202

Goodman RD, Oldroyd BP (1988) Honeybee pollination of strawberries (*Fragaria ananassa* Duchesne). Aust J Exp Agric 28:435–438

Goodwin RM (1985) Honeybee habits under scientific scrutiny. New Zealand Kiwifruit, May

Goodwin RM (1986a) Kiwifruit flowers: anther dehiscence and daily collection of pollen by honey bees. N Z J Exp Agric 14:449–452

Goodwin RM (1986b) Learning more about the honey bee and the flowers we hope it will pollinate. New Zealand Kiwifruit, 9 Feb

Goodwin RM (1989) Pollination hints for Hi-Cane vines. New Zealand Kiwifruit, 16 Sep

Goodwin RM (1995) Afternoon decline in kiwifruit pollen collection. N Z J Crop Hortic Sci 23:163–171

Goodwin RM, Houten T (1991) Poisoning of honey bees (Apismellifera) by sodium flouroacetate (1080) in baits. N Z J Zool 18(1):45–51

Goodwin RM, Steven D (1993) Behaviour of honey bees visiting kiwifruit flowers. N Z J Crop Hortic Sci 21:17–24

Goodwin RM, Ten Houten A, Perry JH (1990) Hydrogen cyanamide and kiwifruit pollination. N Z Kiwifruit Spec Publ 3:14

Goodwin RM, Ten Houten A, Perry JH (1994) Effect of feeding pollen substitutes to honey bee colonies used for kiwifruit pollination and honey production. N Z J Crop Hortic Sci 22:459–462

Goodwin RM, Ten Houten A, Perry JH (1999) Effect of staminate kiwifruit vine distribution and flower number on kiwifruit pollination. N Z J Crop Hortic Sci 27:63–67

Grannet J, Walker MA, Kocsis L, Omer AD (2001) Biology and management of grape phylloxera. Annu Rev Entomol 26:387–412. doi:10.1146/annurev.ento.46.1.387

Greaves AJ, Davys JW, Dow BW, Tomkins AR, Thomson C, Wilson DJ (1994) Seasonal temperatures and the phenologyof greedy scale populations (Hemiptera: Diaspididae) on kiwifruit vines in New Zealand. N Z J Crop Hortic Sci 22:7–16

Grewal P, Ehlers RU, Shapiro – Ilan D (eds) (2005) Nematodes as biological control agents. CABI Publishing, Walling Ford

Gundia M et al (1995) The use of bumble bees, Bombus terrestirs, for pollination of green house strawberry. Hassadeh 75(9):49–50, 71–73

Gupton CL, Spiers JM (1994) Interspecific and intraspecific pollination effects in rabbiteye and southern highbush blueberry. HortSci 29:324–326

Gyan KY, Woodell SRJ (1987) Analysis of insect pollen loads and pollination efficiency of some common insect visitors for four species of woody Rosaceae. Funct Ecol 1:269–274

Hale CR, Jones LT (1956) The pollination of ohanez grapes. J Agr West Austral 5:565–567

Hall IV, Aalders LE (1961) Note on male sterility in the common lowbush blueberry, *Vaccinium angustifolium* Ait. Can J Plant Sci 41:86

Hanna R, Wilson LT, Zalom FG, Flaherty DL (1997) Effects of predation and competition on population dynamics of *Tetranychus pacificus* on grapevines. J Appl Ecol 34:878–888

Harris BM, Maclean B (1999) Spinosad; control of lepidopterous pests in vegetable brassiscas. In: Proceedings of the 52nd plant protection conference (Proceedings of the New Zealand plant protection conference), Wellington, pp 65–69

Hart AJ, Bale JS, Tullett AG, Worland MR, Walters KFA (2002) Effects of temperature on the establishment potential for the predatory mite *Amblyseius californicus* McGregor (Acari: Phytoseiid) in the UK. J Insect Physiol 48:593–599

Haviland D, Hernandez N (2012) Development of management programs for white grubs in California blueberries. Int J Fruit Sci 12(1–3):114–123

Heap IM (1997) The occurrence of herbicide-resistant weeds worldwide. Pest Sci 51:235–243

Hedrick UP (1924) Manual of American grape-growing. The Macmillan Co., New York, 458 pp

Herbert EW, Shimanuki H (1980) An evaluation of seven potential pollen substitutes for honey bees. Am Bee J 120:349–350

Hooper CH (1932) Pollination in relation to cherry orchards. JI S-E Agric Coli Wye 30:244–246

Hopping ME (1982) Kiwifruit Hand Pollination for Size Improvement. New Zealand Ministry of Agriculture and Fisheries Aglink HPP 260

Hopping ME, Hacking NJA (1983) A comparison of pollen application methods for artificial pollination of kiwifruit. Acta Hortic 139:41–50

Hopping ME, Jerram EM (1979) Pollination of kiwifruit (*Actinidia chinensis* Planch.): stigma-style structure and pollen tube growth. N Z J Bot 17:233–240

Houbaert D, Borremans G, Baetc W, Jacobs FJ (1992) Quality improvement of strawberries through optimal pollination. Fruit Letts 5(20):26–29

Howard D, Patterson K, Franklin S, Orchard-Lisle V, Morton J (1985) Treatment of word retrieval deficits in aphasia: a comparison of two therapy methods. Brain 108(4):817–829

Huffaker CB, van de Vrie M, McMurtry JA (1969) The ecology of tetranychid mites and their natural control. Annu Rev Entomol 14:125–174

Hughes HM (1962) Pollination studies. Progress report. M.A.A.F., N.A.A.S., Rep Efford Exp Hort Stn 1961

Husmann GC (1916) Muscadine grapes. U.S Dept Agr Farmers' Bul 709, 28 pp

Husmann GC (1932) Grape districts and varieties in the united states. US Dept Agr Farmers' Bul 1689:33
Husmann GC, Dearing C (1913). The muscadine grapes. U.S Dept Agr Burl Plant Ind Bul 273, 64 pp
Hutson R (1924) Bees in fruit pollination. Gleanings Bee Cult 52:290–292
Hutson R (1925) The honeybee as an agent in the pollination of pears, apples and cranberries. J Econ Ent 18:387–391
Hutson R (1926) Relation of the honeybee to fruit pollination in new jersey. NJ Agr Expt Sta Bull 434:32
Hutson R (1927) The use of honeybees as pollinating agents on Cranberry bogs. Amer Cranberry Growers' Assoc Proc 57th Ann Conv: 10–11
Ifoulis AA, Savopoulou-Soultani M (2004) Biological control of *Lobesiabotrana* (Lepidoptera: Tortricidae) larvae by using different formulations of *Bacillus thuringiensis* in 11 vine cultivars under field conditions. J Econ Entomol 97:340–343
Isaacs R (2010) Insects. http://www.blueberries.msu.edu/. Accessed 6 Sept 2010
Isaacs R, Gajek D (2003) Abundance of blueberry bud mite (Acalitus vaccinii) in Michigan blueberries, and variation in infestation among common highbush blueberry varieties. Integ Plant Protect Orchards – Soft Fruits IOBC/wprs Bull 26:127–132
Isaacs R, Van Timmeren S (2009) Monitoring and temperature-based prediction of the whitemarked tussock moth (Lepidoptera: Lymantriidae) in blueberry. J Econ Entomol 102:637–645
Isaacs R, Schilder A, Miles T, Longstroth M (2008) Blueberry aphid and blueberry shoestring virus, Michigan blueberry facts E-3050. Michigan State University, East Lansing
Isaacs R, Hahn N, Tritten B, Garcia C (2010) Spotted wing drosophila: a new invasive pest in Michigan, MSU extension bulletin E-3140. Michigan State University, East Lansing
Iyer CPA, Randhawa GS (1965) Hybridization studies in grapes: Investigations on the direct influence of pollen on some fruit and seed characters. Indian J Hortic 22(2):107–121
Jay SC (1986) Spatial management of honey bees on crops. Annu Rev Entomol 31:49–65
Jay D, Jay C (1984) Observations of honeybees on Chinese gooseberries ('Kiwifruit') in New Zealand. Bee World 65:155–166
Jay SC, Jay DH (1993) The effect of kiwifruit (Actinidiadeliciosa A Chev) and yellow flowered broom (Cytisusscoparius Link) pollen on the ovary development of worker honey bees (*Apis mellifera* L.). Apidologie 24:557–563
Jaycox ER (1970) Pollination of strawberries. Amer Bee J 110:176–177
Jennings DL (1988) Raspberries and blackberries: their breeding, diseases, and growth. Academic, London
Johansen CA, Hutt R (1963) Encouraging the bumble bee pollinator [bombus mixtus] of Cranberries. Wash Agr Ext Serv EM 2262:2
Johnson DT, Lewis BA, Snow JW (1991) Control of grape root borer (Lepidoptera: Sesiidae) by mating disruption with two synthetic sex pheromone compounds. Environ Entomol 20(3):930–934
Johnston S (1929) Insects aid fruit setting of raspberry. Michigan Agric Exp Station Quart Bull 11:105–106
Kabayashi S (1987) Uniformity of plants regenerated from orange protoplasts. Theor Appl Genet 74:10–14
Kagabu S (1999) Studies on the synthesis and insecticidal activity of neonicotinoid compounds. Pestic Sci 21:231–239
Kang SY (1996) Cutivation of vinylhouse strawberry in Korea (Korean). Chungnam
Kapil RP (1970) Role of honeybees in agriculture. Indian Bee J 32:52–57
Kaplan I, Eubanks MD (2002) Disruption of cotton aphid (Homoptera: Aphididae)-natural enemy dynamics by red imported fire ants (Hymenoptera: Formicidae). Environ Entomol 31:1175–1183
Karban R, English-Loeb G, Walker MA, Thaler J (1995) Abundance of phytoseiid mites on *vitis* sp.: effect of leaf hairs, domata, prey abundance and plant phylogeny. Exp Appl Acarology 19:189–197

Katayama E (1987a) Utilization of honeybees as pollinators for strawberries in plastic greenhouses. Honeybee Sci 8:147–150 (in Japanese)
Katayama E (1987b) Utilization of honeybees as pollination for strawberries in plastic house of strawberry cultivation area by the questionnaire (Korean). Korean J Apicul 4:1–8
Kaya HK, Gaugler R (1993) Entomopathogenic nematodes. Annu Rev Entomol 38:181–206
Kelly CB (1944) The grape in Ontario. Ontario Dept Agr Statis Branch Bull 438:38
Kendall DA, Wilson D, Guttridge GG, Anderson HM (1971) Testing Eristalis as a pollinator of covered crops. Report of Long Ashton Research Station, University of Bristol, pp 120–121
Kevan PG (1993) Pollination in modern agriculture: changing practices, new and hybrid crops. In Asia honey bees and bee mites. Ed. Connor. L.C. Wicwas Press, Connecticut, p 704
Kido H, Flaherty DL, Bosch DF, Valero KA (1984) The variegated leafhopper in the San Joaquin valley. Calif Agric 38(1–2):31–32
King PD, Buchanan GA (1986) The dispersal of phylloxera crawlers and spread of phylloxera infestations in New Zealand and Australian vineyards. Am J Enol Viticult 37:26–33
King MJ, Buchmann SL (2003) Floral sonication by bees: mesosomal vibration by bombus and xylocopa, but not apis (hymenoptera: apidae), ejects pollen from poricidal anthers. J Kans Entomol Soc 76:295–305
Kirchmair M, Huber L, Porten M, Rainer J, Strasser H (2004) *Metarhizium anisopliae*, a potential agent for the control of grape phylloxera. Biocontrol 49:295–303
Kitagawa S (1985) Pollination of strawberries in the greenhouse by honeybees. Apimondia 29:378–380
Knuth P (1908) Handbook of flower pollination, vol 2 (trans: Ainsworth Davis JR). Clarendon Press, Oxford, 705 pp
Krewer G, Bertrand P, Meyers S, Horton D (1986) Kiwi cold injury, other problems be set southeastern plantings. Fruit South 7(5):12–15
Kronenberg HG, Braak JP, Zeilinga AE (1959) Poor fruit setting in strawberries. II. Malformed fruits in Jucunda. Euphytica 8:245–251
Kühn BF (1987) Bestøvning af hindbærrsorten 'Willamette'. Tidsskrift for Planteavl 91:85–88
Lackett JJ, Burkhardt CC (1979) Effects of visits by honeybees, Apis mellifera L., on production and quality of 'Dabreak' and 'Tangi' strawberries Fragaria x ananassa Duch. In: Proceedings 4th international symposium on pollination 1978. Maryland Agricultural Experimental Station Special Miscellaneous Publication 1, pp 137–141
Laiok VD (1953) On pollination of grapes by honey bees. Pchelovodstvo 8:46–48. [in Russian.]
Lang GA, Danka RG (1991) Honey-bee-mediated cross- versus self-pollination of 'Sharpblue' blueberry increases fruit size and hastens ripening. J Am Soc Hortic Sci 116:770–773
Lavrov VV (1956) Interspecific pollination of grapes and the training of bees. In: Krishchunas IV, Gubin AF (eds) Pollination of agricultural plants. Gos, Izd-vo. Selkhoz lit-ry, Moskva, pp 200–203. [in Russian.]
Leclant F, Deguine JP (1994) Aphids. In: Mathews GA, Tunstall JP (eds) Insect pests of cotton. CAB International, Wallingford, pp 285–324
Liburd OE, Arevalo HA (2006) Insects and mites in blueberries. In: Childers NF, Lyrene PM (eds) Blueberries for growers, Gardeners and Promoters. Dr Norman E Childers Publications, Gainesville, pp 99–110
Liburd OE, Alm SR, Casagrande RA (1998) Susceptibility of highbush blueberry cultivars to larval infestation by Rhagoletis mendax (Diptera: Tephritidae). Environ Entomol 27:817–821
Liburd OE, White JC, Rhodes ER, Browdy AA (2007) The residual and direct eVects of reduced-risk and conventional miticides on twospotted spider mites, *tetranychus urticae* (acari: tetranychidae) and predatory mites (acari: phytoseiidae). Florida Entomol 90:249–257
Liburd OE, Sarzynski EM, Arevalo HA, MacKenzie K (2009) Monitoring and emergence of flower thrips species in rabbiteye and southern highbush blueberries. Acta Hortic 810:251–258
Lo PL, Blank RH (1989) A survey of armoured scale species (Herniptera: Diaspididae) in kiwifruit orchards. N Z Entomol 12:1–4

Logan D, Connolly P (2005) Cicadas from kiwifruit orchards in New Zealand and identification of their final instar exuviae (Cicadidae: Homoptera). N Z Entomol 28:37–48
Lomond D, Larson DJ (1983) Honey bees, Apis mellifera (Hymenoptera: Apidae), as pollinators of lowbush blueberry, Vaccinium angustifolium, on Newfoundland coastal barrens. Can Entomol 115:1647–1651
Lyrene PM (1994) Variation within and among blueberry taxa in flower size and shape. J Am Soc Hortic Sci 119:1039–1042
Lyrene PM (2007) Breeding southern highbush blueberries. Fruit Breed Rev 30:353–406
Lyrene PM, Payne JA (1995) Blueberry gall midge: a new pest of rabbiteye blueberries. J Small Fruit Prod 3:111–124
Maas JL (1998) Compendium of Strawberry Diseases. American Phytopathological Society, St Paul, p 98
Macfarlane RP, Ferguson AM (1984) Kiwifruit pollination: a survey of the insect pollinators in New Zealand. In: Proceedings of the 5th international symposium on pollination, Les Colloques del INRA, Versailles. pp 367–373
MacKenzie KE (1994) The foraging behaviour of honey bees (Apis mellifera L) and bumble bees (Bombus spp) on cranberry (*Vaccinium macrocarpon* Ait). Apidologie (Celle) 25:375–383
MacKenzie KE, Javorek S (1997) The potential of alfalfa leafcutter bees (Megachile rotundata L.) as pollinators of cranberry (Vaccinium macrocarpon Aiton). Acta Hortic 437:345–352
Magoon CA, Snyder E (1943) Grapes for Different regions. US Dept Agr Farmers' Bul 1936:38
Mahar AN, Munir M, Gowen SR, Hague NGM, Tabil MA (2004) Biological control of black vine weevil Otiorhynchus sulcatus, Coleoptera: Curculionidae by entomopathogenic bacteria and their cell-free toxic metabolites. J Food Agric Environ 2:209–212
Majek BA (2006) Weeds in blueberries. In: Childers NF, Lyrene PM (eds) Blueberries for Growers, Gardeners and Promoters. Dr Norman E Childers Publications, Gainesville, pp 86–98
Marriott PF (1950) Pollination of table grapes. Victoria Dept Agr J 48(9):391–394
Martinson T (1995) Management of insect pests in organic vineyards. In: Organic grape and wine production symposium, Geneva, New York. Communications Services, New York State Agricultural Experiment Station, Cornell University, Geneva, New York.
Marucci PE (1966) Cranberry pollination. Cranberries 30(9):11–13
Marucci PE (1967a) Cranberry pollination. Cranberries 32(7):8–9
Marucci PE (1967b) Cranberry pollination. Amer Bee J 107:212–213
Marucci PE, Filmer RS (1964) Preliminary cross pollination tests on cranberries. Amer. Cranberry Growers' Assoc Proc 91st–94th Ann Mtg 1961–64: 48–51.
Matsuka M, Sakai T (1989) Bee pollination in Japan with special reference to strawberry production in greenhouses. Bee World 70:55–61
McCutcheon D (1978) Pollination of raspberries. British Columbia Ministry of Agriculture, Apiary Branch, Leaflet 503
McCutcheon D (1983) Blueberry pollination. British Columbia Department of Agriculture, Apiary Branch Bee Notes 507
McGregor SE (1976) Insect pollination of cultivated crop plants. Agr Res Serv, United States Dep of Agric, Handb No. 496
McKenna CE (1998a) The effectiveness of different insecticide programmes for kiwifruit. Proceedings of the 51st New Zealand plant protection conference, pp 184–188
McKenna CE (1998b) Pest monitoring in kiwifruit crops. N Z Kiwifruit J 125:12
McKenna CE, Steven D (1993) Phytotoxicity to kiwifruit of oil sprayed after flowering. Proceedings of the 46th New Zealand plant protection conference, p 75–79
McKenna CE, Stevens PS, Steven D (1995) A new *bacillus thuringiensis* product for use on kiwifruit. Proceedings of the 48th New Zealand plant protection conference, p 135–138
McKenna CE, Stevens PS, Steven D (1997) Phytotoxicity to kiwifruit from sprays of mineral oil. Acta Hortic 444(2):779–784
Mckenna CE, Dobson SJ, Phare JM (2009) The insect pest complex of *Actinidia arguta* kiwifruit. N Z Plant Protect 62:262–267
McKenry M (1999) The replsant problem and its management. Catalina Publication, Fresno

McMurtry JA, Croft BA (1997) Life-styles of phytoseiid mites and their roles in biological control. Annu Rev Entomol 42:291–321

Meeta Y (1978) Comparative studies on the biology of bees of the genus Osmia of Japan, with special reference to their managements for pollination of crops (Hymenoptera: Megachilidae). Bull Tokohu Natl Agr Exp Stn 57:1–221

Meyer JR, Ballington JR (1990) Resistance of Vaccinium spp. to the leafhopper *Scaphytopius magdalensis* (Homoptera:Cicadellidae). Ann Entomol Soc Amer 83:515–520

Moeller FE, Koval CF (1973) Honeybee pollination of strawberries in Wisconsin. University of Wisconsin--Extension. Cooperative Extension Publication, A 2549, 7 p

Mommers J (1961) De bestuiving van aardbeien onder glas. Meded Dir Tuinb 24:353–355

Moore JN (1964) Duration of receptivity to pollination of flowers of the high-bush blueberry and the cultivated strawberry. Amer Soc Hortic Sci Proc 85:295–301

Moore JN (1969) Insect pollination of strawberries. Amer Soc Hortic Sci Proc 94:362–364

Morrissette R, Francoeur A, Perron J-M (1985) Importance des abeillessauvages (Apoidea) dans la pollinisation des bleuetiersnains (*Vaccinium* spp.) en Sagamie, Quebec. Revue Entomologie Québec 30:44–53

Mossler M, Nesheim ON (2003) Florida crop/pest management profile: strawberries. University of Florida, IFAS Cooperative Extension Service, EDIS 129

Munson TV (1899) Investigation in and improvement of American Grapes tex. Agr Expt Sta Bull 56:217–285

Murrell DC, McCutcheon DM (1977) Red raspberry pollination in British Columbia. Am Bee J 117:750

Muttoo RN (1952) Bees produce more fruit. Indian Bee J 1(7):118–119

Nagarkatti S, Tobin PC, Muza AJ, Saunders MC (2002) Carbaryl resistance in populations of grape berry moth (Lepidoptera: Tortricidae) in New York and Pennsylvania. J Econ Entomol 95:1027–1032

Neira MC, Lobos NS, Riveros MG, Carrillo RL, Pessot RZ, Mundaca NB (1997) Insect fauna associated with flowers of raspberries (Rubus idaeus L., cv Meeker) and evaluation of the pollination activity of Apis mellifera L. under the influence of a pheromonal attractant. Revista Chilena de Entomologia 24:37–44

Nicholls CI, Parrella MP, Altieri MA (2000) Reducing the abundance of leafhoppers and thrips in a northern California organic vineyard through maintenance of full season floral diversity with summer cover crops. Agric For Entomol 2(2):107–113

Nicol JM, Stirling GR, Rose BJ, May P, Heeswijck RV (1999) Impact of nematodes on grapevine growth and productivity: current knowledge and future directions, with special reference to Australian viticulture. Aust J Grape Wine Res 5(3):109–127

Nondillo A, Redaelli LR, Pinent SMJ, Botton M (2008) Exigências térmicas e estimativa do número de gerações anuais de Frankliniella occidentalis (Pergande) (Thysanoptera: Thripidae) em morangueiro. Neotropical Entomol 37:646–650

Nougaret RL, Lapham MH (1928) A study of phylloxera infestation in California as related to types of soils. Technical Bulletin No. 2, US Department of Agriculture, Washington, DC

Nye WP, Anderson JL (1974) Insect pollinators frequenting strawberry blossoms and the effect of honey bees on yield and fruit quality. J Amer Soc Hortic Sci 99(1):40–44

Oatman ER, Badgley ME, Platner GR (1985) Predators of the two-spotted spider mite on strawberry. Calif Agr 39(1–2):9–12

Oertel E (1967) Nectar and pollen plants. In beekeeping in the united states. US Dept Agr, Agr Handb 335:10–16

Olmo HP (1943) Pollination of the Almeria grape. Proc Am Soc Hort Sci 42:401–406

Omer AD, Granett J, De Benedictis JA, Walker MA (1995) Effects of fungal root infections on the vigor of grapevines infested by root- feeding grape phylloxera. Vitis 34:165–170

Palmer Jones T, Clinch PG (1974) Observations on the pollination of Chinese gooseberries variety "Flayward". NZ J ExpAgric 2:455–458

Palumbo JC, Horowitz AR, Prabhaker N (2001) Insecticidal control and resistance management for Bemisia tabaci. Crop Prot 20:739–765
Pammel LH, King CM (1930) Honey plants of Iowa. Iowa Geol Surv Bull 7:892–894
Papke AM, Eaton GW, Bowen PA (1980) Airborne pollen above a cranberry bog. HortSci 15:756
Paranjpe AV, Cantliffe DJ (2004) Developing a system to produce organic plug transplants for organic strawberry production. Proc Fla State Hort Soc 117:276–282
Pearson RC, Goheen AC (1988) Compendium of grape diseases. American Phytopathological Society, St. Paul
Pellett FC (1947) American honey plants, 4th edn. Dadant & Sons Inc., Hamilton, 467 pp
Percival MS (1955) The presentation of pollen in certain angiosperms and its collection by Apis mellifera. New Phytol 54:353–368
Perring TM, Gruenhagen NM, Farrar CA (1999) Management of plant viral diseases through chemical control of insect vectors. Annu Rev Entomol 44(1):457–481
Petkov VG (1963) The share of bees in the pollination of strawberry plants. In 19th Internatl. Apic Cong Proc, Prague, p 94–95
Petkov VG (1965a) Studies on the participation of bees in strawberry pollination. Gradinar I Lozar Nauka 2(4):421–431 [In Bulgarian, German and Russian summaries.]
Petkov V (1965b) Contribution of honeybees to the population of strawberries Gradinar. Lozar Nauk 2:421–431
Pinzauti M (1987) Comportamento e ruolo delle api nell'impollinazione della fragola. Colture Protette 16:65–68
Pion S, de Oliveira D, Paradis RO (1980) Agents pollinasateurs et productivite du fraisier 'Redcoat', Fragaria x ananassa Duch. Phytoprotection 61:72–78
Poinar GO Jr (1990) Biology and taxonomy of steinernematide and heterorhabditidae. In: Gaugler R, Kaya HK (eds) Entomopathogenic Nematodes in Biological control. CRC Press, Boca Raton, pp 23–62
Powles SB, Lorraine-Colwill DF, Dellow JJ, Preston C (1998) Evolved resistance to glyphosate in rigid ryegrass (Lolium rigidum) in Australia. Weed Sci 46(5):604–607
Price JF, Kring JB (1991) *Dyscinetus Morator* (Coleoptera: Scarabaeidae) flight activity, food plant acceptance, damage and control in Caladium. Fla Entomol 74:415–421
Priore R, Sannino O (1979) Ruolo dell' ape (*Apis mellifera*) nella coltura della frogola sotto serra. Apicoltore Mod 70:115–119
Pritts MP, Hancock JF (1992) Highbush blueberry production guide. Pub. NRAES-55. Northeast Regional Agricultural Engineering Service. Cooperative Extension Service, Ithaca
Pritts M, Kovach J (2000) 'Strawberry production systems' in greenhouse berried treasures. http://www.hort.cornell.edu/department/faculty/pritts/organic.htm. Accessed 14 Jun 2004
Pyke NB, Alspach PA (1986) Inter-relationships of fruit weight, seed number and seed weight in kiwifruit. N Z Agr Sci 20(3):153–156
Randhawa GS, Negi SS (1965) Further Studies on flowering and pollination in grapes. Indian J Hort 22(3/4):287–308
Randhawa GS, Negi SS, Sharma PL (1960) Studies on flowering and Pollination in grapes. Hort Adv 4:21–37
Raski DJ, Hewitt WB, Goheen AC, Taylor CE, Taylor RH (1965) Survival of Xiphinema Index and reservoirs of fanleaf virus in fallowed vineyard soil. Nematologica 11(3):349–352
Ratti CM, Higo HA, Griswold TL, Winston ML (2008) Bumble bees influence berry size in commercial Vacciniumspp. Cultivation in British Columbia. Can Entomol 140:348–363
Redalen G (1980) Morphological studies of raspberry flowers. Science Report, Agricultural University, Norway 59:1–11
Reimer FC (1910) Selfsterility of rotundifolia grapes. Amer Soc Hort Sci Proc 7th Ann Mtg: 27–32
Reimer FLC, Detjen R (1910) Seef-sterility of the scuppernong And other muscadine grapes. NC Agr Expt Sta Bull 209:23
Reisenzein H, Pfeffer M, Aust G, Baumgarten A (2007) The influence of soil properties on the development of grape phylloxera populations in Austrian viticulture. Acta Hortic 733:13–23
Rhodes EM, Liburd OE (2005) Predatory mite, *Neoseiulus californicus* (McGregor) (Arachnida:Acari:Phytoseiidae). IFAS Extension. IN639. University of Florida, Gainesville

Rhodes EM, Liburd OE, Kelts C, Rondon SI, Francis RR (2006) Comparison of single and combination treatments of *Phytoseiulus persimilis, Neoseiulus californicus*, and Acramite (bifenazate) for control of twospotted spider mites in strawberries. Exp Appl Acarol 39:213–225
Rigby B, Dana MN (1972) Flower opening, pollen shedding, stigma receptivity and pollen tube growth in the cranberry. Hortscience 7:84–85
Ritzinger R, Lyrene PM (1999) Flower morphology in blueberry species and hybrids. HortSci 34:130–131
Roberts RH, Struckmeyer BE (1942) Growth and fruiting of the cranberry. Amer Soc Hortic Sci Proc 40:373–379
Rondon SI, Price JF, Cantliffe DJ (2004) An Integrated Pest Management Approach: Monitoring strawberry pests grown under protected structures. Acta Hortic 659:351–356
Rosati P (1995) Recent trends in strawbeery production and research:an overview. Department Di Biotechnologie Agrarie e Ambientali. Universita – Facolta Di Agraria Ancona, Italy
Safran B (1954) The problem of cross-pollination and defective fruit set in the grapevine (in Herbew). M.Sc thesis, The Hebrew University of Jerusalem, Rehovot, Israel
Sakai T, Matsuka M (1988) Bee-pollination in Japan, with special reference to strawberry production. Honeybee Sci 9:97–101
Sale PR (1972) Oil and Chinese gooseberries do not agree. The Orchardist of N Z 44:97–99
Sale PR (1983) In: Williams DA (ed) Kiwifruit culture. Government Printer, Wellington
Sale PR (1984) Kiwifruit pollination. New Zealand Ministry of Agriculture and Fisheries. Aglink HPP 233
Sale PR, Lyford PB (1990) Cultural, management and harvesting practices for kiwifruit in New Zealand. In: Warrington IJ, Weston GC (eds) Kiwifruit science and management. New Zealand Society of Horticultural Science, Auckland
Sannino G, Priore R (1979) Le role des abeilles (Apis mellifera L.) dans la culture du fraisier sous serre. Fruits 34:503–506
Sarracino J, Vorsa N (1991) Self and cross fertility in cranberry. Euphytica 58:129–136
Sarzynski EM, Liburd OE (2003) Techniques for monitoring cranberry tipworm (Diptera: Cecidomyiidae in rabbiteye and southern highbush blueberries. J Econ Entomol 96:1821–1827
Sato ME, Miyata T, Da Silva M, Raga A, De Souza Filho MF (2004) Selections for fenpyroximate resistance and susceptibility, and inheritance, cross-resistance and stability of fenpyroximate resistance in *Tetranychus urticae* Koch (Acari: Tetranychidae). Appl Entomol Zoo 39(2):293–302
Saxton VP, Creasy GL, Paterson AM, Trought MCT (2004a) Experimental method to investigate and monitor bird behavior and damage in vineyards. Am J Enol Vitic 55(3):288–291
Saxton VP, Hickling GJ, Trought MCT, Creasy GL (2004b) Comparative behavior of free-ranging blackbirds (Turdus merula) and silvereyes (Zosterops lateralis) with hexose sugars in artificial grapes. Appl Anim Behav Sci 85:157–166
Schmid R (1978) Reproductive anatomy of Actinidiachinensis (Actinidiaceae). BotanischeJahrbuecherfuerSystematikPflanzengeschichte und Pflanzengeographie 100:149–195
Schoonover JR, Larson LL (1995) Laboratory activity of spinosad on non target beneficial arthropods. J Arthropod Manage Tests 20:357
Scott-Dupree C, Winston M, Hergert G, Jay SC, Nelson D, Gates J, Termeer B, Otis G (1995) A guide to managing bees for crop pollination. Canadian Association of Professionals Apiculturists. Canadian Department of Agriculture, Aylesford, 41 pp
Settle WH, Wilson LT, Flaherty DL, English-Loeb GM (1986) The variegated leaf hopper, an increasing pest of grape. Calif Agric 40(7–8):30–32
Shapiro-Ilan DI (2004) In: Capinera JL (ed) Encyclopedia of Entomology, vol 1. Kluwer Academic Publishing, Dordrecht, pp 781–784
Sharma D, Abrol DP (2005) Contact toxicity of some insecticides to honey bee Apis mellifera and Apis cerana F. J Asian Pacific Entomol 8(1):113–115
Sharma RM, Yamdagini R (2000) Modern strawberry cultivation. Kalyani Publishers, Ludhiana

Sharma RM, Singh AK, Sharma S, Masoodi FA, Shankar U (2013) Strawberry regeneration and assessment of runners quality in sub-tropical plains. J Appl Hortic 15(3):191–194
Sharples GC, Kuykendall JR, True LF, Tate HF (1961) Improvement of market quality of cardinal grape by Inflorescence apex removal. Amer Soc Hortic Sci Proc 77:316–321
Sharples GC, Todd FE, Mcgregor SE, Milne RL (1965) The importance of insects in the pollination and fertilization of the Cardinal grape. Proc Am Soc Hortic Sci 86:321–325
Shimanuki H, Lehnert T, Stricker M[H] (1967) Differential collection of cranberry pollen by honey Bees. J Econ Ent 60:1031–1033
Shimotori K (1981) Honeybees and strawberry industry in Tochigi. Honeybee Sci 2(2):57–60
Sibert I (1967) The cranberry pollinator. Gleanings Bee Cult 95:281–282
Singh JP (1969) Effect of thiodon on foraging honey bees. Indian Bee J 31:9–10
Singh Y (1979) Pollination activity on strawberry at Jeolikote, District, Nainital, India. Indian Bee J 41:179–19
Skowronek J, Jablonski B, Szklanowska K (1985) Wplyw owadow zapylajacych na owocowanie 6 odmian truskawki (*Fragaria grandiflora* Ehrah). Pszczelnicze Zeszyty Naukowe 29:205–229
Skrebtsova ND (1957) [The role of bees in pollinating strawberries.] Pchelovodstvo. Mosk 34:34–36
Skrebtsova ND (1958) Effectiveness of the saturated pollination of strawberries by bees. Biull. Nauchno-Tekh. Inform (Nauchno-Issled. Inst. Pchelovod.) (3–4): 25–26. [In Russian.]
Smith JB (1890) The insects injuriously affecting cranberries. New Jersey Agric College Expt Sta Special Bull K: 41
Snyder E (1937) Grape development and Improvement. U.S. Dept Agr Yearbook 1937: 631–655
Snyder E, Harmon' FN (1951) The cardinal, calmeria and blackrose grapes for vinifera regions. U.S. Dept Agric Cir 882:8 pp
Solomon MG, Jay CN, Innocenzi PJ, Fitzgerald JD, Crook D, Crook AM, Easterbrook MA, Cross JV (2001) Review: Natural Enemies and Biocontrol of Pests of Strawberry in Northern and Central Europe. Biocontrol Sci Technol 11(2):165–216
Sosunov VI (1953) On Crossed pollination of grapes. Sad I Ogorod 5:26–27. [In Russian.]
Sparks TC, Thompson GD, Larson LL, Kirst HA, Jantz OK, Worden TV, Hertlien MB, Busacca JD (1995) Biological characteristic of the spinosyns: new naturally derived insect control agents. In: Proceedings, beltwide cotton conference, National cotton council, Meinphis. TN, pp 903–907
Staub T (1991) Fungicide resistance: practical experience with antiresistance strategies and the role of integrated use. Annu Rev Phytopathol 29(1):421–442
Steiner FM, Schlick-Steiner BC, Trager JC, Moder K, Sanetra M, Christian E, StauVer C (2006) *Tetramorium tsushimae*, a new invasive ant in North America. Biol Invasions 8:117–123
Steshenko FN (1958) [The role of honey bees in cross-pollination of grape vines.]. Pchelovodstvo 35:37–40, [In Russian.] AA-182/60
Steven D (1990) Entomology in kiwifruit. In: Warrington IJ, Weston GC (eds) Kiwifruit Science and Management. Ray Richards Publisher in association with the New Zealand Society for Horticultural Science, Auckland, pp 362–412
Steven D, Blank RH, Tomkins AR (1991) Monitoring scale to reduce sprays in kiwifruit. N Z Agr Sci 26:4
Stevens PS, McKenna CE (1999) Factors affecting the efficacy of *bacillus thuringiensis* against *Cnephasia jactatana* in kiwifruit. Proceedings of the 52nd New Zealand plant protection conference, pp 89–93. www.hortnet.co.nz/publications/nzpps/
Stevens PS, Steven D, Malcolm C, Holland PT (1993) Reducing pesticide residues on kiwifruit. Proceedings of the 46th New Zealand plant protection conference, p 62–66
Stevens PS, McKenna CE, Steven D (1995) The timing of lepidopteran damage to kiwifruit. In: Proceedings of the 48th New Zealand plant protection conference, pp 130–134
Stevens PS, McKenna CE, Blank R, Tomkins A, Steven D (1997) Compariso n of armoured scale thresholds in kiwifruit. In: Proceedings 50th New Zealand plant protection conference, pp 288–292. www.hortnet.co.nz/publications/nzpps/
Stewart JD (1970) Cranberry pollination in new jersey. In the indispensable pollinators. Ark Agr Ext Serv Misc Pub 127:181–184

Stewart JD, Marucci PE (1970) Honey bees for cranberry pollination. Nj Agr Expt Sta Cir 588(a):4
Stonneveld T, Wainwright H, Labuschagne L (1996) Development of twospotted spider mite (Acari:Tetranychidae) populations on strawberry and raspberry cultivars. Ann Appl Biol 129:405–413
Stricker MH (1953) Blueberry and cranberry pollination. Gleanings Bee Cult 81:268–271
Stubbs CS, Drummond FA (1997) Blueberry and cranberry (Vaccinium spp.) pollination: A comparison of managed and native bee foraging behavior. Acta Hortic 437:341–344
Stumpf N, Nauen N (2001) Cross-resistance, inheritance, and biochemistry of mitochondrial electron transport inhibitor-acaricide resistance in *Tetranychus urticae* (Acari: Tetranychidae). J Econ Entomol 94:1577–1583
Svensson B (1991) The importance of honeybee-pollination for the quality and quantity of strawberries (*Fragaria X ananassa*) in Central Sweden. Acta Hortic 288:260–264
Swenson AA (1958) Bees and cranberries - a winning combination. Am Bee J 98:64
Szendrei Z, Mallampalli N, Isaacs R (2005) Effect of cultivation on abundance of Japanese beetle, Popillia japonica Newman (Coleoptera: Scarabaeidae), in highbush blueberry Þelds. J Appl Entomol 129:258–264
Taha AM (1984) Studies on some pests infesting strawberry plants at Qalubia governorate. M.Sc. Faculty of Agriculture Al-Azhar University, 146 p
Thompson GD, Dutton R, Sparks TC (2000) Spinosad a case study: an example from a natural products discovery program. Pest Manage Sci 56:696–702
Tomkins AR, Thomson C, Wilson D, Greaves AJ (1992) Armoured scale insects (Hemiptera: Diaspididae) on unsprayed kiwifruit vines in the Waikato. N Z Entomol 15:58–63
Torchio PF (1990) Osmia ribifloris, a native bee species developed as a commercially managed pollinator of highbush blueberry (Hymenoptera: Megachilidae). J Kansas Entomol Soc 63:427–436
Trjapitzin SV, Trjapitzin VA (1999) Parasites of mealybugs (Homoptera, Pseudococcidae) on cultivated grapes in Argentina, with description of a new species of the genus *Aenasius* Walker (Hymenoptera, Encyrtidae). Entomol Rev 79:386–390
Tsujikawa Y (1981) Honeybees in green houses, their effects on strawberries and a problem of UV cut film house. Honeybee Sci 2(2):49–56
Vaissière BE, Froissart R (1996) Pest management and pollination of cantaloupes grown under spunbonded row covers in West Africa. J Hortic Soc 71:755–766
Valleau WD (1918) Sterility in strawberry. J Agric Res 12:613–670
van de Vrie M, Price JF (1994) Manual for biological control of twospotted spider mites on strawberry in Florida. Univ Fla Dover Res Rpt DOV 1994–1
van driesche RG, Bellows TS JR (1996) Biological control. Chapman and Hall, New York, 539 pp
van Praagh JP (1988) Die Bestäubung der Brombeere 'Thornless Evergreen' (Rubuslaciniatus) in BeziehungzumFruchtwachstum. Gartenbauwissenschaft 53:15–17
VerCauteren KC, Lavelle MJ, Hygnstrom SE (2006) Fences and deer-damage management a review: a review of designs and efficacy. J Wildl Manag 34:191–200
Walsh DB, Zalom FG, Shaw DV, Larson KD (2002) Yield reduction caused by twospotted spider mite feeding in an advanced-cycle strawberry breeding population. J Am Soc Hort Sci 127:230–237
Walton VM, Pringle KL (2004) A survey of mealybugs and associated natural enemies in the vineyards in the Western Cape Province, South Africa. S Afri J Enol Vitic 25:23–25
Ware GW (2000) The pesticide book, 5th edn. Thompson Publications, Fresno, p 17
Weibelzahl E, Liburd OE (2009) Epizootic of Acalitus vaccinii (Acari: Eriophyidea) caused by Hirsutella thompsonii on southern highbush blueberry in north-central Florida. Florida Entomol 92:601–607
Weibelzahl E, Liburd OE (2010) Blueberry bud mite, acalitus vaccinii (Keifer) on southern highbush blueberry in Florida. EDIS-ENY858/IN844. UF/IFAS Extension, University of Florida, Gainesville, Florida. Available at http://edis.ifas.ufl.edu/in844
Williams IH (1994) The dependence of crop production within the European Union on pollination by honey bees. Agric Zool Rev 6:229–257

Williams D (2000) Twospotted spider mite on ornamental plants. pp. 1–2, Agricultural Notes. State of Victoria, Department of Primary Industries, Victoria, Australia

Williams L III, Martinson TE (2000) Colonization of New ork vineyards by *Anagrus*spp. (Hymenoptera: Mymaridae): Overwintering biology, within-vineyard distribution of wasps, and parasitism of grape leafhopper, *Erythroneura*spp. eggs. Biol Control 18:136–146

Willmer PG, Bataw AAM, Hughes JP (1994) The superiority of bumblebees to honeybees as pollinators – insect visits to raspberry flowers. Ecological Entomology 19:271–284

Winkler AJ, Cook JA, Kliewer WM, Lider LA (1974) Climate and soils. In: General viticulture. University of California Press, Berkeley, pp 58–75

Winston ML, Graf LH (1982) Native bee pollinators of berry crops in the Fraser Valley of British Columbia (Canada). J Ent Soc British Columbia 79:14–20

Wise LC, Gut LJ, Isaacs R, Schilder AMC, Sundin GM, Zandstra B, Beaudry R, Lang G (eds) (2010) Michigan Fruit Management Guide, MSU Extension Bulletin E-154. Michigan State University, East Lansing

Witzgall P, Bengtsson M, Trimble RM (2000) Sex pheromone of grape berry moth (Lepidoptera: Tortricidae). Environ Entomol 29:433–436

Woo KS, Choo HY, Choi KR (1986) Studies on the ecology and utilization of pollinating insects. Korean J Apicul 1(1):54–61

Wood GW (1968) Self-fertility in the lowbush blueberry. Can J Plant Sci 48:431–433

Yamamoto I (1999) General review of the mode of insecticidal action of naturally occurring nicotine. Tokyo Nogyo Daigaku Nogaku Shuho 43(4):169–185

Yamamoto I, Tomizawa M, Saito T, Miyamoto T, Walcott EC, Sumikawa K (1998) Structural factors contributing to insecticidal and selective actions of neonicotinoids. Arch Insect Biochem Physiol 37:24–32

Yang WQ (2005) Blueberry gall midge- a possible new pest in the Northwest: identification, life cycle, and plant injury. OSU Extension Publication. EM 8889

Yokoyama VY (1979) Effect of thrips scars on table grape quality. J Am Soc Hortic Sci Alex 104(2):243–245

Zalom FG, Phillips PA, Toscano NC, Udayagiri S (2001) Strawberry aphids. University of California Department of Agriculture and Natural Resources Publication No. 3339, p 4

Zhi-Qiang Z, Sanderson JP (1995) Twospotted spider mite (Acari: Tetranychidae) and Phytoseiulus persimilis (Acari: Phytoseiidae) on greenhouse roses: Spatial distribution and predator efficacy. J Econ Entomol 88(2):352–357

Chapter 8
Tropical Fruits

The tropics with relatively warm climate occupy nearly 40 % of the earth's surface supporting nearly half of the world's population. The temperature variations are less significant. Majority of the world's biodiversity is adapted to the diverse tropical conditions ranging from wet tropical rainforests to deserts and snow covered high mountains. About half of the plant families are tropical (Crane and Lidgard 1989; Meyer et al. 2000). Most botanical families have at least one species of tropical fruit. Asia has about 500 tropical fruit species, the Indian subcontinent about 300, with about 1200 in Africa. Of these fruits only a few are found in local markets and fewer are exported. Ninety per cent of the export market is made up of citrus, banana and plantain, mango and pineapples. A further 5 % is made up of papaya, avocado and dates. The remainder is made up of more than 20 species, ranging from breadfruit and litchi to mangosteen, passion fruit and coconut. More than 90–95 % of tropical fruits are not exported from the producing country but are consumed locally. Fruit crops while growing under particular climatic conditions develop their own pollination system. This unique system remains as long as the plants are grown under similar conditions. The pollination ecology of different pollinators and their safety from pesticides for some fruit crops such as mango, papaya and banana is discussed below:

8.1 Pollinators

8.1.1 Mango

Mangifera indica L., family Anacardiaceae

Mango is a highly cross-pollinated crop. Mangoes are monoecious, with tiny red-yellowish flowers. Each flower has five sepals and three to nine petals. A single tree can have from 200 to 3000 panicles per tree with each panicle having 500 to 10,000 flowers. Large numbers (30–80 %) of flowers are imperfect or staminate. Perfect flowers have a globular ovary and lateral style together with one to three

D.P. Abrol, *Pollination Biology, Vol. 1*, DOI 10.1007/978-3-319-21085-8_8

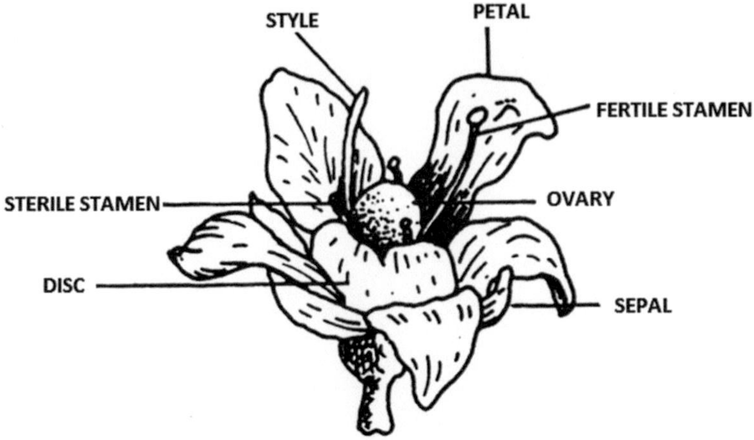

Fig. 8.1 Flower structure of Mango (Singh 1960)

functional stamens. As the style and stamens are of the same length, self-pollination by visiting insects is possible, and a high degree of self pollination has been reported. Only a tiny proportion of perfect flowers actually set fruit (zero to three per panicle) (Fig. 8.1).

The flower opens early in the morning, and the stigma is immediately receptive. However, pollen shedding occurs from about 8 a.m. to noon. The pollen shedding is delayed (0800–1200) which results in inadequate fertilization of the stigma (Spencer and Kinnard 1956). Mango flowers secrete nectar copiously which attracts a large number of insects (Mukherjee 1953); however, relatively little pollen is produced on the anther (Popenoe 1917).

Mango flowers may be pollinated by flies, bees, thrips and other insects, with flies probably the most important. Some fruit set may occur due to wind pollination. About 65 % of the flowers open before 6 a.m. and the rest open during the day. Anther dehiscence occurs within an hour after anthesis, with a maximum between 9 a.m. and noon. Eighty per cent of the pollen is viable. Stigmas are receptive from 1 day before anthesis until 2 days after, with the day of anthesis being the optimum. Mango cultivars are considered to be self-fertile, but self-incompatibility has been reported. Self-pollination produces 0–1.68 % set, while cross-pollination produces 6.2–23.4 %, but there are clear cultivar differences. The pollen grains fall on the base of the ovary and the nectary discs rather than the stigma; hence pollination agents are required. Insects, including bees, wasps and flies, are the principal pollinating agents, as indicated by the sticky pollen, secretion of nectar, colourful corolla and flower scent. Flies seem to be the main pollination agents, with bees and other insects contributing. Wind may play a small role. Pesticide use during flowering affects insect pollination, pollen germination and ovule fertilization.

8.2 Pollination Requirements

There have been diverse opionion regarding pollination requirements of mango. Both wind and insects have been credited as pollinating agents. Young (1942) reported that there was no significant difference between percentages of set in selfed and cross-pollinated mango flowers. Popenoe (1917) found that the mango is self-fertile but cross-pollination increases fruit set. However, Singh et al. (1962) reported that crossed flowers set fruit whereas selfed ones did not, indicating a degree of self-sterility. Wolfe (1962) concluded there is need for pollinating insects to transfer the pollen from anthers to stigma within the cultivar to obtain satisfactory crops of fruit.

8.3 Pollinators

Several agents have been reported as pollinators of mango flowers. Wagle (1929) showed that wind had some role to play in mango pollination, yet insects such as bees, ants, and flies were also recognised important pollinators. Popenoe (1920), Galang and Lazo (1937) and Singh (1969) considered the mango to be an insect-pollinated plant. Recent studies in India (Singh and Sturrock 1969) showed that plants caged to exclude all insects set no fruit but a plant caged with a colony of honey bees set a heavy crop. Fajardo et al. (2008) reported that stingless bees (*Trigona biroi*), calliphorids (*Chrysomya* spp.), syrphids (*Eristalis* spp.) and honeybees (*Apis cerana* and *A. mellifera*) were the important pollinators of mango. That pollination in mangoes is mediated by insects rather than wind was first proposed and demonstrated by Popenoe (1917) although Wester (1920), who has been supported by Davenport and Núñez-Elisea (1997), maintained that wind may be more important than insects in some environments. Free and Williams (1976) found that mangoes were able to set fruit even though insects had been excluded by bagging, thus suggesting that at least some pollination is assisted by wind or gravity. There are numerous reports concerning the insect fauna attending mango flowers and the effect on fruit set of their exclusion. Bhatia et al. (1995) found that on panicles that were bagged to exclude insects fruit set was zero, compared with 4.3 % set on unbagged panicles that allowed insects free access. Similarly, Singh (1997b) recorded zero fruit set on bagged panicles and 1.6 fruits set on unbagged panicles. Galán Saúco et al. (1997), investigating the production of 'Tommy Atkins' mangoes under greenhouse cultivation in the Canary Islands, found that when all insects were excluded no fruit was set but when bees were introduced and other insects had free access, fruit set increased. The make-up of the pollinating fauna of mangoes has been studied in a number of countries. Bhatia et al. (1995) reported Calliphoridae and Syrphidae as the most common visitors to mango flowers in India. Hymenoptera were found to be more prevalent in terms of species in Australia (Anderson et al. 1982), Israel (Dag and Gazit 1996) and South Africa (Eardley and Mansell 1994). Nevertheless, although some species are probably more important than others, there seems to be a consensus that numerous species within the complex of visiting insects contribute to the pollination of mango flowers.

Anderson et al. (1982) in northern Australia showed that wasps and native bees, *Trigona* sp., were more effective pollinators than were large flies. Since mango flowers are generally considered to be unattractive to honeybees, *Apis mellifera* Linnaeus (Free and Williams 1976), and this species is uncommon in northern Australia, Anderson et al. (1982) suggested that *Trigona* sp. might be used in that part of the country to augment the pollinating fauna, since it is common and prevalent on mango blossom and can be hived. *Trigona* spp. are also associated with mangoes in Costa Rica, and although they are regarded as important forest pollinators, they appear to be unimportant in pollinating mangoes. Rather, they are regarded as a nuisance because they chew small pieces of bark from the trees to make their nests. There is also a suggestion that they maybe vectors of the bacterium, *Erwinia* (Jiron 1995). In Thailand, *A. mellifera* is kept for large scale honey production and pollinating longans, but *Apis cerana* Fabricius is preferred for small-scale honey production and for pollinating mangoes (Wongsiri and Chen 1995). Sharma et al. (1998) conducted studies in India to develop in-tree rearing of flies that would assist in mango pollination. Several species, the most numerous of which were *Lucilia* sp. (Calliphoridae) and *Sarcophaga* sp. (Sarcophagidae), were reared from natural populations infesting fish or mutton pieces placed in mesh bags and hung in the lower branches of mango trees.

Singh (1960a) found that flies of the Syrphidae and *M. domestica*, as well as stingless bees of *Trigona* spp., are main pollinators of mango in India. Kumar et al. (2012) recorded six species of insects visiting mango flowers. The abundance of ants, *Camponotus compressus* 33.78 % was high followed by flesh fly, *Chrysomya megacephala* 32.94 % and house fly, Musca domestica 25.44 %. The abundance of *Ropalidia marginata* 0.62 % was lower than rock bee, *Apis dorsata* 3.70 % and little bee, *Apis florea* 3.42 %, respectively. *Chrysomya megacephala* was spending more time in flowers followed by *Musca domestica. Apis dorsata* and *Apis florea* spend equal time in blossoms (Table 8.1).

8.4 Pollination Recommendations and Practices

The mango flowers tend to open in large numbers when more attractive flowers from other plants are available. In order to attract honeybees a heavy concentration of colonies possibly three to six per acre, may be necessary to obtain maximum fruit set. Some agricultural practices adversely affect pollinator diversity. Kremen et al. (2002) observed that intensification brought about a significant reduction in the density of pollinators, including native bees. Pesticide application on mango significantly reduced bee populations. Free (1999) reported that the use of insecticides is probably the most serious threat to wild populations in agricultural areas and provides the greatest deterrent to using bees as pollinators.

Table 8.1 Insect pollinators visiting mango flowers (Source: Sung et al. 2006)

Order	Group	Family/species
Diptera	**Blowflies**	Calliphoridae: *Chrysomya albiceps, C. megacephala, C. pinguis, Lucilia serata.* Important pollinators in Israel and Ghana (this study) As effective as honey bees in Israel "Large flies" reported as effective in Australia. *Chrysomya* species recommended as pollinators
	Houseflies	Muscidae: *Musca domestica.* Not as effective as honeybees or blowflies in Israel. Reported as significant pollinators in India
	Hoverflies	Sirphidae: *Episyrphus balfeatus*, *Melanostoma orientale. M. orientale* one of the most effective pollinators in an Indian study but reported as not alighting on the flowers in Ghana
Hymenoptera	**Wasps**	Sphecidae: *Bembecinus tridens*, *Sceliphron spirifex.* Wasps generally reported as regular visitors to mango flowers in most locations. *B. tridens* specifically mentioned in Israel and *S. spirifex* seen visiting mango flowers in Ghana
	Honeybees	Apidae: *Apis mellifera*, *A. cerana.* Inconsistent reports in literature, but clearly important in some cases (e.g. for Keit varieties in Israel, and in Ghana). Recommended as pollinators despite low pollen loads
	Stingless bees	Apidae: *Meliponinae Trigona*, *Dactylurina.* Frequent reports. *Trigona* effective in Australia and *Dactylurina* seen on flowers in Ghana (this study). Trigona reported as important pollinator in India
	Sweat bees	Halictididae: Hailctus, Lasioglossumm. Recommended pollinators for mango. *Lassioglossum* bees full of pollen on hind femurs
	Ants	Formicidae: *Anoplolepis longipes.* Frequently reported on flowers. Large ants reported as significant in Australia
Coleoptera	**Blister beetles**	Cantharidae: *Cantharis atropoveolatus.* Reported in Israel as frequent visitors to mango flowers but not as effective as flies or honeybees. May be poor cross pollinators as do not tend to move between trees
Coleoptera	**Flour beetles**	Tenrebrionidae: *Omophlus syriacus.* Reported in Israel as frequent visitors to mango flowers but not as effective as flies or honeybees. May be poor cross pollinators as do not tend to move between trees
Lepidoptera	**Butterflies and moths**	Nymphalidae: *Hypolimnas misippu*; Arctiidae: *Euchromia* sp. Butterflies are regularly reported as less abundant floral visitors. Pollinator efficacy not assessed. *H. misippus* and *Euchromia* seen on flowers in Ghana. Hawkmoths reported in Ghana

8.5 Pests of Mango (*Mangifera indica* L., Anacardiaceae)

India is one of the largest producers of mango that in the world. Thailand, Pakistan, Bangladesh, and the Philippines are major exporters of mango in Asia. Quarantine restrictions on fruit flies and stone weevil are the major impediments to international trade of this fruit. Infestation of fruit flies is a major concern in mango cultivation.

More than 492 species of insects and mites infest mango out of which more than 45 % of these have been reported from India (Plate 8.1). The most important pests include fruit flies, stone weevils, mango hoppers, mealy bugs, scale insects and tree shoot borers which cause considerable losses, several secondary pests and a large number of occasional pests in localized areas are also found (Laroussilhe 1980; Tandon and Verghese 1985; Veeresh 1989).

The mango pests have been reported in different parts of the world by several investigators: India (Srivastava 1998; Anonymous 2006), Australia (Anonymous 1989), Pakistan (Mohyuddin 1981), Israel (Wysoki et al. 1993; Swirski et al., 2002), the USA (Peña 1993), West Africa (Vannière et al. 2004), Brazil (Assis and Rabelo 2005), Central America (Coto et al. 1995) and Puerto Rico (Martorell 1975) have also been described.

Of all the mango pests, hoppers are considered as the most serious and widespread pest throughout the Indian subcontinent. Large number of nymphs and adults of *Idioscopus clypealis* and *I. niveosparsus* suck and puncture the sap from tender parts, thereby reducing the vigour of the plants and particularly destroying the inflorescence and causing fruit drop (Tandon and Verghese 1985) (Table 8.2).

8.5.1 Fruit Flies

In case of fruit flies infested fruits, larval feeding renders fruit unmarketable and indirect economic losses, resulting from the presence of fruit flies (Hill 1975; Umeya and Hirao 1975; Yee 1987; Singh 1991; Aluja 1993; Aluja et al. 1996; Vannière et al. 2004; Aluja and Mangan 2008). Approximately 60 species of fruit flies are reported to attack mango and a related species, *Mangifera foetida* (White and Elson-Harris 1992; Aluja et al. 1996; Malavasi and Zucchi 2000; Clarke et al. 2001, 2005; Norrbom 2004; Vayssières et al. 2005). Fruit flies attacking mango belong to the genera *Anastrepha* (*c*.12 species), *Bactrocera* (*c*.33 species), *Ceratitis* (eight species) and *Dirioxa* (two species) (White and Elson-Harris 1992; Vayssières and Kalabane 2000; Lux et al. 2003; Norrbom 2004; Vayssières et al. 2004, 2005; Secretariat of the Pacific Community 2005). Some of these species are referred to as the 'mango fruit fly': *Anastrepha obliqua* Macquart, *Bactrocera frauenfeldi* (Shiner), *Ceratitis cosyra* (Walker) (Aluja 1993; Leblanc and Allwood 1997; Lux et al. 2003; Steck 2003).

Plate 8.1 (**a**) House fly pollinating mango flowers (**b**) Syrphid fly pollinating mango flowers (**c**) Nectar bird plays role in pollination by default (**d**) *Apis dorsata* on guava flowers (**e**) *Apis cerana* foraging on guava flower (**f**) Dorsata and other bees on litchi flowers (**g**) *Apis mellifera* pollinating litchi flowers (**h**) Syrphid fly foraging on litchi

Table 8.2 Some important insect pests of mango (Source Pena et al. 2009)

Common name	Organism	Parts affected, symptoms	Region or country
Mexican fruit fly	*Anastrepha ludens*	Larval damage to fruit	Caribbean
South American fruit fly	*Anastrepha fraterculus*	Larval damage to fruit	Americas
Caribbean fruit fly	*Anastrepha suspensa*	Larval damage to fruit	Caribbean, Florida
Queensland fruit fly	*Batrocera tryoni*	Larval damage to fruit	Australia
Mediterranean fruit fly	*Ceratitis capitata*	Larval damage to fruit	Widespread
Marula fruit fly	*Ceratitis cosyra*	Larval damage to fruit	Africa
Natal fruit fly	*Ceratitis rosa*	Larval damage to fruit	Africa
Oriental fruit fly	*Dacus dorsalis*	Larval damage to fruit	Asia, Hawaii, Philippines
Mango seed weevil	*Sternochetus mangiferae*	Seed	India, Hawaii, Philippines, S. Africa, South-east Asia, Oceania, Caribbean
Mango blossom midge	*Erosomyia indica* *Dasineura mangifera*	Sucking sap from floral parts	India, Hawaii
Mango hopper Oceania, Americas	*Idioscopus* sp.	Sucking sap from flowering shoots	Philippines, India, Africa,
Red-banded thrips	*Selenothrips rubrocinctus*	Sucking on underside of young leaves	Widespread
Coconut bug	*Pseudotheraptus wayi*	Sucking sap from young fruit, watery spot on fruit, fruit drop	Africa

Postharvest treatments specifically related to mangoes and fruit flies have been discussed by Sharp et al. (1988, 1989a, b), Hallman and Sharp (1990), Nascimento et al. (1992), Mangan and Hallman (1998), Shellie and Mangan (2002a, b) and Bustos et al. (2004). Broad regulatory issues were recently reviewed by Follet and Neven (2006). General reviews on biology, ecology and behaviour of economically important and non-pestiferous fruit flies, many of them infesting mangoes, were written or edited by Christenson and Foote (1960), Bateman (1972), Fletcher (1987), Robinson and Hooper (1989), Aluja (1994), Aluja and Norrbom (2000) and Malavasi and Zucchi (2000).

8.5.2 *Anastrepha*

Twelve *Anastrepha* species have been reported associated with mango (Norrbom 2004). *Anastrepha obliqua* is reportedly the most common fruit fly pest in the Americas (Jirón and Hedström 1988; Nascimento et al. 1992). This species is the most common fruit fly pest of mangoes in Mexico, Costa Rica, Honduras and Guatemala

(Soto-Manitiú et al. 1987; Jirón and Hedström 1991; Aluja et al. 1996; Camargo et al. 1996; Sponagel et al. 1996), but it also attacks mangoes in Cuba, Puerto Rico, Jamaica, El Salvador and Venezuela (Norrbom 2004). In Mexico, *A. ludens* commonly attacks mangoes at higher elevations, while *A. obliqua* dominates at lower altitudes (Aluja et al. 1996). In Brazil and Ecuador, mangoes are mainly attacked by *A. fraterculus* (Wiedemann), but *A. turpiniae* Stone, *A. serpentina* Wiedemann, *A. pseudoparallela* (Loew) and *A. zuelaniae* Stone have been reported (Zucchi 1988; Rebouças et al. 1996; Arias and Jines 2004; Norrbom 2004; Barbosa et al. 2005; A. Malavasi, January 2008).

8.5.3 Bactrocera

Bactrocera spp. are serious pests of mango (Hancock et al. 2000; Hollingsworth et al. 2003; Clarke et al. 2005). The most important species include *B. tryoni* (Frogatt), *B. Zonata* (Saunders), *B. dorsalis* (Hendel), *B. neohumeralis* (Hardy), *B. jarvisi* (Tryon), *B. papayae* Drew and *B. frauenfeldi* (Schiner) (Umeya and Hirao 1975; Drew and Hancock 1994; Hollingsworth et al. 2003). Two species, *B. phillippiensis* Drew & Hancock and *B. occipitalis* Bezzi, have been recorded for Palau, Pacific Islands (Secretariat of the Pacific Community 2005), and recently, a new species, *B. invadens* Drew, Tsuruta and White, was reported for West Africa (Kenya, Benin) (Lux et al. 2003; Vayssières et al. 2005). *Bactrocera correcta* (Bezzi), *B. caryeae* (Kapoor), *B. curcubitae* (Coquillett), *B. diversa* (Coquillett) and *B. tau* (Walker) have been reported in India (Australian Government 2004).

8.5.4 Ceratitis

Eight *Ceratitis* spp. have been reported to attack mango fruit. *Ceratitis quinaria* and *C. silvestrii* are considered of economic importance in Benin (Vayssières et al. 2005). *Ceratitis cosyra* is broadly distributed across Africa and causes enormous damage, which can result in total loss of the crop. On average about 20–30 % of mango production is lost due to this fly species in various African countries (Lux et al. 2003).

8.6 Fruit Fly Control

Hendrichs (1996) stated that spread of fruit flies from one orchard to another is reduced if the fruit growers follow a concerted fruit fly management strategy (Aluja 1993, 1996). Aluja et al. (1996) suggested a fruit fly management scheme based on border trapping, enhancement of host-plant resistance through use of plant growth regulators, use of the sterile insect technique and bait stations and augmentative parasitoid releases (Sivinski 1996; Sivinski et al. 1996; Malavasi and Zucchi 2000; Montoya et al. 2000; Tan 2000; Dyck et al. 2005; Mangan and Moreno 2007).

8.6.1 Monitoring and Sampling

Monitoring fruit flies attacking mango serves different purposes: (i) to apply a control or management tactic after the presence of the fruit fly is noticed; and (ii) to verify if fruit fly species will attack mango under natural conditions. In general, thresholds for adult fruit flies are quarantine-mediated (Beers et al. 1993). These thresholds vary from location to location, but depending on the fruit fly species they are typically based on the capture of a single fruit fly. In other fruit crops, a threshold of five flies/trap is recommended resulting in a reduction from four chemical sprays to 1.5 sprays/season (Beers et al. 1993). Sampling for fruit flies in mango is mostly performed using adult traps, because eggs and young larvae are often difficult to see in the fruit and because the primary aim of management programmes is to prevent fruit damage. More recently, human urine has been successfully tested as bait for McPhail and McPhail-type traps for resource-poor farmers in tropical countries (Piñero et al. 2003; Aluja and Piñero 2004). The McPhail trap has provided different results in mango orchards. Due to the low efficiency of the McPhail trap it is being replaced with Multi-Lure® traps, which provide new trap designs. Dry synthetic-food-based lures have also been developed, i.e. BioLure® (Suterra LLC, Inc., Bend, Oregon) (Heath et al. 1995, 1997; Epsky et al. 1999) and Nu-Lure® (Advanced Pheromone Technologies) (Robacker and Warfield 1993; Robacker et al. 1997; Robacker 2001).

The attractiveness of mango compounds is currently being investigated. For example some of the volatiles emitted by 'Tommy Atkins' mangoes, i.e. terpenes (*p*-cymene and limonene), are attractive to *C. capitates* adults (Hernández-Sánchez et al. 2001). Many questions linger with respect to the optimal trap number and time for trap placement in mango groves. In Naru, to produce mango free of *B. frauenfeldi*, 300–400 traps baited with methyl eugenol plus a toxicant were placed every square kilometre and trapping density was increased around mango plantings (Anonymous 2002). Even though large numbers of traps can be utilized to increase detection sensitivity, the cumulative costs and logistical considerations do not make this option practical. Traps with specific and effective lures that can detect the F1 generation at low trap densities (5–10 traps/km^2) would fit the description of a good detection and monitoring device (Tween 1993).

Sampling for earlier fruit fly stages can be used to demonstrate that the fruit is not susceptible to fruit fly attack. For instance, the Caribbean fruit fly, *A. suspensa*, may not attack green mango fruit (Peña et al. 2006b). Through a sequential collection of fruit from fruit-fly infested mango orchards, fruit were dissected for eggs and larvae. At the same time, fruit were stored and held for puparia emergence. In addition fruit were exposed in cages to wild fruit flies and traps were placed to verify the presence of fruit flies. Estimating the time that 'Tommy Atkins' fruit remain immature and therefore non-hosts for fruit flies, may provide a window for mango exports in some fruit fly-infested areas.

8.6.2 Chemical Control

Mango plantations account for major insecticide use in the tropics (Cunningham 1984). From the late 1960s to date, the conventional control of fruit flies was through toxic bait sprays that combine proteinaceous bait (e.g. hydrolysed protein) with an insecticide (López et al. 1969; Soto-Manatiú et al. 1987; Mangan et al. 2006; Mangan and Moreno 2007). For many years the insecticide of choice has been malathion (Peck and McQuate 2000; Burns et al. 2001). Fruit flies are highly susceptible to any insecticide, and other compounds have also been widely used. Vergherse et al. (2004), working on the control of *B. Dorsalis* in India, used a rotation of fenthion (0.05 %), deltamethrin (0.0028 %), carbaryl (0.2 %) and dimethoate (0.06 %) to reduce the risk of resistance development.

Cyromazine, imidacloprid (organochlorinated compound), spinosad (bacteria-derived insecticide) and phototoxic dyes (Phloxine B) have been successfully tested against various fruit fly species (Díaz-Fleischer et al. 1996; King and Hennessey 1996; Peck and McQuate 2000; Vargas et al. 2002; Liburd et al. 2004; McQuate et al. 2005). Despite their success, and as is typical with insecticides intensively applied on a large scale, resistance has already been documented in the case of spinosad (Wang et al. 2005; Hsu and Feng 2006) or collateral damage (Stark et al. 2004). In Pakistan, application of pesticides caused reduction of fruit fly infestations, but their use has created scale insect problems by eliminating their natural enemies (Mohyuddin and Mahmood 1993). Such an effect had been reported by Ehler and Endicott (1984) with pests of olive, citrus and walnut. Another recent development with respect to chemical control of fruit flies has been the refinement of the bait-station concept (Mangan and Moreno 2007).

Decisions on when to apply insecticides are based on the appearance of the first trapped males (trimedlurebaited traps are used). When applying insecticides directly, dimethoate (0.1 %) and fenthion (0.15 %) are used (Khanzada and Naqvi 1985). Pestiferous *Anastrepha* spp. are susceptible to most insecticides (Shaw and Spisshakoff 1958; Shaw 1961).

In Costa Rica, dipterex and malathion are sprayed weekly and reduce mango damage up to 40 % (Soto-Manitiu et al. 1987). In Brazil, malathion with protein and sugarcane bagasse is used (Carvalho and De Queiroz 2002). In Ecuador, Arias and Jines (2004) recommend a spray of malathion (1 %) with protein (4 %) once the fruit fly population reaches 0.14 fruit flies/trap/day (FTD).

8.6.3 Biological Control

8.6.3.1 Parasitoids

Use of parasitoids with mango is hindered by the fact that fruit are very large and therefore provide larvae a refuge from parasitism (López et al. 1999). As a consequence, Aluja (1993) and Montoya et al. (2000) recommended that parasitoids should be released outside the mango orchards to attack fly larvae in their much smaller native hosts and thereby significantly reduce the size of populations entering mango orchards. Several parasitoids, for example *Opius fullawayi* (=*Diachasmimorpha fullawayi* (Silvestri)), *Diachasmimorpha kraussi*, *D.* Fullaway, *D. tryoni* (Cameron), *Opius bellus* Gahan, *Biosteres longicaudatus* Ashmead (=*D. longicaudataB. tryoni* (Couron)) (=*D. tryoni* (Cameron)) and *Biosteres oophilus* Fullaway (=*Fopius arisanus* (Sonan)), parasitize *C. capitata* (Beardsley 1961; Wharton and Marsh 1978). In Kenya, Ghana, Tanzania, Uganda and Cote d'Ivoire, the most important parasitoids obtained from *Ceratitis* spp. emerging from mangoes were *D. fullawayi*, *Fopius caudatus* (Szépligeti), *Psyttalia cosyrae* (Wilkinson) and *Tetrastichus giffardianus* Silvestri (Lux et al. 2003).

8.6.4 Microbial Control

Use of pathogens/disease agents (fungi, bacteria, nematodes) has been attempted with varying degrees of success. For example, *Metarhizium anisopliae* has been evaluated in small-scale mango orchards in Kenya using bait stations laced with the pathogen. Results do not show differences between use of pathogens and use of insecticides (malathion) (Lux et al. 2003). Lezama-Gutierrez et al. (2000) also evaluated isolates of *M. Anisopliae* against larvae of *A. ludens*. They suggested that *M. anisopliae* can cause a 22–43 % reduction in adult emergence, depending on the soil where the larvae pupariates. De la Rosa et al. (2002) evaluated the fungus *Beauveria bassiana* (Bals.) under laboratory conditions and concluded that highest mortality was achieved at the adult stage, while Dimbi et al. (2003) reported on the pathogenicity of *M. anisopliae* and *B. bassiana* on different species of *Ceratitis*. Poinar and Hislop (1981), Lindegren and Vail (1986) and Toledo et al. (2006) have investigated the use of various nematodes, *Heterorhabditis bacteriophora*, *Heterorhabditis heliothidis* (Khan, Brooks and Hirschmann) and *Steinernema feltiae* Filipjev, against *Anastrepha*, *Bactrocera* and *Ceratitis*.

8.6.5 Predators

In addition to parasitoids, the other biocontrol agents include pathogens, nematodes and ants which have been used for control of fruit flies in mango orchards. Peng and Christian (2006) used the weaver ant, *Oecophylla smaragdina* (Fabricius) for control of the Jarvis fruit fly, *B. jarvisi*, in mango orchards in Australia.

8.6.6 Cultural Fruit Fly Control

Fruit bagging is one of the best solutions to prevent fruit fly attack of mango and other tropical fruits (Aluja 1996; Peña et al. 1999; Paderes and Orden 2004). Success with mangoes can be quite high, but Bondad (1985) demonstrated that bagging materials do not always resist the effect of rain/wind. Therefore, while bagged mangoes tend to produce a greater amount of marketable fruit than those not bagged, which requires to determine the type bags to use for different mango varieties and the best time to bag fruit (Love et al. 2003). In India, cultural control practices include removal of fallen fruit and inter-tree ploughing and raking (followed by insecticide cover sprays). Such practices can reduce fruit fly infestation between 77 % and 100 % (Verghese et al. 2004).

A cultural practice that can impinge on the success of management programmes is the widespread use of potassium nitrate (KNO3) sprays to accelerate and synchronize flowering of mango trees. As a result, fruit harvests can be advanced and synchronized. Such a procedure can, under certain circumstances, help control fruit flies, but can also exacerbate the problem. For example, in the case of the Mexican fruit fly (*A. ludens*), a notorius pest of citrus that also attacks mangoes, advancing the mango harvest offers ideal conditions for adult pests to move from citrus groves to mango orchards. This would not occur if mango trees flowered naturally since fruit ripen several months after the citrus harvest.

8.6.7 Host Resistance

All the mango varieties donot suffer similar damage, some varieties show resistance to *B. dorsalis* whereas others are susceptible. For example, Yee (1987) reported that *B. dorsalis* does not attack all mango cultivars to the same extent. Carvalho et al. (1996) reported that 'Espada' showed no infestation by *A. obliqua*, whereas 'Carlota' was highly infested. Jirón and Soto-Manitiu (1987) also observed that susceptibility of mangoes to *A. obliqua* differed among cultivars. 'Rosinha', 'Coquinho' and 'Espada' were resistant to *A. obliqua* attack, whereas 'Smith' and 'Pope' were highly susceptible.

In Brazil, use of gibberellic acid (GA3) reduces the susceptibility of 'Tommy Atkins' mangoes to attack by *C. capitata* based on artificial delay of fruit maturity (De Macedo 1988). Differences on attack by *A. ludens* to mango might be influenced by volatiles from green or yellow fruits (Garcia-Ramirez et al. 2004).

8.6.8 Quarantine Treatments

Quarantine treatments have been reviewed by Johnston and Hofman (2004). Quarantine treatments for harvested mangoes include: irradiation, hot water or hot water followed by immersion cooling are widely used (Sharp et al. 1988, 1989a, b, c; Hallman and Sharp 1990; Nascimento et al. 1992; Mangan and Sharp 1994; Mangan and Hallman 1998; Shellie and Mangan 2002a, b; Bustos et al. 2004; additional references in reviews by Mangan and Hallman 1998 and Follett and Neven 2006).

8.7 Mango Seed Weevils

The mango seed weevil, *Sternochetus mangifereae* (Fabricius), and the mango pulp weevil, *Sternochetus frigidus* (Fabricius) are important pests of mango. They damage flesh of ripe fruit which may limit plant propagation in nurseries (Johnson 1989). Severe infestation may cause early fruit drop (Subramanian 1925). The mango seed weevil has wide spread occurrence throughout the world (Balock and Kozuma 1964; Shukla and Tandon 1985; Johnson 1989; Schotman 1989).

8.7.1 *Sampling*

Shukla et al. (1988) reported the intra-tree distribution of eggs of *S. Mangiferae* on 'Baganpalli' mango; the highest number of eggs per fruit occurred on fruit in the lower region of the tree. With increasing tree height, egg deposition on fruit decreases. No statistical differences on fruit infestation were observed on north, south, east or west directional quadrats of the tree.

8.8 Management of Mango Seed Weevils

In India, field sanitation has been reported to reduce infestation of the mango nut weevil, *Sternochetus gravis* (Fabricius), by 22 % (De and Pande 1987). However, In Hawaii USA, field sanitation failed to reduce infestation rates (Hansen and Amstrong 1990). Organophosphate fenthion has been found to be most effective to reduce field infestation. Spot application of diazinon on tree trunks was recommended. Verghese et al. (2004) reported that azadirachtin was not effective for management of *S. mangiferae* in India. Selection of resistance cultivars is another most effective method for control of mango seed weevils (Bagle and Prasad 1984; Hansen et al. 1989). Mango weevils have few natural enemies. *A baculovirus* has been reported that affects the larvae of *S. mangiferae* (Shukla et al. 1984).

8.9 Mango Seed Borer (Lepidoptera: Pyralidae)

8.9.1 *Distribution and Biology*

The red banded mango caterpillar or mango seed borer, *Deanolis sublimbalis* Snellen (Waterhouse 1998), is an important pest of mangoes in the Philippines (Anonymous 1984), India (Zaheruddeen and Sujatha 1993), Vietnam (Nguyen et al. 1998; van Mele et al. 2001), China (Li et al. 1997), Thailand, Indonesia and Papua

New Guinea (Cunningham 1984). The oval white eggs are laid in clusters at the fruit apex. They hatch in 3–4 days. The larvae develop through five instars in 14–20 days and they pupate in cocoons in the soil. The period from egg to adult requires 28–40 days. The insect apparently prefers mango, but *M. odorata* and *M. minor* have also been recorded as hosts. Adults are generally nocturnal and spend most of the day under leaves on the tree (Waterhouse 1998).

8.9.2 Damage

Infested fruit split and rot, and fall to the ground (Anonymous 1984). In the Guimaras Islands, the Philippines, Golez (1991) recorded 12.5 % fruit infestation and in serious outbreak years, 40–50 % yield reductions are possible. Waterhouse (1998) considered that since *D. sublimbalis* is capable of causing such levels of damage, it might be a more important pest of mangoes than has generally been realized. Van Mele et al. (2001) suggested that damage caused by *D. sublimbalis* in the Mekong Delta has been wrongly attributed to fruit flies; however, Waterhouse (1998) states that soon after boring by *D. sublimbalis*, secondary infestations with fruit flies (*Bactrocera ferrugineous*, *B. frauenfeldi*) occur, together with infections by bacteria and fungi.

8.9.2.1 Biological Control

According to Waterhouse (1998) no parasitoids were detected in Java, Indonesia. However, in the Guimaras Islands of the Philippines, the vespid wasp, *Rychium attrisimum*, preys on the larvae as they exit the fruit to pupate. Larvae are used to stock the wasps' nests as food for their young. The egg parasitoids *Trichogramma chilonis* Ishii and *Trichogramma chilotreae* attack the pest in Luzon (Golez 1991). Leefmans and van der Vecht (1930) reported that an entomopathogenic fungus infected the larvae in Indonesia.

8.9.2.2 Monitoring and Control

Infestation in fruits can be detected by the presence of a dark brown ring and frass at the point entry (CAB International 2003). Infested fruit should be removed from trees before the larvae can leave them to attack neighbouring fruit. Fruit should be wrapped in protective bags at 55–65 days after pollination, and fallen fruit should be destroyed (Anonymous 1984). Other Lepidoptera that can attack fruit have been reported in India and the Philippines (Australian Government 2004). The pomegranate fruit borer, *Deudores isocrates* (Fabricius) (Lepidoptera: Lycaenidae), which is also a pest of loquat, lychee, guava and pear, could attack mango by laying single eggs on shoots; the emerging larva bore into the fruit (Srivastava 1998).

8.9.3 Fruit Spotting Bugs (Hemiptera: Coreidae)

The yellowish green coreid bugs, *Amblypelta lutescens lutescens* (Distant) and *Amblypelta nitida* Stål occur along the coast of Queensland, Australia, and attack most of the tropical and subtropical fruit crops there (Waite and Huwer 1998). They prefer to feed on young, green fruit, but *A. l. lutescens* also damages the terminals of a number of hosts. In tropical north Queensland, *A. l. lutescens* is the dominant species and feeds on young fruit causing black lesions to develop and the fruit to fall. It also feeds on the terminals and leaf petioles, causing wilting and dieback (Cunningham 1989). In the subtropical south, both species attack mango, but *A. nitida* confines its attention to the fruit, while *A. l. lutescens* also attacks fruit and terminal growth (Waite 1998). The bugs breed in natural rainforest areas, and fly into the orchards to feed on the fruit and terminals. Female bugs lay individual, opalescent green eggs under the leaves. There are five nymphal instars and a generation takes *c.*40 days.

The main predators of fruitspotting bugs are spiders, particularly members of the family Thomisidae. Several species of egg parasitoids have been recorded. In north Queensland, *Ooencyrtus* sp. (Encyrtidae), *Anastatus* sp. (Eupelmidae) and *Gryon* sp. (Scelionidae) parasitized 37.5–91.6 % of eggs of *A. l. lutescens* (Fay and Huwer 1994). In south Queensland, *Anastatus* sp. And *Gryon meridianum* (Dodd) parasitize eggs of *A. nitida* and *A. l. lutescens* to a similar extent (Waite and Petzl 1994).

Because fruitspotting bugs continuously migrate into orchards, more than one insecticide spray may be required to protect the young fruit. However, the fruit are safe from attack once they are *c.*50 mm long, and two or three sprays of endosulfan at intervals of 2 weeks are generally sufficient to protect them. The coconut bug, *Pseudotheraptus wayi* Brown, was first recorded on mangoes in South Africa in 1977, and now also attacks guavas, pecans, macadamias, avocados and loquats. It causes damage similar to that of *Amblypelta* spp. (De Villiers 1990). *Helopeltis* spp. (Miridae) are minor pests of mango, cashew and cacao in the Philippines and in northern Australia, where they feed on fruit and cause superficial corky blemishes. Insecticides are used to control them, but in the Philippines, bagging is also effective (Anonymous 1984).

Plant bugs within the Lygaeidae and Pyrrhocoridae, i.e. the Indian milkweed bug, *Spilostethus pandurus* (Scopoli) and the red cotton bug, *Dysdercus koenigri* (Fabricius), can injure fruit, inflorescences and leaves of mangoes in India (Australian Government 2004).

8.10 Thrips

The citrus thrips, *Scirtothrips aurantii* Faure and the red banded thrips *Selenothrips rubrocinctus* (Giard) are the only thrips that caused lesions on fruit in South Africa (Grove et al. 2001). However, Grove et al. (2001) reported that former species was of more economic importance than the later.

8.11 Blossom Pests

Midges, caterpillars, hoppers, thrips and mites are the most important pests attacking mango inflorescences.

8.11.1 Midges

Barnes (1948) recognized nine gall midges from mango; two of these, *Asynapta* sp. and *E. mangiferae*, are from the West Indies. Butani (1979) reported five cecidomyiid species on mango blossoms, including *Erosomya indica* (Grover and Prasad). *Dasyneura mangiferae* (Felt) was reported in Hawaii USA (Vannière et al. 2004). In recent times, Gagne and Etienne (2006) reported the species *Gephyraulus mangiferae* (Felt), n.comb. infesting mango flowers on the island of Guadeloupe, French West Indies. Male adults of *E. mangiferae* are 1.61 mm and females 1.32 mm long. Eggs are deposited in folds between sepals and petals of flower buds. The larval stage has four instars. The second generation then infests very young fruit, which drop before the marble stage. Sampling of mango midges needs to include affected tissue, different trapping devices, pheromones, etc. On citrus, use of coloured sticky traps placed in the tree canopy provides a more efficient method of sampling the citrus midge, *Prodiplosis longifila* Cagné, than ground emergence traps and collection of larval samples (Peña and Duncan 1992). In a survey of parasitoids of cecidomyiid pests of mango in India, Grover (1986b) reported that *Platygaster* sp., *Systasis* sp. and *Eupelmus* sp. were associated with *Dasineura* sp., and *Tetrastychus* sp. was associated with *E. indica*.

8.11.2 Mango Hoppers

More than 18 species of leaf hoppers have been reported as pests of mango, of these, *Idioscopus clypealis* Leth., *Idioscopus niveosparsus* Leth., *Idioscopus magpurensis* Pruthi and *Amritodus atkinsoni* Leth., are important (Viraktamath and Viraktamath 1985; Viraktamath 1997; Fletcher and Dangerfield 2002). The females deposit their eggs in panicles or midribs of tender leaves. The adults and nymphs preferentially feed on young leaves and flowers or shoots, and excrete honeydew upon which sooty mould develops (Ahmed et al. 1981).

This interferes with photosynthesis, adversely affecting plant growth and yield (Godase et al. 2004). Affected inflorescences turn brown, become dehydrated and fruit set does not occur. *Amritodus brevistilus* and *I. niveosparus* populations increase from February to peak in March–April in Sri Lanka, while peaks of *I. clypealis* occur in March and September (Kudagamage et al. 2001). *Idioscopus clypealis* populations peak in south-eastern India during March and April (Tandon et al. 1983). *Idioscopus nivesoparus* and *I. clypealis* peaks coincided with major and minor flowering periods while population peaks for *A. brevistilus* coincide with the occurrence of vegetative flushing (Kudagamage et al. 2001). Azizur Rahman and Singh (2004) demonstrated that *A. atkinsoni* populations on panicles of 'Langra' mango were negatively correlated with high RH; whereas no significant relationships were observed with rainfall, sunshine and wind velocity.

Sampling Very few sampling studies have been reported for hoppers on mango. A sequential sampling plan for mango hoppers was recommended by Verghese et al. (1985) in India. Tandon et al. (1989) reported that distribution of *I. niveosparsus* was aggregated and was best explained by Iwao's patchiness regression. To assess damage, they recommended a sampling size of 59–98 panicles/tree. Verghese et al. (1985) developed a sequential sampling plan classifying infestations of adults and nymphs of *I. clypealis* as light, moderate and severe.

Biological Control Several natural enemies have been described from west and South-east Asia. In Pakistan, several species of egg parasitoids, *Gonatocentrus* sp., *Miurfens* sp. nr. *mangiferae* Viggiani and Hayat, *Centrodora* sp. nr. *scolypopae* Valentine, *Aprostocetus* sp. and *Quadrastichus* sp., and the adult ectoparasitoid *Epipyrops fuliginosa* Tames have been reported by Mohyuddin and Mahmood (1993) for the control of mango hoppers. In India, Sadana and Kumari (1991) studied the efficacy of the lyssomanid spider, *Lyssomanes sikkimensis* on *I. clypealis.* Classical biological control of mango hoppers has not been attempted. The entomopathogens, *Verticillium lecanii* (Zimmerman) Viegas, *Beauveria bassiana* Balsamo (Vuillemin) and *Isaria tax*, infect *I. clypealis* in India (Kumar et al. 1993; Srivastava and Tandon 1986) while the effectiveness of *Metarhizium anisopliae* var. *anisopliae* was tested under laboratory conditions against *A. atkinsoni* (Vyas et al. 1993).

Chemical Control Several pesticides have been evaluated for controlling mango hoppers (Tandon and Lai 1979; Yazdani and Mehto 1980; Shah et al. 1983; Shukla and Prassad 1984; Islam and Elegio 1997; Kudagamage et al. 2001). Jhala et al. (1989) considered that sprays of carbaryl during the off-season maintained the hopper population at low-density levels. Godase et al. (2004) demonstrated that sprays of 0.05 % monocrotophos at the first panicle emergence and a second spray 15 days later are essential to prevent yield loss. Kudagamage et al. (2001) found that imidacloprid (Admire SL 200) controlled mango hoppers if applied just after flowering and again 10 days later.

8.12 Lepidoptera

The lepidopteran flower feeders are the second most serious pests of mango. They include *Chloropteryx glauciptera* Hampson and *Oxydia vesulia* (Cramer) (Whitwell 1993). The infestation increase with the flowering period. The eggs are laid on or near the inflorescences or new leaves. The larvae of both species feed on the axis of the inflorescence, dried fallen flowers are webbed together and fastened to flower clusters to form nests (Patel et al. 1977). The Lepidoptera complex attacking mango flowers in Australia consists of several species from the families Geometridae, Lymanthridae, Noctuidae, Pyralidae and Tortricidae. In Brazil, Barbosa (2005) and Barbosa et al. (2005) reported *Pleuroprucha asthenaria* (Walker) (Lepidoptera: Geometridae) and *Cryptoblabes gnidiella* (Milliere) (Lepidoptera: Pyralidae) affecting mango inflorescences. In Brazil, the pesticides *Bacillus thuringiensis*, trichlorfon and lambdacyhalothrin provided 66–59 % mortality (Barbosa 2005). Classical biological control of lepidopteran insects attacking mango in Dominica was initiated with the introduction of the wasps, *Aleiodes* sp. And *Euplectrus* sp., and the fly *Blepharella lateralis* Macquart. In Brazil, *C. gnidiella* is parasitized by *Brachymeria pseudoovata* Blanch (Hymenoptera: Chalcididae).

8.13 Thrips

The western flower thrips, *Frankliniella occidentalis* (Pergande) has been reported to damage flowers and fruit in Israel (Wysoki et al. 1993). In Florida the thrips complex consisting of *Frankliniella bispinosa* (Morgan) and *F. Kelliae* (Sakimura) is the most frequently observed blossom pest on flowers and causes damage by ovipositing in the panicle and feeding on the floral nectarines and anthers, which may result in premature loss of pollen.

Sampling In India, Verghese et al. (1985) determined that the lower mango canopy is better for sampling, and recommended a sample size of 55 panicles/tree for surveying. Verghese et al. (1988) indicated that distribution of *Thrips palmi* (Karny) on mango panicles is better explained by Iwao's patchiness regression, which indicated an aggregated distribution and suggested that the lower canopy should be sampled. Sample sizes should be 55 panicles/tree for control and survey studies, with a 20 % error of the mean and 92 panicles/tree to obtain a lower (5 %) percentage error of the mean. In Florida, Peña et al. (2006a) showed that a cumulative number of 400–700 thrips/panicle during 4 weeks causes 33–50 % yield reduction of 'Keitt' mangoes.

Chemical Control Peña et al. (2006a) tested the efficacy of different pesticides such as acetamiprid, fenpropatrin, milbemectin and zeta-cypermethrin, novaluron against mango flower thrips. All treatments, except milbemectin, reduced thrips densities 5 days after application of the first spray. Danitol and zeta-cypermethrin had the lowest numbers of thrips 12 and 20 days after the first spray. All treatments had lower thrips densities compared to the control at 28, 35 and 43 days after the second application.

Biological Control Several parasitoids and predators, such as *Ceranisus menes* (Walker) in Israel (Rubin and Kuslitizky 1992) and the predators, *Orius* sp., *Anystis* *agilis* Banks and *Hypoaspis aculifer* (Canestrini) (Loomans et al. 1995), are potential biocontrol agents for *F. occidentalis,* while *Dasyscapus parvipennis* Gahan is considered to have good potential for the biological control of flower thrips in Puerto Rico (Bartlet 1938).

8.14 Pests of Buds and Leaves

The most destructive mango leaf feeders are thripsmidges, mites, scales, whiteflies, mealybugs, weevils, ants, locusts and caterpillars (Jeppson et al. 1975; Bhole et al. 1987; Jadhav and Dalvi 1987; Tigvattnanont 1988). Formation of leaf galls in India is caused by the Eurytomidae, *Mangoma spinidorsum* (Subba Rao 1986), but there is little information on their importance as foliage pests.

8.14.1 Thrips

The thrips *Scirtothrips mangiferae* Priesner causes the young leaves to curl along the midrib, distorting their shape and leading to premature drop (Wysoki et al. 1993). The twigs of infested shoots are much shorter than those of uninfested ones. The population of the thrips is low during winter, increases in early spring and reaches its peak during summer (Wysoki et al. 1993). Yellow sticky traps can be used for monitoring thrips densities. According to Yee (1987), the thrips are controlled by malathion (25 % v/w). The weaver ant, *O. smaragdina* (Hymenoptera: Formicidae) is considered an effective biological control of *S. rubrocinctus* in the Northern Territory of Australia (Peng and Christian 2004).

8.14.1.1 Midges

Midges belonging to family Cecidomyiidae are serious pests of mango in Asia and the Caribbean region. Two genera, *Procontarinia* Kieffer and Cecconi and *Erosomyia* Felt, are particularly associated with mango (USDA 1981; Schreiner 1991; Harris and Schreiner 1992; Uechi et al. 2002; Gagne and Medina 2004). Gall formation intensity is more during rainy season because RH improves larval and pupal survival. No differences were observed in gall densities collected from lower and top portions of the tree. Askari and Radjabi (2003) observed overlapping generations of *Procontariniamatteiana* in Iran related to the different leaf flushing patterns and found that the optimum pest temperatures were 10–26 °C. Differences on susceptibility of mango cultivars to *P. matteiana* might indicate that susceptible cultivars should not be grown in areas infested by this gall midge (Jhala et al. 1987; Githure et al. 1998). Daneel et al. (2000) suggested that products to control *P. matteiana* should be applied after harvest, coinciding with the first major flush and a second spray 6 weeks later. Austin (1984) and Sankaran and Mjeni (1989) have reported several platygastrid species parasitizing *Procontarinia* spp. and their prospects for biological control of the pest.

8.14.1.2 Mites

The mango bud mite, *Aceria mangiferae* (Sayed) (Acari: Eriophyidae), attacks buds and inflorescences (Keifer et al. 1982; Ochoa et al. 1994), while another species, *Metaculus mangiferae* (Attiah) (Acari: Eriophyidae) causes russeting of terminal leaves, buds and inflorescences. The latter is an important pest in Egypt, India, Palestine and Angola (Jeppson et al. 1975). The puncture wounds of several acarines (Acari: Tetranychidae) cause serious damage to leaves, which may dry and fall. The main pest in Mauritius, India, Egypt, Israel and Peru is *Oligonychus mangiferus* (Rahman and Sapra); in Israel, the spider mite *Tetranychus cinnabarinus* (Boisduval), which lives on the underside of the leaves, causes bronzing around the puncture wounds. If the tetranychid mites are sufficiently abundant, infested leaves may drop. There is little or no information on sampling techniques or for their economic thresholds. *Aceria mangiferae* occurs wherever mango is grown (Denmark 1983; Doreste 1984). However, *A. mangiferae* does not cause mango malformation, but may be a carrier of *Fusarium mangiferae*, which is recognized as the causal agent of mango malformation (Varma et al. 1974; Freeman et al. 2004).

Damage According to Keifer (cited in Jeppson et al. 1975), *A. mangiferae* infestation of buds results in arrested growth, with the stunted, short, young stems close together at the terminal branch. When the leaves fall, the overall effect is scanty growth of twiggy branches, with a few stubby, short branchlets and discoloured buds.

Sampling Sampling small arthropods such as is operationally difficult and often time consuming. To ease this burden, presence-absence, or binomial, sampling was tested in place of complete counts for estimating or classifying densities of these organisms (Peña et al. 2005).

Biological Control The phytoseiid *Amblyseius swirski* Athias Henriot is associated with *A. mangiferae* (Abou-Awad 1981). In Florida USA, several unidentifi ed phytoseiids occur on buds infested with *A. mangiferae.* Tenuipalpid (*Brevipalpus phoenicis*), Tydeid and Tarsonemid (*Tarsonemus confuses* (Ewing)) mites also inhabit mango buds. Therefore, it is difficult to determine the mite species that is the host prey.

Chemical Control Osman (1979) reported that applications of four full coverage sprays of dichlorvos were effective for controlling *A. mangiferae* in Egypt. Rai et al. (1966) cautioned that chemical control should be directed to apparently healthy and not malformed tissues.

Plant Resistance In Florida USA, the densities of *A. mangiferae* were measured on 22 mango cultivars from December 1997 to June 1998. 'Keenan', an unknown cultivar, 'cv. 9819', 'Brander' and 'Bombay Green' had significantly more mites than 'Joellen', 'Duncan', 'Red Itamaraca', 'Smith', 'Wally' and 'Hindi' (Peña et al. 2005).

8.14.1.3 Scales

Armored Scales At least 26 species of diaspidids attack mangoes worldwide (Chua and Wood 1990). In India, *Aspidiotus destructor* Signoret causes serious damage, while *Parlatoria pergandii* Comstock and *Lepidosaphes gloverii* (Packard) damage 3-year-old plants (Chua and Wood 1990). *Radionaspis indica* (Marlatt) (=*Leucaspis indica*) encourages growth of black mould, which covers young branches (Dekle 1976). Van Halteren (1970) concluded that *A. mangiferae* development is completed in 35–40 days for females and 23–28 days for males.

Soft Scales They include *Coccus viridis* (Green), *Coccus longulus*, *Ceroplastes actiniformis*, *Philephedra tuberculosa* Nakahara and Gill and the mango shield scales *Milviscutulus mangiferae* (Green) and *Viusonia stellifera* (Westwood). These coccids are generally polyphagous, attacking different genera and species. They are mobile and injure mango because of the production of honeydew and the subsequent accumulation of sooty mould on the honeydew (Escalante 1974; Silva and Cavalcante 1977). Most of these scales can be suppressed at sub-economic levels, either by application of selective pesticides (i.e. oils) or by biological control agents.

Sampling Because of their small size, it is laborious to sample all stages of scales. Pheromones in tent-style traps have been used with other fruit crops, as well as monitoring crawler movement using sticky bands close to infested leaves. Either double-sided sticky tape, or tape coated with Vaseline® is effective for trapping crawlers. Bands are removed and examined under a microscope to determine crawler numbers. Dark-coloured tape provides a better contrast to detect crawlers (Beers et al. 1993).

8.14.2 Biological Control

Labuschagne (1993) determined that the predatory thrips *Auleurodothrips fasciapennis* Franklin and the parasitoid *Aspidiotiphagus citrinus* (Crawford) are the most important biocontrol agents of *A. tubercularis*. In South Africa, Joubert et al. (2000) obtained 46 % parasitism of *A. tubercularis* using an unidentified species of the parasitoid, *Aphytis* sp. Arias et al. (2004a) observed *Coccidophilus* spp. (Coleoptera: Coccinellidae) and *Chrysopa* spp. preying on *A. tubercularis* in Ecuador; the exotic predator *Cybocephalus nipponicus* (Coleoptera: Nitidulidae) was introduced to supplement predation of the former scale (Arias et al. 2004b).

8.14.3 Whiteflies and Blackflies

Two species of whitefly, *Aleurodicus disperses* Russel and the blackfly, *Aleurocanthus woglumi* Ashby are of great economic value. They suck cell sap from leaves, which wilt if populations are high. The infestations can blacken entire trees and reduce photosyntheitic area if populations are high (Angeles et al. 1971; Peña 1993). Some parasitoids *Encarsia opulenta* (Silvestri) and *Amitus hesperidus* (Silvestri) attacking immature stages of white flies can provide good control.

8.14.4 Mealybugs

The mango mealy bug, *Drosicha mangiferae* Green is one of the serious insect pests of mango in the Northern and the Central parts of India. According to Miller et al. (2002) 158 species of mealybugs are recognized as pests worldwide. Approximately 22 % of the mealybug pests are polyphagous, 20 % occur on grasses, 16 % on citrus and tropical fruits, and 6 % on coffee.

8.14.5 Sampling

Narasimham and Chacko (1991a, b) determined that densities of *R. invadens* Williams, *Rastrococcus iceryodes* (Green) and *Rastrococcus mangiferae* (Green) were significantly higher on the abaxial than the adaxial surface. Mealybug density was also higher from ground level to 2 m compared with >2 m; they also observed that spatial distribution of *R. iceryodes* did not differ among internal and external canopies, whereas densities of *R. invadens* and *R. mangiferae* were higher in the external canopy. They also determined that there were statistical differences causing some inter-tree variation, but did not determine if the mealybugs followed a random or contagious distribution.

8.14.6 Host Ranges of Different Types of Mealy Bug

Dietz and Harwood (1960) reported the host range and damage by the Grass mealy bug, *Heterococcus graminicola* on 67 species representing 23 genera. They have recorded 64 mealy bug species from Brazil. Tandon and Varghese (1985) reported that *D. mangiferae* is dangerous for mango crop. It is not only pest of mango but it attacks more than 70 other host plants (Tandon and Lal 1978). Though this insect is mainly a pest of mango tree, however in the areas of heavy populations, it has the tendency to attack a variety of other fruit trees like peach (*Prunus persica*), plum (*P. domestica*), papaya (*Carica papaya*) and all citrus species etc. (Khan 2001).

Kannan and Rao (2006) determined the occurrence of pests during the new flush, twig expansion, matured leaf and fruit maturity stages of 0–5, 5–15 and >=15 year-old mango cv. Neelum trees. Trees 15 years old and above were preferred by the pests *Procontarinia matteiana, Amradiplosis ecinogalliperda, Orthaga exvinacea, Amritodus atkinsoni, Idioscopus niveosparsus, Thrips hawaiiensis, Drosicha man-*

giferae, Sternochaetus mangiferae, Bactrocera dorsalis, Conogethes punctiferalis and *Batocera rufomaculata*.

Sharma and Arora (2009) identified six mealy bug species, viz. *Planococcus citri* (Risso), *Planococcus lilacinus* (Cockerell), *Ferrisia virgata* (Cockerell), *Nipaecoccus viridis* (Newstead), *Maconellicoccus hirsutus* (Green) and *Drosicha mangiferae* (Green) active on various fruit crops in Punjab. *P. citri* attacked citrus, guava, pomegranate, sapota, jamun, aonla and ber fruit plants. Its infestation was found on leaves, twigs and branches of all these fruit crops. *D. mangiferae* was the most serious mealy bug species of mango from February to April and its attack was also observed on sapota and litchi. *P. lilacinus*, *F. virgata*, *N. viridis* and *M. hirsutus* are reported for the first time on fruit crops in Punjab. Chlorpyriphos (0.075 %), acephate (0.075 %), thiamethoxam (0.008 %) and profenophos (0.075 %) were found effective against mealy bugs on citrus, guava and grapes. Wih and Billah (2012) identified nine fruit fly species and four mealybug species (*Pseudococcus longispinus*, *Paracoccus marginatus*, *Rastrococcus invadens* and *Icerya* sp.) during the survey. While mango was dominated by *R. invadens*, the ornamental plants were mostly affected by *Icerya* sp., papaw by *P. marginatus*, and *Jatropha* species infested by *P. longispinus*. The mealybug species were fairly common in the region. In certain cases, other pest species such as aphids and whiteflies were found in close association (in complex mixtures) with the mealybugs.

8.15 Management of Mango Mealy Bug

The management of mealy bug is difficult owing to its polyphagous and hardy nature. Many workers after carrying out detailed investigations have recommended maneurvability of cultural practices for destruction of egg masses, application of banding materials around the tree trunks for arresting the ascending and descending nymphs, insecticidal sprays and biological methods for the containment of the pest. Various control methods for the management of mealy bugs are given below:

8.15.1 Raking and Scrapping of Soil

The digging of the soil around the trees has been found to help destruction of eggs at Pusa (Fletcher 1917) and elsewhere in Delhi (Dutt 1925). Destruction of eggs of mango mealy bug by digging them out with spades from the soil is not an encouraging practice (Rahman and Latif 1944), whereas this way of destroying the eggs is an effective practice as reported by (Singh 1947; Mohyuddin and Mahmood 1993). Similarly the use of burlap band, burning of gravid females and removal of soil contaminated with eggs of mango mealybug gave complete control of mango mealy bug without the use of pesticides (Sial 1999) and should be destroyed in autumn (Xu et al. 1999).

The raking of soil and exposure of the eggs to sun, birds, ants and other predators has been found to yield encouraging results Sen (1955) and Pruthi and Batra (1960), Butani (1964) advocated scrapping of soil to a depth of about 150 mm and placement of the scrapped soil as litre, underneath the cattle to ensure complete destruction of the eggs lying within the scrapped soil. Singh (1978) has demonstrated effective control by the exposure of the *rak*ed soil to sun during May and June. Maximum control of insects has been achieved by Chandra et al. (1989) by raking the soil followed by irrigating the fields in the months of May, June, August and October. The sticky bands with grease material or slippery bands with alkathene or plastic sheets around the trunk at about one meter above ground level in 2nd week of December could also prevent the upward movement of nymphs.

The turning over of the soil in mango orchards in Jammu in the months of May–June with a furrow turning plough has been suggested to expose most of the egg masses for destruction by summer heat (SKUAST 1993). Lakra et al. (1980) reported that sand was also used as barrier for upward migrating nymphs of *D. mangiferae.* Srivastava et al. (1982) suggested a band width of 25–30 cm ideally suiting in preventing ascend of nymphs.

8.15.2 Evaluation of Banding Materials

Different types of banding materials have been tried for its control but with varying success. Stebbing (1903) after evaluation of grease coaltar, grease, coaltar, coaltar glue and coaltar rosin bands concluded that none of these are effective in stopping the ascend of nymphs statifactorily. Sticky bands of rosin and castor oil have also proved effective in checking the upward movement of the insect at Pusa (Fletcher 1927). Similarly, rosin + Toria oil ban 1:1 is, recorded b Chopra (1928) to be very promising. Effectiveness of rosins neem oil and Vaseline band. Rehman and Latif (1944) obtained satisfactory results after using "Namhar' band at Layallpur. The ordinary sannhemp fibre or hessian cloth dipped in equal quantities of crude oil emulsion and coaltar has also been proved a successful method of control. Lal (1952) tried grease band and castor oil + rosin at Delhi, and Chaudhary and Mazid (1954) also observed the ostico band yielding encouraging results. Amongst various types of bands tried at Punjab and Bihar ostico band has been rated as the best for checking the ascend of the nymphs (Sen 1955). Sen and Prasad (1956) and Atwal (1963) confirmed the superiority of ostico band for effective arrest of the ascend of nymphs. Bindra et al. (1970) at Punjab has ratified the view of above mentioned workers. Bindra (1967) while working at Hissar observed the coaltar + grease band effective for 2 weeks. The results on the evaluation of different types of stickly coaltar 1:2 as the most effective treatment.

Srivastava (1980) observed the maximum number of nymphs congregated at the level (L) in case of alkathene band while comparing the previously used alkathene band with presently used ostico and Esso fruit grease band. Randhawa (1974) has suggested that Esso fruit tree grease of slippery alkathene sheet does help in checking the upward movement of crawlers. Superiority of alkathene has also been determined at Delhi (Srivastava, 1976) and Punjab (Lakra et al. 1980). Polythene band dusted with Methyl parathion has proved equally effective in Punjab (Sandhu et al. 1979a). Srivastava et al. (1972) suggested a band width of 25–30 cm ideally suiting in preventing ascend of nymphs. Bindra and Sohi (1974) recommended tree banding with 30 cm wide polythene sheet over a mud plaster.

Polyethelene bands of 400 mil gauge and 25 cm width fastened around the tree trunk have given good results. The sticky bands with grease material or slippery bands with alkathene or plastic sheets around the trunk at about 1 m above ground level in 2nd week of Dec. could also prevent the upward movement of nymphs (Atwal 1963). Similarly Tandon and Verghese (1987) suggested that exposure of eggs to sun, removal of alternative host plants and conservation of natural enemies by using garlic oil or neem seed extract around the trunk of trees and application of alkathane bands could also eliminate mango mealybug population.

Bajwa and Gul (2000) reported that *Drosicha stebbingii* could be managed through destruction of eggs, banding of trees and spray of insecticides v.z., Bulldock 25EC, Endon 35EC and Mepra 50EC on trees of *Paulownia tomentosa* and *Paulownia fortunei.*

8.15.3 Evaluation of Various Insecticides and Combination Strategy

Chemical control methods for mealy bug and fruit fly have been inefficient (Tandon and Lal 1980; Yousuf and Ashraf 1987). Srivastava (2008) tested the field efficacy of 3, 2 and 1 % ether and alcoholic extracts of neem products for 3 years against nymphs of mealy bug *D. mangiferae* Green during 1997–1998 to 1999–2000 on c.v Mallika and revealed that all the concentrations were found significantly superior over the control in the effective pest management.

The spray of fish soil rosin soap emulsion (240z in 4 gal of water) was suggested earlier at Delhi (Dutt 1925) and Punjab (Rehman 1940) for control of this insect. Morrill and Otanes (1947) obtained 83 % kill of the insect with a single spray of DDT emulsion comprising 20 % DDT, 65 % xylene and 10 % triton×−100 whereas the second spray on the plants showing infestation offered cent per cent of the mealy bugs in phillippines. Sen (1955) while determining the effectivity of BHC 50 %, DDT 10 % Toxaphene 10 %, Chlordane 2.5 %, methyl parathion dust 0.5 % potassium permagnate, kerosene oil emulsion and calomel as soil applicants observed that none of these pesticides could offer my satisfactory control of the nymphs of mealy bug. Sen and Prasad (1956) observed that percentage of mortality of the first instar nymphs was 100 % with folidol E- 605 (0.03 %) treatment, 56.6 % with pyrocolloid and 61 % with BHC (Hortex) treatment within 6 days. The percentage of mortality of second instar nymphs was 93 % of Folidol E- 605 (0.04 %), 26.3 % in case of pyrocolloid (0.25 %) and 29.7 % in case of BHC (0.1 %)

within the above stated period and none of these chemicals was found to be much effective against the third instar nymphs.

Pradhan et al. (1956) carried out bioassay tests in the laboratory and found melathion to be more toxic than P'-P DDT and within the commercial preparations. Parathion, Demeton and Diazinon proved more toxic than DDT. Trehan, Gangoli et al. (1957) advocated BHC dust for its effective control. Singh (1957) recorded 98.98 % check in the population ascent of first instar nymphs when Diazinon 0.33 % was applied on trunk and application of the same chemical at same concentration on foliage yielded 96.2–100 % mortality of second instar nymphs. Singh (1960) recommended a spray of the infested trees with a mixture of nicotine sulphate 0.1 % sesame oil 1.25 % soft soap 0.25 % washing soda 0.3 % in water and ethylalcohal 0.3 % for effective control of the pest. He has further argued that Diazinon 0.06 % has also shown promise.

Atwal (1963) observed that first and second instar nymphs could be effectively killed in Punjab by spraying the tree with Endrin or Parathion 0.04 % or dusting with 10 % BHC or DDT and the third instar nymphs with Diazinon 0.04 % sprays. Birat (1963) after carrying out comparative efficacy studies of organophosphorous insecticides has found Diazinon 0.06 % superior to melathion and parathion. Bhatia and Gauba (1964) have also reported Diazinon to be effective against the pest. Atwal et al. (1969) found 0.1 % Fomothion or methylparathion most promising and these were followed by parathion 0.05 %, 0.10 % Formothion or Zectran in their effectively for the control of first and second nymphal instars. They also obtained equally successful control of third instar nymphs with parathion (0.05 %, 0.10 %) formothion (0.10 %) Methyl parathion (0.10 %), trichlorophan, (0.1 %) and Zectran (0.1 %).

Reddy (1968) experimented spraying the trees with Diazinon 0.04 % or Malathion 0.08 % and concluded that all the three yield encouraging results for controlling the nymphs and adults. Sexena and Rawat (1968) obtained effective control at Jabalpur when they sprayed the infested trees with parathion 0.06 % emulsion or melathion 0.07–0.08 % emulsion. Pradhan (1969) recommended strong organophosphrous insecticidal sprays against young nymphs. Vevai (1969) found dimethoate 30 % 12–20 ml with 50 g sulphur (80 W.P.) in 15–20 l of water and spray mango fruits and flowers only. Cent percent mortality of mealy bugs collected below the bands and sprayed with malathion emulsion (30 ml of 50 EC in 10 l of water was observed by (Bindra and Sohi 1974). The nymphs and adults are also reported to be satisfactorily controlled by spraying 0.03 % Monocrotophos or Dimecron, Endosulfan 0.05 % and Dimethoate (0.03 %) (Srivastava et al. 1973).

Tandon (1969) obtained maximum reduction of mealy bug population with monocrotophos (0.05 %) carabaryl (0.1 %) and synthetic garlic oil (0.75 %) which behaved at par whereas the second best was methyl parathion and quinalphos, each applied at 0.05 % concentration.

Pushpa et al. (1973) recorded the mealy bug (*Maconellicoccus hirsutus* Green) on mesta (*Hibiscus cannabinus*) were controlled effectively through the application of roger (dimethoate) or metasystox (methyl demeton). High volume application of insecticides was very important to ensure coverage and reach the mealy bugs in crevices and *craks* on the bark, for example diazinon, quinalphos and parathion-methyl have been shown to suppress the pest quite effectively both at nymph and adult stages.

Lakra et al. (1980) reported that chemicals like diazinon, quinalphos and parathion-methyl (methyl parathion) were highly effective against 1st instar nymphs that were gathered below the bands. Dalaya et al. (1983) observed that fenitrothion (0.1 %) is the most effective the followed by phosolone (0.07 %), Quinalphos (0.05 %), Monocrotophos (0.04 %), Methyl parathion (0.05 %) and Bromophos ethyl (0.07 %) and Phosphomidon (0.03 %) and the chemicals viz. Phenthoate (0.05 %) dimethoate (0.03 %) and Melathion (0.1 %) and Melathion (0.1 %) proved less effective against mango mealy bug.

Dhingra (1990) evaluated ten insecticides as the alternative chemical for successful control of the mango mealy bug. He further observed that lamda cyhalothrin, alphamethrin, decamethrin, cypermethrin, methyl parathin, fenvalerate, monoctophos, endosulfan whose LC_{50} value was calculated as 0.0024, 0.0196, 0.0209, 0.0292, 0.0385, 0.0454, 0.2357 and 0.9303 respectively offered promise result. He further noted that lindane and malathion proved ineffective even at as high as 1 % concentration.

Karar et al. (2010b) evaluated 11 formulation of insecticides for the control of mango mealy bug under field conditions and found that the maximum mortality of 1st instar mango mealybug was observed in those treatments, where Mospilan were applied with 80 %, 85 % and 91 % after 24, 72 and 168 h of spray. However, in case of 2nd and 3rd instar, Decis and Curacron gave maximum mortality 71 and 70, 24 h after spray. After 72 and 168 h Mospilan proved best with 78 % and 81 % mortality. Supracide the most effective insecticides for the control of adult female at all the post treatment intervals i.e., 60 %, 72 % and 73 % mortality under field conditions.

Srivastava (1976) recommended the spraying of diazinon, monocrotophos and Chloropyriphos for achieving good control. Prasad and Singh (1976) opined that the protions of the infested tree trunks above the applied tree bands, and the foliage if sprayed with chemicals viz. Melathion 0.04 % (3–4 sprays) or Dimethoate 0.03 % or Phosphomidon 0.04 % or Diazinon 0.04 % or Thilometon 0.04 % or Diazinon 0.04 % or Thiometon 0.04 % (3 sprays) during January on tree trunks can offer a very satisfactory control of mealy bugs. Stem injection is also suggested as a means of effective control of the insect at its earlier stages. The monocrotophos injected in the trunk has proved more effective than fenitrothion (Sandhu et al. 1979). Monocrotophos (0.05 %), carbaryl (0.1 %) and synthetic garlic oil (0.75 %) have been rated at par but effective for the control of insects at their nymphal stages (Tandon et al. 1989). Tandon (1969) obtained maximum reduction of mealy bug population with monocrotophos (0.05 %) carabaryl (0.1 %) and synthetic garlic oil (0.75 %) which behaved at par whereas the second best was methyl parathion and quinalphos, each applied at 0.05 % concentration.

Batra et al. (1980) revealed that Quinalphos was the most effective insecticide on terminals followed by monocrotophos but when the chemicals were sprayed on branches and stem only 25 % and 35 % morality up to 7 days was recorded respectively and amongst the dusts methyl parathion offered best results. Lakra et al. (1980) obtained 97.04 %, 93.50 % and 93.80 % mortality of the clustering second instar nymphs on mango trees with Quinalphos (0.75 %) Diazinon (0.05 %) and methyl parathion (0.05 %) respectively

Siddique and Mahur (1982) noted that diazinon (0.04 %) methyl demeton (0.1 %), phosphomidon (0.04 %) and ethyl parathion (0.04 %) offered 99 % insect kill in the laboratory studies whereas the chemicals viz. diazinon (0.06 %) methyl-demeton (0.15 % and 0.1 %) phsophamidon (0.06 %) and ethyl parathion (0.06 %) under field studies offered insect mortality of 98 %. Dalaya et al. (1983) after undertaking a field trial in Maharashtra observed that fenitrothion (0.1 %) is the most effective the followed by phosolone (0.07 %), Quinalphos (0.05 %), Monocrotophos (0.04 %), Methyl parathion (0.05 %) and Bromophos ethyl (0.07 %) and Phosphomidon (0.03 %) and the chemicals viz. Phenthoate (0.05 %) dimethoate (0.03 %) and Melathion (0.1 %) and Melathion (0.1 %) proved less effective.

Nayer et al. (1986) advocated application of aldrin (5 % dust around the base of the tress and spraying of aerial parts with Parathion (0.05 %) or Melathion (0.05 %) for the effective control of early instar nymphs. Methyl demeton (0.05 %) Dimethoate (0.05 %) and Dichloros (0.1 %) have been proved equally effective when used as foliar sprays. Tandon (1988) observed the spray of Monocrotophos (0.05 %) or Chloropyrphos yielding good results when bugs settle on inflorescene panicles. Khurana and Verma (1988) at Sirsa evaluated the efficiency of Phosalne, Dicrotophos each at (0.05 %) Carbaryl, Tetrachlorvenphos, Trichlorphan each at (0.1 %) and found these effective against second instar nymphs under field conditions.

Karar (2010) tested a combination of mounds on the plastic sheet, Haider's band and application of acetamiprid and found that the combination of these treatments to be the most effective resulted in 98 % reduction of first instars of mango mealybug. It is further stated that the Haider's band was the most effective and cheaper which was a new addition in the mechanical control management of mango mealybug on mango trees. The males of mango mealy bug were attracted to mercury light and no males were attracted to yellow, green, red, blue lights. Male preferred to pupate in wet places near the 'kacha' (mud) water which can be exposed to sunlight by hoeing. This research project demonstrates the complete management programme for the control of mango mealy bug under field condition for mango growers.

8.16 Pests of Trunks, Twigs and Roots

8.16.1 Coleopterans and Scales

Stem-boring insects are of serious concern in mango growing areas. They include Hypocryphalus *mangiferae* (Stebbing), *Apate monachus* Fabricius and *Batocera rubus* L. The tunnelling results in serious reduction in yields and decline in mango production. In addition to borers, the infestations by mango scale *Radionaspis indica* (Marlatt), and plumose scale, *Morganella longispina* occur on the trunks, branches and buds. The severe infestation may result in cracking of bark, exudation of sap and decline of upper brances. The insects prefer trees that have been weakened by pathogens, wind, etc., but after a population has been established, the infestation spreadsto healthy trees (van Wyk et al. 2007).

8.16.1.1 Termites

Mango orchards are becoming more common in dry and semi-arid areas with vast termite populations. Mango growing in infested areas often results in plant growth suppression as a result of reduced root establishment, invasion and pruning of roots (Rogers et al. 1999). For example, six termite species (*Odontotermes indicus* Thakur, *O. lokanandi* Chatterjee and Thakur, *O. obesus* (Rambur), *O. giriensis* (Roonwal and Chhotani), *O. bhagwatii* Chatterjee and Thakur and *Microtermes obesi* Holm) were recorded from mango orchards in India (Srivastava and Singh 2004). More species of termites were observed in winter than during the summer and rainy season. Veeresh et al. (1989) observed that *O. wallonensis*, *O. horni* and *O. obesus* constructed earthen sheeting on the stem of small mango trees. Singh (1960), cited by Srivastava (1998), reports *Eutermes* (*Nasutitermes*) *costali*, *Calutermes* (*Cryptotermes*) *brebis*, *Heterotermes tenuis*, *Coptotermes gestroi*, *Neotermes* (*Kelatermes*) *bosei* (Gardneri) Synder, *Microcereoutermes peroffinis*, *Calotermes* (*Neotermes*) *greeni* and *Coptotermes curvignathus* also affecting mango in India. According to Srivastava (1998), the most important termite species affecting mango are *O. Obesus* and *O. wallonensis*; *O. wallonensis* nests in the root zone in Uttar Pradesh, India. The workers feed on roots, stems and branches.

In Florida, urban dwellings infested with the Formosan termite, *C. formosanus* Shiraki, were treated with baits containing 0.5 % weight/weight novifl umuron (Cabrera and Thoms 2006). These baits might be useful when mango orchards are planned for areas infested with termites. In India, termite infestations are controlled with a combination of monthly irrigation, hoeing and application of neem cake (Singh and Singh 2003).

In general, most mango pests also occur on other fruit crop species. Fruit flies, scales, mites, thrips, lepidopteran flower feeders, mirids, weevils and beetles are mostly generalists, and some of their management schemes need to be implemented with this in mind. In the case of fruit flies, Aluja (1996) suggests surveying vegetation adjacent to infested mango orchards as populations are sustained and multiplied in these locations and from them adult fl ies move into commercial orchards to attack ripening fruit (Aluja et al. 1996). Management tactics that can be improved include the following:

1. Selective pesticides.

Pesticides that are used in integrated pest management programmes must have selective toxicity. Cunningham (1989) suggested that oils could be utilized for control of scales in mango; however, most of the recommendations are based on highly toxic or illegal, non-registered persistent chemicals (Singh 1991; van Mele et al. 2001; de Bie 2004). In South Africa, Joubert et al. (2004) tested kaolin, which is the active ingredient of Surround®, a non-toxic natural clay mineral, against the mango seed weevil, mango scale and citrus thrips. Surround® was effective against citrus thrips, mango weevil and coconut bug, *P. wayi*, but caused outbreaks of mango scale and long-tailed mealybug, *Pseudococcus longispinus*. However, producers of export fruit rely on calendar-based chemical control when trees are heavily infested with mango scale (Joubert et al. 2004).

Biological control. Biological control has great potential as a tactic for regulating pest populations in integrated pest management programmes in mango orchards. Biological control should be highly effective for indirect pests. Indeed, numerous studies have been conducted in many mango-producing countries to promote the use of parasites and predators for this type of pest (Cunningham 1989; Mohyuddin and Mahmood 1993; Moore and Cross 1993; Whitwell 1993; Wysoki et al. 1993; Labuschagne et al. 1995).

2. Host plant resistance. Cultivars of mango resistant to pests should be evaluated in established germplasm collections. After the initial selection has been made, evidence for the pattern of resistance must be established and changes in the environment, whether geographic or temporal, should not disrupt or decrease the resistance to any great extent.
3. Pheromones and trapping devices. Synthesis of sex pheromones have resulted in their possible use for pest management in mango orchards (Chu et al. 1994; Khan et al. 2002, 2005; Sheikh et al. 2008). Trapping techniques can be utilized to reduce pesticide use by improving timing of sprays as a result of better monitoring of pest populations.
4. Cultural and physical control. Use of cultural and physical techniques (i.e. pruning, bagging, etc.) d pends on costs of control, availability of technical assistance and market purposes. Paderes and Orden (2004) observed that bagging and pruning of 'Manila' in the Philippines is mostly influenced by the availability of technical assistance for growers.

8.17 Pollination in Banana Flowers

Banana flowers grow in a whorl around a stalk. The numerous cultivated banana species (*Musa spp.*) were developed from wild species that bear seedy, inedible bananas. The flowers of cultivated bananas do not require pollination. The wild banana species do need pollination and their seeds need to be spread. About 10–15 months after planting, a long, tapering, oval-shaped bud clad in purple emerges from the tip of the true stem. The purple covering of the bud opens, revealing slim, tubular white flowers. Whorled clusters of these flowers grow in double rows along the stalk. Each cluster is covered by a thick, waxy hood that has a deep red interior and purple exterior. The first 5–15 rows of flowers are female. As the shaft grows longer, sterile female and male flowers appear. A day after the flowers open, the

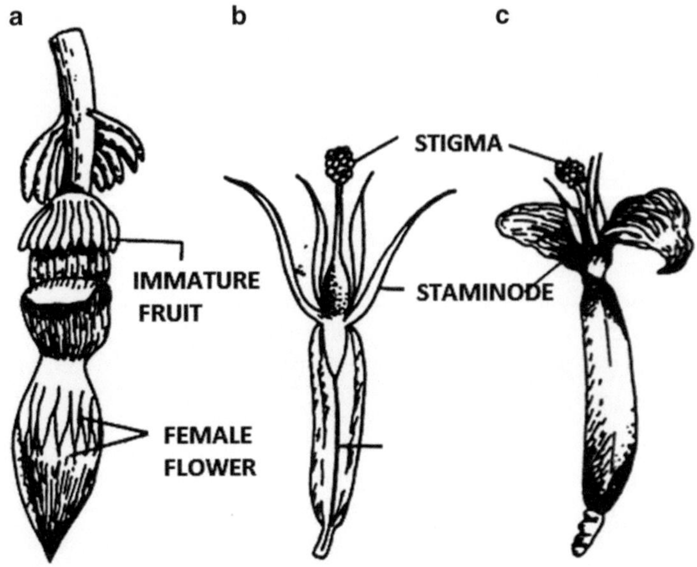

Fig. 8.2 Flower structure of Banana. (**a**) Banana inflorescence, (**b**) Vertical section of Banana flower, (**c**) Banana flower with developing fruit

male flowers fall off. Female flowers produce the bananas. The first female flowers are rich in nectar and have ovaries in the shape of tiny bananas. In cultivated species of bananas the flowers develop fruit without pollination. Cultivated bananas are also seedless (Fig. 8.2).

Banana flowers are visited by a variety of insects but characterstics of the inflorescence suggest that flowers are bat pollinated such as nocturnal opening of flowers, characteristic odour, strong, often pendent inflorescences, accessible nectar, dull flower colour, and flowers exposed freely below the foliage (Simmonds 1962; Nur 1976; Liu et al. 2002). A number of flying foxes such as *Pteropus* has been reported to forage on *Musa* flowers (Nur 1976; Fujita and Tuttle 1991; Liu et al. 2002). The hair on the heads of the flying foxes is modified with hooks which can entrap pollen. The majority of these flying foxes feed during the night within a radius of 30 km from and thus regarded as long distance pollinators (Eby 1995). Honeybees and birds are also regarded as pollinators of *Musa* in different parts of the world (Ortiz and Crouch 1997). The sunbird *Arachnothera longirostris* (family Nectariniidae) pollinates *Musa itinerans* in southwestern China (Liu et al. 2002). The maximum number of fruit that develop in a bunch depends upon climatic conditions occurring during the very early formation of the flowers at the time when the last 3–4 leaves are developing (Simmonds 1959).

8.18 Pests of Bananas (*Musa* spp., Musaceae)

India is the world's largest producer of bananas. Besides, bananas are also widely grown in South and Southeast Asia and the Pacific Islands. Of several pests, the banana weevil is one of the most serious pests of bananas throughout Asia, but the banana leaf roller and the scab moth are also common pests in Southeast Asia. Even though the direct damage caused by banana aphid is minimal, it is a serious pest because of its ability to vector banana bunchy top virus disease.

8.18.1 Banana Weevil Borer

Cosmopolites sordidus (Germar) (Coleoptera: Curculionidae). It is a serious pest in all banana growing areas of the world (Waterhouse and Sands 2001). Its hosts are *Musa* spp. and *Ensete* spp. (Banana, plantain, and Manila hemp are severely affected.

Life Cycle Nocturnal adults live up to 2 years and lay up to 100 eggs per year. The eggs are white, ovoid, 2 mm long and are laid singly on the banana corm or pseudostem in small cavities gnawed out by the female near the ground surface. The egg stage lasts about 8 days, and larvae reach maturity in about 20 days. Larvae tunnel deep into the corm, return near the corm surface, and pupate in an oval chamber. The life cycle varies from 1 to 6 months depending on the temperature. Dispersal is mostly through infested plant material (Gold et al. 2002).

Damage Most damage is caused by larvae tunneling in the corm. Severe damage results in premature death of leaves, development of poor root systems, and topplingand breaking of the plants. Larval tunneling also may provide entry points for fungi and bacteria that cause rott ing (Smith 1977).

Natural Enemies The histerid beetle, *Plaesius javanus*, is one of the important natural enemy of banana weevil and is distributed throughout the tropical world. The other natural enemies such as histerid beetle, *Plaesius laevigatus*, the staphylinids *Belonuchus ferrugatus* and *Priochirus unicolor*, a cucujid beetle, *Cathartus* sp., and the Similarly, a predatory hydrophilid beetle, *Dactylosternum hydrophiloides*, from Malaysia, a histerid, *Hololepta quadridentata*, from Trinidad, and the ant, *Tetramorium guineense*, in Cuba, are known to attack larvae of *C. sordidus* (Waterhouse 1998; Waterhouse and Sands 2001). Braithwaite (1958) reported a blue planarian worm, *Caenoplana coerulea*, attacking *C. sordidus* in Australia.

8.18.2 Pseudostem Borer

Odoiporus longicollis (Olivier) Coleoptera: Curculionidae. It is distributed in India, Nepal, Myanmar, Sri Lanka, Thailand, Indonesia, and the Philippines.

Hosts Banana and Manila hemp.

Life Cycle Adults are slightly larger than those of the banana weevil. They live 6–24 months. Eggs are laid singly at the rate of one to six per day in the leaf sheaths made by the female rostrum. The egg stage lasts from 3 to 12 days. There are four or five instars and the larval stage lasts 2–10 weeks. Prepupal and pupal stages last 3–9 and 3–20 days, respectively.

Damage The larvae bore through the inner leaf sheaths. Larval tunneling results in pseudostem rott ing and in some cases the internal shoot is killed. Leaves of infested plants wither and fruits are undersized.

Natural Enemies No information is available.

8.18.3 Banana Aphid

Pentalonia nigronervosa Coquerel Hemiptera: Aphididae. This aphid occurs in almost all countries where banana is grown.

Its host are *Musa* spp. Foottit et al. (2010) have restricted *P. nigronervosa* to banana feeding populations and restored the name *Pentalonia caladii* van der Goot to the populations that feed on Zingiberaceae and Araceae.

Life Cycle Apterous and alate forms coexist. This aphid reproduces parthenogenetically throughout the year. Males have not been observed and reproduction is viviparous. There are four instars. Adults are about 1.5 mm in length and vary in color from reddish to dark brown. This pest completes 21–26 generations per year.

Damage The preferred sites for feeding are within the whorl of the growing shoot, base of the leaves of young plants, and immature fruits. This aphid produces copious amounts of honeydew which att racts ants. This aphid transmits bunchy top virus which makes it a serious pest. Direct damage by the aphid is minimal. Bunchy top virus disease is established in Asia, some parts of Africa, the Mariana Islands, and Hawaii.

Natural Enemies Several parasitoids and generalist predators have been recorded for this aphid. Some of the parasitoids recorded are *Aphidius colemani*, *Aphidius* sp., *Lysiphlebus testaceipes,* and *Pseudendaphis maculans*. The predators recorded were *Cryptogonus orbiculus nigripennis*, *Micraspis discolor*, *Menochilus sexmaculatus*, *Coccinella septempunctata brucki,* and *Allograpta nasuta*. The fungi *Paecilomyces farinosus* and *Cephalosporium crassum* were also reported to infect this aphid (Waterhouse and Norris 1987).

8.18.4 Banana Scab Moth

Nacoleia (*Lamprosema*) *octasema* (Meyrick) Lepidoptera Pyralidae. It is distributed throughout Northeast Australia, Solomon Islands, Fij i, New Caledonia, Tonga, Samoa, Vanuatu, American Samoa, Wallis and Futuna Islands, New Guinea, Indonesia, and Malaysia (Paine 1964; Gold et al. 2002).

Hosts *Musa* spp., *Heliconia* spp., *Nypa fruticans*, and *Pandanus* spp.

Life Cycle Eggs are oval, scalelike, laid either singly or in clusters, hatch in 4–6 days. The life cycle is completed in about 28 days.

Damage Larvae feed on the surface of the young developing banana fruits causing scarring and lesions that develop into black scabs. Severe damage may result in distorted to hollow fruits. Banana is the preferred host, but *Heliconia*, *Nipa*, and *Pandanus* serve as alternate hosts.

Natural Enemies The parasitoids *Agathis* sp., *Apanteles* sp., *Macrocentrus* nr. *trimaculatus*, *Meteorus octasemae*, *Meteorus trichogrammae*, and *Chelonus striatigenas* were reported from Flores, and the latt er was introduced from Flores to Java in 1959–1960. An egg parasitoid, *Trichogramma* sp., was recorded from Malaysia. About ten different parasitoids were recorded from the South Pacifi c Islands, but none were effective. The ant *Tetramorium bicarinatum* was observed att acking larvae in North Queensland (Waterhouse and Norris 1987).

8.18.5 Banana Skipper, Banana Leaf Roller

Erionota thrax (Linnaeus) Lepidoptera Hesperidae. It is distributed through Southeast Asia, Guam, Saipan, Hawaii, Papua New Guinea, Northeastern India, Andaman Islands, and Mauritius.

Hosts Bananas, *Musa textiles, Calamus trachycoleus, Caryota* sp., and nipa

Life Cycle The female butterfly lays eggs singly or in groups on the lower surface of the leaves. Eggs hatch in 5–8 days; fi rst instars are pale green. Larvae move to the outer margin of the leaf where they start feeding and begin to roll the leaf. As the larvae grow, they cut and roll the leaf along the midrib. There are five instars, and all but the fi rst instar are covered with white waxy powder. The larval stage lasts 23–30 days. Pupation occurs in the leaf roll and lasts from 8 to 12 days

Damage Heavy infestations may result in severe defoliation leaving only the leaf midrib with numerous rolls.

Natural Enemies Egg parasitoids, *Agiommatus* sp. nr. *sumatraensis*, *Anastatus* sp., *Leurocerus hongkongensis*, *Ooencyrtus erionotae*, and *Ooencyrtus papilionis*; larval parasitoids, *Cotesia* (*Apanteles*) *erionotae*, *Bessa remota*, *Echthromorpha*

agrestoria, *Pediobius* sp., *Scenocharops* sp., and *Sympiesis* sp.; and pupal parasitoids, *Brachymeria albotibialis*, *Brachymeria euploeae*, *Brachymeria lasus*, *Brachymeria marginata*, *Brachymeria* sp. nr. *marginicollis*, *Xanthopimpla regina*, and *Pediobius* sp., are eff ective in reducing populations of *E. thrax* (Waterhouse and Norris 1989; Gold et al. 2002). A larval parasitoid, *Elasmus philippinensis*, was reported on *Erionota torus* in Taiwan.

8.18.6 Rose Beetles, Chafer Beetles

Adoretus compressus (Weber), *Adoretus sinicus* Burmeister Coleoptera: Scarabaeidae. It is distributed in Southeast Asia.

Hosts Polyphagous: banana, rose, grapes, okra, cott on, sweet potato, rambutan, coffee, cocoa, tea, oil palm, and others

Life Cycle Adults (10–12 mm long) lay eggs in the soil and developing grubs feed on decaying organic material. There are fi ve instars, and pupation takes place in the soil.

Damage The adults feed at night and can cause extensive damage by shredding margins of leaves. Because the beetles are nocturnal and hide during the day, farmers are oft en puzzled as to what is causing the damage.

Natural Enemies The parasitoids *Micromeriella* (*Campsomeris*) *marginella modesta* and *Tiphia segregata* were introduced to Hawaii and Guam to control this pest, but they were not effective.

8.18.7 Glasshouse Thrips, Black Tea Thrips

Heliothrips haemorrhoidalis (Bouché) Thysanoptera: Thripidae. It has a Cosmopolitan distribution.

Hosts Mango, banana, taro, rose, cocoa.

Life Cycle Adults are 1.5–2.0 mm long and dark brown with greenish white legs and wings. Nymphs are greenish white. Reproduction is asexual, and males are not known to occur. Each female lays about 25 eggs. Immature stages oft en retain a globule of feces at the distal end of the abdomen. The life cycle is completed in 20–30 days.

Damage The species is polyphagous and feeds mostly on the lower surface of leaves causing bleached patches covered with black spots (feces). Heavily infested plants become weak with discolored leaves.

Natural Enemies Heavy parasitism by *Thripobius javae* (=*Thripobius semiluteus*) and predation by the mite *Bdella distincta* on the larvae of *H. haemorrhoidalis* have been reported from Guam (Muniappan et al. 2000). The parasitoids *Goetheana*

shakespearei (in Indonesia) and *Megaphragma mymaripenne* (in Hawaii) were reported from these thrips. *Thripobius javae* was imported from Italy into New Zealand and became established (Jamieson et al. 2008). Reported predators of these thrips include *Franklinothrips vespiformis* in California and Trinidad, *Franklinothrips tenuicornis* in Trinidad, and *Franklinothrips megalops* and *Chrysoperla carnea* in Israel (Watson et al. 1995).

8.18.8 Banana Lace-bug

Stephanitis typica Distant Hemiptera: Tingidae. It is distributed in India, Palau, Thailand, Malaysia, Indonesia, Vietnam, the Philippines, Bangladesh, Myanmar, China, Hong Kong, Japan, Korea, Nepal, Sri Lanka, Maldives, Taiwan, and Papua New Guinea.

Hosts Banana, coconut, cardamom, turmeric, manila hemp, Alpinia.

Life Cycle Eggs are embedded in the undersurface of leaves. Nymphs are gregarious and feed underneath leaves. There are fi ve nymphal instars. Development from egg to adult takes about 33 days (Kalshoven 1981). Adults (4 mm in length) are black and remain mostly on the lower surface of the leaves.

Damage Feeding by this polyphagous species causes yellowish or whitish spots on the leaves.

Natural Enemies A predatory bug, *Stethoconus praefectus*, has been reported to feed on this bug. Other generalist predators almost certainly prey on the nymphs and adults.

Other major pests are:

Coconut scale, *Aspidiotus destructor*, Spiraling whitefly, *Aleurodicus dispersus* and Nettle caterpillar, *Parasa philepida* (*lepida*)

8.19 Papaya

Carica papaya L., family Caricaceae

8.19.1 Pollination of Papaya

The papaya is a dioecious or hermaphrodite herbaceous plant. The papaya plant has separate male (pollen-bearing) and female (fruit-bearing) flowers, along with some flowers that are both male and female (hermaphroditic). The floral biology and fruit

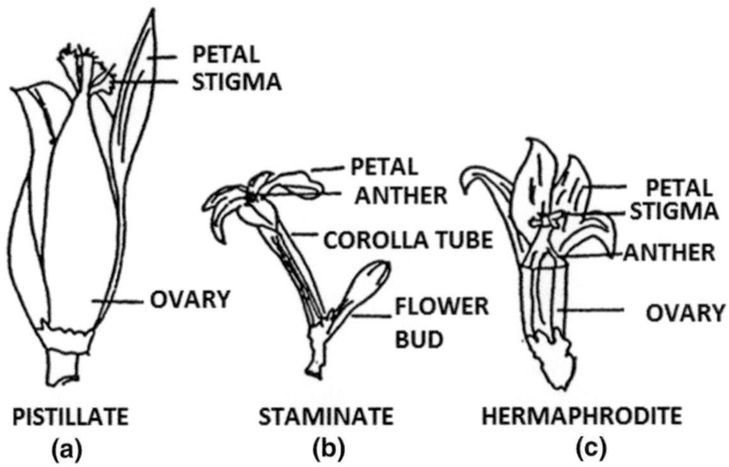

Fig. 8.3 Papaya flower. (**a**) Vertical section of male flower, (**b**) Vertical section of female flower, (**c**) Vertical section of hermaphrodite flower (Nicholls and Holland 1940)

development of papaya is quite variable (Ronse Decraene and Smets 1999). In general, the female flowers account for the majority of saleable fruit while hermaphroditic flowers can self-pollinate to produce smaller fruits that ripen less readily, but can be cooked and eaten. Cross-pollination of these female flowers produces higher fruit set of better quality. Only male flowers, as a rule, actually offer nectar to pollinators. Female flowers have no nectar, and pollinators only visit them because they are attracted to stands of papaya trees by the cues advertised to pollinators by male flowers. This has two important implications: the first is that papaya requires cross pollination by animal pollinators to set good, saleable fruit. The second is that if farmers remove male trees, knowing they are not particularly productive, there will be a shortage of pollen to cross fertilize the female flowers, and the papaya trees will be less attractive to pollinators (Fig. 8.3).

8.19.2 Papaya Pollinators

Papaya flowers produce a sweet fragrance after the sun sets in the evening. The flowers are white which stand out in the dark, suggesting that the flowers are attractive to moths that fly at night. A number of insect visitors are attracted to papaya flowers which include: Bluebottle flies, Diptera: Calliphoridae. Fruit Flies, Honeybees, Hymenoptera: Apidae: *Apis mellifera.,* Skipper butterfly, Lepidoptera:Hesperiidae*:* *Ceoliades* sp., Hawkmoth, Lepidoptera: Sphingidae: *Hippotion celerio.* Hawkmoth, Lepidoptera: Sphingidae: *Herse convolvuli.* Hawkmoth, Lepidoptera: Sphingidae: *Macroglossum trochilus.* Hawkmoth, Lepidoptera: Sphingidae: *Daphnis nerii.* Hawkmoth, Lepidoptera: Sphingidae:

Nephele comma. Moth, Lepidoptera: Noctuidae: *Sphingomorpha chlorea.* Bayless (1931) reported that honeybees visit staminate and hermaphrodite flowers for pollen and nectar from the pistillate and hermaphrodite flowers. He further stated that the corolla tube of the staminate flower is too narrow to permit the proboscis of honeybee to reach to the nectar secreted at the base of the corolla. However, humming birds and sphinx moths can reach this nectar (Brooks 1936; Stambaugh 1960; Traub et al. 1942). In general, honeybees were most active around papaya flowers (Malan 1964)

Parthenocarpic fruit set in C. *papaya* has often been reported from various geographical locations of tropical and subtropical distribution such as South Africa (Hofmeyr 1938; Allan et al. 1987), the United States of America (Traub et al. 1942; McGregor 1976), Israel (Cohen et al. 1989), the Philippines (Ordonyo 1959) and Australia (Prest 1955). Seedlessness generally results in underdeveloped fruit which in most instances reach walnut size, tum yellow, shrivel and abscise (Traub et al. 1942; Prest 1955). Parthenocarpic fruit ripens rarely (Agnew 1954; Ordonyo 1959) and the fruit is of small size, thinly fleshed and has an oblong pointed appearance (Foster 1943). Seedlessness has also been associated with poor flavour. Pollination with non-viable pollen during winter can initiate parthenocarpic fruit set, which indicates that pollination has a stimulating effect on fruit set without the necessity for fertilization (Cohen et al. 1989).

Prest (1955) suggested that C. *papaya* is wind-pollinated, however, various other observations suggest (at least circumstantially) the involvement of insects. Scattered observations of floral characteristics and studies of potential pollinators suggest entomophilic pollen transfer for the following reasons: the timing of anthesis, the emission of scent, the production of nectar of staminate flowers and the presence of brightly coloured petals (Baker 1976). Records of pollinating insects are diverse but remain inconclusive, except for that of Baker (1976). The possibility of insect-mediated pollination has been suggested by various authors based on: (a) observed insect activity on solely staminate flowers (b) insect presence within the orchards and (c) floral characteristics. The following examples indicate the nature of some of these studies:

Some workers condidered wind as the primary pollinating agent (Prest 1957, Agnew 1941), whereas Stambaugh (1960) found sphinx moths, Brooks (1936) and Allan (1963) honey bees and Traub et al. (1942) hummingbird as the main pollinating agents.

"Insects were rarely seen visiting female flowers, but nevertheless one must still conclude that they affect pollination" (Allan 1963). Garrett (1995) observed honeybees visiting staminate flowers, recommending these insects as pollination vectors; observations at dusk confirmed that…each male tree was being visited by about three Skipper butterflies, *Perichares philetesphiletes* (Gmelin). "These butterflies also visited hermaphrodite flowers. Their numbers and behaviour indicate that they would be effective pollinators if the need arose" (Free 1975). "But even though moths (Sphingidae) were not observed on female flowers …these are probably the pollinators of the species" (Bawa 1980). Traub et al. (1942) as cited by Baker (1976) based their conclusions of insect visits exclusively on observations of staminate flowers. In general there is a lack of rigour in these studies. Overall, observations of the activity of insects concentrate on staminate flowers and only on rare occasions have insects been observed visiting pistillate flowers (Allan 1963; Baker 1976; Caron 1985). Baker (1976) noted the presence of mosquitoes and midges on the stigmas of pistillate flowers. The mosquitoes (Culicidae: Toxorhynchites) were also found inside the corolla tube of staminate flowers, although the frequency of this observation was not recorded. The same author documented the involvement of hawkmoths (Sphingidae) in the transfer of papaya pollen, suggesting a species of *'Hyles'* visiting male and female flowers at dusk.

This observation was made in Costa Rica within the area of origin of C. *papaya.* Moths were present over a half hour period at dusk, and collected nectar whilst in hovering flight, visiting staminate flowers and making short 'mistake' visits to the pistillate papaw flowers. Baker observed ten such 'mistake' visits over a 25 min period but did not indicate the number of sphingids involved. Sphingids also have been mentioned as the pollinating vectors of C. *papaya* in the south east of the United States of America (Traub et al. 1942). However, the authors based their observations only on staminate flowers. Based on Traub's et at. (1942) conclusion, Stambaugh (1960) later issued the generalization that sphingidae are the sole the pollinating agents of papaya. Allan (1963) examined the pollination of papaws in South Africa and concluded that the size of the potential pollinators had to be larger than 16 mm square mesh, since this prevented female trees from setting fruit. He suggested that honey bees where involved, despite the fact that bees were only observed foraging on staminate flowers. So called 'mistake' pollination or pollination by 'deceit' has often been resorted to in attempts to explain papaya pollination (Baker 1976). A number of such 'mistake' plant-pollinator interactions have been documented, including various bee species (Bawa 1977) and sphingid moths (Baker 1976) amongst others. Insects, in these cases, are supposedly misled into visiting pistillate flowers, either by similar scent emission and petal colouring or misconception of intersexual flower morphologies. No obvious reward is offered to the pollinator (Bawa 1977, 1980; Bawa and Bullock 1981). According to Bawa (1980), a pollination mechanism based on mimicry of floral parts of non-reward offering in either dioecious or papaws is reduced during the season in to the extent that some instances it does not occur at data are available from Israel (Cohen et al 1989), South Africa (Malan 1964; Allan 1976; Allan et al. 1987), the United States of America (Stambaugh 1960; Agnew 1968). Dioecious papayas tolerant to subtropical conditions than bisexual clones, which are of purely tropical distribution.

Generally, plants have pollinators belonging to more than one species and sometimes more than one order, and the effectiveness of pollination varies accordingly (Bertin 1989). However, in the moist tropics, where *C. originates*, the large number of animal species is coupled with greater specialization in plant-pollinator relationships (Heinrich and Raven 1972). Generally insects visit a more diverse range of plants than plants receive pollinators (Heinrich and Raven 1972). In summary the literature documents a confusing and conflicting record of insect pollination in papaw.

8.19.3 Pollination Requirements

Pollen transfer is must from the staminate flowers to the pistillate for development of seeded fruits. Some varieties are parthenocarpic depending upon pollinating agents. Cheema and Dani (1929) and Traub et al. (1942) showed that flowers bagged to exclude pollen set fruit, but quality was reduced.

8.19.4 Pollination Recommendations and Practices

There is no recommendation for the use of pollinating agents for papaya fruit production, except observations by Malan (1964) and Allan (1963) who advocated the use of honeybees for quality fruit production. The placement of bee hives around papaya have been reported to result in maximum quality fruit. The Best practices for papaya pollination management include (1) maintenance of male trees on the farms, (2) protection and encouragement of alternative nectar sources for pollinators, (3) planting of larval food plants in hedgerows, and (4) protection of trees in surrounding bush and woodland areas to preserve the indigenous biodiversity of native plants and hawkmoths.

8.20 Pests of Papaya (*Carica papaya* L., Caricaceae)

Papaya is of tropical American origin and is grown throughout the tropics. Papaya is cultivated for the fresh fruit market, green fruits that are used in cooking (especially in Asia), and papain collection. Fruit fl ies are constraints in export trade and require disinfestation treatment. Papaya mealybug originated in Mexico; it has spread progressively to the Caribbean, tropical South America, the Pacific, Asia, and, recently, Africa, causing devastation of papaya cultivation. However, implementation of a classical biological control program has eff ectively controlled this pest in countries where it invaded. Aphids, *Aphis gossypii* and *Myzus persicae*, are important as vectors of papaya ring spot virus.

Papaya Mealybug *Paracoccus marginatus* Williams and Granara de Willink Hemiptera : Pseudococcidae. It is distributed in Caribbean, Central and South America, Florida, Guam, Palau, Tinian, Hawaii, Bangladesh, Thailand, Malaysia, India, Indonesia, Cambodia, Philippines, Pakistan, Oman, Réunion Island, Ghana, Benin, Nigeria. This pest has a wide host range of over 60 species of plants including economically important ones such as *Annona squamosa, Carica papaya, Hibiscus rosasinensis, Ipomoea* sp., *Manihot esculenta*, *Plumeria* sp., and *Solanum melongena*. Adult females are yellow and covered by a white waxy secretion. In general, populations of mealybugs build up in the dry season and decline in the rainy season. Heavy infestations kill papaya trees if no control measures are implemented.

Natural Enemies Five species of parasitoids, *Apoanagyrus californicus*, *Anagyrus loecki*, *Acerophagus papayae*, *Pseudaphycus* sp., and *Pseudleptomastix mexicana,* were collected from *P. marginatus* in Mexico. Of these, *A. loecki, A. papayae*, and *P. Mexicana* were introduced to various countries for biological control of *P. marginatus.* The coccinellid beetle, *Cryptolaemus montrouzieri*, is a common predator on this mealybug.

8.20.1 Asian Papaya Fly

Bactrocera papayae Drew and Hancock, Diptera: Tephritidae. It is distributed in Thailand, Malaysia, Singapore, Indonesia, Papua New Guinea

Hosts Papaya, citrus, banana, guava, *Syzygium*, apple, mango, carambola, cashew, all other edible fruits and fleshy vegetables

Life Cycle This is a polyphagous pest that can lay eggs in green papaya, young banana, and citrus. Females have a long ovipositor that allows them to penetrate beyond the sap layer of green fruit. The egg stage lasts for 2–3 days, and maggots develop through 3 instars inside the fruit, feeding on the pulp. Mature larvae exit the fruit and pupate in the soil within a puparium. The pupal stage lasts for about 10 days.

Damage Fruit that is att acked by this fruit fly will rot. It is a serious quarantine pest. This fly was found in Cairns, Northern Australia in 1992; it has since been eradicated by using male annihilation and protein bait spraying, at an expense of AUD 35 million.

Natural Enemies Some of the parasitoids recorded in Southeast Asia are *Biosteres persulcatus, Fopius arisanus*, *Diachasmimorpha longicaudata*, *Fopius vandenboschi*, *Psytt alia incisi*, and *Psytt alia* (*Opius*) *fl etcheri*. The Asian weaver ant, *Oecophylla smaragdina,* is known to prey on fruit fly larvae.

Other remarks: Other fruit fl ies that att ack papaya in Asia are *Bactrocera dorsalis*, *Bactrocera cucurbitae, Bactrocera melanotus*, and *Bactrocera trilineola.*

8.20.2 Coconut Scale

Aspidiotus destructor Signoret Hemiptera Diaspididae. It is distributed throughout the tropics

Hosts This scale is a pest of many tropical crops including coconut, mango, banana, avocado, breadfruit, cassava, cott on, guava, and papaya.

Life Cycle Adult females are circular, sessile, and covered by a translucent, flexible scale which is fl at, whitish, and waxy. Males have a pair of wings, well-developed eyes, legs, and antennae, but no mouthparts. Eggs are laid under the scale at an average of 90 per female. Eggs hatch into nymphs, known as crawlers, in about 8 days. These crawl and sett le on a suitable spot, insert their mouthparts, and begin feeding. They secrete wax which covers them. They develop and molt within this waxy shield. The period from egg to adult is about 32 days.

Damage In severe infestations, and in the absence of natural enemies, this scale can kill coconut palms. The removal of sap by the sucking result in leaves turning yellow and eventually drying up.

Natural Enemies Classical biological control programs have been successfully adopted in several countries to control the coconut scale. Several hymenopterous parasitoids and coccinellid beetles are known to att ack this scale insect. Natural enemies utilized include *Aleurodothrips fasciapennis*, *Aphytis chrysomphali*, *Aphytis lingnanensis*, *Bardylis parvipennis*, *Chilocorus cacti, Chilocorus malasiae, Chilocorus nigritus, Chilocorus politus*, *Comperiella bifasciata*, *Comperiella unifasciata*, *Cryptogonus orbiculus nigripennis, Cryptognatha gemellata*, *Cryptognatha nodiceps*, *Cryptolaemus montrouzieri*, *Delphastus* spp., *Encarsia citrina*, *Pentilia insidiosa*, *Pseudoazya trinitalis*, *Pseudoscymnus anomalus*, *Rhyzobius satelles*, *Scymnus* sp., *Spaniopterus crucifer*, *Telsimia nitida*, *Zagloba aeneipennis*, and *Zaomma microphagum*.

Other major pests are: Cotton aphid or Melon aphid, *Aphis gossypii* (See Pests of Eggplant), Green peach aphid, *Myzus persicae* (See Pests of Pepper) Spiraling whitefl y, *Aleurodicus dispersus* (See Pests of Guava) Spider mites, *Tetranychus* spp. (See Pests of Eggplant) Broad mite or Yellow tea mite, *Polyphagotarsonemus latus* (See Pests of Pepper)

Other pests of regional importance are:

Palm scale, *Aspidiotus orientalis* Newstead, Oriental scale, *Aonidiella orientalis* (Newstead) (Hem.: Diaspididae), Red and black fl at mite, *Brevipalpus phoenicis* (Geij skes) (Acari: Tenuipalpidae), and *Thrips parvispinus* Karny (Thysanoptera: Thripidae).

8.21 Pests of Pineapple (*Ananas comosus* Merr., Bromeliaceae)

The pineapple is indigenous to South America but is widely grown on a commercial scale in Asia. Pineapple ranks third aft er banana and mango in fruit crops harvested in the tropics and subtropics (Petty et al. 2002). It is marketed as fresh fruit, canned, and made into juice, jam, and other products. Thailand, India, Indonesia, and the Philippines are the major producers of pineapple in Asia. The pineapple mealybug is by far the most important pest of this fruit crop. Petty et al. (2002) have reviewed pests of pineapple on a global scale.

8.21.1 *Pineapple Mealybug*

It has widespread distribution wherever pineapple is grown.

Hosts The preferred host is pineapple, but this pest also occurs on soybean, coff ee, rice, and other crops.

Life Cycle The adult female is pinkish with a thick waxy covering and short, conical, waxy projections. It is parthenogenetic and ovoviviparous. About 250 crawlers are produced by a female. The female has three nymphal instars and the male has two. Both have prepupal and pupal stages. The adult male lives 1–3 days and the female 17–49 days.

Damage This mealybug transmits a virus that causes stunting, reddening, and wilting known as pineapple mealybug wilt.

Natural Enemies The parasitoid, *Anagyrus ananatis*, eff ectively controls this mealybug in the absence of ants (Gonzalez-Hernandez 1995). It was introduced to Hawaii from Brazil in 1934–1935 (Waterhouse 1998). Other parasitoids introduced to Hawaii are *Euryrhopalus propinquus*, *Hambletonia pseudococcina*, *Leptomastix dactylopii*, *Pseudaphycus dysmicocci*, and *Pseudaphycus* sp. The coccinellids, *Scymnus apicifl avus* and *Sticholotis quatrosignata*, and seven species of unidentifi ed hymenopteran parasitoids have been recorded in Malaysia. *Cryptolaemus montrouzieri* and *Rhyzobius ventralis* have been reported from the Pacific.

8.21.2 *Pineapple Scale*

Diaspis bromeliae (Kerner) Hemiptera: Diaspididae. It Occurs wherever pineapple is grown

Hosts Agave, pineapple, bromeliads

Life Cycle Eggs are laid within the scale cover, and the egg stage lasts for 7 days. There are three larval stages and a pupal stage. It takes about 2 months to complete the life cycle.

Damage Pineapple scale mostly infests lower leaves. Ratoon crops are more subjected to this scale att ack as these plants are denser. A yellow spot develops at the place of feeding of the scale due to the injection of toxic saliva. In heavy infestations, fruit cracking may occur.

Natural Enemies The parasitoids, *Ablerus elegantulus*, *Encarsia* sp., and *Coccidencyrtus ochraceipes* (in South Africa), and *Aphytis* sp. and *Aspidiotiphagus* sp. (in Australia) have been recorded. The ladybird beetle, *Rhyzobius lophanthae*, is an important predator of this scale (Petty et al. 2002).

8.21.3 Pineapple Fruit Mite

Steneotarsonemus ananas (Tyron) Acari: Tarsonemidae. It occurs wherever pineapple is grown

Host Pineapple

Life Cycle Female mites lay oval and opaque white eggs singly. Larvae are white, and adults are creamy brown. The life cycle is completed in 1–2 weeks.

Damage These mites feed on basal leaf tissue, flower bracts, and sepals, causing light brown spots (Paull and Duarte 2011).

Natural Enemies A predatory thrips, *Podothrips lucasseni,* has been recorded.

8.21.4 Pineapple Flat Mite, Pineapple Red Mite

Dolichotetranychus floridanus (Banks) Acari: Tenuipalpidae. It occurs wherever pineapple is grown. Pineapple, cardamom, grass, bamboo

Life Cycle Eggs are light orange, larvae are amber, nymphs are orange, and adults are red. The life cycle is completed in 10 days.

Damage These mites cause severe damage when fruits mature under drought conditions. They feed on the basal tissue of leaves, especially of the crown (Paull and Duarte 2011).

Natural Enemies *Amblyseius benjamini* is a common predatory mite recorded on *D. floridanus.*

Other major pests are:

Oriental fruit fly, *Bactrocera dorsalis,* Onion thrips, *Thrips tabaci*

8.22 Aonla

*8.22.1 Aonla (*Emblica officinalis *Gaertn.)*

Aonla or Indian gooseberry (*Emblica officinalis* Gaertn.) is grown and known in India for last more than 3500 years. Sushruta, the father of ancient medicine (during 1500 BC–1300 BC), mentioned about its effect in 'Ayurveda' in great detail. Owing to its significant medicinal and nutritive value, it finds a prominent place in ancient Indian mythological literatures like Vedas, Askandhpuran, Shivpuran, Padmapuran, Ramayana, Kadambari, Charak Shanghita, Sushrut Shanghita, etc. and considered as Amrit Phal (life giving fruit). The plant and fruit of aonla are regarded sacred by Hindus and have great mythological significance

8.22.2 *Origin*

The aonla tree are of common occurrence in the mixed deciduous dry forests of India, ascending from two sea level (western and eastern ghats, Aravali and Vindhyan hills) to 1300 m asl, from northwest Himalayas (Jammu & Kashmir, Himachal Pradesh, Uttranchal) to eastern Himalayas in Assam, Meghalaya, Mizoram, Manipur and Tripura.

8.22.3 *Pollination*

8.22.3.1 Flowering and Fruiting

As per flowering behaviour of aonla, the male flowers appear first at basal portion of determinate shoot, while female flowers appear at the distal portion. After pollination and fertilization, the developing embryo goes under dormancy during summer months (April–June). During this period, vegetative growth continues. On the onset of monsoon, the fruit development takes place, and fruits mature during autumn (November–December) and plants again go in dormancy during winter (December–February). Thus, as per phenology, the flower bud differentiation in aonla starts in first week of March and flowering takes place from last week of March to mid-April.

8.22.3.2 Inflorescence

Inflorescence is racemose type, flowers minute, unisexual, with short pedicel. Male flowers appear first in cluster; perianth 6, yellowish green or deep pink in colour with valvate aestivation. Androecium consists of 3 stamens, each profusely branched, and filament attachment is basi-fixed or innate type, short, and cohesion of anther is syngenesious. Female flowers have tiny green perianth and number of segments varies from 5 to 7 but commonly six. Ovary hypogynous, carpels 3–4, three chambered placentation axile, two ovules per locule, margin straight to crescent shaped and ovarian chamber shallow to deep

8.22.4 *Pollination in Aonla (***Emblica officinalis***)*

Aonla flowers have been reported to benefit from insct pollination. If the pollination does not occur in the first 3 weeks of flowering more than 70 % of the flowers shed. In northern India, flowering takes place from March to May while as

Southern India, the tree blooms in June–July and again in February–March, the second flowering producing only a small crop. Allemullah and Ram (1990) found that the use of pollinators (honey bees) and pollinizers in aonla orchards is necessary for increasing the fruit yield. Singh et al 1998 found that the maximum fruit-retention was recorded under open pollination followed by sibbing and geitonogamy, whereas bagging resulted in no-fruit retention. Improvement in fruit retention under sibbing and open pollination compared with the geitonogamy might be due to the effect of foreign pollen grains having high pollination stimulus. Poor-fruit retention under geitonogamy and bagging indicated the possibility of self-incompatility in all the aonla cultivars/genotypes. Aonla flowers are good source of nectar for honeybees. Tewari et al. (2010) found *Apis dorsata* foraging on Aonla flowers for nectar.

8.22.5 Insect Pests of Aonla

Aonla was supposed to be free from major insect and disease menace. With lapse of time and intensification of its cultivation, number of insects and diseases have started feeding and infecting at various stages and causing considerable damage (Masarrat et al. 2000; Rani et al. 2006; Parkash 2012). Lal et al. (1996), Haseeb et al. (1990, 2000), and Jhala et al. (2003) reported occurrence of fruit borer, midge and leaf cutting weevil for the first time in northern and western India (Plate 8.2). Bharpoda et al. (2009) studied insect pests and natural enemies at Anand Gujarat, India. They found that Aonla fruit trees mainly attacked by different insect pests belonging to four major insect groups viz., Lepidoptera, Homoptera, Coleoptera and Isoptera. In a similar study, Singh and Singh (2012) studied the arthropod diversity on aonla (*Emblica officinalis* Gaertn) and management of its major sucking insect-pests and found the association of a total of 31 arthropods with aonla plant. Among these, 5 belong to sap feeder, 6 defoliator, 2 each gall former and borer, 12 natural enemies, 3 pollinators and one other group of arthropod. Among sap feeders, defoliators, gall formers and borers, respectively, aphid (*S. bougainvilliae*), leaf binder (*T. ziziphy*), shoot gall maker (*B. stylophora*), bark eating caterpillar (*Indarbela sp.*) as pest and among natural enemies, lady bird beetles were important. Three species of honeybees were observed visiting on aonla plants as pollinator. Out of 9 treatments including untreated check evaluated against aphid (*S. bougainvilliae*), imidacloprid @ 40 g a.i./ha was found most effective, however, it was at par with acephate 350 g a.i./ha, ethion 40 % + cypermethrin 5EC 0.03 % and profenofos 50SL, 0.07 % after 3 and 10 days of application.

Plate 8.2 (**a**) Deudorix larvae damaged the aonla fruits (**b**) Gall maker larva inside the twig gall of aonla (**c**) Damage symptoms produced by bark eating caterpillar on aonla tree trunk (**d**) Aonla leaves infestation by aphids (**e**) Pomegranate fruit damaged by fruit borer (**f**) Adult of fruit borer, Deudorix laying eggs on flower of pomegranate (**g**) Curled pomegranate leaf by thrips infestation (**h**) Infestation of mango mealybugs on pomegranate leaves and twigs

Bibliography

Abbas S (1985) Studies on the mango inflorescence midge, *Erosomyia indica* Grover. Acta Horticult 231:593–596

Abou-Awad BA (1981) Ecological and biological studies on the mango bud mite, *Eriophyes mangiferae* (Sayed), with description of immature stages Eriophyoidea: Eriophyidae. Acarologia 22:145–150

Acounk ED, Agricol AU, Bokonon-Gant HAA (1988) *Rustrococcur invudens* Williams (Hemiptera: Pseudococcidae), a serious exotic pest of fruit trees and other plants West Africa. Bull Entomol Res 78:695–702

Agnew GWJ (1941) Notes on the papaw and its improvement in queensland. Queensland Agric J 56(5):358–373

Agnew GWJ (1954) The papaw. Queensland Agric J 79:77–90

Agnew GWJ (1968) Growing quality papayas in Queensland. Queensland Agric J 94:24–36

Agricola U, Agounkè D, Fischer HU, Moore D (1989) The control of *Rastrococcus invadens* Williams (Hemiptera: Pseudococcidae) in Togo by the introduction of *Gyranusoidea tebygi* Noyes (Hymenoptera: Encyrtidae). Bull Entomol Res 79:671–678

Ahmed MU, Ahmed M, Baluch MA, Naheed R (1981) Sooty mold on mango plants and its relationship with leafhoppers and climatic factors in Karachi-Pakistan during 1978–79. Pak J Sci Ind Res 24:140–144

Ali SM (1961) A note on Nipacoccus vestator Mask. Indian J Ent 23:304–305

Ali M (1980) Control of mango mealy bug, *Drosicha stebbingi* (Gr.) (Homoptera: Coccidae). M.Sc (Hons.), thesis, Department of Agricultural Entomolology, University of Agriculture Faisalabad

Allan P (1963) Pollination of papaws. Farm South Africa 38(11):13–15

Allan P (1976) Out of season production of pawpaws (*Carica papaya* L.) in cool subtropical areas. Acta Horticult 57:97–103

Allan PJ, McChlery M, Biggs D (1987) Enviromental effects on clonal female and male Carica papaya L. plants. Sci Hortic 32:221–232

Allemullah M, Ram S (1990) Causes of low fruit set and heavy fruit drop in Indian gooseberry (*Emblica officinalis* Gaertn. Indian J Hortic 47(3):270–277

Aluja M (1993) *Manejo Integrado de la Mosca de la Fruta*. Editorial Trillas, Mexico Distrito Federal, Mexico

Aluja M (1994) Bionomics and management of *Anastrepha*. Annu Rev Entomol 39:155–178

Aluja M (1996) Future trends in fruit fly management. In: McPheron B, Steck G (eds) Fruit fly pests: a world assessment of their biology and management. St Lucie Press, Delray Beach, pp 309–320

Aluja M (1999) Fruit fly (Diptera: Tephritidae) research in Latin America: myths, realities and dreams. Anais da Sociedade Entomologica do Brasil 28:565–594

Aluja MR, Liedo PF (1986) Perspectives on future integrated management of fruit flies in Mexico. In: Mangel M, Carey JR, Plant RE (eds) Pest control: operations and systems analysis in fruit fly management, vol G11, NATO ASI series. Springer, Berlin, pp 9–42

Aluja M, Mangan RL (2008) Host status determination for pestiferous fruit fly (Diptera: Tephritidae) species: conceptual, methodological and regulatory considerations. Annu Rev Entomol 53:473–502

Aluja M, Norrbom AL (eds) (2000) Fruit flies (Tephritidae): phylogeny and evolution of behavior. CRC Press, Boca Raton

Aluja M, Pinero J (2004) Testing human urine as a low-tech bait for *Anastrepha* spp. (Diptera: Tephritidae) in small guava, mango, sapodilla and grapefruit orchards. Fla Entomol 87:41–50

Aluja M, Cabrera M, Guillen J, Celedonio H, Ayora F (1989) Behavior of *Anastrepha ludens*, *A. obliqua* and *A. serpentina* (Diptera: Tephritidae) on a wild mango tree (*Mangifera indica*) harbouring three McPhail traps. Insect Sci Appl 10:309–318

Aluja M, Celedonio-Hurtado H, Liedo P, Cabrera M, Castillo F, Guillen JA, Rios E (1996) Seasonal population fluctuations and ecological implications for management of *Anastrepha* fruit flies

(Diptera: Tachinidae) in commercial mango orchards in southern Mexico. J Econ Entomol 89:654–667

Aluja M, Pinero J, Jacome I, Diaz-Fleischer F, Sivinski J (2000) Behavior of flies in the genus *Anastrepha* (Trypetinae: Toxotrypanini). In: Aluja M, Norrbom A (eds) Fruit flies (Tephritidae): phylogeny and evolution of behavior. CRC Press, Boca Raton, pp 375–406

Aluja M, Sivinski J, Rull J, Hodgson PJ (2005) Behavior and predation of fruit fly larvae (*Anastrepha* spp.) (Diptera: Tephritidae) after exiting fruit in four types of habitats in tropical Veracruz, Mexico. Environ Entomol 34:1507–1516

Aluja M, Sivinski J, Ovruski S, Guillen L, Lopez M, Cancino J, Torres-Anaya A, Gallegos-Chan G, Ruiz L (2009) Colonization and domestication ofseven species of native New World Hymenopterous larval-prepupal and pupal fruit fly (Diptera: Tephritidae) parasitoids. Biocontrol Sci Technol 19:49–79

Anderson DL, Sedgley H, Short JRT, Allwood AJ (1982) Insect pollination of mango in northern Australia. Aust J Agric Res 33:541–548

Andrews K, Poe S (1980) Spider mites of El Salvador, Central America (Acari: Tetranychidae). Fla Entomol 63:502–505

Andrews KL, Quezada JR (1989) Manejo Integrado de Plagas Insectiles en la Agricultura: Estado Actual y Futuro. Escuela Agrícola Panamericana, Honduras, 623 p

Angeles DE (1991) *Mangifera atissima*. In: Coronel RE, Verheij EWM (eds) Plant resources of south-east Asia no.2: edible fruits and nuts. Pudoc-DLO, Wageningen, pp 206–207

Angeles N, Requena J (1966) Moth injury of mangoes in Venezuela. Plant Prot Bull FAO 14:86–87

Angeles N, Dedordy N, Paredes P, Requena J (1971) Citrus blackfly in Venezuela. Agronomia Trop 21:71–75

Anonymous (1984) Major insect pests and diseases of mango and their control. In: The Mango Technical committee. The Philippines recommends for Mango. Philippines Agriculture and Resources Research Foundation, 124 pp

Anonymous (1989) Mango pests and disorders. Information series QI89007. Queensland Department of Primary Industries, Brisbane

Anonymous (1996) Nationally coordinated research project on mango crises. First annual report. Department of Horticulture, University of Agriculture Faisalabad, pp 2–3

Anonymous (2002) Fruit flies eradicated from Nauru. 3pp. Available at: http://www.Pacifly.org/Success_stories/Nauru_eradl.htm. Accessed 10 Sept 2008

Anonymous (2005) Malathion (CAS Number 121-75-5). Toxicology information about insecticides used for eradicating mosquitoes. Available at: http://www.atsdr.cdc.gov/consultations/malathion.html. Accessed 20 July 2007

Anonymous (2006) Importation of fresh mango fruit (*Mangifera indica*) from India into the United States, a qualitative pest risk assessment. Animal and Plant Health Inspection Service, Plant Protection Quarantine, Raleigh

Arias M, Jines A (2004) Manejo Integrado de Moscas de la Fruta en el Litoral Ecuatoriano

Arias M, Jines A, Bustos P, Pluas M, Carrera C (2004a) Enemigos Naturales de Aulacaspis Tubercularis (Homoptera: Diaspididae) en Mango. Instituto Nacional Autonomo de Investigaciones Agropecuarias (INIAP), Boliche, Guayas

Arias M, Jines A, Maldonado E, Pluas M (2004b) *Cybocephalus nipponicu* Endrody-Younga (Coleoptera: Nitidulidae) un Predator Exotico Para el Control del Aulacaspis Tubercularis en Mango de Exportacion. Instituto Nacional Autonomo de Investigaciones Agropecuarias (INIAP), Boliche, Guayas

Arif MI, Rafiq M, Ghaffar A (2002) Host plants of cotton mealy bug (*Phenacoccus solenopsis*): a new menace to cotton agroecosystem of Punjab, Pakistan. Int J Agric Biol online:1814–9596. http://www.fspublishers.org

Armstrong JA (1979) Biotic pollination mechanism in the Australian flora-a review. NZ J Bot 17:467–508

Ashfaq M, Khan RA, Khan MA, Rasheed F, Hafeez S (2005) Complete control of mango mealy bug using funnel type slippery trap. Pak Entomol 27(1):45–48

Askari M, Radjabi G (2003) Study on the biology and population fluctuations of mango midge gall *Procontarinia mattiana* (Diptera: Cecidomyiidae) in Homozogan province. Appl Entomol Phytopathol 70:121–135

Assis E, Rabelo F (2005) Pragas da Mangueira, Monitoramento, Nivel de Acao e Controle. Empresa Brasileira de Pesquisa Agropecuaria, Embrapa Semi-Arido, Ministerio da Agricultura, Pecuariae Abastecimiento, Petrolina

Atwal AS (1963) Insect pests of mango and their control. Punjab Hortic J 3(2–1):235–58

Atwal AS (1976) Agriculture pests of India and South East Asia. Kalyani Publishers, Ludhiana, pp 224–227

Atwal AS, Bhatti DS, Sandhu GS (1969) Some observations on the ecology and control of mango mealy bug *D. mangiferae* Green (Coccidae: Hemiptera). J Res (Punjab Agricultural University, Ludhiana) 6(1):107–114

Austin AD (1984) New species of Platygastridae (Hymenoptera) from India which parasitise pests of mango, particularly *Procontarinia* spp. (Diptera: Cecidomyiidae). Bull Entomol Res 74:549–557

Australian Government (2004) Mangoes from India, draft revised import policy. Department of Agriculture, Fisheries and Forestry, Canberra

Azizur Rahman SK, Singh G (2004) Population dynamics of mango hopper (*Amritodus atkinsoni*) on 'Langra' mango (*Mangifera indica*) and its relationship with abiotic factors. Indian J Agric Sci 74:566–569

Babu LB, Maheswari TU, Rao NV (2002) Seasonal incidence and biology of themango hopper *Amritodus atkinsoni* Lethier (Homoptera: Cicadellidae). Entomon 27:35–42

Bagle B, Prasad V (1984) Varietal incidence and control of stone weevil *Sternochetus mangiferae* Fabricius. Indian J Entomol 46:389–392

Bajwa GA, Gul H (2000) Some observations on insect species of *Paulownia* species at Pakistan Forest Institute campus. Peshawar Pak J Forestry 50(1–2):71–80

Baker HG (1976) 'Mistake' pollination as a reproductive system with special reference to the Caricaceae. In: Burley J, Styles BT (eds) Tropical trees— variation breeding and conservation. Linnean society symposium, series no. 2. 161–169. Proceedings of an international symposium held at Oxford University, England. Academic Press, London/New York/San Francisco, Apr 1975

Bakshi P, Jasrotia A (2012) Amla: an important fruit of Rainfed areas. http://www.krishisandesh.com/general/amla-fruit-rainfed-areas

Balock JW, Kozuma T (1964) Notes on the biology and economic importance of the mango weevil, *Sternochetus mangiferae* (Fabricius), in Hawaii (Coleoptera: Curculionidae). Proc Hawaiian Entomol Soc 8:353–364

Balock JW, Lopez DF (1969) Trapping for control of the Mexican fruit fly in mango and citrus groves. J Econ Entomol 62:53–56

Barbosa FR (2005) Arthropodes-pragas associados a cultura da mangueira no Brasil e seu controle. In: Menezes E, Barbosa FR (eds) Pragas da Mangueira, Embrapa Semi-arido. Ministerio da Agricultura, Pecuaria e Abastecimiento, Petrolina, pp 1–50

Barbosa FR, de Goncalves ME, Moreira WA, de Alencar JA, de Souza EA, da Silva CSB, Souza A, da Miranda I (2005) Arthropodes-Pragae Predatores (Arthropoda) Associados a Cultura da Manguerira no Vale do Sao Francisco, Nordeste do Brasil. Neotrop Entomol 34:471–474

Barclay HJ (1987) Models for pest control: complementary effects of periodic releases of sterile pests and parasitoids. Theor Popul Biol 32:76–89

Barnes HF (1948) Gall midges of economic importance. Vol: 3: Gall midges of fruit. Crosby Lockwood and Sons, London

Bartlet K (1938) The introduction and colonization in Puerto Rico of *Dasyscapus parvipennis* Gahan, a parasite of thrips. Agric Notes Puerto Rico Exper Station 87:6

Bateman MA (1972) The ecology of fruit flies. Annu Rev Entomol 17:493–518

Batra RC, Sohi AS, Bhalla JS, Sandhu GS (1980) Soil application of dusts/granules and foliar/bark sprays for control of mango mealy-bug *Drosicha mangiferae*. Punjab Hortic J 20(1/2):80–83

Bawa KS (1980) Mimicry of male by female flowers and intreasexual competition for pollinators in Jacaratia dolichaula (D. Smith) Woodson (caricaceae). Evolution 34:467–476

Bayless B (1931) Papayas. Fla State Hortic Soc Proc 44:86–89

Beardsley JW (1961) A new review of the Hawaiian Braconidae (Hymenoptera). Proceed Hawaiian Entomol Soc 27:333–366

Becker S (1958) The production of papainñan agricultural industry for tropical america. Econ Bot 12:62–79

Beers E, Hull L, Jones V (1993) Sampling pest and beneficial arthropods of apples. In: Pedigo L, Buntin D (eds) Handbook of sampling methods for arthropods in agriculture. Preface. CRC Press, Boca Raton, pp 383–416

Berti Filho E, Flechtman C (1986) A model of ethanol trap to collect scolytidae and platypodidae (Insecta, Coleoptera). Instituto de Pesquisas e Estudios Florestais 34:53–56

Bertin RI (1989) Pollination biology. In: Abrahamson WG (ed) Plant–animal interactions. McGraw-Hill, New York, pp 23–83

Bess HA, van den Bosch R, Haramoto FH (1961) Fruit fly parasites and their activities in Hawaii. Proc Hawaiian Entomol Soc 27:367–378

Bhagat KC (2004) Mango mealy bug, *Drosicha mangiferae* (Green) (Margarodidae: Hemiptera) on Ashwagandha – a medicinal plant. Insect Environ 10(1):14

Bharpoda TM, Koshiya DJ, Korat DM (2009) Seasonal occurrence of insect-pests on aonla (*Emblica officinalis* Geartn) and their natural enemies. Karnataka J Agric Sci 22(2):314–318

Bhatia LJ, Gauba HR (1964) Efficiency of different insecticides for the control of *Drosicha mangiferae* Green. Plant Protect Bull 16(1–6):13–15

Bhatia R, Gupta D, Chandel JS, Sharma NK (1995) Relative abundance of insect visitors on flowers of major subtropical fruits in Himachal Pradesh and their effect on fruit set. Indian J Agric Sci 65:907–912

Bhole S, Jadhav V, Dumbre R, Dalvi C (1987) Seasonal incidence and chemical control of mango nursery pests in the Konkan region. J Maharashtra Agric Univ 12:387–388

Bindra OS (1967) Fighting pests of commercial fruits. Indian J Entomol 43(2):148–152

Bindra OS, Sohi BS (1974) A note on control of mango mealybug in Punjab. Indian J Hortic 31(1):102–103

Bindra OS, Varma GC, Sandhu GS (1970) Studies on the relative efficacy of banding materials for the control of mango mealy bug *Drosicha stebbingi* (Green). J Res, Ludhiana 7(4):491–494

Birat RBS (1963) Comparative efficacy of organohosphorous insecticides for the control of *Drosicha mangiferae* Green. Indian J Entomol 25(4):377–399

Birat RBS (1964) Mango's mealy bug menace. Indian Farm 14(1):14–15

Birke A, Aluja M, Greany P, Bigurra E, Perez-Staples D, McDonald R (2006) Long aculeus and behavior of *Anastrepha ludens* renders gibberellic acid ineffective as an agent to reduce 'Ruby Red' grapefruit susceptibility to the attack of this pestiferous fruit fly in commercial citrus orchards. J Econ Entomol 99:1184–1193

Boavida C, Neuenschwander P, Schulthess F (1992) Spatial distribution of *Rastrococcus invadens* Williams (Homoptera: Pseudococcidae) in mango trees. J Appl Entomol 114:381–391

Boavida CP, Neuenschwander P, Herren HR (1995) Experimental assesment of the impact of the introduced parasitoid Gyransoidea tebgi Noyes on the mango mealy bug *Rastrococcus invadens* william by exclusion. Biocontrol 5:99–103

Bondad ND (1985) Mango bagging. Anim Hus Agric J June Issue, 11–14

Bonkon-Ganta AH, Neuenschwander P (1995) Impact of the biological control agent *Gyranusoidea tebygi* Noyes on mango mealy bug *Rastrococcus invadens* Williams in Benin. Biocontrol 5:95–97

Braithwaite BM (1958) Ground spray treatments for control of the banana beetle borer (Cosmopolites sordidus) (Germ.). J Aust Inst Agric Sci 24:27–34

Broadley R, Rigden P, Chay- Prove P, Daniells P (2004) Subtropical banana growers handbook. Queensland Department of Primary Industries, Brisbane, pp 1–206

Brooks JR (1936) The papaya. Fla State Hortic Soc Proc 49:134–136

Burns RE, Harris DL, Moreno DS, Eger JE (2001) Efficacy of spinosad bait sprays to control Mediterranean and Caribbean fruit flies (Diptera: Tephritidae) in commercial citrus in Florida. Fla Entomol 84:672–678

Bustos ME, Enkerlin W, Reyes J, Toledo J (2004) Irradiation of mangoes as a postharvest quarantine treatment for fruit flies (Diptera: Tephritidae). J Econ Entomol 97:286–292

Butani DK (1964) Pests that paster your mango crop. Indian Hortic 9(1):28
Butani DK (1979) Insects and fruits. Periodical Expert Book Agency, Delhi
CAB International (2003) Crop protection compendium. CAB International, Wallingford
Cabrera BJ, Thoms E (2006) Versatility of baits containing noviflumuron for control of structural infestations. Fla Entomol 89:20–31
Cabrera H, Villanueva J, Ortega D (1993) Fruit fly infestation timing on 'Manila' mango fruit at Veracruz, Mexico, Miami, USA (Abstract). IV international mango symposium, p 118
Camargo CA, Odell E, Jiron LF (1996) Interspecific interactions and host preference of *Anastrepha obliqua* and *Ceratitis capitata* (Diptera: Tephritidae), two pests of mango in Central America. Fla Entomol 79:266–268
Carvalho PJ, De Queiroz AC (eds) (2002) A Cultura da Mangueira. CIP-Brasil, Brasilia
Celedonio-Hurtado H, Aluja M, Liedo P (1995) Adult population fl uctuations of *Anastrepha* species (Diptera: Tephritidae) in tropical orchard habitats of Chiapas, Mexico. Environ Entomol 24:861–869
Chandra A, Bhati DPS, Singh KM (1987) Bionomics of mango mealy bug, *Drosicha mangiferae* Green. Bull Entomol 28(2):145–152
Chandra A, Bhati DPS, Singh KM (1989) Egg laying behavior of mango mealy bug, *Drosicha mangiferae* Green. Indian J Entomol 51(1):101–104
Chandra A, Bhat DPS, Singh KM (1991) Mechanical control of mango mealy bug, *Drosicha mangiferae* (Green) by tree trunk banding. Indian J Entomol 53(1):76–79
Chaudhary S, Mazid S (1954) Hand book of plant protection. Department of Agriculture, Assam, p 312
Cheema GS, Dani PG (1929) Seedlessness in papayas. Agric J India 26(3):206–207
Chen WS, Chen SK (1997) Mass-rearing of oriental latrine fly (*Chrysomya megacephala* (Fabricius)). Proceedings of a symposium on enhancing competitiveness of Fruit Industry (I), Taichung DARES. Series 38:189–196 (in Chinese)
Chopra RL (1928) Annual report of the entomologist to the Government of Punjab, Layallpur for the year 1925–26. Report Department of Agriculture Punjab 1:67–125. (fide Srivastava RP (1980) Indian J Entomol 42(1):10–5
Christenson LD, Foote RH (1960) Biology of fruit flies. Annu Rev Entomol 5:171–192
Chu Y, Lee KT, Tseng YH (1994) Occurrence of melon and oriental fruit fly in Republic of Naru. Plant Protect Bull 36:131–140
Chua TH, Wood BJ (1990) Other tropical fruit trees and shrubs. In: Rosen D (ed) Armored scale insects, their biology, natural enemies and control, 4B. Elsevier, New York, pp 543–552
Clarke AR, Allwood A, Chinajariyawong A, Drew RAI, Hengsawad C, Jirasurat M, Kong-Krong C, Kritsaneepaibon S, Vijaysegaran S (2001) Seasonal abundance and host use patterns of seven *Bactrocera* Macquart species (Diptera: Tephritidae) in Thailand and Peninsular Malaysia. Raffles Bull Zool 49:207–220
Clarke AR, Armstrong KF, Carmichael AE, Milne JR, Raghu S, Roderick GK, Yeates DK (2005) Invasive phytophagous pests arising through a recent tropical evolutionary radiation: the *Bactrocera dorsalis* complex of fruit flies. Annu Rev Entomol 50:293–319
COA (2004) Index of agriculture production. Council of Agriculture, Executive Yuan, Taiwan
Cohen E, Lavi U, Spiegel-Roy P (1989) Papaya pollen viability and storage. Sci Hortic 40(4):317–324
Coto D, Saunders JL, Vargas CL, King ABS (1995) Plagas invertebradas de cultivos tropicales con enfasis en America central. Centro Agronomico Tropical de Investigacion y Ensenanza. Serie Tecnica, Manual Tecnico 12. CATIE, Turrialba
Crane PR, Lidgard S (1989) Angiosperm diversification and paleolatitudinal gradients in cretaceous floristic diversity. Science 246(4930):675–80
Culik MP, dos Martins DS, Entura JA (2003) Indice de Artropodes Pragas do Mamoeiro (*Carica papaya* L.). INCAPER, Vitoria, 48 pp
Cunningham IC (1984) Mango insect pests. In: Australian mango research workshop, Commonwealth Scientifi c and Industrial Research Organization (CSIRO), Melbourne, pp 211–224

Cunningham IC (1989) Pests. In: Bagshaw J (ed) Mango pests and disorders. Department of Primary Industries Information Series, Q189007. Queensland Department of Primary Industries, Brisbane, pp 10–21
da Carvalho RS, Nascimento A, Morgante J, Fonseca N (1996) Susceptibility of different mango varieties to the attack of the fruit fly, *Anastrepha obliqua*. In: McPheron B, Steck G (eds) Fruit fly pests: a world assessment of their biology and management. St Lucie Press, Delray Beach, pp 325–331
da Silva RC (2006) Comportamento e comunicacao quimica da broca da mangueira (*Hypocryphalus mangiferae* Stebb) (Coleoptera: Curculionidae: Scolytinae). Universidade Estadual do Norte Fluminense, Camposdos Goytacazes, Rio do Janeiro, Theses, 59 pp
da Silveira R (1960) Contribucao ao studo do *Hypocryphalus mangiferae* (Coleoptera: Scolytidae). PhD thesis, Escuela Agricultura de Sao Paulo, Sao Paulo
Dag A, Gazit S (1996) Mango pollinators in Israel. Alon HaNotea 50:458–465
Dag A, Gazit S (2000) Mango pollinators in Israel. J Appl Hortic 2:39–43
Daneel MS, de Jager K, Steyn W, Husselman J (2000) Effi cacy of different insecticides against gall fly on mangoes. South African Mango Growers' Assoc Yearb 19:20
Davenport TL, Núñez-Elisea R (1997) Reproductive physiology. In: Litz RE (ed) The mango – botany, production and uses. CAB International, Wallingford, pp 69–146
De K, Pande YD (1987) Evaluation of certain non-insecticidal methods of reducing infestation of the mango nut weevil, *Sternochetus gravis* (F.) in India. Trop Pest Manag 33:27–28
De K, Pande YD (1988) Bionomics and some behavioural aspects of the mango stone weevil, *Sternochetus gravis* (Fabricius) (Coleoptera: Curculionidae). Entomon 13:17–24
de Bie CA (2004) The yield map of mango in Phrao, Thailand, as investigated through comparative performance evaluation. Sci Hortic 102:37–52
De la Rosa W, Lopez FL, Liedo P (2002) *Beauveria bassiana* as a pathogen of the Mexican fruit fly (Diptera: Tephritidae) under laboratory conditions. J Econ Entomol 95:36–43
de Laroussilhe F (1980) Le Manguier. G.P. Maisonneuve and Larouse, Paris
De Macedo FP (1998) Efeitos da aplicacao de acido giberelico em manga (*Mangifera indica* L.) cultivar Tommy Atkins: aspectos horticulturais e entomologicos. Tesis de doctorado, Universidade de Sao Paulo, 120 pp
De Villiers EA (1990) Coconut bugs in avocados. Farming in South Africa. Avocados H.5, 2 pp
Dekle GW (1976) Florida armored scale insects. Arthropods of Florida and neighboring land areas, Florida Department of Agriculture and Consumer Services, Gainesville, 3, 345 pp
Denmark H (1983) *Eriophyes mangiferae* (Sayed) a pest of mango (Acarina: Eriophyidae). Entomology Circular Florida Department of Agriculture and Consumer Services 254
Dhingra S (1990) Alternative insecticides for the control of mango mealy bug *Drosicha mangiferae* Green to overcome the failure of usually available insecticides. J Entomol Res 14(2):139–141
Diaz-Fleischer F, Aluja M (2003) Clutch size in frugivorous insects as a function of host firmness: the case of the tephritid fly *Anastrepha ludens*. Ecol Entomol 28:268–277
Diaz-Fleischer F, Toledo J, Enkerlin W, Fernandez J (1996) Cyromazine: effects on three species of *Anastrepha*. In: McPheron B, Steck G (eds) Fruit fly pests: a world assessment of their biology and management. St Lucie Press, Delray Beach, pp 336–337
Dietz SM, Harwood RF (1960) Host range and damage by the Grass Mealy bug, *Heterococcus graminicola*. J Econ Entomol 53(5):737–740
Dimbi S, Maniania NK, Lux S, Ekesi S, Mueke JK (2003) Pathogenicity of *Metarhizium anisopliae* (Metsch.) Sorokin and *Beauveria bassiana* (Balsamo) Vuillemin, to three adult fruit fly species: *Ceratitis capitata* (Wiedemann), *C. rosa* var. *Fasciventris* Karsch and *C. cosyra* (Walker) (Diptera: Tephritidae). Mycopathologia 156:375–382
Dinesh AS, Venkatesha MGG (2011) Predation of the apefly, *Spalgis epius* (Lepidoptera: Lycaenidae) on citrus mealybug, *Planococcus citri* (Hemiptera: Pseudococcidae). Biocontrol Sci Technol 21(5):523–533
Dolinski C, Lacey LA (2007) Microbial control of arthropod pests of tropical tree fruits. Neotrop Entomol 36:161–179

Doreste E (1984) Acarologia. Instituto Interamericano de Cooperacion para la Agricultura, San Jose

Downie DA, Gullan PJ (2004) 2004 Phylogenetic analysis of mealybugs (Hemiptera: Coccoidea: Pseudococcidae) based on DNA sequences from three nuclear genes, and a review of the higher classification. Syst Entomol 29:238–259

Drew RAI (1989) The tropical fruit flies (Diptera: Tephritidae: Dacinae) of the Australasian and Oceanian regions. Mem Qld Mus 26:521

Drew RAI, Hancock DL (1994) The *Bactrocera dorsalis* complex of fruit flies (Diptera: Tephritidae: Dacinae) in Asia. Bull Entomol Res Suppl Ser 2:68

DuToit AP (1994) Pollination of avocados, mangoes and litchis. Plant Prot News 35:4–5

DuToit AP, Swart D (1993) Pollination of mango in the Letsitele Valley during the 1992 flowering season: first report. Yearb South African Mango Growers Assoc 13:129–130

Dutt GR (1925) The Giant mealy bug and its control. Bull Entomol Res 16:155–158

Dwivedi SC, Kuldeep, Singh SM, Katiyar RR (2003) Seasonal incidence of insect pests associated with mango crop. Department of Entomology, C.S.A. Univ. Agri. & Tech., Kanpur 208 002, India. Ann Plant Protect Sci 11(1):159–160

Dyck VA, Hendrichs J, Robinson AS (eds) (2005) Sterile insect technique. Principles and practice in area-wide integrated pest management. Springer, Dordrecht

Eardley CD, Mansell MW (1994) Report on the natural occurrence of insect pollinators in a mango orchard. Yearb S Afr Mango Grow Assoc 14:65–66

Eby P (1995) The biology and management of flying foxes in NSW. Species management report number 18. NSW National Parks and Wildlife Service, Hurstville

Ehler IE, Endicott PC (1984) Effect of malathion-bait sprays on biological control of insect pests of olive, citrus and walnut. Hilgardia 52:1–47

Ekesi S, Nderitu PW, Rwomushana I (2006) Field infestation, life history and demographic parameters of the fruit fly *Bactrocera invadens* (Diptera: Tephritidae) in Africa. Bull Entomol Res 96:379–386

Emery R (2002) Mango seed weevil and mango pulp weevil. Available at: http://www.agric.wa.gov.au/ento/Su…%20seed%20and%20pulp%20weevil.html. Accessed 20 July 2007

Enkerlin WR (2007) Impact of fruit fly control programmes using the sterile insect technique. In: Dyck VA, Hendrichs J, Robinson AS (eds) Sterile insect technique. Principles and practice in area-wide integrated pest management. Springer, Dordrecht, pp 651–676

Epsky N, Hendrichs J, Katsoyannos B, Vasquez A, Ros J, Zumreolu A, Bakri A, Seewooruthum S, Heath R (1999) Field evaluation of female-targeted trapping systems for *Ceratitis capitata* (Diptera: Tephritidae) in seven countries. J Econ Entomol 92:156–164

Escalante JA (1974) Insects of economic importance in Quillabamba, Cusco. Revista Peruana de Entomologia 17:51–53

Etienne J (1966) Lutte contre les mouches des fruits. Rapporte de Institute de Recherches Agronomiques Tropicales Reunion. Institute National de la Recherche Agronomique, Linne de Paradis, Reunion, pp 112–113

Etienne J (1968) Lutte contre les mouches des fruits. Rapporte de Institute de Recherches Agronomiques Tropicales Reunion. Institute National de la Recherche Agronomique, Linne de Paradis, Reunion, pp 117–124

Faih HA, Huwer R (1994) Egg parasitoids collected from *Amblypelta lutescens lutescens* (Distant) (Hemiptera: Coreidae) in north Queensland. J Aust Entomol Soc 32:365–367

Fajardo AC Jr, Medina JR, Opina OS, Cervancia CR (2008) Insect pollinators and floral visitors of mango *(Mangifera indica* L. cv.Carabao). Philippine Agric Sci 91(4):372–382

Fasih M, Srivastava RP (1990) Parasites and predators of insect pests of mango. Integrated Pest Control 32:39–41

Fasih M, Srivastava RP, Abbas S, Sharma S (1989) Outbreak of *Orgyia postica* Walker (Lymantriidae: Lepidoptera), a new pest on mango in Uttar Pradesh. Curr Sci 58:1258–1260

Fay HA, Huwer R (1994) Egg parasitoids collected from *Amblypelta lutescens* (Distant) (Hemiptera: Coreidae) in north Queensland. J Aust Entomol Soc 32:365–367

Fletcher TB (1917) Proceedings second entomological meeting Pusa 1917 (Fide Rehman, K.A., Latif, A. 1944. Bull Entomol Res 35(2):197–209
Fletcher TB (1927) Report of the imperial entomologist. Science Reporter Agriculture Research Institute Pusa. 1926–7 (fide Srivastava, R.P. 1980. Indian J Entomol 42(1):110–115
Fletcher BS (1987) The biology of dacine fruit flies. Annu Rev Entomol 32:115–144
Fletcher MJ, Dangerfield PC (2002) *Idioscopus clypealis* (Lethierry), a second new leafhopper pest of mango in Australia (Hemiptera: Cicadellidae: Idiocerinae). Aust J Entomol 41:35–38
Follett P, Gabbard Z (2000) Effect of mango weevil (Coleoptera: Curculionidae) damage on mango seed viability in Hawaii. J Econ Entomol 93:1237–1240
Follett P, Neven LG (2006) Current trends in quarantine entomology. Annu Rev Entomol 51:359–385
Foottit RG, Maw HEL, Pike KS, Miller RH (2010) The identity of *Pentalonia nigronervosa* and *P. caladii*van der Goot (Hemiptera: Aphididae) based on molecular and morphometric analysis. Zootaxa 2358:25–38
Fortescue JA, Turner DW (2005) Pollen and fertility in *Musa*: viability in cultivars grown in Southern Australia. Aust J Agr Res 55:1085–1091
Foster LT (1943) Morphological and cytological studies on Carica papaya. Bot Gaz 105:116–126
Franco JC, Zada A, Mendel Z (2009) Novel approaches for the management of mealybug pests. In: Ishaaya I, Horowitz AR (eds) Biorational control of arthropod pests: application and resistance management. Springer, New York, pp 233–278
Fraser S (1927) American fruits, their propagation, cultivation, harvesting and distribution. Orange-Judd, New York, 829 pp
Free JB (1975) Observations on the pollination of papaya (Carica papaya L.) in Jamaica. Trop Agric 52:275–279
Free JB (1999) Pollination in the tropics. Beekeeping Dev 50:10–11; 51:6–7; 53:4–5
Free JB, Williams IH (1976) Insect pollination of *Anacardioum occidentale* L., *Mangifera indica* L., *Blighia sapida* Koening and *Persea americana* Mill. Trop Agric 53:125–139
Freeman S, Klein-Gueta D, Korolev D, Sztenjnberg A (2004) Epidemiology and survival of *Fusarium mangiferae*, the causal agent of mango malformation disease. Acta Hortic 645:487–491
Fujita M, Tuttle MD (1991) Flying foxes (Chiroptera: Pteropodidae) threatened animals of key ecological and economic importance. Conserv Biol 5:455–463
Gaffar SA (1989) Chemical control of mealy bug, *Drosicha dalbergiae* Stebbing on almond in Kashmir. Indian J Plant Protect 17(1):31–40
Gagne RJ, Etienne J (2006) *Gephyraulus mangiferae* (Felt) N. Comb. (Diptera: Cecidomyiidae): a mango pest from India newly recorded from the western hemisphere. Proc Entomol Soc Wash 108:930–937
Gagne RJ, Medina CD (2004) A new species of *Procontarinia* (Diptera: Cecidomyiidae), an important new pest of mango in the Philippines. Proc Entomol Soc Wash 106:19–25
Galang FG, Lazo FD (1937) The setting of carabo mango fruits as affected by certain sprays. Philipp J Agric 8(2):187–210
Galán-Saúco V (1990) Los Frutales Tropicales en los Subtropicos. MundiPrensa, Madrid, pp 83–84
Galán-Saúco V, Fernández Galván D, Hernández Conde JC, Morales Navarro A (1997) Preliminary studies on fruit-set of mango cultivar Tommy Atkins under greenhouse cultivation in the Canary Islands. Acta Hortic 455:530–537
Gangoli SR, Singh R, Katyal SL, Singh D (1957) The mango. IARI, New Delhi, p 530
Ganz S, Hamieri Y, Nakach Y (1990) Control trials on the mango thrips. Annual Report of the Eden Experimental Station Israel. Institute of Plant Protection, Bet She'am, Israel
Garcia-Medel D, Sivinski J, Diaz-Fleischer F, Ramirez-Romero R, Aluja M (2007) Foraging behavior by six fruit fly parasitoids (Hymenoptera: Braconidae) released as single- or multiple-species cohorts in fi eld cages: influence of fruit location and host density. Biol Control 43:12–22

Garcia-Ramirez MJ, Cibraian-Tovat J, Arzufi-Barrera R, Lopez-Collado J, Soto-Hernandez M (2004) Preferencia de *Anastrepha ludens* (Loew) (Diptera: Tephritidaepor volatiles de frutos verdes o amarillos de mango y naranja. Agrociencia 38:423–430

Garrett A (1995) The pollination biology of Papaw (Carica papaya L.) in Central Queensland. Ph.D. Thesis, Central Queensland University

Gautam A, Tesfaye RD (2002) Potential of green lacewing *Chrysoperla carnea* (Stephens) in crop pest management. New Agric 13(1/2):147–158

Ge XJ, Liu MH, Wang WK, Schaal BA, Chiang TY (2005) Population struvture ogf wild bananas, *Musa balbisiana* in China determined by SSR fingerprinting and cpDNA PCR- RFLP. Mol Ecol 14:933–944

Geiger CA, Daane KM (2001) Seasonal movement and distribution of the grape mealybug (Homoptera: Pseudococcidae): developing a sampling program for San Joaquin Valley vineyards. J Econ Entomol 94:291–301

Gibbs DMH, Muirhead IF (1998) The economic value and environmental impact of the Australian beekeeping industry. [Internet] Australian Honeybee Industry Council. From: http://www.honeybee.org.au/economics.htm. [Cited Apr 2006]

Githure CW, Schoeman AS, McGeoch MA (1998) Differential susceptibility of mango cultivars in South Africa to galling by the mango gall fly, *Procontarinia matteiana* Kieffer & Cecconi (Diptera: Cecidoyiidae). Afr Entomol 6:33–40

Godase SK, Bhole SR, Shivpuje PR, Patil BP (2004) Assessment of yield loss in mango (*Mangifera indica*) due to mango hopper (*Idioscopus niveoparvasus*) (Homoptera: Cicadelidae). Indian J Agric Sci 74:370–372

Godse SK, Bhole SR (2003) Mango mealy bug fauna of Konkan Region of Maharashtra. Insect Environ 9(4):157

Gold CS, Okech SH, Nokoe S (2002) Evaluation of pseudostem trapping as a control measure against banana weevil, *Cosmopolites sordidus* (Coleoptera: Curaulionidae). Bull Entomol Res 92:35–44

Golez HG (1991) Bionomics and control of the mango seed borer, *Noorda albizonalis* Hampson (Pyralidae: Lepidoptera). Acta Hortic 291:418–424

Gonzalez-Hernandez H (1995) The status of the biological control of pineapple mealybugs in Hawaii. Ph.D. Thesis, The University of Hawaii at Manoa, Honolulu, 128 pp

Green EE (1908) Remarks of Indian scale insects (*Coccidae*) part-III with a catalogue of all species hitherto recorded from the Indian continent. Memo Dept Agric India (Entomol Ser) 2:15–46

Greenway PJ, Wabbace MM (1953) The papaw, its botany, cultivation, diseases, and chemistry. Tanganyika Dept Agr Pam 52:32

Grove T, Pringle K (2000) A sampling system for estimating population levels of the citrus thrips, *Scirtothrips aurantii faure* (Thysanoptera: Thripidae), mango orchards. Afr Entomol 8:223–226

Grove T, Giliomee JH, Pringle K (2001) *Thrips* (Thysanoptera) species associated with mango trees in South Africa. Afr Entomol 9:153–162

Grover P (1986a) Integrated control of midge pests. Cecidol Int 7:1–28

Grover P (1986b) Population fluctuation of *Erosomya indica* and *Dasineura amaramanjarae* and co-related extent of damage. Cecidol Int 7:43–57

Hala N, Kehe M, Allou K (2004) Incidence of the mango mealy bug *Rastrococcus invadens* Williams, 1986 (*Homoptera*; Pseudococcidae) in the Ivory Coast. Agron Afr 16(3):29–36 [French]

Hallman GJ, Sharp JL (1990) Mortality of Caribbean fruit fly (Diptera: Tephritidae) larvae infesting mangoes subjected to hot-water treatment, then immersion cooling. J Econ Entomol 83:2320–2323

Hancock DL, Hamacek EL, Lloyd AC, Elson-Harris MM (2000) The distribution and host plants of fruit flies (Diptera: Tephritidae) in Australia. Department of Primary Industries, Brisbane

Hansen JD (1993) Dynamics and control of the mango seed weevil. Acta Hortic 341:415–420

Hansen JD, Amstrong JW (1990) The failure of field sanitation to reduce infestation by the mango nut weevil, *Cryptorhynchus mangiferae* (F.) (Coleoptera: Curculionidae). Trop Pest Manag 36:359–361

Hansen JD, Armstrong JW, Brown S (1989) The distribution of the mango weevil, *Cryptorhynchus mangiferae* (Coleoptera: Curculionidae), in Hawaii. Proc Hawaiian Entomol Soc 29:31–39
Haq KA, Akmal M (1960) Mango mealy bug and its control. Pb Fruit J 23(82–83):199–202
Harkness RW (1967) Papaya growing in florida. Fla Agr Expt Sta Cir S-180:15 pp
Harris KM, Schreiner IH (1992) A new species of the gall midge (Diptera: Cecidomyiidae) attacking mango foliage in Guam, with observations on its pest status and biology. Bull Entomol Res 82:41–48
Harris EJ, Vargas RI, Gilmore JE (1993) Seasonality in occurrence and distribution of Mediterranean fruit fly (Diptera: Tephritidae) in upland and lowland areas of Kauai, Hawaii. Environ Entomol 22:404–410
Haseeb M, Abbas SR, Srivastava RP, Sharma S (1990) New records of insect pest of aonla. Narendra Dev J Res 5(1):116–119
Haseeb M, Abbas SR, Srivastava RP, Sharma S (2000) Studies on insect pests of aonla (*Emblica officinalis*). Plant Proc Sci 8(1):85–88
Hayat M (2009) Description of a new species of Aenasius Walker (Hymenoptera: Encyrtidae), parasitoid of the mealybug *Phenacoccus solenopsis* Tinsley (Homoptera: Pseudococcidae) in India. Biosystematica 3(1):21–26
Heath RR, Epsky ND, Guzman A, Dueben BD, Manukian A, Meyer WL (1995) Development of a dry plastic insect trap with food-based synthetic attractant for the Mediterranean and Mexican fruit flies (Diptera: Tephritidae). J Econ Entomol 88:1307–1315
Heath RR, Epsky ND, Dueben BD, Rizo J, Jeronimo F (1997) Adding methyl substitute ammonia derivates to food-based synthetic attractant on capture of the Mediterranean and Mexican fruit flies (Diptera: Tephritidae). J Econ Entomol 90:1584–1589
Heinrich B, Raven PH (1972) Energetics and pollination ecology. Science 176:497–602
Hendrichs J (1996) Action programs against fruit flies of economic importance: session overview. In: McPheron B, Steck G (eds) Fruit fly pests: a world assessment of their biology and management. St Lucie Press, Delray Beach, pp 513–519
Hendrichs J (2001) Use of the sterile insect technique against key insect pests. Food and Agriculture Organization (FAO)/International Atomic Energy Agency Divisionof Nuclear techniques in Food and Agriculture. Available at: http://www.sustdev.org/journals/edition.02/download/sdi2_2_2.pdf. Accessed 20 July 2007
Hennesey M, Schnell RJ (2001) Resistance of immature mango fruits to Caribbean fruit fly (Diptera: Tephritidae). Fla Entomol 84:318–319
Hernández-Sánchez G, Sanz-Barbosa I, Casaña-Giner V, Primo-Yúfera E (2001) Attractiveness for *Ceratitis capitata* (Wiedemann) (Dipt., Tephritidae) of mango (*Mangifera indica*, cv. Tommy Atkins) airborne terpenes. J Appl Entomol 125:189–192
Herrera AJ, Vinas LE (1977) Fruit flies (Diptera: Tephritidae) on mangoes in Chulucanas, Piura. Revista Peruana de Entomoogia 20:107–114
Higgins JE, Hobt VS (1914) The papaya in hawaii. Hawaii Agr Expt Sta Bull 32:44 pp
Hill D (1975) Agricultural insect pests of the tropics and their control. Cambridge University Press, New York
Hofmeyr JDJ (1938) Genetic studies of Carica papaya L. S Afr J Sci 35:300–304
Hollingsworth RG, Drew R, Allwood AJ, Roming M, Vagalo M, Tsatsia F (2003) Host plants and relative abundance of fruit fly (Diptera: Tephritidae) species in the Solomon Islands. Aust J Entomol 42:95–108
Howell DJ (1974) Bats and pollen: physiological aspects of syndrome chiropterophily. Comp Biochem Physiol 48A:263–276
Hsu J-C, Feng H-T (2006) Development of resistance to Sinosad in oriental fruit fly (Diptera: Tephritidae) in laboratory selection and cross-resistance. J Econ Entomol 99:931–936
Huang T-Ch, Tze-chung H, Cheng EY, Kao Ch, Hwang Y, Chiang M (2006) Area-wide control of the oriental fruit fly and melon fly in Taiwan. Available at: www.agnet.org/activities/sw/2006/954198798/paper-868546184.pdf. Accessed 20 July 2007
Hung SC (1997) Mass rearing, preserving and release methods of *Chrysomya megacephala* (Fabricius). Proceedings of a symposium on enhancing competitiveness of Fruit Industry (I), Taichung DARES. Series 38:177–188 (in Chinese)

Husain MA, Pruthi HS (1924) A short note on the life history of the mango hopper (*Idiocerus* spp.) in the Punjab. In: Report proceedings V of the entomology meeting, Pusa, pp 252–260

Instituto Nacional de Investigaciones Agropecuarias – Programa de Modernizacion de Servicios Agropecuarios, Guayaquil, Equador.

Ish-Am G, Barrientos-Priego F, Castañeda-Vildozola A, Gazit S (1999) Avocado (*Persea Americana* Mill.) pollination in its region of origin. Revista Chapingo Ser Hort 5:137–143

Ishaq M, Usman M, Asif M, Khan IA (2004) Integrated pest management of mango against mealy bug and fruit fly. Int J Agric Biol 6:452–454

Islam MS, Elegio DT (1997) Effect of time and frequency of insecticide spraying on the control of mango leafhoppers (*Idioscopus* spp.). Bangladesh J Entomol 7:93–99

Iwahashi O (1984) The control project of the oriental fruit fly in Okinawa. Chin J Entomol 4:107–120 (in Chinese)

Jadhav V, Dalvi C (1987) Bionomics of mango nursery leaf eating caterpillar. J Maharashtra Agric Univ 12:384–385

Jamieson LE, Froud K, Edwards R, Stevens PS (2008) Establishment of Thripobius javae (=Semiluteus) in New Zealand. N Z Plant Protect 61:17–23

Jeppson L, Keiffer H, Baker E (1975) Mites injurious to economic plants. University of California Press, Berkeley

Jhala RC, Patel Z, Shah A (1987) Studies on the relative occurrence of the leafgall midge (*Procontarinia matteiana* Keiffer and Cecconi) on different varieties of mango in south Gujarat, India. Trop Pest Manag 33:277–279

Jhala RC, Shah AH, Patel ZP, Patel RL (1989) Studies on population dynamics of mango hopper and scope of spraying in integrated pest management programme. Acta Hortic 231:597–601

Jhala RC, Bharpoda TM, Patel MG, Patel VB, Patel JR (2003) Survey on insect pest and their natural enemies occurring in aonla orchard under middle Gujarat condition. Paper presented in national seminar on production and utilization of aonla, organized by All India Aonla Growers Association, Salem

Jia XY, Ma GY, Wang LG, Liang W, Wen H (2001) [Integrated control of walnut pests. Chin Fruits 1:39–40 (In Chinese)

Jiron LF (1993) Plagas del follaje y de paniculas asociadas con el mango en Costa Rica. In: Memorias Ier Seminario Nacional de Mango. Punta Arenas, pp 22–26

Jiron LF (1995) Opciones al uso de insecticidas en mango. In: Garcia J, Fuentes G, Monge J (eds) Opciones al uso Unilateral de Plaguicidas en Costa Rica, vol 2. Editorial Euned, San Jose, pp 129–155

Jiron LF, Hedstrom I (1988) Occurrence of fruit flies of the genera *Anastrepha* and *Ceratitis* (Diptera: Tephritidae), and their host plant availability. Fla Entomol 71:62–73

Jiron LF, Hedstrom I (1991) Population fluctuations of economic species of *Anastrepha* (Diptera: Tephritidae) related to mango fruiting phenology in Costa Rica. Fla Entomol 74:98–105

Jiron LF, Soto-Manitiu J (1987) Las Moscas de las frutas (Diptera: Tephritidae) en Costa Rica: situacion actual, analisis y comentario. Agronomia Costarricense 11:255–261

Joel DM (1980) Resin ducts in the mango fruit: a defense system. J Exp Bot 30:1707–1718

Johnson PJ (1989) Mango seed weevil (*Sternochetus mangiferae* F.): identification, biology and management. Univ Idaho Pip Tip 6:3–7

Johnson GI, Hofman PJ (eds) (2004) Agri-product supply-chain management in developing countries. Australian Centre for International Agricultural Research (ACIAR) Proceedings 119. ACIAR, Canberra

Joubert PH, Daneel MS, Grove T (2000) Progress towards integrated pest management (IPM) on mangoes in South Africa. Acta Hortic 509:811–817

Joubert P, Grove T, de Beer MS, Steyn WP (2004) Evaluation of kaolin (Surround WP) in an ipm program on mangoes in South Africa. Acta Hortic 645:493–499

Juliano JB, And Cuevas NL (1932) Floral morphology of the mango (*mangifera indica* l.) with special reference to the pico variety. Philipp Agric J 21:449–472

Jyothi JVA (1994) Visitation frequency and abundance of *Apis cerana indica* F. on mango (*Mangifera indica* L.) at Bangalore India. Indian Bee J 56:35–36

Kalshoven LGE (1981) Pests of crops in Indonesia. P.T. Ichtiar Baru-Nun Hoeve, Jakarta, 701 pp

Kannan M, Rao NV (2006) Influence of age and stage of the host plant on insect pests of mango (*Mangifera indica* L). Int J Agric Sci 2(2):351–353

Karar H (2010) Bio-ecology and management of mango mealybug (Drosicha mangiferae (Green) in mango orchards of Punjab, Pakistan. Ph.D. Thesis, University of Agriculture, Faisalabad

Karar H, Arif MJ, Saeed S, Abbas G (2007) Mechanical control of mango mealybug *Drochica mangifera* Green in mango orchards. In: XI international symposium on scale insect studies ISSIS 2007 in Oeiras, 24–27 Sept, pp 46

Karar H, Syyed AH, Arif MJ, Ashfaq MA, Aslam M (2010a) Integration of cultural and mechanical practices for management of mango mealy bug *Droschia mangiferae* G. Phytoparasitica 38:233–229

Karar H, Arif MJ, Sayyed HA, Ashfaq M, Khan MA (2010b) Comparative efficacy of new and old insecticides for the control of mango mealybug (*Drosicha mangiferae*) in mango orchards. Int J Agric Biol 12(3):443–446. 19

Karar H, Arif MJ, Ali A, Hameed A, Abbas G, Abbas Q (2012) Assessment of yield losses and impact of morphological markers of various genotypes on mango (*Mangiferae indica*) genotypes on mango mealybug (*Droschia mangiferae* Green) (Homoptera: Margarodidae). Paki J Zool 44:1643–1651

Keifer HH, Baker EW, Kono T, Delfinado M, Styer W (1982) An illustrated guide to plant abnormalities caused by Eriophyid mites in North America. US Dept Agric Agric Handb 573:74–75

Kenneth EF (1952) The value of some organic phosphate insecticides in control of grape mealy bug. J Econ Entomol 45(2):340–341

Kfir R, Rosen D (1980) Parasites of soft scales (Homoptera: Coccidae) in Israel: an annotated list. J Entomol Soc South Africa 43:113–128

Khan MA (1989) Control of insect pests of mango. In: Proceedings of international mango workshop 27th Feb–1st Mar. Direct Agric. Multan region. Punjab, p 224

Khan MA (2001) Control of insect pests of mango. In: Proceedings of the international mango workshop 27 February–1 March. Direct. Agric-Multan Region, Punjab, p 224

Khan RA, Ashfaq M (2004) Funnel type slippery trap: a mechanical device to control mango mealybug. SAIC Newslett Bangladesh 14(3):2–9

Khan MA, Ashfaq M, Khaliq A, Ahmad A (2002) Effect of pheromone traps on the population capture of fruit fly species and their infestation in mango orchards at Multan. Pak Entomol 24:153–157

Khan MA, Ashfaq M, Akram W, Lee J-J (2005) Management of fruit flies (Diptera: Tephritidae) of the most perishable fruits. Entomol Res 35:79–84

Khanzada AG, Naqvi KM (1985) The optimum time for the control of mango hopper *Idioscopus* spp. (Hemiptera: Cicadellidae). Proc Entomol Soc Karachi 14(15):149–156

Khurana AD, Verma AN (1988) Laboratory and field evaluation of insecticides against nymphs and gravid females of mango mealy bug, *Drosciha mangiferae* Green. Indian J Entomol 50(3):319–326

King JR, Hennessey MK (1996) Spinosad bait for the Caribbean fruit fly (Diptera: Tephritidae). Fla Entomol 79:526–531

Kosztarab M (1996) Scale insects of north eastern North America, identification, biology and distribution. Virginia Museum of Natural History, Martinsville

Kremen C, Williams NM, Thorp RW (2002) Crop pollination from native bees at risk from agricultural intensification. Proc Natl Acad Sci U S A 99:16812–16816. doi:10.1073/pnas.262413599

Kring TJ, Yearian WC, Steward VB (1987) *Pleuroprucha insulsaria* (Lepidoptera: Geometridae): a potential pest of grain sorghum. J Kansas Entomol Soc 60:476–477

Kudagamage C, de Rajapakse HLZ, Rajapakse RH, Ratnasekara D (2001) The population dynamics and insecticidal control of three leaf hoppers, *Amrytodus brevistilus* Leth., *Idioscopus niveosparus* Leth., and *Idioscopus clipealis* Leth. (Homoptera: Cicadellidae) in mango. J Entomol Res 25:121–125

Kumar S, Rai A, Patel C, Bhatl R (1993) Studies on spatial dsitribution of *Amritodes atckinsoni* (Leth.) infesting mango in south Gujarat, India. (Abstract). IV international mango symposium, p. 122

Kumar P, Baskaran S, Sundaravadivelan C, Kuberan T, Anburaj J (2012) Influence of environmental factors on insect pollination activity of *Mangifera indica* Linn. Asian J Plant Sci Res 2(6):692–698

Kuruvilla KM, Radhakrishnan VV, Madhusoodanan KJ (1995) Small cardamom plantations – floristic calendar and bee pasturage trees. J Environ Res 3:32–33

Labuschagne TI (1993) Progress with integrated control of the mango scale *Aulacaspis tubercularis* Newstead. South African Mango Growers' Assoc Yearb 13:134–135

Labuschagne TI, Steyn WP, De Beer MS (1995) Voorkoms en moontlike alternative beheermetodes van vrugtevliee opp mango's. South African Mango Growers' Assoc Yearb 15:102–105

Lakra RK, Kharub WS, Singh Z (1980) Pest Management for the mango mealy bug Drosicha mangiferae Green. A polyphagous pest of fruit trees in Haryana. Indian J Entomol 42(2):153–165

Lal MN, Joshi GC, Singh SS (1996) Studies on the incidence of *Indarbela* spp., *Cerciaphis emblica* and *Betousa stylophora* on aonla in eastern. Uttar Pradesh Prog Hortic 28:66–71

Lal MM (1952) Insect pests of fruit trees grown in plains of U.P. Agric Anim Hus J 3(1, 2, 3):55–7

Lassoudiere A (1969) The papaya iv. Description of inflorescences and flowers of 'solo' papaya. Fruits 24(3):143–151 [in french]

Latif A (1940) Short noted and exhibits. Indian J Entomol 2:93

Latif A (1941) Punjab Fruit J 5–19 (fide Rehman KA, Latif MA (1944) Bull Entomol Res 35(2):197–199

Law BS (2001) Diet of the common blossom bat (*Syconycteris australis*) in upland tropical rainforest and the importance of riperian areas. Wildl Res 28:619–626

Law BS, Lean M (1999) Common blossom bat (*Syconycteris australis*) as pollinators in fragmented Australian tropical rainforest. Biol Conserv 91:201–212

Leblanc L, Allwood AJ (1997) Mango fruit fly (*Bactrocera frauenfeldi*): why so many in federated states in Micronesia? In: Allwood AJ, Drew RAI (eds) Management of fruit flies in the Pacific, vol 76, Australian Centre for International Agricultural Research (ACIAR) proceedings. ACIAR, Canberra, pp 125–130

Leblanc L, Leweniqila L, Tau D, Tumukon T, Kassim A, Hollingsworth R (1997) Can fruit flies be controlled in a village with a mixed orchard? Pacific Islandexperiences. In: Allwood AJ, Drew RAI (eds) Management of fruit flies in the Pacific, vol 76, Australian Centre for International Agricultural Research (ACIAR) proceedings. ACIAR, Canberra, pp 187–191

Leefmans S, van der Vecht J (1930) The red-ringed mango caterpillar (in Dutch, English summary). Korte Mededeelingen van het lustituut voor Plantenziekten, 14 pp

Leyva JL, Browning H, Gilstrap F (1991) Development of *Anastrepha ludens* (Diptera: Tephritidae) in several host fruit. Environ Entomol 20:1160–1165

Lezama-Gutierrez R, Trujillo A, Molina-Ochoa J, Rebolledo-Dominguez O, Pescador A, Lopez-Edwards M, Aluja M (2000) Virulence of *Metarrizium anisopliae* (Deuteromycotina: Hypomycetes) on *Anastrepha ludens* (Diptera: Tephritidae): laboratory and field trials. J Econ Entomol 93:1080–1084

Li LY, Wang R, Waterhouse DF (1997) The distribution and importance of major insect pests and weeds of agriculture and plantation forests in southern China. Chinese Academy of Agricultural Sciences, Guangdong Entomological Institute and ACIAR, Beijing/Canberra, 185 pp

Liburd OE, Holler TC, Moses AL (2004) Toxicity of imidacloprid-treated spheres to Caribbean fruit fl y, *Anastrepha suspensa* (Diptera: Tephritidae) and its parasitoid *Diachasmimorpha longicaudata* (Hymenoptera: Braconidae) in the laboratory. J Econ Entomol 97:525–529

Lieu MF (1999) Species, introduction and breeding. In: Yang SJ, Wu YL, Huang YM, Cheng CS (eds) Integrated pest management of mango. TACTRI Press, Taichung, pp 5–10 (in Chinese)

Lindegren J, Vail P (1986) Susceptibility of Mediterranean fruit fly, melon fly and oriental fruit fly (Diptera: Tephritidae) to the entomogenous nematode *Steinernema feltiae* in laboratory tests. Environ Entomol 4:465–468

Lindgreen S, Borden J, Gray D, Lee P, Palmer D, Chong L (1982) Evaluation of two trap log techniques for ambrosia beetles (Coleoptera: Scolytidae) in timber processing areas. J Econ Entomol 75:577–586

Liu AZ, Li DZ, Wang H, Kress WJ (2002) Orinthphilous and cheriropterophilous pollination in *Musa itinerans* (Musaceae) a pionerr species in tropical rainforestf of Yunnan, Southwestern China. Biotropica 34:254–260

Loomans AJM, van Lenteren JC, Tommasini MG, Maini S, Riudavets J (1995) Biological control of thrips pests, vol 95, Wageningen Agricultural University papers. Wageningen Agricultural University, Wageningen

Lopez F, Chambers D, Sanchez M, Kamasaki H (1969) Control of the Mexican fruit fly by bait sprays concentrated at discrete locations. J Econ Entomol 62:1255–1257

Lopez M, Aluja M, Sivinski J (1999) Hymenopterous larval-pupal and pupal parasitoids of *Anastrepha* fl ies (Diptera: Tephritidae) in Mexico. Biol Control 15:119–129

Love K, Bowen R, Paull R (2003) Increasing marketable production of exotic tropical fruit with protective covering. Western Region, Sustainable Agriculture Research and Education, Project Report, Farmer/Rancher, Sustainable Agriculture Research and Education, Utah State University, Captain Cook, Hawaii

Lux SA, Ekesi S, Dimbi S, Mohamed S, Billah M (2003) Mango-infesting fruit flies in Africa: perspectives and limitations of biological approaches to their management. In: Neuenschwander P, Borgemeister C, Langewald J (eds) Biological control in IPM systems in Africa. CAB International, Wallingford, pp 277–293

Maban EE (1964). Papaws in South Africa. South Africa Dept Agr Tech Serv Bull 375:12 pp

Maher WJ (1992) Breeding biology of yellow-bellied sunbird Nectarinia jugularis in northern queenland. Emu 92:57–60

Malavasi A, Zucchi RA (eds) (2000) Moscas-Das-Frutas de Importancia Economica no Brasil: Conhecimento Basico e Aplicado. Holos Editora, Ribeirao Preto

Mangan RL, Hallman GH (1998) Temperature treatments for quarantine security: new approaches for fresh commodities. In: Hallman GH, Denlinger DL (eds) Temperature sensitivity in insects and application in integrated pest management. Westview Press, Boulder, pp 201–234

Mangan RL, Moreno DS (2007) Development of bait stations for fruit fly population suppression. J Econ Entomol 100:440–450

Mangan RL, Sharp JL (1994) Combination treatments. In: Sharp JL, Hallman GH (eds) Quarantine treatments for pests of food plants. Westview Press, Boulder, pp 239–247

Mangan RL, Moreno DS, Thompson GD (2006) Bait dilution, spinosad concentration, and efficacy of GF-120 based fruit fly sprays. Crop Prot 25:125–133

Mani M, Krishnamoorthy A, Pattar GL (2007) Suppression of pink mealy bug (*Maconellicoccus hirsutus*) on ber (*Zizyphus mauritiana*). Indian J Agric Sci 77(2):135

Manning R (1995) Honeybee pollination technical data. Western Australia Department of Agriculture, Bulletin No 4298. Department of Agriculture, South Perth

Mansour R, Grissa Lebdi K, Rezgui S (2010) Assessment of the performance of some new insecticides for the control of the vine mealybug *Planococcus ficus* in a Tunisian vineyard. Entomol Hell 19(1):21–33

Marin Acosta JC (1969) Insects in relation to the papaya in Venezuela. Trop Agron 19(4):251–267 [in spanish]

Marino PC, Cornell HV (1993) Adult feeding and oviposition of *Phytomyza ilicicola* Loew (Diptera: Agromyzidae) in response to leaf and tree phenology. Environ Entomol 22:1294–1301

Martorell LF (1975) Annotated food plant catalog of the insects of Puerto Rico. Agricultural Experiment Station, University of Puerto Rico, Rio Piedras

Masarrat, H., Abbas, S. R., Srivastava, R. P. and Sharma, S., 2000. Studies on insect pests of aonla (*Emblica officinalis* Linn.). *Annual Pl. Prot. Sci.*, 8: 84–88.

Matokot L, Reyd G, Malonga P, Ru BI (1992) Population dynamics of *Rastrococcus invadens* (Homoptera: Pseudococcidae) in the Congo; influence of accidental introduction of the Asiatic parasitoid *Gyranusoidea tebygi* (Hymenoptera: Encyrtidae). Entomophaga 37:123–140

McGregor SE (1976) Insect pollination of cultivated crop plants. Agriculture handbook 496. United States Department of Agriculture, Agricultural Research Service, Washington, DC

McKenzie HL (1967) Mealy bug of Califirnia with taxonomy, biology, and control of North American species (Homoptera: Cooccoidea: Pseudococcidae). Uniersity of California Press, Berkeley, 526 pp

McQuate GT, Peck SL, Barr PG, Sylva CD (2005) Comparative evaluation of spinosad and phloxine B as toxicants in protein baits for suppression of three fruit fly (Diptera: Tephritidae) species. J Econ Entomol 98:1170–1178

Messing R (1999) Managing fruit flies on farms in Hawaii. Cooperative Extension Service, College of Tropical Agriculture and Human Resources, University of Hawaii at Manoa, Hawaii

Meyer N, Mittermeier RA, Mittermeier CG, da Fonseca GAB, Kent J (2000) Biodiversity hotspots for conservation priorities. Nature 403:853–858

Mibey RK (1997) Sooty moulds. In: Ben-Dov Y, Hodgson CJ (eds) Soft scale insects: their biology. Natural Enemies Control 7A:275–290. Elsevier, Amsterdam/New York

Michener CD (2000) The bees of the world. The Johns Hopkins University Press, Baltimore/London, 913 pp

Miller DR, Davidson J (2005) Armored scale insect pests of trees and shrubs (Hemiptera: Diaspididae). Comstock, Ithaca

Miller DR, Miller GL, Watson GW (2002) Invasive species ofmealy bug (Hemiptera: Pseudococcidae) and their threat U.S. agriculture. Proc Entomol Soc Wash 104:825–836

Miller DR, Miller GL, Hodges GS, Davidson JA (2005) Introduced scale insects (Hemiptera: Coccoidea) of the United States and their impact on U.S. agriculture. Proc Entomol Soc Wash 107:123–158

Mohindru B, Jindal V, Dhawan AK (2009) Record of parasitoid on mealy bug *Phenacoccus solenopsis* in Tomato. Indian J Ecol 36(1):101–102

Mohyuddin AI (1981) A review of biological control in Pakistan. In: Proceedings of the second Pakistan congress of zoology, Tandojam, pp 31–79

Mohyuddin AI (1989) Integrated pest management of three mango pests. In: Proceedings of international mango workshop, Multan, pp 137–141

Mohyuddin AI, Mahmood R (1993) Integrated control of mango pests in Pakistan. Acta Hortic 341:467–483

Montoya P, Liedo P, Benrey B, Cancino J, Barrera JF, Sivinski J, Aluja M (2000) Biological control of *Anastrepha* spp. (Diptera: Tephritidae) in mango orchards through augmentative releases of *Diachasmimorpha longicaudata* (Ashmead) (Hymenoptera: Braconidae). Biol Control 18:216–224

Moore D (2004) Biological control of *Rastrococcus invadens*. Review article. Biocontrol News Inf 25(1):17N–27N. www.sel.barc.usda.gov/Coccoidea/scale2004.htm

Moore D, Cross AE (1992) Competition between two primary parasitoids, *Gyranusoidea tebygi* Noyes and *Anagyrus mangicola* Noyes, attacking the mealybug *Rastrococcus invadens* Williams and the influence of a hyperparasitoid *Chartocerus hyalipennis* Hayat. Biocontrol Sci Technol 2(3):225–234

Moore D, Cross AE (1993) Biological control of the fruit tree mealy bug, *Rastrococcus invadens* Williams. Acta Hortic 341:433–440

Morin C (1967) Cultivo de Frutales Tropicales. Librerias ABC, Lima, pp 225–229

Morrill AW, Otanes FQ (1947) DDT emulsion to control mealy bugs and scales. J Econ Entomol 40:599–600

Morton JF (1987) Banana Musaxparadisiaca. In: JF Morton (ed) Fruits of warm climates. Distributed by Creative Resource Systems, Miami, pp 29–46. Available online at http://www.hort.purdue.edu/newcrop/morton/banana.html

Moznette G (1922) Insects injurious to the mango in Florida and how to combat them. Farm Bull 1257

Mukherjee SK (1953) The mango, its botany, cultivation, uses and future improvement. Econ Bot 7(2):130–162

Muniappan R, Marler T, Hirsh H (2000) *Trioza suavis* galls on leaves of *Elaeocapus joga*. Micronesica 33:167–169

Murray RC (1991) Arthropod pests of mango in the Caribbean Islands. Review of crop protection. University of the West Indies, Kingston
Nachiappan RM, Baskaran P (1986) Field evaluation of certain insecticidal sprays against mango leaf-hoppers. Pesticides 20:41–44
Nafus D (1991) Biological control of *Penicilllaria jocosatrix* (Lepidoptera: Noctuidae) on mango on Guam with notes on the biology of its parasitoids. Environ Entomol 20:1725–1731
Naik KC, Rao MM (1943) Studies on the blossom biology and pollination in mangoes (*Mangifera indica* l.). Indian J Hortic 1(2):107–119
Narasimham AU, Chacko MJ (1991a) The distribution of some *Rastrococcus* spp. (Homoptera: Pseudococcidae) on mango in India. Bull Entomol Res 78:703–708
Narasimham UA, Chacko MJ (1991b) The distribution of some *Rastrococcus* spp., (Homoptera: Pseudococcidae) on mango in India. Bull Entomol Res 81:445–448
Narasinham M (1959) Control of mango malformation disease. Curr Sci 28:254
Nascimento A, Malavasi A, Morgante J, Duarte AL (1992) Hot-water immersion treatment for mango infested with *Anastrepha fraterculus*, *A. obliqua*, and *Ceratitis capitata* (Diptera: Tephritidae) in Brazil. J Econ Entomol 85:456–460
Nascimento do AS, da Carvalho S, da Costa Mendonca M, Braga R (2002) Pragas e seu controle. In: de Carvalho PJ, de Queiroz Pinto AC (eds) A Cultura da Mangueira. Embrapa Informacao Tecnologica, Brasilia, pp 279–297
Nayer NK, Anantha Krishna TN, David BV (1986) Generall and applied entomology. Tata McGraw Hill, Bombay, p 589
Neuenschwander P (1993) Human interactions in classical biological control of Cassava and Mango mealybugs on subsistence farms in tropical Africa. In: Altieri MA (ed) Crop protection strategies for subsistence farmers. Westview studies in insect biology. Westview Press, Boulder, pp 143–177 [Neuens1993]
Neuenschwander P (1996) Evaluating the efficacy of biological control of three exotic Homopteran pests in tropical *Africa*. Entomophaga 41:405–424
Nguyen TN, Nguyen V, Le van D (1998) Insect pests on citrus, durian, longan and mango in selected provinces of the Meking Delta of Vietnam. In: Nguyen VH, van Mele P (eds) Proceedings of the first symposium on fruit production in the Mekong delta focusing on integrated pest management. Cautho, Vietnam, pp 71–76 (in Vietnamese)
Nicholls HA, Holland JH (1940) A textbook of tropical agriculture. Macmillan, London
Norrbom AL (2004) Host plant database for Anastrepha and Toxotrypana (Diptera: Tephritidae: Toxotrypanini). Diptera Data Dissemination Disk (CD-ROM), North American Dipterist's Society, Washington, DC
Nur N (1976) Studies on pollinationin Musaceae. Ann Bot 40:167–177
Ochoa R, Aguilar H, Vargas C (1994) Phytophagous mites of Central America: an illustrated guide. Centro Agronomico de Investigacion y Ensenanza Technical Service 6, Turrialba, Costa Rica
Orankanok W, Chinvinijkul S, Thanaphum S, Sutantawong M, Enkerlin WR (2007) Area-wide integrated control of oriental fruit fly (*Bactrocera dorsalis*, Hendel) and Guava fruit fly (*B. correcta*, Bezzii) in Thailand using the sterile insect technique (SIT). In: Vreysen MJB, Robinson AS, Hendrichs J (eds) Areawide control of insect pests: from research to field experimentation. Springer, Dordrecht, pp 517–526
Ordonyo MV (1959) A artificial production of parthenocarpic fruits. Pillip Agric 35:115–129
Ortiz R, Crouch JH (1997) The efficiency of natural and artificial pollinators in plantain (Musa spp. AAB group) hybridization and seed production. Ann Bot 80:693–695
Osman AA (1979) Notes on the control of the bud mite *Aceria mangiferae* (Sayed) in Egypt (Acarine: Eriophyidae). Bull Entomol Soc Egypt Econ Serv 9:119–126
Ovruski S, Aluja M, Sivinski J, Wharton R (2000) Hymenopteran parasitoids on fruit infesting Tephritidae (Diptera) in Latin America and the southern United States: diversity, distribution, taxonomic status and their use in fruit fly biological control. Integr Pest Manag Rev 5:81–107
Paderes A, Orden ME (2004) Factors determinating the adoption of pruning and bagging by major mango producers in the Philippines. Available at: http://www.kmitl.ac.th/science/Journal/jan2004/MANGOAdopt.pdf. Accessed 12 Mar 2005.

Paine RW (1964) The banana scab moth, Nacoleia octasema (MEYRICK): its distribution, ecology and control, Australia technical paper 145. South Pacific Commission Canberra, Nouméa, pp 1–70
Pandey AK, Rajesh Kumar KS (2009) Population dynamics of mango mealy bug, *Drosicha mangiferae* Green from Jhansi, Uttar Pradesh. Biol Forum 1(2):66–68
Papadopoulos NK, Carey JR, Katsoyannos BI, Kouloussis NA (1996) Overwintering of the Mediterranean fruit fly (Diptera: Tephritidae) in Northern Greece. Ann Entomol Soc Am 89:526–534
Parvathi C, Ramachandra YL (2005) Relation between flower density in small cardamom (*Elettaria cardamomum* L. Maton) and pollinator species diversity. Indian Bee J 67:78–82
Parvathi C, Ramachandra YL (2006) Pollination biology of small cardamom (*Elettaria cardamomum* L. Maton). Indian Bee J 67:41–46
Patel RK, Patel SR, Shah S (1977) Biology of mango hopper, *Amritodus atkinsoni* (Leth.) (Jassidae: Hemiptera) in South Gujarat. Indian J Entomol 37:150–153
Paull RE, Duarte O (2011) Tropical fruits, vol 1. CABI, Wallingford
Peck SL, McQuate GT (2000) Field tests of environmentally friendly malathion replacements to suppress wild Mediterranean fruit fly (Diptera: Tephritidae) populations. J Econ Entomol 93:280–289
Pena JE (1993) Pests of mango in Florida. Acta Hortic 341:395–406
Pena JE, Duncan R (1992) Sampling methods for *Prodiplosis longifila* (Diptera: Cecidomyiidae) in limes. Environ Entomol 21:996–1001
Pena JE, Duncan R, Vasquez T, Hennessey M (1999) Guava arthropod seasonality and control of fruit flies in south Florida. Proc Fla State Hortic Soc 112:206–209
Pena JE, Palevsky E, Otero-Colina G, Ochoa R, Meister CW (2005) Mango bud mite, *Aceria mangiferae* bionomics and control under Florida conditions. Proc Fla State Hortic Soc 118:228–234
Pena JE, Duncan R, Meister C (2006a) Chemical control of flowering thrips infesting tropical fruits. Proc Fla State Hortic Soc 119:21–24
Pena JE, Gould W, Hennessey MK, Hallman G, Crane JH (2006b) Laboratory and field infestation studies on immature green 'Tommy Atkins' and 'Keitt' mangoes to determine host status to the Caribbean fruit fly (Diptera: Tephritidae). Proc Fla State Hortic Soc 119:16–20
Pena JE, Aluja M, Wyoski M (2009) Pests. In: Litz RE (ed) The Mango, botany and used, 2nd edn. Cab International, Wallingford, pp 317–366
Peng RK, Christian K (2004) The weaver ant, *Oecophylla smaragdina* (Hymenoptera: Formicidae), an effective biological control agent of the red banded thrips *Selenothrips rubrocinctus* (Thysanoptera: Thripidae) in mango crops in the Northern Territory of Australia. Int J Pest Manag 50:107–114
Peng R, Christian K (2005a) Integrated pest management in mango orchards in the Northern Territory Australia, using the weaver ant *Oecophylla smaragdina* (Hymenoptera: Formicidae) as a key element. Int J Pest Manag 51:149–155
Peng RK, Christian K (2005b) The control efficacy of the weaver ant, *Oecophylla smaragdina* (Hymenoptera: Formicidae), on the mango leafhopper, *Idioscopus nitidulus* (Hemiptera: Cicadellidae) in mango orchards in the Northern Territory. Int J Pest Manag 51:297–304
Peng RK, Christian K (2006) Effective control of Jarvis's fruit fly *Bactrocera jarvisi* (Diptera: Tephritidae), by weaver ant, *Oecophyla smaragdina* (Hymenoptera: Formicidae), in mango orchards in the Northern Territory of Australia. Int J Pest Manag 52:275–282
Petty GJ, Stirling GR, Bartholomew DP (2002) Pests of pineapple. In: Peña JE, Sharp JL, Wysoki M (eds) Tropical fruit pests and pollinators. CABI Publishing, Wallingford, pp 157–195
Pimentel D (2007) Area-wide pest management: environmental, economic and food issues. In: Vreysen MJB, Robinson AS, Hendrichs J (eds) Area-wide control of insect pests: from research to field implementation. Springer, Dordrecht, pp 35–47
Pimentel D (2009) Pesticides and pest control. In: Peshin R, Dhawan AK (eds) Integrated pest management innovation development, vol 1. Springer, Dordrecht, pp 83–88
Pinero J, Aluja M, Vazquez A, Equihua M, Varon J (2003) Human urine and chicken feces as fruit fly (Diptera: Tephritidae) attractants for resource-poor fruit growers. J Econ Entomol 96:334–340

Poinar GO Jr, Hislop RG (1981) Mortality of mediterranean fruit fly adults (*Ceratitis capitata*) from parasitic nematodes (*Neoaplectana* and *Heterorhabditis* spp.). Institute for Research and Cognitive Science, Medical Science. Microbiol Parasitol Infect Dis 9: 641
Pope WT (1930) Papaya culture in Hawaii. Hawaii Agr Expt Sta Bull 61:40 pp
Popenoe W (1917) The pollination of the mango. U.S. Department of Agriculture, Bulletin No. 542, Washington, DC, 20 pp
Popenoe W (1920) Manual of tropical and subtropical fruits. The Macmillan, New York, 474 pp
Pradhan S (1969) Insect pests crops. National Book Trust of India, New Delhi, pp 1–280, pp 208
Pradhan S, Jotwani MG, Rai BK (1956) Bioassay of insecticides. VI comparative toxicity of modern insecticides to *Drosicha mangiferae* Green. Indian J Entomol 18(4):315–317
Prakash O (2012) IPM schedule for aonla pests. Horticulture Year, 2012 extension bulletin No. 5. National Horticulture Mission Ministry of Agriculture Department of Agriculture & Cooperation Krishi Bhawan, New Delhi, pp 1–19.
Prasad SN (1971) The mango midge pests. Cecidological Society, Allahabad
Prasad V, Singh RK (1976) Prevalence and control of mango mealy bug *Drosicha stebbingii* (Green) in Bihar. Indian J Entomol 38:214–224
Prest RL (1955) Unfruitfulness in papayas. Queensland Agric J 81:144–148
Prest RL (1957) Unfruitfulnes in pawpaws. Qld Agric J 81(3):144–148
Pruthi HS, Batra HN (1960) Important fruit pests of North West India. Indian J Agric Res Bull 80:61–64
Pushpa V, Rao GPRM, Appa P, Rao P (1973) A note on the biology and control of mealybugs (*Maconellicoccus hirsutus* Green) on mesta. Jute-Bull 36(1–2):25–28
Rahman KA (1940) Important insect pests of the mango and how to control them. Punjab Fruit J 3:6
Rai B, Verma S, Kumar K (1966) Evaluation of pesticides for control of mango bud mite, *Aceria mangiferae* Sayed (Acarina: Eriophyidae). Indian J Entomol 28:176–180
Rai B, Shah A, Patel R (1988) Biology of *Oligonychus mangiferus* (Tetranychidae: Acarina), a pest of mango in Gujarat. Gujarat Agric Univ Res J 14:5–10
Rajendran TP (2009) Integrated pest management-policy directions in the context of climate change. In: Ramamurthy VV, Gupta GP, Puri SN (eds) Proceedings of the national symposium on IPM strategies to combat emerging pests in the current scenario of climate change, Pasighat, 28–30 Jan 2009, pp 8–13
Randhawa GS (1974) Importance of pest control. Pestic Ann 8:85–87
Rani BU, Kalyanasundaram M, Suresh K (2006) Major sucking pests of Aonla and their management. Indian J Arecanut Spices Med Plants 8(3):108–110
Reboucas A, Boas IV, Martins J, Magalhaes O (1996) Manga, Tecnologia de Producao e Mercado. Universidade Estadual do Sudoeste da Bahia, Vitoria da Conquista
Reddy DB (1968) Plant Protection in India. Allied Publishers, Calcultta, p 454
Rehman KA (1940) Insect pests of the Punjab. Punjab Agric College Mag 7:5–7
Rehman KA, Latif MA (1944) Description, bionomics and control of gaint mealy bug *Drosicha stebbingi*, Green (Homoptera: coccid ac). Bull Entomol Res 35:197–209
Richards PB, Sharma HN (1934) Bulletin Department of Agriculture U.P. Fruit series No 33
Ricketts T (2004) Tropical forest fragments enhance pollinator activity in nearby coffee crops. Conserv Biol 18:1262–1272
Robacker DC (2001) Roles of putrescine and 1-pyrroline in attractiveness of technicalgrade putrescine to the Mexican fruit fly (Diptera: Tephritidae). Fla Entomol 84:679–685
Robacker DC, Warfield WC (1993) Attraction of both sexes of Mexican fruit fly, *Anastrepha ludens*, to a mixture of ammonia, methylamine, and putrescine. J Chem Ecol 19:2999–3016
Robacker DC, Martinez AJ, Garcia JA, Diaz M, Romero C (1996) Toxicity of *Bacillus thuringiensis* to Mexican fruit fly (Diptera: Tephritidae). J Econ Entomol 89:104–110
Robacker DC, DeMilo AB, Voaden DJ (1997) Mexican fruit fly attractants: effects of 1-pyrroline and other amines on attractiveness of a mixture of ammonia, methylamine, and putrescine. J Chem Ecol 23:1263–1280

Robinson AS, Hooper G (eds) (1989) Fruit flies. Their biology, natural enemies and control, vols A and B. Elsevier, Amsterdam

Rogers L, French J, Elgar M (1999) Suppression of plant growth on the mounds of the termite *Coptotermes lacteus* Froggatt (Isoptera: Rhinotermitidae). Insect Soc 46:366–371

Ronse Decraene LP, Smets EF (1999) The floral development and anatomy of Carica papaya (caricaceae). Can J Bot 77:582–598

Roubik DW (1989) Ecology and natural history of tropical bees. Cambridge University Press, Cambridge, NY, 514 pp

Rubin A, Kuslitizky W (1992) First record of *Ceranisus menes* (Hymenoptera: Eulophidae) in Israel. Phytoparasitica 20:123–124

Ruehle GD, Ledin RB (1955) Mango growing in Florida. Fla Agr Expt Sta Bull 574:90 pp

Sadana GL, Kumari M (1991) Predatory potential of the lyssomanid spider, *Lyssomanes sikkimeri* Tikader on the mango hopper, *Ideoscopus clipealis* (Lethierry). Entomon 16:283–285

Said Sajap A, Amit S, Welker J (2000) Evaluation of hexafl umuron for controlling the subterranean termite *Coptotrermes curvignathus* (Isoptera: Rhinotermitidae) in Malaysia. J Econ Entomol 93:429–433

Sakagami SF (1989) Taxonomic notes on a Malesian bee *Lasioglossum carinatum*, the type species of the subgenus *Ctenonomia*, and its allies (Hymenoptera: Halictidae). J Kans Entomol Soc 62:496–510

Sakimura K (1981) A review of *Frankliniella bruneri* Watson and description of *F. kelliae*, n. sp. (Thysanoptera: Thripidae). Fla Entomol 64:483–490

Sandhu GS, Sohi AS, Batra RC (1979a) Improved polythene band for the control of mango mealy bug in Punjab. Punjab Hortic J XIX(3 & 4):174–178

Sandhu GS, Sohi AS, Batra RC, Bhalla JS (1979b) Stem infection for the control of mango pests with particular reference to mango mealy bug *D. mangiferae* Green (Hemiptera: Margarodidae). In: Proceedings of mango worker meeting, Goa, pp 248–249

Sandhu GS, Batra RC, Sohi AS, Bhalla JS (1980) Control mango mealybug, *Drosicha mangiferae* (Gr.) by tree trunk banding. Indian J Entomol 42(2):225–229

Sankaran T, Mjeni AM (1989) Recent studies on the mango leaf-midge *Procontarinia matteiana* Kieffer and Cecconi (Diptera: Cecidomyiidae) and its parasites in India and on prospects for biological control of the pest in Oman. Acta Hortice 231:587–592

Santa-Cecília LVC, Reis PR, Souza JC (2002) About the nomenclature of coffee mealy bug species in Minas Gerais and Espírito Santo States, Brazil. [Sobre a nomenclatura das espécies de cochonilhas-farinhentas do cafeeiro nos Estados de Minas Gerais e Espírito Santo]. Neotrop Entomol 31:333–334

Saxena DK, Rawat RR (1968) Chemical control of *Drosicha mangiferae* Green a serious pest of citrus species in Madhya Pradesh. Indian J Entomol 30(1):66–68

Sayed M (1946) *Aceria mangiferae*. Bull Soc Entomol 30:7–10

Sazo L, Pizarro E, Araya JE (2006) Effect of the form of application of imidacloprid on control of the long-tailed mealy bug *Pseudococcus lingispinus* (Targioni & Tozzetti) on avocado and its impact on *Neoseiulus californicus* (McGregor) in Chile. Boletin-de-Sanidad-Vegetal 32(4):483–494

Schaefer CW, Ahmad I (2000) Cotton strainers and their relatives (Pyrrhocoroidea: Pyrrhocoridae and Largidae). In: Schaefer CW, Panizzi AR (eds) Heteroptera of economic importance. CRC Press, Boca Raton, pp 271–307

Schotman L (1989) FAO/RLAC plant quarantine programme. Data sheet on mango seed weevil. FAO accession XF8988239, 23 pp. http://www.fao.org/agris/search. Accessed 1 Sept 2008

Schreiner I (1987) Mango shoot caterpillar control on mango flowers, 1985. Insect Acaricide Tests 12:94

Schreiner I (1991) *Procontarinia* sp. nov.: a mango pest long misdiagnosed as anthracnose. In: Agricultural Development American Pacific, Crop Protection Conference Proceedings 1989, Hawaii Institute Tropical Agriculture and Human Resources, Research Extension Series 134, 53–55

Secretariat of the Pacific Community (2005) Fruit flies in Palau. Pest Advisory Leaflet No. 44, Palau

Sen AC (1955) Control of mango mealy bug in Bihar. Indian J Entomol 17(1):129.32

Sen AC, Prasad D (1956) Biology and control of the mango mealy bug. *Drosicha mangiferae* Green. Indian J Entomol 18(2):127–140

Sfemanthani B (1965) Sex expression in certain inbred selections of papaya (*Carica papaya* linn.). So Indian Hortic 13(1/2):15–19

Shah S (1980) Bionomics and control of some lesser known insect pests in Sabah. In: Proceedings of international conference on cocoa and coconuts, Incorporated Society of Planters, Kuala Lumpur, pp i–xviii, 1–766, 295–307

Shah AH, Jhala RC, Patel GM, Patel SH (1983) Evaluation of effective dose of monocrotophos for the control of mango hopper *Amritodus atkinsoni* Lethierry (Cicadellidae: Homoptera) by injection method and its comparison with foliar application. Gujarat Agric Univ Res J 9:14–18

Sharma DR, Arora PK (2009) Status and control of mealy bugs on fruit crops in Punjab. J Insect Sci 22(2):156–161

Sharma S, Abbas SR, Shukia RP, Sharma S (1998) An easy and quick method of breeding flies for pollination of mango blossoms. Insect Environ 4:76–77

Sharp JL, Ouye MT, Thalman R, Hart WG, Ingle SJ, Chew V (1988) Submersion of 'Francis' mango in hot water as a quarantine treatment for the West Indian fruit fly and the Caribbean fruit fly (Diptera: Tephritidae). J Econ Entomol 81:1431–1436

Sharp JL, Ouye MT, Hart WG, Ingle SJ, Hallman G, Gould W, Chew V (1989a) Immersion of Florida mangoes in hot water as a quarantine treatment for Caribbean fruit fly (Diptera: Tephritidae). J Econ Entomol 82:186–188

Sharp JL, Ouye MT, Ingle SJ, Hart WG (1989b) Hot water quarantine treatment for mangoes from Mexico infested with Mexican fruit fly and West Indian fruit fly (Diptera: Tephritidae). J Econ Entomol 82:1657–1662

Sharp JL, Ouye MT, Ingle SJ, Hart WG, Enkerlin E, Celedonio H, Toledo J, Stevens L, Quintero E, Reyes J, Schwarz A (1989c) Hot-water quarantine treatment for mangoes from the State of Chiapas, Mexico, infested with Mediterranean fruit fly and *Anastrepha serpentina* (Wiedemann) (Diptera: Tephritidae). J Econ Entomol 82:1663–1666

Shaw J (1961) Airplane applications of malathion bait spray for Mexican fruit fly control. J Econ Entomol 54:600–601

Shaw J, Spisshakoff LM (1958) Insecticidas para el control de la mosca Mexicana de la fruta en citricos y mango. Memorias Primer Congreso Naciona de Entomologia Fitopatologia Escuela Nacional Agricola Chapingo. Escuela Nacional Agricola de Chapingo, Texcoco, Mexico, pp 205–217

Sheikh AD, Bashir A, Jilani G (2008) Incidence and effects of fruit fly on mango quality in Punjab. Technology and Transfer Institute, Faisalabad, 3 pp

Shellie KC, Mangan RL (2002a) Hot water immersion as a quarantine treatment for large mangoes: artificial versus cage infestation. J Am Soc Hortic Sci 127:430–434

Shellie KC, Mangan RL (2002b) Cooling method and fruit weight: efficacy of hot water quarantine treatment for Mexican fruit fly (Diptera: Tephritidae) in mango. HortScience 37:910–913

Shu ZH, Yen CR, Ke LS, Wang DN, Lin TS, Liu MF, Shiesh CC (2000) Mango production in Taiwan. Acta Hortic 509:87–94

Shukla RP, Prassad VG (1984) Evaluation of some insecticides against mango hopper, *Idioscopus clypealis* (Lethierry) (Homoptera: Cicadellidae). Indian J Agric Sci 54:677–681

Shukla RP, Tandon PL (1985) Bio-ecology and management of the mango weevil, *Sternochetus mangiferae* (Fabricius) (Coleoptera: Curculionidae). Int J Trop Agric 3:293–303

Shukla RP, Tandon PL, Singh SJ (1984) Baculovirus – a new pathogen of mango nut weevil, *Sternochetus mangiferae* (Fabricius) (Coleoptera: Curculionidae). Curr Sci 53:593–599

Shukla RP, Tandon PL, Suman CL (1988) Intra-tree distribution of eggs and diapausing adults of the stone weevil. Acta Hortic 231:566–570

Sial MI (1999) Integrated pest management of mango insedt pests. M.Sc. (Hons.) thesis, Department of Agricultural Entomology, University of Agriculture, Faisalabad

Siddiqi MS, Mahur KP (1982) Comparative efficiency of some insecticides against the females of *Drosicha mangiferae* Green. Progress Hortic 14(1):63–66

Silva QM, Cavalcante RD (1977) *Vinsonia stellifera* (Westwood 1871), (Hom. Coccidae), on various plants in the state of Ceara, Brazil. Fitossanidade 2:25–26

Simao S, Maranhao ZC (1959) Insects pollinating mango. Anais esc Sup Agr 'Luiz queiroz' 16:299–304 [in portuguese, english summary]

Simmonds NW (1953) The development of banana fruit. J Exp Bot 4:87–105

Simmonds NW (1959) Banana. Longmans, Green & Co., London

Simmonds NW (1962) The evolution of bananas. Longmans, Green & Co., London

Singh RN (1946) Factors affecting development and time of hatching of eggs of *Drosicha stebingii* Green. Indian J Entomol 8(1):132

Singh RN (1947) Short notes and exhibits. Indian J Entomol 8(1):132

Singh RN (1954a) Studies on floral biology and subsequent production of fruits in the mango (*Mangifera indica* L.) varieties Dushehari and Langra. Indian J Hortic 11:69–88

Singh S (1954b) Horticultural crops as bee pasture. Indian J Hortic 11(2):49–52

Singh SM (1957) Control of Mango mealy bug. *Drosicha mangiferae* Green. Hortic Adv 1:38–47

Singh LB (1960a) Pollination. The mango: botany, cultivation, and utilization, chap 3. Interscience Publishers, New York, pp 42–43

Singh LB (1960b) The mango a handbook. Leonard Hill (Book)/ICAR, London/New Delhi, p 438

Singh SN (1961) Studies on the morphology and viability of the pollen grains of mango. Hortic Adv 5:121–144

Singh LB (1969) Mango. In: Ferwerda FP, Wit F, Veenman FH, Zonen NV (eds) Outlines of perennial crop breeding in the tropics. Veenen and Zonen, Wageningen, pp 309–327

Singh SP (1978) Propagation of a Coccinellid beetle for the biological control of citrus and coffee mealybugs. Scientific conference CPA. Indian Institute of Horticultural Research, Bangalore

Singh G (1988) Insect pollinators of mango and their role in fruit setting. Acta Hortic 231:629–632

Singh G (1991) Loss assessment, ecology and management of mango fruit fly, *Dacus* sp. Acta Hortic 291:425–436

Singh G (1997) Pollination, pollinators and fruit setting in mango. Acta Hortic 455:116–123

Singh SK, Singh GB (2003) Management of termite infestation in mango orchard by cultural practices and organic amendments. Indian J Agric Res 37:148–150

Singh RP, Singh HM (2012) Arthropod diversity and management of major sucking insect pests of Aonla (*Emblica officinalis* Gaertn). LAP Lambert Academic Publishing, 2012 Nov 15, pp 156

Singh LB, Sturrock D (1969). Mango. In: Ferwerda FP, Wit F (eds) Outlines of perennial crop breeding in the tropics. H. Veenman and zonen, n. V. Wageningen, pp 309–327

Singh RN, Majumdar PK, Sharma DK (1962) Self incompatibility in mango (*Mangifera indica* l.) Var. Dashehari. Curr Sci 31(5):209

Singh HK, Srivastava AK, Dwivedi R (1998) Pollination and fruit set behaviour of aonla (*Emblica officinalis*). Indian J Agric Sci 68(4):204

Singh D, Bhatnagar P, Niwas HOR (2010) Impact of weather variations on the incidence of mealy bug *Drosicha mangiferae* (Green) and leaf hopper *Amritodes atkinsoni* (Lethier) in mango. Environ Ecol 28(4A):2482–2484

Sinu PA, Shivanna KA (2007) Pollination ecology of cardamom *(Eletteria cardamomum)* in the Western Ghats, India. J Trop Ecol 23:493–496

Sivinski JM (1996) The past and potential of biological control of fruit flies. In: McPheron BA, Steck GJ (eds) Fruit fly pests. A world assessment of their biology and management. St Lucie Press, Delray Beach, pp 365–375

Sivinski JM, Calkins CO, Baranowski RM, Harris D, Brambila J, Diaz J, Bums RE, Holler T, Dodson D (1996) Suppression of Caribbean fruit fly (*Anastrepha suspensa* (Loew) Diptera: Tephritidae) population through releases of the parasitoid *Diachasmimorpha longicaudata* (Ashmead) (Hymenoptera: Braconidae). Biol Control 6:177–185

Sivinski J, Aluja M, Lopez M (1997) Spatial and temporal distributions of parasitoids of Mexican *Anastrepha* species (Diptera: Tephritidae) within the canopies of fruit trees. Ann Entomol Soc Am 90:604–618

Sivinski J, Pinero J, Aluja M (2000) The distributions of parasitoids (Hymenoptera) of *Anastrepha* fruit flies (Diptera: Tephritidae) along an altitudinal gradient in Veracruz, Mexico. Biol Control 18:258–269
SKUAST (1993) Package of practices for fruit crops. Shere-Kashmir University of Agricultural sciences & Technology, Jammu and Kashmir, p 80
Slater P, Slater P, Slater R (1986) The slate field guide to Australian birds. Welden Publishing, Willoughby
Smith RI (1977) Cosmopolites sordidus Germ. In: Kranz J, Schmutterer H, Koch W (eds) Diseases, pests and weeds in tropical crops. John Wiley & Sons, New York, pp 406–410
Sohi AS, Sohi AS (1990) Mango leafhoppers (Homoptera: Cicadellidae) – a review. J Insect Sci 3:1–12
Soto-Manitiu J, Jiron L, Hernandez R (1987) Chemical control and ecological observations of fruit fl ies of the genus *Anastrepha* Schiner (Diptera: Tephritidae) on mango. Turrialba 37:245–251
Spencer JL, Kennard WC (1956) Limited stigmatic receptivity may contribute to low fruit set in the mango (*Mangifera indica* l.). Am Soc Horticut Sci Proc 67:287–289
Sponagel K, Diaz F, Cribas A (1996) Las moscas de la fruta (Diptera: Tephritidae) y su importancia en la agricultura de Honduras. Fundacion Hondureña de Investigacion Agricola, La Lima, Honduras, 75 pp
Srivastava RP (1976) Studies on feasibility of integrated control of fruit pests, mango mealy bug *Drosicha mangiferae*, *Papilio demoleus*. PhD thesis IARI, New Delhi
Srivastava RP (1980) Relative effectiveness of different mechanical bands in preventing ascent of mango mealy bug. Indian J Entomol 42(1):110–115
Srivastava RP (1981) Comparative efficacy of various insecticidal dusts against mango mealybug eggs. Indian J Entomol 43:225–229
Srivastava RP (1998) Pests of mango. In: Srivastava RP (ed) Mango cultivation. International Bood Distributing, Charbagh, pp 175–299
Srivastava RP (2008) Field efficacy of neem kernel extracts against mango mealy bug nymphs, *Drosicha mangiferae* Green. J Eco-friendly Agric 3(2):150–152
Srivastava RP, Butani DK (1972) A method to prevent the mango mealy bug *Drosicha mangiferae* Green damage. Entomol News Lett 2(5):35
Srivastava S, Singh NB (2004) Survey of termites infesting mango orchards at four locations in Bareilly, Uttar Pradesh. Biol Mem 30:48–51
Srivastava R, Tandon P (1986) Natural occurrence of two entomogenous fungi pathogenic to mango hopper, *Idioscopus clypealis* Leth. Indian J Plant Pathol 4:121–123
Srivastava AS, Singh YP, Pandey RC, Awasthi BK (1973) Bionomics and control of mango mealy bug. World Crops 25(2):87–88
Stambaugh SV (1960) Forty years of papaya development. Fla State Hortic Soc Proc 73:311–314
Stark JD, Vargas R, Miller N (2004) Toxicity of spinosad in protein bait to three economically important tephritid fruit fly species (Diptera: Tephritidae) and their parasitoids (Hymenoptera: Braconidae). J Econ Entomol 97:911–915
Stebbing EP (1903) Notes on insect pests from the entomological section, Indian Museum. Insect pests of fruit trees. Indian Mus Notes 6:63–67
Steck GJ (2003) Mango fruit fly, marula fruit fly, *Ceratitis cosyra* (Walker) (Insecta: Diptera: Tephritidae). Available at: http://edis.ifas.ufl.edu/IN563. Accessed 20 July 2007
Steiner LF, Lee RK (1955) Large area tests of male annihilation method for oriental fruitfly control. J Econ Entomol 48:311–317
Steiner LF, Mitchell WC, Harris E, Kozuma T, Fujimoto M (1965) Oriental fruit fly eradication by male annihilation. J Econ Entomol 58:961–964
Storey WB (1937) The primary flower types of papaya and the fruit types that develop from them. Am Soc Horticult Sci Proc 35:80–82
Storey WB (1941) The botany and sex relationships of the papaya. Part 1. In papaya production in the Hawaiian islands. Hawaii Agr Expt Sta Bull 87:64 pp.
Storey WB (1958) Modification of sex expression in papaya. Hortic Adv 2:49–60
Storey WB (1969) Papaya. In: Ferwerda FP, Wit F (eds) Outlines of perennial crop breeding in the tropics. H. Veenman and Zonen, N. V. Wageningen, pp 389–408

Sturrock TT (1944) Notes on the mango. Sturart Daily News, Sturart, 122 pp
Sturrock TT (1966) The mango inflorescence. Fla State Hortic Soc Proc 79:366–369
Subba Rao BR (1986) *Mangoma spinidorsum* gen. et sp. n. (Hymneoptera: Eurytomidae) associated with leaf galls. Bull Entomol Res 76:389–392
Subramanian CK (1925) A note on the life history of *Cryptorhynchus mangiferae* Fab. Madras Agricultural Department yearbook. Madras Agricultural Department, Madras, pp 29–36
Summanwar A, Raychoudhury S (1968) The role of the eriophyid mite (*Aceria mangiferae*) in the causation of mango malformation. Indian Phytopathol 21:463–464
Sung IH (2005) The biology, foraging activities and resource partition between Taiwanese stingless bee *Trigona ventralis hoozana* and eastern honeybee *Apis cerana*. Doctoral dissertation, Graduate School of Science and Engineering, Ibaraki University, Mito, 136 pp. (in Japanese)
Sung IH, Yamane S, Ho KK, Chen WS (2006) Geographic distribution and nesting sites of the Taiwanese stingless bee *Trigona ventralis hoozana* and Asian honeybee *Apis cerana* on the island of Taiwan (Hymenoptera, Apidae). Jpn J Entomol New Ser 9:33–45 (in Japanese)
Sweet MH (2000) Seed and chinch bugs (Lygaeoidea). In: Schaefer CW, Panizzi A (eds) Heteroptera of economic importance. CRC Press, Boca Raton, pp 143–264
Swirski E, Lzhar Y, Wysoki M, Gurevitz E, Greenberg S (1980) Integrated control of the long-tailed mealybug, *Pseudococcus longispinus* [Hom.: Pseudococcidae], in avocado plantations in Israel. Entomophaga 25:415–426
Swirski E, Wysoki M, Izhar Y (2002) Subtropical pests of Israel. Fruit Board of Israel and Institute of Plant Protection, The Volcani Center, Tel Aviv, 284 pp
Syed RA, Ghani MA, Murtaza M (1970) Studies on the *Trypetides* and their natural enemies in West Pakistan III. Tech Bull Common Wealth Inst Biol Control 13:1–16
Tan K-H (2000) Area-wide control of fruit flies and other insect pests. Penerbit Universiti Sains Malaysia, Pulau Pinang
Tandon PL (1969) New pests of mango. In: Proceedings of mango worker meeting, Goa, pp 251–252
Tandon PL (1988) Present status of pests and their management, insect pests of tropical fruit trees and their management. Indian Institute of Horticultural Research, Bangalore, pp 278–293
Tandon PL, Lai B (1979) Studies on the chemical control of the mango hopper, *Idioscopus clypealis* Leth. (Cicadellidae: Homoptera). Int Pest Control 21:6–9
Tandon PL, Lal B (1978) The mango coccid, *Rastrococcus iceryoides* Green (Homoptera: Coccidae) and its natural enemies. Curr Sci 13:46–48
Tandon PL, Lal B (1979) Comparative efficacy of synthetic garlic oil with some modern insecticides against *Drosicha mangiferae* Green. Progress Hortic 12(3):62–65
Tandon PL, Lal B (1980) Control of mango mealy bug *Drosicha mangiferae* Green by soil application of insecticides. Entomon 5(1):67–69
Tandon PL, Lal B (1981) Effectiveness of different banding materials for preventing migration of *D. mangiferae* Green. Indian J Entomol 43(2):189–191
Tandon PL, Srivastava RP (1979) New pests of mango. In: Proceedings of mango worker meeting, Goa, pp 278–279
Tandon PL, Verghese A (1985) World list of insect, mite and other pests of mango. Technical document No 5. IIHR, Banglore, p 22
Tandon PL, Verghese A (1987) New insect pests of certain fruit crops. Indian J Hortic 44:120–121
Tandon PL, Lal B, Srivastava RP (1978) New records on additional hosts of mango mealy bug, *Droschia mangiferae* Green (Margarodidae: Hemiptera). Indian J Hortic 35:281–282
Tandon PL, Lal B, Rao GS (1983) Prediction of mango hopper *Idioscopus clypealis* (Leth.) population in relation to physical environmental factors. J Entomol 8:257–261
Tandon PL, Verghese A, Prasada GS (1989) Spatial distribution, sampling plan and appropriate transformation of the mango hopper, *Idioscopus niveoparsus* (Lethierry) (Homoptera: Cicadellidae). G Ital Entomol 4:235–242

Tanwar RK, Jeyakumar P, Monga D (2007) Mealybugs and their management. Technical Bulletin 19. National Centre for Integrated Pest Management, New Delhi, p 12. Available online: http://www.ncipm.org.in/MealybugPDFs/Bulletin-Mealybugs%20

Tenakanai D (1997) Fruit fly fauna in Papua New Guinea. In: Allwood AJ, Drew RAI (eds) Management of Fruit Flies in the Pacific, vol 76, Australian Centre for International Agricultural Research (ACIAR) proceedings. ACIAR, Canberra, pp 87–94

Thomas P, Kannan A, Degwekar VH, Ramamurthy MS (1995) Non-destructive detection of seed weevil-infested mango fruits by X-ray imaging. Post Harvest Biol Technol 5:161–165

Thomson FC (ed) (2005) BioSystematic database of world diptera, version 7.5. Available at: http://www.diptera.org/biosys.htm. Accessed 5 Nov 2007

Tigvattnanont S (1988) Biological and autoecological studies of the mango leaf-cutting weevil, *Deporaus marginatus* Pascos (Coleoptera: Attelabidae). Khon Kaen Agric J 16:51–62

Tiwari P, Tiwari JK, Ballabha R (2010) Studies on sources of bee-forage for rock bee (*Apis dorsata* F.) from Garhwal Himalaya, India: a melissopalynologaical approach. Nat Sci 8(6):5–15

Toledo J, Lara J (1996) Comparison of the biology of *Anastrepha obliqua* reared in mango (*Mangifera indica* L.) and in mombin (*Spondias mombin*) infested under field conditions. In: McPheron B, Steck G (eds) Fruit fly pests: a world assessment of their biology and management. St Lucie Press, Delray Beach, pp 359–362

Toledo J, Liedo P, Williams T, Ibarra J (1999) Toxicity of *Bacillus thuringiensis* β-exotoxin to three species of fruit flies (Diptera: Tephritidae). J Econ Entomol 92:1052–1056

Toledo J, Rasgado MA, Ibarra JE, Gomez A, Liedo P, Williams T (2006) Infection of *Anastrepha ludens* following soil applications of *Heterorhabditis bacteriophora* in a mango orchard. Entomol Exp Appl 119:155–162

Traub HP, Robinson TR, Stevens HE (1942) Papaya production in the United States, U.S. Department of agriculture circular 633. U.S. Department of Agriculture, Washington, DC, 36 pp

Tween G (1993) Fruit fly control and eradication program management: factors influencing action criteria and program design. In: Aluja M, Liedo P (eds) Fruit flies: biology and management. Springer, New York, pp 308–310

Uechi N, Kawamura F, Tokuda M, Yukawa J (2002) A mango pest, *Procontarinia mangicola* (shi) comb. Nov. (Diptera: Cecidomyiidae), recently found in Okinawa, Japan. Appl Entomol Zool 37:589–593

Umeya K, Hirao J (1975) Attraction of the jackfruitfly, *Dacus umbrosus* F. (Diptera: Tephritidae) and lace wing, *Chrysopa* sp. (Neuroptera: Chrysopidae) to lure traps baited with methyl eugenol and cuelure in the Philippines. Appl Entomol Zool 10:60–62

United States Department of Agriculture (USDA) (1981) Plant Pest News 1

US Department of Agriculture (USDA) 87, 6

van Dine DL (1906) The mango weevil (*Cryptorhynchus mangiferae* Falor.). Hawaii Agric Exper Station Bull 17:1–11

van Halteren P (1970) Notes on the biology of the scale insect *Aulacaspis mangiferae* Newst. (Diaspididae: Hemiptera) on mango. Ghana J Agric Sci 3:83–85

van Mele P, Cuc TT, van Huis A (2001) Farmer's knowledge, perceptions and practices in mango pest management in the Mekong Delta, Vietnam. Int J Pest Manag 47:7–16

van Mele P, Vayssieres J-F, van Tellingen E, Vrolijks J (2007) Effects of an African weaver ant, *Oecophylla longinoda*, in controlling mango fruit flies (Diptera: Tephritidae) in Benin. J Econ Entomol 100:695–701

van Whervin LW (1968) Citrus weevils of Jamaica and some of their parasites. Technical bulletin No 1. Citrus Research Unit, University of the West Indies, St Augustine

van Wyk M, Al Adawi A, Khan I, Deadman M, Al Jahwari A, Wingfield B, Ploetz R, Wingfield M (2007) *Ceratocystis manginecans* sp. nov., causal agent of adestructive mango wilt disease in Oman and Pakistan. Fungal Divers 27:213–230

Van Zyl E, Kotze JM, Steyn PL (1988) Isolation of *Xanthomonas campestris* pv. *Mangifera indica* from gall fly induced lesions on mango leaves. Phytophylactica 20:89–90

Vanniere H, Didier C, Rey JY, Diallo TM, Keita S, Sangare M (2004) La mangue en Afrique de l'Ouest francophone: les systemes de production et les itineraries techniques. Fruits 59:383–398

Vargas RI, Miller NW, Prokopy RJ (2002) Attraction and feeding responses of Mediterranean fruit fly and a natural enemy to protein baits laced with two novel toxins, phloxine B and spinosad. Entomol Exp Appl 102:273–282

Varma A, Lele VC, Raychoudhuri SP, Ram A, Sang A (1974) Mango malformation: a fungal disease. Phytopathologische Zeitschrift 79:254–257

Vayssieres J-F, Kalabane S (2000) Inventory and fluctuations of the catches of Diptera Tephritidae associated with mangoes in Coastal Guinea. Fruits 55:259–270

Vayssieres J-F, Sanogo F, Noussourou M (2004) Inventaire des especes de mouches des fruits (Diptera: Tephritidae) infeodees au manguier au Mali et essays de lutte raisonnee. Fruits 59:3–16

Vayssieres J-F, Goergen G, Lokossou O, Dossa P, Akponon C (2005) A new *Bactrocera* species in Benin among mango fruit fly (Diptera: Tephritidae) species. Fruits 60:371–377

Veeresh GK (1989) Pest problems in mango – world situation. Acta Hortic 231:551–565

Veeresh GK, Rajagopal D, Kumar NG (1989) Management of termites in mango orchard. Acta Hortic 231:633–638

Verghese A, Rao G, Prasada-Rao G (1985) Sequential plan for mango leaf hopper, *Idioscopus clypealis* Lethierry. Entomology 10:285–290

Verghese A, Tandon P, Prasada-Rao G (1988) Ecological studies relevant to the management of *Thrips palmi* Karny on mango in India. Trop Pest Manag 34:55–58

Verghese A, Tandon PL, Stonehouse JM (2003) Economic evaluation of integrated management of oriental fruit fly *Bactrocera dorsalis* (Diptera: Tephritidae) in mango in India. Crop Prot 23:61–63

Verghese A, Nagaraju DK, Jayanthi PDK, Vasudev V, Madhura HS (2004) Is azadirachtin useful in the management of the mango stone weevil (*Sternochetus mangiferae* Fabricius)? J Food Agric Environ 2:213–216

Vevai EJ (1969) Know your crop, its pest problems and control-15. Pesticides 15:22–31

Viraktamath CA (1997) A revision of the Idiocerine leafhopper genus *Amritodus* (Hemiptera:Cicadellidae) breeding on mango. Entomon 22:111–117

Viraktamath S, Viraktamath CA (1985) New species of *Busoniominus* and *Idioscopus* (Homoptera: Cicadelidae: Idiocerinae) breeding on mango in south India. Entomon 10:305–311

Vyas RV, Patel JJ, Godhani PH, Yadav DN (1993) Evaluation of green muscardine fungus (*Metarrhizium anisopliae* var. *anisopliae*) for control of mango hopper (*Amritodus atkinsoni*). Indian J Agric Sci 63:602–603

Waddell WJ, Crooks NH, Carmichael PL (2004) Correlation of tumours with DNA adducts from methyl eugenol and tamoxifen in rats. Toxicol Sci 79:38–40

Wagle PV (1929) A preliminary study of the pollination of the alphonso mango. Agric J India 24(14):259–263

Waite GK (1998) Integrated pest management in tropical and subtropical fruits in Queensland. In: Proceedings of the international conference of IPM, Guangzhou, p 135

Waite GK, Huwer RK (1998) Host plants and their role in the ecology of the fruitspotting bugs *Amblypelta nitida* Stal and *Amblypelta lutescens* (Hemiptera: Coreidae). Aust J Entomol 37:340–349

Waite GK, Petzl N (1994) A search for egg parasites of *Amblypelta* spp. (Hemiptera: Coreidae) in South-East Queensland. In: Menzel C, Croswell A (eds) Maroochy Research Station report No 7, pp 13–14

Wang XG, Jarjees EA, McGraw BK, Bokonon-Ganta AH, Messing RH, Johnson MW (2005) Effects of spinosad-based fruit fly bait GF-120 on tephritid fruit fly and aphid parasitoids. Biol Control 35:155–162

Waterhouse DF (1998) Biological control of insect pests: South-East Asian prospects. Australian Centre for International Agricultural Research Monograph No 51, Canberra, 548 pp.

Waterhouse DF, Norris KR (1987) Biological control: Pacific prospects. Inkata Press, Melbourne, 454 pp

Waterhouse DF, Norris KR (1989) Biological control: Pacific prospects – supplement 1. Australian Centre for International Agricultural Research, Canberra, 123 pp

Waterhouse GF, Sands DPA (2001) Classical biological control of arthropods in Australia, Australian centre for international agricultural research (ACIAR) monograph no 77. Australian Centre for International Agricultural Research (ACIAR), Canberra, 560 pp

Watson JR (1917) Thysanoptera of Florida. Fla Buggist 1:53–77

Watson GE, Ooi PAC, Girling DJ (1995) Insects on plants in the Maldives and their management. International Institute of Biological Control, Ascot, 124 pp

Wen HC, Lee HS (1978) Bionomics observation and control of mango brown leafhopper (*Idiocerus niveosparsus* Leth.). J Agric Res China 27:47–52 (in Chinese)

Wester PJ (1920) The mango. Philipp Bur Agric Bull 18:3–70

Wharton RA (1978) Classical biological control of fruit-infesting Tephritidae. In: Robinson AS, Hooper G (eds) Fruit flies. Their biology, natural enemies and control, vol B. Elsevier, Amsterdam, pp 303–313

Wharton RA, Marsh PM (1978) New world Opiinae (Hymenoptera: Braconidae) parasitic on Tephritidae (Diptera). J Wash Acad Sci 68:147–165

White IM, Elson-Harris M (1992) Fruit flies of economic significance. CAB International, Wallingford

Whitwell AC (1993) The pest/predator/parasitoid complex on mango inflorescences in Dominica. Acta Hortic 341:421–432

Wih K, Billah MK (2012) Diversity of fruit flies and mealybugs in the Upper-West Region of Ghana. J Dev Sustain Agric 7(1):39–45

Williams DJ, Granara de Willink MC (1992) Mealy bug of Central and South America. CAB International, London, 635 pp

Wolfe HS (1962) The mango in Florida 1887–1962. Fla State Hortic Soc Proc 75:387–391

Wolfe HS, Lynch SJ (1940) Papaya culture in Florida. Fla Agric Expt Sta Bull 350:35 pp.

Wongsiri S, Chen PP (1995) Effects of agricultural development on honey bees in Thailand. Bee World 76:3–5

Woodruff RE (1985) Citrus weevils of Florida and the West Indies: preliminary report on systematics, biology and distribution (Coleoptera: Curculionidae). Fla Entom 68:370–379

Woods B, Lacey IB, Brockway CA, Johnstone CP (2005) Hosts of Mediterranean fruit fly *Ceratitis capitata* (Wiedemann) (Diptera: Tephritidae) from Broome and the Broome Peninsula, Western Australia. Aust J Entomol 44:437–441

Wu WJ, Lin TM (1994) [Investigation of pollinators of mango and rearing experiment.] Proceedings of a symposium on the practical aspects of some economical fruit trees in Taiwan, Taichung DARES. Series 33, pp. 13–20. (in Chinese)

Wysoki M, Ben-Dov Y, Swirski E, Izhar Y (1993) The arthropod pests of mango in Israel. Acta Hortic 341:452–466

Xu Q, Sun XH, Wu DH, Zhang ZP (1999) Biological characteristics and control methods of *Drosicha corpulenta* (Kanawa). J Jiangsu Forestry Sci Techonol 26(1):52–54

Yadav, J .L. Singh, S. P. and Kumar, R . 2004. The population density of the mango mealy bug (*Drosicha mangiferae* G.) in mango. Progressive Agriculture. 4: 1, 35–37.

Yang JF (1959) The origin varieties, breeding, cultivation and management in mango fruits. J Chinese Soc Hortic Sci 5:74–89 (in Chinese)

Yang SR, Lee KC, Chang WN (1990) Physiological studies of sex expression of mango. Res Bull Tainan DAIS 25:31–50 (in Chinese)

Yazdani SS, Mehto DN (1980) A note on the control of mango hopper, *Amritodus atkinsoni* Leth. at Dholi. Indian J Entomol 42:275–276

Yee W (1958) The mango in Hawaii. Hawaii Agric Ext Sen Cir 388:26 pp

Yee W (1987) The mango in Hawaii. Cooperative Extension Service, University of Hawaii Circular 388:19–22

Young TW (1942) Investigations of the unfruitfulness of the haden mango in Florida. Fla State Hortic Soc Proc 55:106–110
Young TW (1955) Influence of temperature on growth of mango pollen. Fla State Hortic Soc Proc 68:308–313
Yousuf M (1993) Mango mealybug control with polyethylene bands. Pak Entomol 15:129
Yousuf M, Ashraf M (1987) Effect of some organophosphates on major insect pests of mango by stem injection. Pak Entomol 9:9–12
Yousaf M, Gagan M (1993) Prospis species in the arid and semi-arid zones of India. In: Proceedings of the conference held at Arid zone research institute, Jodhpur, 21–23 Nov 1993, p 134
Zaheruddeen SM, Sujatha A (1993) Record of *Deanolis albizonalis* (Hampson) (Pyralidae: Odontinae) as mango fruit borer in Andhra Pradesh. J Bombay Nat Hist Soc 90:528
Zucchi RA (1988) Moscas-das-frutas (Diptera: Tephritidae) no Brasil: Taxonomia, Distribuicao Geograficae Hospedeiros. Fundacao Cargill, Campinas
Zucchi RA (2000) Especies de *Anastrepha*, sinonimias, plantas hospedeiras e parasitoides. In: Malavasi A, Zucchi RA (eds) Moscas-das-frutas de Importancia Economica no Brasil: Conhecimento Basico e Aplicado. Holos Editora, Ribeirao Preto

Chapter 9
Subtropical Fruits

The subtropics are the geographical areas which lie between the tropics of cancer and tropics of Capricorn at latitudes 23.5°N and 23.5°S. The fruits grown in subtropical areas require mild temperatures throughout the year, but can survive a light frost. The important fruit crops of subtropics include Avocado, Citrus, Grape, Litchi, Loquat, Mango, ber, Olive, Passion fruit, Persimmon and Pomegranate.

9.1 Citrus Pollination

Citrus comprises a group of fruits which are collectively placed in the family Rutaceae. The most commonly grown citrus types are *Citrus sinensis* Osbeck (sweet orange), *C. aurantifolia* Swing (Lime), *C. reticulata* Blanco (mandarin,tangerine), *C.aurantium* L.(Sour orange), *C.limon* (Lemon) *C.medica* L.(citron), *C.grandis* Osbeck (pummelo), *C. paradisii* Macf. (grape fruit), *Fortunella margarita* (lour) Swingle and *Poncirus trifoliata* (trifoliate orange). On account of existence of very high degree of cross-fertility, both between the species of citrus and among different genera of citrus i.e. *Fortunella* and *Poncirus,* provides enough scope for the development of both simple and multiple hybrids.

A citrus flower is usually botanically perfect (both male and female sex organs are in the same flower). The female portion of the flower is known collectively as the pistil and is comprised of the stigma, style and ovary (Fig. 9.1). During the sexual process, pollen is deposited on the sticky surface of the stigma, germinates and grows down the inside of the style into the ovary. The pistil is surrounded by a whorl of stamens (the male portion of the flower) which consists of long filaments with anthers at the top. The anthers are hollow and contain the pollen necessary for the process of sexual fertilization. Exterior to the stamens and pistil are the petals (usually five) and then the sepals which are fused into a alp-like structure (the "calyx") and like the other flower parts, attached to the receptacle.

D.P. Abrol, *Pollination Biology, Vol. 1*, DOI 10.1007/978-3-319-21085-8_9

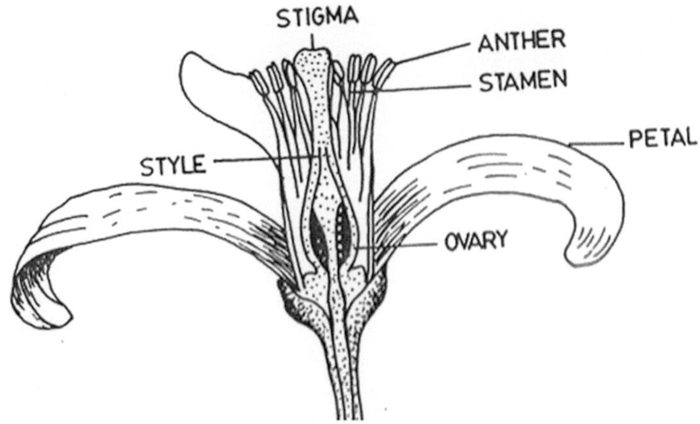

Fig. 9.1 (**a**, **b**) Diagrammatic representation of open citrus flower. (**c**, **d**) Vertical section of citrus flower (Hume 1957)

Citrus flowers occur either singly or in clusters. They have pleasant fragrance to attract the pollinating insects. The flowers are white or pinkish in colour. They are mostly hermaphrodite, releasing pollen only ehen stigmas are receptive. Citrus flowers produce very good amount of copious nectar and nectar secretion continues for at least 48 h after opening the flowers (Vansell et al. 1942). Vansell et al. (1942) further found that some flowers produced 20 microliters of nectar, compared to 0.8–2.4 microliters per blossom for an alfalfa flower.

The flower opening in citrus starts in the early hours of the day and continues till in the afternoon (8 AM to 4 PM), with peak anthesis in the noon. As large number of cultivars set substantial fruits without the aid of the pollinating, the citrus fruits in general presents wide variation in their pollination requirements. However, Webber (1946) observed self pollination in many cultivars which took place mainly due to the fact that the anthers touched stigma or when pollen fell on the stigma.

9.2 Pollination Requirements

Citrus pollen is heavy and sticky and depends upon contact of flower parts or insects to effect pollination and cannot be carried by wind (Krezdorn 1986a; Sanford 1992; Schneider 1968). Therefore, the presence of pollinating insects is necessary for quality and quantity fruit production. Pollination requirements of citrus crops such as orange (*Citrus sinensis*), tangerine (*C. reticulata*), grapefruit (*C. paradisi* Macf.), and lemon (*C. limon*) are so diverse that no generalizations can be made.

In citrus, breeding systems are quite variable among and within the species (Cameron and Frost 1968; Crane and Walker 1984; Free 1993; McGregor 1976; Roubik 1995). In some there is almost complete self-sterility, in addition to apomixis, polyembryony and parthenocarpy (Soost 1969, 1987; Mcgregor 1976; Soost and Cameron 1975). The self-incompatibility has been genetically demonstrated in citrus. The tetraploids show reduced fertility and are slow in growth compared to their diploid counterparts (Soost and Cameron 1975; Soost 1987). Earlier, citrus crop was said to have little or no need of pollination. However, Krezdorn (1970) reported better fruit set and higher yields through better pollination. Some sweet oranges are completely or partially self-incompatible (Cameron and Frost 1968; McGregor 1976; Sanford 1992) depending upon insects for pollination and for fruit set (Webber 1946; Krezdorn 1986a; Roubik 1995). However, in some cultivars, most seeds are produced by asexual means (McGregor 1976). McGregor 1976 found that in 'Marsh' grapefruit, open-pollinated flowers set about twice as many seeds and four times as many fruits as spontaneously, self-pollinated flowers. Similar results were reported by Crane and Walker (1984) who found that insect pollination increases fruit set in grapefruit.

9.3 Role of Honeybees in Citrus Pollination

Honeybees have been reported as the main pollinators of citrus collecting both nectar and pollen (Plate 9.1). Fahn (1949) established a positive correlation between the quantum of nectar secreted and the nectar size in 11 different citrus species. Vansell et al. (1942) studied the bee activity in relation to orange nectar and pollen. The investigation showed an initial sugar percentage ranging between 13 % and 17 % which later scaled to 31 % in older flowers. The bee visitation was more on older than on younger flowers, but Loper et al. (1976) observed decrease in nectar (*C. macrophylla, C. depressa*, Fairchild tangerine) content in older flowers. Wafa and Ibrahim (1960) and Hassanein and Ibrahim (1959) found honeybees to be most important amongst insect visitors for citrus pollination in Egypt. They indicated that honeybee visits were objective based specifically either for nectar collection, pollen collection or both. They further found that pollen collectors spent less time per flowers (5–8 s) than the nectar collectors (15–20 s). The nectar collection activity was more intense during early hours of the day with peak between 0900 and 1500 h and pollen collectors made their visits during later part of the day. In Yucatán, Mexico, *A. mellifera* was most abundant comprising more than 98 % of the flower visitors in lemon and sweet orange orchards (Grajales-Conesa et al. 2013)

Dhankar (1999) reported number of insect visitors frequenting citrus flowers which included *Apis dorsata* F., *Apis florea* F., *Apis mellfera* L., *Eumenes petiolata* F., *Polistes hebraeus* F., *Xylocopa aestuans* L., *Xylocopa amethystina* F., *Xylocopa* sp., *Utethesia pulchella* L., *Utethesia pulchelloides* Hampson, *Danais chrysippus* L., *Thysanoplusia orichalcea* F., *Papilio demoleus* L., *Papilio polytes* L., *Tros aristolochiae* F., *Delias eucharis* Drury, *Huphina* sp., *Ceryx godarti* Boisduval, *Psichotoe duvauceli* Boisduval and *Coccinella septempunctata* L.

Kumar and Lenin (1998) found that in orange, *Citrus reticulata Trigona iridipennis* constituted 41.3 % of insect visitors, *A. florea* 30.3 %, *A. mellifera* 3.2 % and *A. cerana* 3.1 %. The last 2 species were attracted by several other species that were flowering in

Plate 9.1 (**a**) Swellotail butterfly on persimmon flower (**b**) Pieridae butterfly foraging on persimmon flower (**c**) Ceratina on citrus flower (**d**) Bumblebee on citrus flowers (**e**) *Apis cerana* on citrus flower (**f**) *Apis mellifera* on citrus (**g**) *Apis florea* on cherry flower (**h**) Sweet and trigona bee pollinating cherry flowers

the area. Bees visiting the orange flowers foraged for nectar from 07.00 to 17.00 h with a peak at 08.00 to 10.00 h; peak of pollen-collection was at 07.30 to 09.00 h. Average fruit set was 3.8 %, but on flowers bagged to exclude insects it was only 1.3 %.

Manzoorul-Haq et al. (1978) studied the impact of insect pollinators such as *Apis dorsata* and *A. florea* on the fruit production and quality characteristics of Kinnow mandarin (*Citrus reticulata*) and found that fruit set, fruit size, juice content and number of seeds were significantly higher in open pollinated branches than those isolated from insect visits.

Chacoff and Aizen (2007) studied the insect visitors to grapefruit flowers and found that of the 50 species of insects visiting grape fruit flowers Africanized honeybees comprised more than 95 % of the flower visitors, followed by stingless bees and solitary bees. Abrol (2013) reported that in case of citrus, 18 species of insects belonging to 3 orders, 9 families and 13 genera were found to visit the sweet orange (*Citrus sinensis* L. Osbeck,) flowers. These included honeybees *Apis florea* F. *Apis cerana, Apis dorsata* F. *Apis mellifera* L., Carpenter bees *Xylocopa fenestrata, X. pubescens*. Besides, *Musca spp, Eristalis spp, Syrphid spp*, and butterflies were other insect visitors observed on citrus flowers. Of all these insects, honeybees were most abundant and comprised more than 63 % of the flower visitors. Studies on commencement and cessation of flight activities of honeybees revealed that *Apis dorsata* commenced field activities much earlier than *A. cerana* followed by *A. mellifera* and *A. florea*.similarly, activities of *A. dorsata* also ceased earlier than *A. mellifera, A. florea* and *A. cerana*, respectively. On an average, *A. cerana* had a higher foraging duration in the field followed by *A. dorsata, A. mellifera* and *A. florea*. The data further revealed that peak foraging periods also varied fron one species to another. Peak foraging occurred between 0900 and 1400 in *A. cerana*, 1100–1300 in *A. mellifera*, 1000–1100 in *A. dorsata* and 1300–1500 in *A. florea* similarly foraging rates, duration of foraging trips and population dynamics varied from one bee species to another depending upon their adaptiability.

Souza and Cauto (2002) studied pollination by insects in sweet orange (*Citrus sinensis* L. Osbeck) cultivars. They found that the uncovered flower buds produced more fruits than the covered ones. The flower buds covered in the morning produced less fruits than the ones covered in the afternoon. It was also observed that fruits derived from covered treatment were smaller and having smaller number of seeds.

Toledo et al. (2013) reported that Orange (*Citrus sinensis* L. Osbeck) that pollination results in production of fruits in greater quantity and better quality. They found that in Brazil, pollination by Africanized honey bees in *C. sinensis* can result in sweeter fruits, besides increasing production of Pera-Rio orange by about 30 %.

9.4 Pollination Recommendations and Practices

There is no uniformity in recommendations on the number of honeybee colonies to be placed in citrus orchards. This is most likely due to studies conducted in different locations with different cultivards which have variable breeding systems. Baldwin

(1916) recommended five colonies per acre. Van Horn and Todd (1954) recommended one colony per acre of 'Clementines'. Similarly, Robinson and Krezdorn (1962) recommended a minimum of one strong colony of honey bees per acre of 'Orlando' tangelos. Soost (1963) considered 2 hives per acres sufficient for optimum fruit production." In general 1–4 hives/ha (McGregor 1976) could result in optimum yield, although small variations are expected for different cultivars in different locations". The colonies probably should be present throughout the citrus flowering period for maximum benefit. The colonies should be placed in orchards when roughly 5–10 % of the flowers have blossomed so that the bees focus on target crops. The colonies should be placed in groups of 4–8 at intervals of 150 m throughout the orchard.

9.5 Citrus Pests

Citrus is an important fruit crop cultivated in tropical and subtropical areas of the world. Citrus ecosystem hosts a wide diversity of organisms. More than 300 insect species, 24 mites, 17 nematodes and several vertebrate have been recorded that attack citrus crop. Most of these insects are not a serious threat, but they develop at least a part of their life cycle in citrus ecosystems and occasionally cause damage due to environmental modifications. In certain cases outbreaks of secondary pests occur due to indiscriminate use of pesticides (Hardin et al. 1995; Dutcher 2007). Invasive pests are also cause of concern and in some areas more than 80 % of the key pests of citrus are of exotic origin (Franco et al. 2006). Integrated pest management of citrus requires a basic knowledge of environmental conditions in which the orchard is located, so that suitable agricultural practices such as appropriate cultivar, tillage, weeds management and pruning etc. are adopted. Information should be available on relation between the injurious economic levels and the population of the pests, so that the interventions in most appropriate manner using all available control methods (mechanical, biological and chemical) be made. National Research Centre for Citrus Nagpur, India has listed several species of insect pests attacking citrus (Plate 9.2). They include "Citrus blackfly (Aleurocanthus woglumi Ashby), whitefly (Aleurolobus spp., Dialeurodes citri), psylla (Diaphorina citri Kuwayama), leaf-miner (Phyllocnistis citrella Stainton), lemon butterfly (Papilio demoleus), aphids (Aphis gossypii Glover, Aphis spiraecola Patch, Toxoptera spp.), thrips (Aeolothrips sp., Scirtothrips ologochaetus (Karny), Thrips hawaiiensis (Morgan), Scirothrips dorsalis Hood), mealy bugs (Planococcus citri (Risso), Maconellicoccus hirsutus, Nipaecoccus viridis (Newstead), Ferrisia virgata), scales (Coccus hesperidus Linn),bark eating caterpillar (Indebela spp.), fruit piercing moths (Eudocima materna (Linnaeus), Eudocima homaena (Hubner), Parallellia stuposa Fabricius, Chalciope mygdon (Cramer), Ericeia inangula (Guenee), Trigonodes hyppasia (Cramer) Eudocima spp., Achaea janata L.), fruitfly (Bactrocera correcta (Bezzi), Bactrocera dorsalis (Hendel), Bactrocera invadens Drew, Tsuruta and White, Bactrocera zonata (Saunders)) Bactrocera diversa (Coquillett) Bactrocera

Plate 9.2 (**a**) Syrphid fly on pear flower (**b**) Indian bee on plum flower (**c**) European bee on plum (**d**) *Apis florea* on plum flowers (**e**) Syrphids and sweet bee on apple flower (**f**) Pieridae butterfly on apple flower (**g**) *Apis cerana* on apple flower (**h**) Syrphids and tachanid fly pollinating apple flowers

cucurbitae (Coquillett), Bactrocera tau (Walker), Bactrocera vishnu Drew and Hancock, Dacus (Callantra) discophorus Hering and citrus mites (Eutetranychus orientalis Klien, Brevipalpus pheonicis and Polyphagotarsonemus latus (Banks)) are recorded on various citrus cultivars". The information on important pests of citrus such as insects, mites and nematodes is given below:

India occupies fifth position in the citrus production in the world. The other important citrus producing countries are Spain, USA, Israel, Morocco, South Africa, Japan, Brazil, Turkey and Cuba. In India alone, more than 250 species of insects and mites have been reported infesting different species of citrus. The important ones include Citrus butterfly, leaf miner, blackfly, whitefly, psylla, scales etc. that cause severe damage to the citrus crop. A brief description of important pests is given below:

9.5.1 Bark Borer, Inderbela tetraonis, Inderbela quadrinotata (Cossidae: Lepidoptera)

This pest is distributed in India, Bangladesh, Burma and Srilanka. The caterpillar makes holes into the branches and weakens the tree. Presence of webby mass of chewed wood particle and excreta of larvae conspicuously plastered on tree trunks is the clear indication of damage by these borers. The feeding of these pests causes interruption in translocation of cell sap, which adversely affect the growth and fruiting of the plant. The pest can be controlled by removing the webbing followed by plugging the holes with cotton wool soaked in the 2–5 ml of Dichlorvos 76EC solution (20 ml/10 l of water) or inject kerosene oil. The holes are then sealed with mud. The best time for its management is September-October and the operation should be repeated in January-February.

9.5.2 Citrus Whitefly, Dialeurodes citri; Aleurocanthus woglumi (Aleurodidae: Hemiptera)

The pest is distributed in India, China, Vietnam, America, Srilanka, Bangladesh, Japan, Mexico, Guatemala, south America, and south France. The nymph and adults suck the sap of the plants and during this process, secrete honeydew due to which sooty mould develops on the leaves. Sometimes, the infestation is so severe that the whole orchard looks black. Fruits turn black in colour and have insipid taste. The attack is more intense on the shady side of the tree. Drenching the trees with sufficient solution (preferably 8–10 l for fully-grown tree) of either Monocrotophos (1.5 ml/litre of water) or Acephate (0.8 ml/litre of water) controls the pest effectively.

9.5.3 *Aphids, Toxoptera citricida, Myzus persicae and Aphis gossypii (Aphididae: Hemiptera)*

On citrus, 18 aphid species are recorded worldwide (Blackman and Eastop 2000). Of these, only *Aphis spiraecola* Patch, *A. gossypii* Glover, *Toxoptera citricidus* (Kirkaldy) and occasionally *T. aurantii* (Boyer de Fonscolombe) are to be considered as pests both for direct or indirect damage (Barbagallo et al. 1997), while all remaining species are of minor economic importance. Nymphs and adults suck sap from tender leaves and shoots. Affected leaves turn yellow, get curled, deformed and dry up. Growth of young shoots is adversely affected. Plant growth is stunted. Sooty mould is produced on honeydew excreted by aphids. Infestation of aphids at the time of flowering is associated with reduced fruit set. Black aphids and peach green aphids are vector of **citrus tristeza virus** (CTV). Numerous natural enemies (lacewings, ladybirds, syrphid flies and hymenopterous parasitoids) have detrimental effects on colonies and greatly contribute to reduce infestations of aphids. Population aphids can be easily controlled by spraying the plants with Parathion (0.03 %) and Malathion (0.03 %). Also Monocrotophos (0.025 %) effectively controls this pest. Spraying should be immediately undertaken as soon as the pest is observed. Conservation of coccinellid populations.

9.5.4 *Citrus or Lemon Butterfly, Papilio demoleus (Papilionidae: Lepidoptera)*

It is widely distributed from north Australia to Arabia including Iran, Pakistan, India, Srilanka, Burma, Bangladesh, China, Taiwan, south-east Asia, Indonesia, Africa, Philippines, Japan, Formosa. It is the most destructive pest in nurseries. The caterpillars feed on the young foliage at the nursery stage and also on young flushes of grown up trees. The caterpillar feed voraciously on leaf lamina leaving behind only the midrib. In case of severe infestation, entire tree is defoliated. *Trichgramma evanescens* West., *Petromalus luzonensis* and *Telonomus sp.* are parasite on eggs of this pest, while *Erycia nymphalidaephaga* Bar., *Charops sp., Brachymaria sp.* parasitize the larvae. The fully grown caterpillars are green in colour. Hand picking of the various stages of the pest and their destruction in nurseries and new orchards has been suggested to suppress the population of this pest. Spraying with Monocrotophos (1.5 ml/litre of water) is effective in managing the pest. Spraying of the entomogenous fungus, Bacillus thuringiensis Berliner or nematode DD-136 strain also gives very high mortality.

9.5.5 *Citrus Psylla, Diaphorina citri (Psyllidae: Hemiptera)*

The pest is distributed in India, China, Formosa, Japan, Burma, Srilanka, East Indies, New Guinea, Pakistan, Philippines, Indonesia. The pest is active throughout the year but development is slow in winter. Both the adults and nymphs suck the sap from the tender parts of the buds, leaves, branches and injects a toxic substance into them. The nymphs excrete white crystalline waxy pellets on which black sooty mould may develop which reduces the photosynthetic area. In case of severe attack leaves get distorted, curled up and ultimately fall resulting in complete defoliation of the plant. This pest acts as a vector for spreading the 'citrus greening' disease. A number of natural enemies such as species of coccinellids and ladybird beetles attack the nymphs of citrus psylla. In areas where these insects are already established, biological control may give satisfactory results in containing their populations and various parasitoids are known to live on these psyllids. In particular, the eulophid wasps *Tamarixia radiata* (Waterston) and the encyrtid wasp *Diaphorencyrtus aligarhensis* (Shafee, Alam and Agarwal) to control *D. citri* (Gómez-Torres et al. 2012), species originating from Punjab (India) and Taiwan, respectively, provided good control of the psyllid (Hoy and Nguyen 2001; Qureshi et al. 2009) in various parts of the worlds. Discrete possibilities of natural control are provided by *T. dryi* (Waterston) on *T. erytreae* (van den Berg and Greenland 2000). In the field, when a new infestation site is found, contact insecticides must be timely and vigorously applied. The pest can be controlled by spraying the plants with Parathion (0.025 %). Also application of Monocrotophos (0.025 %) or Malathion (0.03 %) or Dimethoate (1.5 %) is useful during March or with the appearance of the pest and again in the first week of September.

9.5.6 *Fruit Sucking Moths, Eudocima fullonia and Eudocima maternal (Noctuidae: Lepidoptera)*

They are distributed throughout India. The moths are nocturnal in habit. During the daytime, they hide in fallen leaves and in weeds and become active at dusk and swarm in large numbers when citrus fruits are about to ripen. The moths continue feeding throughout the night and cause colossal damage. They pierce the ripened fruits and suck the juice from them. Such fruits are exposed to the secondary infections of diseases and infestation of flies. The affected fruit usually falls within a few days. Elimination of alternate hosts plants from the vicinity of the orchards (especially Tinospora cardifolia) and collection and destruction of affected fruits reduce the pest population. Uses of poison baits have proved useful. Effective bait may be prepared by mixing 15 g lead arsenate and 450 g molasses in 10 l of water. A little vinegar may be added to it and the bait should be suspended from trees in shallow wide-mouthed containers. Bagging of fruits at small scale is effective. Creating smoke in the orchards after sunset. Fallen fruits should be disposed off as they attract the moths. Spray trees with 1 kg of Sevin 50 WP (carbaryl) in 500 l of water per acre at the time of maturity of fruits.

9.5.7 Citrus Leaf Miner, Phyllocnistis citrella (Gracillaridae: Lepidoptera)

This pest is found all over the orient from Africa to Australia including India and Pakistan. It causes damages both in nursery and in grown up stages of the citrus plant. The larvae attack tender leaves and feed in the epidermal layers of the leaf by making serpentine mines in which air gets trapped and gives them silvery appearance. The affected leaves turn pale yellow, get distorted and crumpled. Such leaves gradually dry and die away. The attack of this pest also encourages the development of **citrus canker** disease. To keep the pest population under check, pruning of all the affected parts during winter should be done. Spraying the plants with Methyl Demeton (0.03 %) at the emergence of new leaves is highly beneficial. Application of Phorate 10 G (2.5 kg a.i./ha) applied 1 day before planting is effective in reducing the larval population of leaf miner.

9.5.8 Fruit Flies, Bactrocera zonata and Bactrocera dorsalis (Tephritidae: Diptera)

In the genus *Bactrocera* there are several species that may be very destructive for citrus. Most of them are native from Asia and afterwards some species have spread in other areas of the world. The female adult fruit fly punctures the ripening fruits by penetrating its needle like ovipositor and lays the eggs inside. On hatching, the maggots feed on pulp. Fruits at colour break stage are more prone to its damage. Infested fruit shows many dark green depressions due to punctures caused by insertion of ovipositor by female fly. Later on, the damaged area around the punctures become enlarged and yellow. On squeezing the infested fruit, a number of jets of juice come out, as there are many holes on a single fruit. Rotting of the fruit occurs due to fungal and bacterial infection through the puncture hole and due to feeding by maggots, resulting in premature fruit fall. Collection and destruction of infested fruits reduces the insect population. Use of flytraps containing 1 % Methyl Eugenol and 0.5 % Malathion mixed with sugar syrup 2 months prior to harvesting is effective.

9.5.9 *Cottony Cushion Scale,* **Icerya purchasi** *Maskell*

It is a serious pest of citrus in tropical and subtropical areas of the world. It is a polyphagous pest but citrus spp. are the main hosts. Full grown larvae are broadly oval (3.0 × 1.5 mm), reddish brown to brick red in colour and adult females are flat oval (4.5 × 3.3 mm), brown to reddish brown soft bodied scale. The most conspicuous part of the of the insect is the large, white flutted egg–sac which is secreted by the female. The adult males are slender, about 3 mm long, reddish purple in colour. The pest is active throughout the year with maximum multiplication during the hot dry months. As the males are rare, the scales reproduce parthenogenetically. The female lays up to 700 eggs in the ovisac held behind the body. Young nymphs red in colour hatch out from the eggs within 24 h during hot months and may take several weeks when the weather is cold. The newly emerged nymphs move a little away before fixing them on leaves and twigs for feeding. After moulting three times, they become full adults females. Depending upon weather conditions the life cycle is completed in 46–240 days.

The leaves and twigs are infested with large white fluted scales. The infested leaves turn pale and the fall prematurely. The lady bird beetle, Rodolia cardinalis (Mulsant) introduced from Australia could effectively control this pest. Application of Monocrotophos (0.025 %) or Malathion (0.03 %) or Dimethoate (1.5 %) is useful during March or with the appearance of the pest and again in the first week of September.

9.5.10 *Mealybug, Planococcus citri (Pseudococcidae: Hemiptera)*

Citrus mealy bugs are found in all the citrus growing areas of the world. Nymphs and females suck sap from leaves, twigs, tender shoots, and from fruits at the base near the fruit stalk. Besides they also excrete honeydew on which sooty mold fungus grows, giving the plant a blackish look. Flowers and fruit dry up and fall. They also damage citrus nursery seedlings. Black ants are associated with mealy bug infestation and number of ants can be seen moving up and down on the trees infested with mealy bugs. Do not grow tall crops like cotton, bajra, maize, berseem, okra and creeper like vegetables in the orchards. Maintain proper sanitation of the orchards. Drench spray 1875 ml 20 EC (chlorpyriphos) in 500 l of water first on the appearance of the pest and repeat the spray 15 days after spray.

9.5.11 *Citrus Thrips, Scirtothrips citri (Thripidae: Thysanoptera)*

All thrips stages may attack the fruits, but the second instars larvae is the most damaging because they tend to refuge under the sepals (Webster et al. 2006). They cause damage to flowers, leaves and young as well as grown-up fruits by lacerating, rasping and sucking the cell sap. Damage caused by thrips reduces the quality and the fresh market value of fruits. Chemical treatments if necessary, non-toxic insecticides be used to protect bees and other beneficial insects. Thrips damage in nursery cause stunted growth of plant. Repeated chemical applications be avoided which may lead to develop resistance in thrips population. Spray 1000 ml Metasystox 25 EC (oxydemeton methyl) or 1000 ml Fosmite 50 EC (ethion) twice in 500 l of water. First spray is given in mid March and second in mid April.

NRC Nagpur (2014) has identified different biocontrol agents for control of citrus insect pests which include, "Mallada boninensis, Serangium paracesetosum, spiders (predators), Encarsia bennetti, E. opulent, Eretmocerus gunturiensis on citrus Blackfly, Chrysoperla crassinervis, M. boninensis, Syrphids, Coccinellids, Spiders (predators), Tamarixia radiata, Diaphorencyrtus aligharensis for citrus psylla, M. boninensis, M. sexmaculatus, Preying Mantid(predators), Cirrospilus quadristriatus, Citrostichus phyllocnistoides (parasitoids) for leaf miner were identified. Among the predominant bio-agents in the citrus ecosystem, a chrysopid predator, Mallada boninensis against blackfly, psylla, leaf miner, aphids and mealy bugs; hymenopteran encyrtid parasitoid, Encarsia sp on blackfly".

9.5.12 Whiteflies

The Woolly whitefly, *Aleurothrixus floccosus* (Maskell), is one of the most important citrus whiteflies. In the genus *Aleurocanthus*, the Orange spiny whitefly, *A. spiniferus* (Quaintance), and the Citrus blackfly, *A. woglumi* Ashby are serious pests, both having adults with orange-reddish body and dark blue to grey wings, and black nymphs covered with many stout dorsal spines. They occur in many citrus growing countries of Asia, Africa and the Austro-Oriental-Pacific region (both species) as well as in North, Central and South America (*A. woglumi*) (Martin 1987; Nguyen and Hamon 1993; EPPO/CABI 1997).

Several natural enemies attack these insects and provide partial to complete biological control when undisturbed by insecticide treatments. Successful biological control programs, almost leading to suppression of the whitefly populations, are documented for the Citrus whitefly, by means of *Encarsia lahorensis* (Howard), the Woolly whitefly, especially with *Cales noacki* (Howard) and secondarily *Amitus spiniferus* (Brethes), the Citrus blackfly, by means of *Encarsia opulenta* (Silvestri) and *Amitus hesperidum* Silvestri (the latter species well performing to reduce high densities of the whitefly and the former one better adapt to maintain the balance when the population density of the whitefly is low), or the Orange spiny whitefly, with *Encarsia smithi* (Silvestri) and *A. spiniferus*.

Chemical applications are usually not effective, since they only ensure a temporary suppression of whitefly populations. Thus, they should be considered only when very heavy infestations occur and it is necessary to timely control the whitefly before trying to rebuilt the balance with their natural enemies. When the need of a chemical treatment occurs, the use of simply spray oils should be preferred as far as possible, at a concentration insuring 1 % actual oil in the spray mixture.

9.5.13 Mite Pests of Citrus

Several species of mites attack citrus trees. The Eriophyidae *Phyllocoptruta oleivora* (Ashmead) and *Aculops pelekassi* (Keifer) attack leaves, twigs and fruits, destroying epidermal cells by injecting saliva and sucking out their contents. As a result, infested parts become darker and often, their surface roughens too. In the same family, the Citrus bud mite, *Aceria sheldoni* (Ewing), frequently attacks lemon (especially in humid, coastal areas) which cause typical deformations to flowers and fruits. The Broad mite, *Polyphagotarsonemus latus* (Banks), of family Tarsonemidae is another pest of citrus. The mite attacks especially terminal leaves, which are twisted, hardened distorted and discoloured. In severe cases premature fruit drop may occur. In the family Tenuipalpidae, more than 20 species are known to live also on citrus, yet they are usually considered as secondary pests. The Citrus red mite, *Panonychus citri* is one of the most important mite pests on citrus and has spread almost worldwide during recent years.

Integrated control of mites injurious to citrus should be primarily based on cultural practices and biological control. Predaceous mites, such as phytoseiids of the genera *Amblyseius*, *Euseius*, *Galendromus*, *Typhlodromus*, *Typhlodromalus* or *Typhlodromina* are highly important. Among predacious insects, the thrips [such as *Scolothrips sexmaculatus* (Pergande) or *Aelothrips fasciatus* (L.)], the dustywings [such as *Conwentzia barretti* (Banks), *C. psociformis* (Curtis) or *Coniopteryx vicina* Hagen], the staphylinid beetles (such as *Oligota pygmea* Solier), the cecidomyid flies (such as *Therodiplosis persicae* Kieffer) are worth of being mentioned, most of them having almost the same predacious habits as the black lady beetles. Pathogens are also frequent mortality factors on citrus mites, such as non-inclusion virus disease and especially fungi (mainly of the genera *Entomophtora* or *Hirsutella*), whose incidence is particularly relevant when high mite populations occur during rain summer periods or foggy autumn weather (Weiser and Muma 1966; McCoy and Selhime 1974).

Despite the relevant number of natural enemies, biological control is not always satisfactory and chemical control is often required (Vacante 2010). Treatments with oil sprays are usually effective against most citrus mites, while controlling also other pests (especially scale insects). Eriophyds and Tenuipalpids are also susceptible to sulphur (both sprays and dusts) and application with selective miticides could be also evaluated anyway against these and other citrus mites. In any case, the correct use of chemicals should be preceded by a precise monitoring of the mite population, based on innovative sampling methods and updated threshold values (Smith and Peña 2002; Song et al. 2003; Martinez-Ferrer et al. 2006; Childers et al. 2007; Hall et al. 2007; Vacante 2010).

9.5.14 Nematode Pests of Citrus

NRC Nagpur (2014) has reported eleven nematode genera, "Tylenchulus, semipenetrans, Hoplolaimus, Criconematid, Paralongidorus, Helicotylenchus, Xiphinema, Tylenchus, Aphelenchus, Longidours, Pratylenchus, Rotylenchulus and fourteen nematodes species were found associated with the citrus orchards in Nagpur, Amravati and Wardha districts of Maharashtra, India . Tylenchulus semipenetrans was the most predominant nematodes out of all the nematodes species recorded during survey". Nematode species attacking citrus are Citrus nematode, *Tylenchulus semipenetrans* Cobb, and the Burrowing nematode, *Radopholus similis* (Cobb) (Duncan and Cohn 1990; Duncan 1999, 2005). In soils infested by *T. semipenetrans*, slow plant growth may represent a serious economic loss, especially in regions where the Citrus Tristeza Virus occurs and the fruit production of new orchards should start as fast as possible. In areas where nematodes are regular problem, only resistant rootstocks and plants free of nematode infestations free plants should be used. This simple practice has been successful in controlling nematodes (Lee et al. 1999; Verdejo-Lucas et al. 2003), Controlling the introduction of infested soils or rootstocks by applying quarantine measures is equally useful (Inserra et al. 2005). Moreover, agronomic practices should be aimed at encouraging the development of various and sometimes abundant populations of antagonists organism of nematodes (bacteria, fungi and predacious nematodes) which are naturally present in the soil.

Chemical control of nematodes is difficult and frequently the results are not completely satisfactory. A useful recommendation is to apply pesticides (both in granules or liquid formulation) when nematode populations reach the numeric peak. Pesticides should be spread in the soil around the tree and incorporated on it mechanically. Local irrigation by a drip system, allows to nematocide a uniform distribution into the first 30–40 cm of soil. The treatment must be repeated 2–3 times to reduce population densities of parasites and its efficacy, in terms of yield and fruit quality increment, can be observed after 2–3 years (Van Gundy 1982).

9.6 Pollination

*9.6.1 Guava (*Psidium guajava *L.)*

Guava (*Psidium guajava* L.) an important fruit crop of India and ranks fourth only after banana, mango and citrus both in terms of area under cultivation and fruit production. The flowers are white and slightly protandrous with five petals and numerous stamens. The flowers are highly attractive to honeybees producing large amount of pollen and nectar. *Psidium guajava* is known to reward bees with pollen and nectar both being essential for brood growth and colony development. Guava is highly cross pollinated crop and pollination by honeybees results in higher fruit yield (Malo and Campbell 1968; Kumar et al. 1996, Nalawadi et al. 1973) studied the floral biology of Guava and emphasized the role of cross pollination.

The guava flowers are borne singly or in clusters in leaf axils. They are white in colour, fragrant, bisexual measuring about an inch. The flowers has superior calyx (1–1.5 cm long) with five lobes, 6–10 white petals (1–2 cm long) arranged in a

Fig. 9.2 Vertical section of guava flower (Rendle 1925)

single or two in whorls. The stamens are numerous and one to two cm long carry bilobed anthers which are closely packed together and produce a large number of pollen grains (Fig. 9.2).

9.6.1.1 Pollinators

Guava flowers a visited by a wide variety of insects. In all the studies, honeybees have been recognized as most important pollinators and comprised of very high population in all the locations. Lakshmi and Mohan Rao (1998) reported 26–43 % pollination due to honeybees resulting in significant increase in fruit set and improvement in quality. The other flower visitors included solitary bees, syrphids and wasps. In Assam, India, (Anonymous 2011), the honeybee *Apis cerana* was found to be the most important pollinator visiting guava flowers. The bee pollinated plots gave significantly higher yield (88.74 %, 86q/ha) over self pollinated ones deprived of bee visits(88.74 %, 86q/ha). Lakshmi and Rao (1998) observed all the four species of honeybees such as Apis dorsata, Apis florea, A.mellifera and A.cerana visiting guava flowers. However, the A. cerana was most predominant of all the flower visitors. The other insect visitors were solitary bees and butterflies. Castro and Araújo (1998) found that *Apis mellifera* (21.6 %) was most abundant followed by, *Nannotrigona punctata* (19.8 %), *Trigona spinipes* (19 %) and

Melipona scutellaris (12.9 %) visiting guava flowers. Rajagopal and Eswarappa (2005) reported the pollination potential of honey bees in increasing productivity of guava in, Allalasandra, Bangalore India. *Apis dorsata, A. cerana, A. mellifera, A. florea* and *Trigona iridipennis* (06.00–11.00 h) foraged on the flowers for 4–6 h and other insects such as dipterans, hemipterans and coleopterans observed on the flowers (08.00–11.00 h) for 4 h. The per cent proportion of different pollinators on the flowers revealed that the honey bees accounted for 92.86 % compared to 7.14 % with other pollinators. Among the honey bees, *A. dorsata* (46.58 %) and *A. mellifera* (37.99 %) visited the flowers very frequently on the natural trees and on caged trees, respectively, compared to the rest of the bee species. Among the different species of honey bees caged for the pollination of trees, the percent flower drop was maximum in the untreated mode and caged trees with *T. iridipennis*. The percent fruit set (88.0) and number of seeds per fruit (446.50) were maximum in trees caged with *A. cerana* compared with the untreated mode and the trees caged with other bee species. The fruits harvested from the trees caged with *A. mellifera* recorded the maximum weight (89.25 g) and the fruits harvested from trees caged with *A. florea* recorded the maximum total soluble solids (18.75 %). The different modes of pollination in guava showed complete flower drop in self-pollination indicating high degree of self incompatibility. The percent fruit set, fruit weight, number of seeds per fruit and test weight were significantly higher in hand pollination and also in guava orchard treated with five hives of *A. cerana* compared to guava orchard without bee hives. The total soluble solids of the fruits ranged from 15.62 % to 17.0 %.

The pollination of guava (*Psidium guajava* Linn.) was found to be effected by four insects viz. *A. c. indica, A. florea, Melipona fasciata* and *Musca domestica* (Plate 9.3). The relative abundance, pollen depletion, pollen deposition and time spent at flowers revealed that the honeybees, *A. c. indica* and *A. florea* were the major pollinators of guava. The amount of pollen deposited by the pollinators during the successive visits was found to decrease because of the decreasing stigma receptivity with time. The amount of pollen depleted and deposited was found to be more at 0800 h due to more pollen stickiness and stigma receptivity (Prakash et al. 1993). Further studies on pollination of guava (*P. guajava* Linn.) by honey bees (*A. c. indica* F.) by Jyothi (2004) at East Godavari district of Andhra Pradesh came to a conclusion that fruit set was increased by honey bee pollination.

The number of domesticated honeybee colonies in India is decreasing due to huge number of viral diseases and pests. In order to attract the pollinating bees, attractants may be used (Delaplane and Mayer 2000). Among the tested bee attractants based on queen mandibular pheromone (QMP) (Mayer et al. 1989; Elmstorm and Maynard 1991; Winston and Slessor 1993; Ambrose et al. 1995; Higo et al. 1995), fruit boost and Bee-Q (based on carbohydrate rich) have the most promising effect (Currie et al. 1992; Naumann et al. 1994). The attractants have been reported to enhance pollination and yield in cucumber (Viraktamath and Anagoudar 2002; Pateel and Sattigi 2007), onion (Kalmath and Sattigi 2002) and radish (Chandrashekar and Sattigi 2009).

Plate 9.3 (**a**) Citrus psylla infestation on citrus leaves (**b**) Citrus psylla nymph with crystalline exudates (**c**) Citrus leaf miner larva inside the tunnel (**d**) Symptoms of leaf mines on citrus e. Damaged leaf by citrus caterpillar (**f**) Adult of citrus caterpillar, Papilio demoleus (**g**) Aphids infestation o citrus plant (**h**) Red ants, Oecophylla smargdulla on citrus plant

9.6.1.2 Pollination Requirements

Based on various studies made by Hamilton and Seagrave-Smith (1954) and Malo and Campbell 1968 guava has been established a self fertile plant but higher yields are obtained if pollinated by insects. Ray and Chhonkar 1981 studied pollination and fruit development in three cultivars of guava namely Habashi, Harisha and Local seedless by different methods like local, self and cross pollination, setting by hand or emasculation and no pollination. The fruit set in all the cultivars ranged between 62 % and 82 % under open pollination which was significantly more than self pollination. Similar results were obtained by Ojha 1985 in case of two cultivars of guava namely Allahabad and Sardar where open pollination proved superior over self pollination. Sehgal (1961) recorded a high percentage of fruit set under natural conditions in diploid varieties of guava (80–86 %) as compared to triploid varieties (54 %).

9.6.1.3 Pollination Recommendations and Practices

Studies on pollination recommendations are not documented in literature because of the limited studies made on the crop. The available information, however, clearly demonstrates that pollination by insects is at least highly beneficial for maximum production.

9.6.1.4 Insects

More than 45 species of insects and 6 species of mites have been reported attacking guavas (Table 9.1). These aphids, thrips, scales, mealy bugs, beetles, moth larvae, false spider, eriophyid and spider mites. Besides, there are number of parasitic wasps and predators that keep scale insects and mealy bugs under control. Fruit flies cause serious fruit damage and fruit. Fruit bagging can significantly reduce the problem and produce high-quality fruits. A review of guava pests in South Africa included bird and bat control (Villiers and Grove 2000). Bats frequently attack ripening fruit in Cuba. Thrips can be troublesome, causing leaf silvering and fruit scarification. In the Amazon region, *Anastrepha striata* (fruit fly), the lepidopteran *Timocratica albella* (a stem and branch borer) and *Conotrachelus psidii* (guava weevil) are potentially serious enemies (Vasquez et al. 2002). The guava weevil spends its larval time in the fruit and feeds on the seeds, producing petrifi cation and premature fruit ripening. It is important to stress that insecticides should be applied in the afternoon to protect honey-bees; thus, better pollination and harvests will be obtained.

Table 9.1 Important insect and nematode pests of guava (Source: Paull and Duarte 2012)

Common name	Organism	Part(s) affected	Region
Mediterranean fruit fly	*Ceratitis capitata*	Fruit	Hawaii, South Africa
Natal fruit fly	*Ceratitis rosa*	Fruit	South Africa
Oriental fruit fly	*Dacus dorsalis*	Fruit	Hawaii
Melon fly	*Dacus cucurbitae*	Fruit	Hawaii
Caribbean fruit fly	*Anastrepha striata*	Fruit	Caribbean, American tropics Amazon area
Guava weevil	*Conotrachelus Psidii3*	Fruit	South American tropics
Guava stem moth	*Timocratica albella*	Stem, branches	Amazon area
No common name	*Monolepta australis*	Leaf A	ustralia
Red-banded thrips	*Selenothrips rubrocinctus*	Leaf, fruit	Worldwide
Bark-eating caterpillar	*Indarbela quadrinotata*	Bark	India
Root-knot nematode	*Meloidogyne incognita*	Root	Australia
	M. arenaria	Root	Caribbean
	M. acrita	Root	Caribbean
Tea mosquito bug	*Helopeltis antonii* Signoret	Nymphs and adults feed on tender leaves	India, Srilanka, S.E. Asia
Tea mosquito bug	*Helopeltis febrioculosa* Bergroth	Nymphs and adults feed on tender leaves	India
Citrus black fly	*Aleurocanthus woglumi* Ashby	Infest foliage	India, china
Cotton aphid	*Aphis gossypii* Glover	Infest foliage	India, Malaysia, Laos, Philippines
Olive soft scale	*Saissetia oleae (Bernard)*	Infest foliage	India, China
Red banded thrips	*Selenothrips rubrocintus* (Giard)	Infest foliage	India, china
Guava stem borer	*Microcolona leucosticta* Meyrik	Larvae bore shoots	India
Grey weevil	*Myllocerus blandus* Faust	Adult eat leaves	India
Grey weevil	*M. undecimpustulatus* Desbrocher	Adult eat leaves	India
Pomegranate butterfly	*Virachola isocrates* (Fab.)	Larvae feed inside fruits	India
Castor semi-looper	*Achaea janata* (L.)	Larvae feed on leaves	India
Cherry stem borer	*Aeolesthes holoserica Fab.*	Larvae bore stems	India, S.E. Asia

Common Guava Pests

Some common guava pests cause damage to the leaves and others cause damage to the fruits. All important pests of guava cause significant economic loss. The implementation of an Integrated Pest Management (IPM) program can provide a good control of pest damage.

Fruit Fly

The guava is a prime host of fruit flies, especially *Anastrepha sp.* in Belize. The ripe fruits infested with larvae are unmarketable. In order to protect from fruit fly damage, bagging of fruit and field cleaning is strongly recommended. Management and control–Fruit flies can be effectively controlled by bagging of young fruit under 60 days, use of baited traps, and use of chemical and ethological control measures. Collect all the fallen and infested fruits and destroy them. Ploughing around the tree, kill the pupae in the soil by exposing them to the sun's heat another predators. Poison baiting with protein hydrolysate,Malathion and water (1:10) has been suggested. When chemical control measures are implemented be sure to observe the recommended pre-harvest interval.

Mealy Bug

Nymphs and the adult female mealy bug suck up the plant saps thus impeding proper leaf development resulting in deformed leaves. Mealy bugs secret honeydew on which secondary fungi may grow. Management and control –Mealy bugs can be controlled by the use of recommended systemic insecticide and a fungicide to control the secondary fungal disease for example sooty mold.

White Fly

White flies cause direct damage to plants by sucking plant sap and weakening the plants. They produce large quantity of honeydew that leads to the growth of sooty mold on plant leaves. White flies also serve as vectors of viral diseases.

Management and control: (1) Monitoring white fly population with yellow sticky traps, (2) Good field sanitation and clear weed or vegetation surrounding the field; (3) In a severe case, apply chemical control measures.

Thrip

Thrips suck sap from the plant tissues and the leaves give mottled appearance, and also cause fruit damage. In severe cases, young fruits, terminal buds, die back as a result of severe physical damage. Management and control: (1) Monitoring thrip population with blue sticky traps, (2) Good field sanitation and clear weed or vegetation surrounding the field; (3) In a severe case, apply chemical control measures.

Chinch (Stink Bug)

Leptoglossus stigma; *L. Zonatus* Stink bugs are sucking insects. The female stink bug lays its eggs on foliage and on branches where the nymphs hatch and live as groups in this stage of their development. Both adult and nymphs will suck sap from flower buds and fruits. Affected or damaged flower buds will fall off. Fruits that have been sucked by stink bugs will have black sunken specks on the fruit unsuitable for the market. The adult stink bug is able to penetrate through the paper or plastic barrier around the bagged fruit. Management and Control – In severe cases apply Neem oil at 7 days interval – three applications/cycle. Integrated pest management helps to minimize the use of pesticides to protect beneficial organism (e.g. a natural predator) that occur naturally in the area and is recommended practice in guava production.

9.6.2 Litchi

9.6.2.1 Pollination

Litchi is one of the important fruit crop grown extensively in China, India, Southeast Asia, and South Africa (Menzel 1984). The litchi inflorescence comprises of a branched panicle, with large number of small, greenish-yellow flowers in the terminal clusters about a foot long (Fig. 9.3). The individual of L.chinensis has a cup

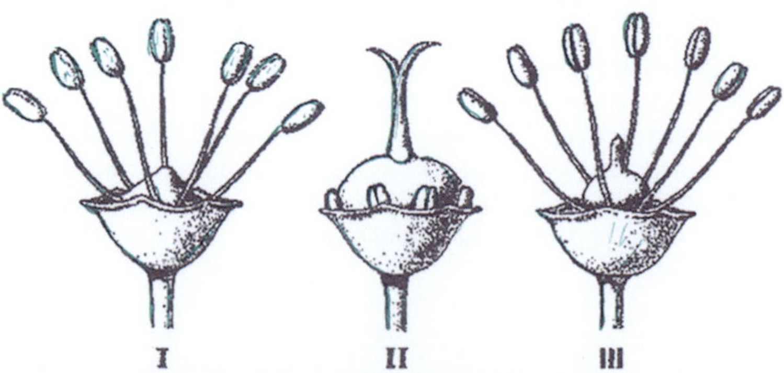

Fig. 9.3 Litchi flower. 1. Staminate flower with long filaments. 2. Staminate flower with short filaments. 3. Hermaphrodite flower functioning as female. 4. Hermaphrodite flower functioning as male with long filaments. 5. Hermaphrodite flower functioning as male with short filaments

shaped calyx with stamens and pistils inserted on it. The flowers are apetalous, two lobed stigma, an ovary on short stalk with one ovule in each of its two or three sections (Menzel 1991). Butcher (1957a) reported three types of flowers in litch which bloom in stages: Male or staminate flowers (M_1) with no functional ovaries bloom first; female or pistillate flowers (F) with anthers that do not dehisce comprise the second stage; and imperfect hermaphrodite flowers (M_2), also lacking functional ovaries, bloom last (Stern and Gazit 1996). Each phase of bloom consists of flowers of the same type. The onset and duration of anthesis is highly variable among cultivars (Menzel and Simpson 1992b). Anther dehiscence occurs on day 2–5 and continues throughout the day and night, attaining peak at around 10 a.m. The duration of flowering continues from 20 to 45 days in a given tree.

Nectary is present as a large fleshy crenulate gland within the calyx which secretes nectar in the morning (Khan 1929). Nectar is highly attractive to honey bees and flies.

9.6.2.2 Pollination Requirements

Although, self-pollination can occur, but the litchi flowers are generally recognized as self-sterile and require insects pollinating insects for transfer of pollen from anthers to stigmas setting of fruits (King et al. 1989; Stern and Gazit 1996). Chaturvedi (1965) reported that open pollinated branches set 43 % fruit whereas zero percent on branches bagged with muslin cloth. Similar results were reported by Das and Choudhury (1958) who found no set of fruit on bagged panicles. Pandey and Yadava (1970) reported open pollinated flowers set 0.7–11.2% fruit whereas those deprived of insect visits set 0.03–0.10% of fruit. Butcher (1957a, b) also reported that no fruit set was found on trees caged to exclude insect pollinators. These studies clearly reveal that Litch requires insect pollinators for setting of fruits.

Since lychee flowers are functionally unisexual, fruit set can occur only if pollen from male flowers (M1 and M2) is transferred to the stigma of F flowers. Lychee flowers are entemophilic and pollination is usually performed by insects, one of which is the honeybee (Free 1993; King et al. 1989; McGregor 1976). A large number of insects, especially bees, visit lychee flowers, mainly during the morning hours (Free 1993; Pivovaro 1974). Adequate pollination is ensured by the fact that the flowering wave occurs between two pollen-releasing waves (M1 and M2), and by the abundance of pollinating agents (Butcher 1956, 1957a; DeGrandi-Hoffman 1987; Groff 1943; McGregor 1976; Pandey and Yadava 1970), provided that weather conditions do not limit bee activity (Menzel 1984). When litchi flowers are bagged to deny access to insects, fruit set is reduced. In India, Kumar et al. (1996) recorded an average of 1.4 fruit on panicles when insects were excluded, 8.9 on panicles caged with *A. mellifera*, and 14.9 on panicles with free access to all pollinators. An average of six fruit per panicle was set in South Africa when honeybees had access to flowers, compared with two per panicle when the panicles were bagged (du Toit 1994). Several studies have indicated that bee pollination resulted in significantly higher fruit yield with better quality. Badiyala and Garg (1990) reported that litchi inflorsecences exposed to open pollination by honeybees resulted in 2–3 higher yield than bagged to exclude them. du Toit (1994) in south Africa, also obtained three times higher fruit set when inflorescences were left open to honeybees. Fruit yield per hectare was 70 tons in bee pollinated plots, compared to 36.8 tons in self pollinated plots (Mohanta and Rahman 1997).

9.6.2.3 Pollinators

Litchi inflorescences are a rich source of nectar, and are highly attractive to insects. Female flowers had greater nectar volumes and sugar concentrations than M2 flowers, which in turn were higher than M1 flowers (Stern and Gazit 1996). A typical 'Mauritius' inflorescence has over 100 female flowers open at the same time, and a typical tree has hundreds of such inflorescences. Anther dehiscence takes place throughout the day and night, mostly from morning to mid-afternoon (Das and Choudhury 1958; Chadha and Rajpoot 1969; Wang and Qiu 1997). Pollen is not attractive to most of the pollinators. Pandey and Yadava (1970) found that honeybees collect pollen mainly at the beginning and end of flowering.

Bees are the most important pollinators (Butcher 1956, 1957a; McGregor 1976; Free 1993). Pollination by honeybees has been reported to increase yields by 2.5–2.9 times compared with orchards without bees (Chen 1993). Only a few of the several *Apis* species in China are involved in pollination, with the Chinese honeybee, *Apis cerana cerana*, being the main pollinator. This species was domesticated 2000 years ago and is used for honey production in litchi-growing areas. The European honeybee, *A. mellifera*, was introduced to China about 100 years ago (Crane 1990; Chen 1993), and also pollinates litchi. The European honeybee is the most important species in Florida, South Africa, Israel and Australia, with hives moved to orchards in bloom. It is a very effective pollinator, with the bees touching the stigmas on almost every visit to the female flowers (King et al. 1989).

A bee can easily fill its stomach with nectar from just a few litchi flowers (Free 1993), so both the quality and quantity of nectar can have a large bearing on the efficiency of honey production. Stern and Gazit (1996) found a correlation between bee density on the M1, F and M2 flowers and sugar in the nectar, but a negative correlation between bee density and nectar volume for M2 and F flowers. The proportion of glucose (43 %), fructose (39 %) and sucrose (18 %) in the nectar was similar for the three flower types.

The amount of pollen found on bees sampled from F and M2 flowers was twice as high as that found on bees sampled during M1 bloom, reflecting the pronounced differences in the quantity and quality of the nectar present in the different flower types (Stern and Gazit 1996). Bees visiting M1 blooms will be 'disappointed' with the meagre yield, and will abandon them, whereas bees visiting M2 flowers will be rewarded, will visit many flowers, and will be dusted with plenty of pollen.

Pollination has been reported to range from 0 % to 10 %, notwithstanding the large number of bees visiting during the first days of female bloom (Stern and Gazit 1996). The few bees visiting M1 flowers resulted in little pollen adhering to the bees (less than 10 pollen grains per bee). When the M2 bloom commenced, pollination increased to 20 % or 30 %, reaching 80–90 % within a few days. The greater attractiveness of M2 flowers resulted in hundreds of pollen grains adhering to each bee. Consequently, the number of germinating pollen grains per stigma increased from one or two to about 20 (Stern and Gazit 1996). McConchie and Batten (1991) provided information on the relationship between fruit set and pollination. They found that hand pollination of all the F flowers on a 'Bengal' inflorescence (225 flowers) produced only nine fruit, whereas pollination of only 47 or 48 flowers and removal of all other female flowers yielded 13 or 23 fruit for early- and late-opening flowers, respectively. The greater fruit set of the late flowers may reflect a higher fertility rate, or the use of the more potent M2 pollen.

Du Toit and Swart (1995) concluded that litchi has a limited ability for self-pollination and that insects are necessary to ensure optimum fruit set. In India, Singh and Chopra (1998) recorded 16 species of bees, wasps, flies and other insects visiting litchi flowers. Most common were the honeybees, *A. cerana* Fabricius, *A. dorsata* Fabricius, *A. mellifera* Linnaeus and *A. florea* Fabricius, and the syrphid flies *Melanostoma univittatum* Wiedemann and *Episyrphus balteatus* (De Geer). Pollination by these insects resulted in a 387 % increase in fruit set, and a 505 % increase in fruit retention compared with panicles denied access by pollinators.

Of the bees, *A. dorsata* has been considered the most important, although *A. mellifera* is also a regular visitor to litchi flowers, and most probably has a significant impact on pollination (Kitroo and Abrol 1996; Kumar et al. 1996; Abrol 1999). In Thailand, *A. mellifera* is preferred as a large-scale producer of honey and for the pollination of longans, while *A. cerana* is preferred for small-scale honey production and the pollination of litchis, rambutans and mangoes (Wongsiri and Chen 1995). *Trigona iridipennis* Smith also contributes to pollination in both litchis and longans in India and Thailand (Boonithee et al. 1991; Kumar et al. 1996). Eardley and Mansell (1996) concluded that of the 38 insect species visiting litchi flowers near Ofcolaco in South Africa, most were ineffectual as pollinators, and that honeybees and indigenous bees, such as *Plebeina denoiti* (Vachal), *Meliponula erythra junodi* (Friese), *Ctenoceratina moerenhouti* (Vachal), *Ctenoceratina rufigaster* (Cockerell) and *Braunsapis facialis* (Gerstaecker), contribute significantly to pollination. In contrast, litchis at Tzaneen were visited mostly by honeybees.

9.6.2.4 Pollination Recommendations and Practices

Placing honeybee colonies in litchi orchards is an important practice ensuring adequate pollination and fruit set. du Toit (1990) recommended 4 hives per hectare for effective pollination and harvest of surplus honey. Kumar and Kumar (2014) recommended that a 5–10 honey bee colonies per ha of litchi cultivation should be provided for optimum fruit production. To obtain higher yield and better quality of fruits 2–3 colonies of *Apis mellifera* and 4–6 colonies of A. cerana should be placed per ha in litchi orchards (Abrol 2013).

9.6.2.5 Pests of Litchi

About 42 insects and mite species reported attacking trees and fruits of litchi at different stages of growth (Singh and Singh 1954; Hameed et al. 1992; Waite and Hwang 2002). Earlier, only two species, namely, erineum mite and bark eating caterpillar were reported causing serious damage to litchi trees (Mathur and Tandon 1974; Butani 1977). Recently, litchi fruit borer and litchi leaf roller have acquired the status of major pests and now, litchi looper, litchi bug and bag worm are emerging pests of litchi (Hameed et al. 2001; Kumar et al. 2011; Choudhary et al. 2013; Kumar et al. 2014a). Besides, few other insect pests viz. *Lepropus lateralis* (Coleoptera: Curculionidae), *Thalassodes* sp. (Lepidoptera: Geometridae), *Statherotis* sp. (Lepidoptera: Tortricidae) were also identified as minor pests in litchi ecosystem (Table 9.2).

Litchi Mite *Aceria litchi* (Acari: Eriophyidae)

Litchi mite has been reported from all the litchi growing countries of the world including India, where litchi mite has been found in almost all litchi growing areas. Litchi mite is a destructive pest, causing curling of leaves and chocolate-brown growth on the ventral surface of leaves at later stage (Singh and Raghuraman 2011). Both nymphs and adults damage the leaves, inflorescence and young developing fruits by puncturing and lacerate the tissues with their stout rostrum and suck the cell sap (Lall and Rahman 1975). The infestation generally begins from lower portion of the trees and gradually extends upwards. The incidence of litchi mite can be observed round the year on litchi trees; however, severe infestation has been noticed from March to September and goes under hibernation in extreme cold (Mishra 1912; Mishra 1980). However, Lall and Rahman (1975) reported active period of mite is January to October and hibernation in adult stage during November to December under surface of leaf. The highest population (18.1 per cm^2 leaf surface) of mite was observed during March, while lowest population (1.4 per cm^2 leaf surface) during November. Sharma et al. (1986) also reported litchi mite during April-May and November- December but mentioned that the population was significantly influenced by temperature and relative humidity. The other mites include the tea red spider mite, *Oligonychus coffeae* (Nietner) (Tetranychidae), occasionally infests litchi leaves in Queensland, but is never a problem. In Queensland and other regions, *Polyphagotarsonemus latus* (Banks) (Tarsonemidae) may occasionally damage individual terminals on orchard trees. However, it is most often seen on nursery trees, where it can be easily controlled with dicofol, sulphur or endosulfan.

Table 9.2 Taxonomic position and economic status of insect pests in litchi

S. No.	Common name	Scientific name	Order: family	Economic status
1.	Litchi fruit & shoot borer	*Conopomorpha cramerella* Snellen	Lepidoptera: Gracillariidae	Major
2.	Bark eating caterpillar	*Indarbela tetraonis* Moore	Lepidoptera: Metarbelidae	Major
3.	Leaf roller	*Pletypeplus aprobola* Meyer	Lepidoptera: Tortricidae	Major
4.	Leaf roller	*Statherotis* sp.	Lepidoptera: Tortricidae	Minor
5.	Ash weevil	*Myllocerus undecimpustulatusn* Faust	Coleoptera: Curculionidae	Major
6.	Red weevil	*Apoderus blandus*	Coleoptera: Curculionidae	Major
7.	Ash weevil	*Lepropus lateralis*	Coleoptera: Curculionidae	Minor
8.	Bag worm	*Eumeta crameri* Westwood	Lepidoptera: Psychidae	Major
9.	Looper	*Perixera illepidaria*	Lepidoptera: Geometridae	Major
10.	Looper	*Thalassodes* sp.	Lepidoptera: Geometridae	Minor
11.	Litchi bug	*Tessaratoma javanica* Thunberg	Hemiptera: Pentatomidae	Major
12.	Litchi mite	*Acerya litchi* Keifer	Acari: Eriophyidae	Major
13.	Tobacco caterpillar	*Spodoptera litura*	Lepidoptera: Noctuidae	Minor/ Sporadic
14.	Hairy caterpillar	*Spilarctia* sp.	Lepidoptera: Arctidae	Minor/ Sporadic
15.	Mango mealy bug	*Drosicha mangiferae*	Hemiptera: Coccidae	Minor

Litchi Fruit and Shoot Borer: *Conopomorpha cramerella* Snellen (Lepidoptera: Gracillariidae)

Litchi fruit borer known as the litchi stem-end borer in China and the Litchi fruit borer in Thailand, is one of the major pest of litchi. Besides India, pests occur all litchi growing countries viz. China, Taiwan, Thailand etc. The pest shows a preference for litchis over longan but in both crops, damaged fruit may fall from the tree. Fruit loss and leaf infestation has been estimated from 24 % to 32 % and 7 % to 70 %, respectively by this pest (Lall and Sharma 1976; Hameed et al. 2001). In Taiwan, 96.1–100 % litchi fruit get damaged by *C. sinensis* Bradley, a closely related species in untreated orchards (Huang et al. 1994). Bhatia et al. (2000) observed 13.6–64.9 % incidence of *C. cramerella* in Himanchal Pradesh.

Leaf Roller : *Platypeplus aprobola* (Meyrick) (Lepidoptera: Tortricidae)

The leaf roller is a cosmopolitan pest and has been reported in various fruit crops in Hawaii, South Africa, Australia and China. Leaf roller is a very serious litchi pest in Bihar and surrounding areas. The incidence of leaf roller can be observed on litchi trees throughout the year, however more population has been observed from July-February (Singh 1971). The symptoms of leaf injury by the larvae are manifested through rolling of tender leaves and feeding inside. According to Lall and Mallik (1976) leaf injured by the leaf roller varied between 16.70 % and 71.60 % while trees infestation varied 12.88–53.54 % during August to February.

Leaf Cutting Weevils: *Myllocerus* Sp and *Apoderus blandus* (Coleoptera: Curculionidae)

Grey weevil (*Myllocerus* Sp,) and red weevil (*Apoderus blandus*) are polyphagous pest therefore, besides litchi its damage other important crops (Singh 1974; 1978). Adult has long snout with grey colour, though poor flier but very active feeder on the leaves of litchi. It attacks leaves, shoot and flower. Adult weevils congregate on the tender leaves and nibble irregular holes on the leaves and sometimes consume the entire leaf leaving the midrib only (Singh et al. 2012). The damage of red weevil is more severe at the time of shoot emergence as it prefers newly leaves therefore; newly established orchard/ nurseries are more vulnerable for pest attack (Kumar et al. 2011).

Litchi Looper: *Perixera illepidaria* (Lepidoptera: Geometridae)

Looper attacks tender leaves in mass and defoliate the new shoots. In severe attack, it completely defoliates the newly emerged flush. Besides litchi so many host plants viz. longan, rambutan, mango, and castor have been reported for this pest

(Schreiner and Nafus 1992; Kumar et al. 2014b). Incidence of this pest has been observed from July to December however, highest population recorded during September- October.

Litchi Bug: *Tessaratoma* sp (Hemiptera: Pentatomidae)

Litchi bug has been emerged as major pest in all litchi growing countries (Choudhary et al. 2013). There are many species of bugs viz. *Tessaratoma javanica* Thunberg, *T.papillosa* Dury, *T. quadrata* Distant, *T. nigripes* Dallas and *T. malaya* Stal that attack litchi plant, out of which, *T. javanica* Thunberg is the most destructive in Australia, China, India, Myanmar and Thailand (Han et al. 1999; Leksawasdi and Kumchu 1991; Lu et al. 2006). In India, pest is also noticed in all litchi growing areas particularly, Jharkhand, Bihar, U.P., Punjab, H.P. and J&K. Both adults and nymph suck cell sap mostly on tender plant parts such as growing buds, leaf petioles, fruit stalks and tender branches of litchi tree. In case of severe infestation drying of growing buds and tender shoots has been observed resulting into poor fruit set. The bugs when crowd on the developing fruit, it causes the fruits to fall a couple of week later. Longan, rambutan, kusum, pummelo, castor, pomegranate, eucalyptus, loquat and rose are major host plants of this bug ((Singh et al. 2009),). In the north-western part of India, the insect appears on litchi from the last week of April and disappears from the orchard after the last week of August and undergoes hibernation in adult stages (Kumar et al. 2008), however, in eastern India incidence of this bug has been observed from February to September (Choudhary et al. 2013).

Bag Worm: *Eumeta crameri* Westwood (Lepidoptera: Psychidae)

Bag worm larvae infest the older leaves and bark. Young caterpillars construct silken bags covering with bark and dry twigs and by living inside scrap the leaf surface (Agrawal and Pati 2002). Coffee, tea, maize, pomegranate, tamarind, castor, sandal, Casuarina, Cinnamon, *Shorea robusta*, are the alternate host of bagworm (Agrawal and Pati 2003). Incidence of this pest has been noticed throughout the year on litchi with maximum population during the month of September-October.

Bark Eating Caterpillar: *Indarbela tetraonis* Moore and *Indarbela quadrinotata* Walker (Metarbelidae: Lepidoptera)

There are two species of bark eating caterpillars, (*Indarbela tetraonis* Moore and *Indarbela quadrinotata* Walker) and both are polyphagous in nature with wide range of host plants including, longan, rambutan, mango, mulberry aonla, ber, citrus, phalsa guava, jack-fruit, jamun, loquat, pomegranate and rose (Verma and Khurana 1974; Khurana and Gupta 1972.). The pest is distributed throughout Indian subcontinent, mainly U.P., M.P., Bihar, Rajasthan, Haryana, Orissa, A.P., T.N. and Maharashtra. Particularly in Bihar, *Indarbela tetraonis* Moore is a very serious pest of litchi, guava, mango and ber (Sharma and Kumar 1986). Damage is caused by caterpillar, which bore into trunk, main stems and thick branches of litchi tree and destroying xylum tissues resulting into poor growth and fruiting of the tree. Caterpillars remain within the tunnel inside the stem during day, come out in night and feed upon the bark. Older and uncared trees are more affected by the pest. Infestation may be noticed by the presence of ribbons of wood chips, frass and silken thread over the bark surface (Hameed et al. 2001; Atwal and Dhaliwal 2008).

Fruit Sucking Moths

Fruit-piercing moth include *Eudocima phalonia* (*Eudocima materna*, *Eudocima ancilla*, *Eudocima salaminia, Platyja umminia, Pericyma cruegeri, Ercheia dubia, Serrodes mediopallens,* and *Oraesia emarginata*) which attacks many fruit crops including litchi in night (Muniappan et al. 2002). Adult moths cause damage whereas the larvae are not harmful. Once the moth has punctured the skin of the fruit, it feeds upon the juices of the fruit. Damage caused by this pest is not direct but indirectly fungal and bacterial infections also develop at the wound site. When moths are abundant, premature ripening and dropping of fruits occur. Collection and destruction of moths at sunset using flash lights is a effective method of their control. Bagging of fruits may be done by covering the fruits. Furthermore smoking orchards could mask the odour of riening fruits that attract moths. Smoking of the orchard masks the odour of mature and ripening fruit that attracts the moth. The effectiveness of chemical control is variable and remains ineffective in most of the cases.

Gall Midges

The Gall midge, *Litchiomyia chinensis* Yang and Luo has been described infesting litchi orchards (Yang and Luo 1999). The larvae of the midge overwinter in the galls resulting from feeding on leaves. They pupate in the soil and the adult flies start emerging in March–April to initiate seven or eight overlapping generations per year. The female midge prefers moist places with dense foliage on which they lay eggs. When the larvae hatch they mine the leaf, on which galls develop. As with the litchi erinose mite, infested leaves can be removed after harvest, and destroyed. In the spring, 75 kg/ha of 2.5 % methyl parathion can be distributed on the ground under the tree, or isofenphos (0.001 %) sprayed on the ground just prior to the expected emergence of the adult flies. In autumn, isocarbophos (0.001 %) should be sprayed twice over a period of 14 days during a leaf flush (Zhang et al. 1997).

Fruit Flies

Several species of fruit flies viz. *Ceratitis capitata* (Weidemann) and *Ceratitis* (*Pterandrus*) *rosa* Karsch have been reported infesting litchi orchards in South Africa and Réunion, and *C. capitata*, *Bactrocera dorsalis* Hendel and *Bactrocera cucurbitae*

(Coquillett) from Hawaii (Vayssieres 1997). All of the above-mentioned species are capable of laying their eggs in the crevices of skin of litchi, yet in some cultivars the thickness of the skin may prevent successful oviposition. Most often, the presence of fruit fly eggs and the occasional larva in litchi fruit can be attributed to oviposition through prior damage inflicted by other pests such as *Cryptophlebia* spp. Only in South Africa are fruit flies considered a real problem, despite the relatively low damage levels recorded. *C. rosa* is thought to be responsible for most of the losses (Grové et al. 1999a, b). The physical damage thus caused, as well as some damage inflicted on the flesh by early larval instars, initiates fruit rot and fermentation (de Villiers 1992b). A study conducted by Gould et al. (1999) in Florida found that, despite the presence of large numbers of the Caribbean fruit fly, *Anastrepha suspensa* (Loew), in litchi and longan orchards, neither crop was attacked by the pest. Numerous parasitoids, especially *Opius* spp. and *Biosteres* spp., have been recorded attacking the fruit fly species that damage litchis. Despite the activity of such parasitoids, fruit flies continue to flourish and often require chemical control in susceptible crops.

In South Africa, the use of pheromone baited traps is recommended for monitoring fruit fly populations around litchi orchards. Control is achieved with bait sprays of protein hydrolysate mixed with trichlorfon or mercaptothion. Alternatively, panicles may be protected with paper bags applied just after the November fruit drop (de Villiers 1990a, b). Grové et al. (1999c) found that quarantine cold treatment killed *C. rosa* in litchi fruit. In other countries, no specific action is recommended because the problem is not serious, although quarantine protocols may change this in the future.

Mealybugs

The citrus mealybug, *Planococcus citri* (Risso), is widely distributed throughout the world on numerous hosts. It sporadically attacks litchis in Taiwan, where it may contribute to the production of sooty mould.

Thrips

In India, *Dolichothrips indicus* Hood and *Megalurothrips distalis* Karny attacked the flowers and leaves of litchi, respectively (Ananthakrishnan 1971). The damage caused by these species was not quantified, but phosphamidon and dimethoate were recommended for control (Butani 1977). The tea yellow thrips, *Scirtothrips dorsalis* Hood, infests litchis and longans in China. All life stages feed on the shoots, causing malformation of the new leaves. The crimped, yellow leaves have a mosaic appearance and eventually lose their sheen, and may fall. The thrips are most numerous from August to October, especially when it is dry. Sprays of isocarbophos, omethoate or dimethoate are recommended, along with orchard management procedures that encourage uniform flushing (Zhang et al. 1997). The plague thrips, *Thrips imaginis* Bagnall, often infests litchi flowers in Queensland. As its name implies, it may be present in extraordinary numbers, and it appears to feed on the florets. However, experiments have indicated that the insect has little, if any, effect on fruit set (Waite 1992a).

Other pests of regional importance are fruit borer, *Conopomorpha sinensis* Bradley, in China and Thailand, leaf-miner, *Conopomorpha litchiella* Bradley, in Thailand, shot hole borer, *Xyleborus fornicatus* (Eichhoff) (Col.: Scolytidae), in India, trunk borer, *Anoplophora maculata* (Thomson) (Col.: Cerambycidae), in Taiwan, and fruit fl ies, *Bactrocera* spp. (Dip.: Tephritidae), in Asia (Muniappan et al. 2002).

9.7 Integrated Pest Management (IPM)

As with all crops, the ultimate aim in the protection of litchis and longans from insect pests isto implement a viable integrated pest management (IPM) system. This system will have been developed for each production region, accounting for the variations in geography and latitude that determine the timing of plant growth, and the local insect and mite fauna.

The tactics adopted against each pest depend on how well they fit in with the overall management strategy for an orchard. Some pests may be effectively controlled by natural enemies, and the options considered for the control of other pests must always take into account the possible side-effects on these, and the possible induction of problems caused by the injudicious use of insecticides. It should be noted, however, that when pest infestations start to cause economic losses, they should be controlled, but with consideration of the impact on the orchard ecosystem.

In China, where litchis have been cultivated for thousands of years, the need to control pests to produce high-quality fruit has been well recognized over the years. There is a strong belief in the concept of biological control combined with sound orchard management practices, especially the pruning of infested branches or leaves, as the basis for IPM. In addition, many studies have been conducted on the effect of orchard floor management on pest incidence. As well as providing a suitable habitat for natural enemies, suitable ground covers may improve the orchard micro-climate (Liang and Huang 1994; Liu and Tan 1999). When farm communes existed, such systems were widely used. However, since their demise and with the availability of more effective insecticides and spray equipment, adherence to this approach has waned. For the Chinese crop, Zhang et al. (1997) divided the year into phenological stages so that the recommended management strategies can be implemented to match the crop's susceptibility to particular pests. A disease management system is also included. Although various biological controls are detailed in the preamble, the commercial recommendations listed do not include them.

During the winter flushing period, suppression of leaf flushes not only promotes flowering but also suppresses some of the overwintering insect and mite pests. In the spring flush/ flowering period, trichlorfon, for the control of litchi stinkbug and suppression of erinose mite and leaf gall midge, is applied in combination with chlorbenside for downy blight disease. Two sprays of the latter chemical may be required during prolonged rainy periods at flowering.

During the fruit let period, pest targets are stem-end borer and litchi stinkbug, which are controlled with a mixture of cypermethrin or chlorpyrifos plus trichlorfon. Ridomil-MZ or Sandofan may be applied for downy blight control. The stem-end borer and downy blight are targeted in the maturing fruit period, with sprays as required. The autumn flush is protected from stem-end borer, erinose mite, leaf midge and various lepidopterous caterpillars by two sprays of isocarbophos or acephate, applied 10–15 days apart. Presumably, releases of *Anastatus* could be substituted for the stinkbug sprays, provided that they are correctly timed and integrated with the stem-end borer sprays. The use of *A. japonicus* in combination with *Ooencytus phongi* has been adopted in Thailand, where the stinkbug is controlled on litchis and longans by mass releases of these egg parasitoids at flowering. If necessary, carbaryl can be applied at fruit set to control excessive numbers of nymphs, if egg parasitism is low. The subsequent control of *Conopomorpha sinensis* using permethrin has no effect on biological control, since the oviposition period for the bugs, and hence the critical period of protection for the parasitoids, has passed by the time the small fruit become susceptible to the borer. In Thailand, pruning to remove infested leaves and fruit, and restriction of tree size so that sprays can be applied effectively, are recommended. Fruit bagging, while not always providing perfect pest control, provides the bonus of enhanced fruit colour in litchi and longan.

Waite (1992a) detailed the procedure that should be adopted for implementing IPM in Queensland litchis. This system follows the phenological cycle of the trees, and monitoring at least monthly during autumn and winter, and weekly during flowering and fruit production in spring and summer. Using this approach, the more important pests can be detected, and timely controls applied. In this way, conservation biological control is employed for pests such as the soft scales and erinose mite, although the latter may still require occasional chemical control. Leaf-eating loopers are tolerated until excessive foliage is consumed, and *Bacillus thuringiensis* is applied if necessary. In the past, carbaryl has been applied up to five times, or azinphos-methyl up to three times, during fruit growth to control the macadamia nut borer. Tebufenozide offers the opportunity for better control of this pest, with fewer sprays required and less disruption of natural enemies. The complex of pests attacking litchis and longans throughout the world invariably includes at least one species of fruit borer. In the absence of completely effective natural enemies of these pests, chemicals will remain a necessary part of IPM. These will generally be the key determinants of the eventual fruit yield and quality, as well as the viability of the whole IPM system.

9.8 Ber, Ziziphus jujube

The common name "jujube" refers to several species of fruit trees in the genus *Ziziphus* (Family Rhamnaceae). Jujubes are predominantly cultivated in several Asian countries, including India, China, Korea, Thailand, and Vietnam. There are minor acreages in Russia, the Middle East, southern Europe, Australia, and the USA (William 2002; Meyer 1995; Nguyen 2008; Wanichkul and Noppapun 2009). Jujube trees can tolerate harsh conditions such as drought and high temperatures (Pareek et al. 2007). The fruits are rich in vitamins C, A, and the B complex, and are available fresh, dried, and candied. In addition, the leaves may be fed to cattle and goats (William 2002). Some parts of the plant (flowers, seeds, fruits) can be used for medicinal purposes. The wood is used in the production of musical instruments because it is smooth and strong. Flowers are a source of nectar for honeybees, and the honey is reported to be as good as sunflower honey (William 2002).

The use of the common name "jujube" varies between countries. In Vietnam, the jujube species grown commercially is *Ziziphus mauritiana* which is locally called "Táo gai" (Vu 2000); the same species in India is usually called "Ber". In contrast, in China the main species grown commercially is *Ziziphus jujuba* (Liu and Zhao 2009). In China, jujubes have been cultivated for ~4000 years, with current annual fruit production of ~450,000 tonnes (William 2002). In India, jujube is one of the most ancient and common crops, with high economic returns in part due to the low cost of cultivation.

The flowers of *Zizyphus* are borne singly or in inflorescences of up to 10 flowers in leaf axils. The central flower of the inflorescence is the first to open; the outermost flower may open as much as 2 weeks later. The greenish-yellow flowers are only 0.5–0.6 mm diameter, with five triangular sepals, five petals, five sheath-enclosed stamens, a two- to three- celled ovary terminating in two to six styles (Fig. 9.4). The anthers dehisce soon after anthesis; but stigma receptivity and nectar production appears to be delayed until the 2nd day (Ackerman 1961). Pollination needs differ with the cultivar concerned. Ackerman (1961), in California, USA, found that self-pollination by hand gave from 0.5 % to 11.5 % set. Cross-pollination was generally more effective. When two trees were caged together with honeybee colonies the sets ranged from 10.6 % to 15.3 %. Fruits resulting from self-pollination were usually smaller than normal, contained many aborted seeds, and tended to drop prematurely. Honeybees, *Musca domestica,* and *Coccinella novem-*

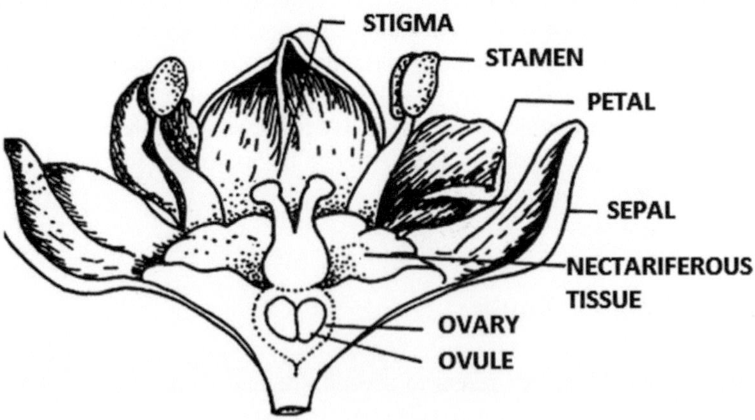

Fig. 9.4 Vertical section of Ber flower (McGregor 1976)

notata were the most abundant visitors to jujube in California, but when the latter two species were introduced onto flowering branches covered in plastic sleeves no fruit was produced indicating they failed to carry much, if any, jujube pollen on their bodies.

In Florida, USA, Lyrene (1983) reported that the introduced cultivars either flower between sunrise and 08.00 h, or between 14.00–17.00 h, and because they are largely self-incompatible and protandrous, plantations need to contain both types for effective pollination. He stated that evidence suggests some cultivars are parthenocarpic. He observed the flowers were visited by numerous bees, wasps and flies.

The trees are visited by several species of insects, particularly honeybees (*A. cerana*) and flies (Singh 1984). A deficit of pollinators and poor cultivation practices may contribute to suboptimal fruit production. Knowledge of pollinators and pollination biology may be important for increasing jujube yields. Jujube flowers attract many insect visitors including honeybees (*A. cerana*; considered the primary insect pollinator), and several taxa of Diptera including *Physiphora* spp. (Ulidiidae), house flies (*Musca domestica*), and other flies (Singh 1984; Devi et al. 1989).

Teaotia and Chauhan (1964) reported that *Zizyphus mauritiana* Lam, Indian jujube, was self unfruitful, and that some pairs of cultivars were cross-incompatible. Anthesis occurs either between 07.30 and 09.00 h or from 12.00 to 13.30 h, according to the cultivar concerned (Vashistha and Pareek 1979). In the Punjab, *Polistes hebraeus* was the most abundant visitor to jujube flowers; it visited about 18 per minute (Dhaliwal 1975). In Jodhpur, Singh (1984) observed seven species of Diptera and 11 of Hymenoptera visiting the flowers, the commonest being *Musca domestica* and *Apis cerana.* The former was active over a longer period each day, and responsible for pollination early in the season as *A. cerana* did not visit the crop until it was in full flower. *M. domestica* and *A. cerana* were mostly present on the lower and upper branches respectively of a jujube tree. Galil and Zeroni (1967) studied 200 wild plants *of Zizyphus spinachristi* L. Willd in Israel. About half were early flowering types (midnight-04.00 h) and the rest late flowering (10.00–12.00 h). Flowers of both types were protandrous, the anthers dehiscing about 12 h before the stigmas became receptive. Hand self-pollination and cross-pollination gave 2 and 29 % fruit set respectively.

Self-pollination is reported to occur in 88 % *of Z. jujuba* cultivars in China, but cross-pollination was reported to increase fruit set up to 600 % (Liu et al. 2009). Fruit that develops from self-fertilized flowers are usually smaller than normal and prone to drop prematurely (Ackerman 1961). In contrast, several cultivars of *Z. mauritiana* failed to set fruit without cross-pollination (Pareek et al. 2007).

Jujube nectar is indispensable for *A. cerana* bees in the Red River Delta during August to October, a time of year when few other nectar sources are available for honeybees. Many professional and semi-professional beekeepers with *A. cerana* colonies place their hives near jujube orchards for honey production. Jujube honey is highly valued by the Vietnamese due to its fragrant odour, light yellow colour, and slow granulation. Jujube also provides pollen for *A. cerana* foragers, although the amount is insufficient for good development of bee colonies.

Ziziphus mauritiana (Jujube, Ber) is visited by Bees (*A. cerana, A. florea, Trigona* spp., *Ceratina* spp.); wasps (*Rhopalidia spatulata*); flies (*Chrysomya megacephala*); flies (*Sarcophaga* sp., *Musca domestica, Stomorhina discolour, Eristalinus arvorum*) and ants (*Camponotus compressus*) (Crane and Walker (1984); Singh (1984); Devi et al. (1989); Jothi and Tandon (1993); Klein et al. (2007) Bạch (2008).

Several species of flies and *A. cerana* are the most important pollinators of jujubes (Singh 1984). Without *A. cerana*, jujubes may not be pollinated sufficiently to attain high yields of fruit (Devi et al. 1989). Sihag and Abrol (1986) recorded *Apis florea* as important pollinator of *Ziziphus mauritiana Lamk.* in Hisar,India. Mishra et al. (2004) reported that 86 species of insects visting ber flowers. Nadia et al. (2007) observed that floral visitors of ber were wasps, bees and flies.

According to Nuru et al. (2012) one *Ziziphus* tree is estimated to produce 3.6 kg of honey (range 2.2–5.2 kg), equivalent to about 900 kg of honey/ha (range 550–1300 kg). These figures indicate the high potential value of the plant for honey production. Nectar secretion was positively correlated with temperature, indicating the adaptation of the tree to hot climates. The study showed that *Ziziphus spina-christi* is one of the high-potential honey source plants in arid and semi-arid climatic conditions. It is worthwhile to plant or to conserve the multipurpose *Ziziphus* trees both for honey production and for environmental value.

9.8.1 Insect Pests of Ber

The major problem with different cultivars of *jujube* is the attack of insects on early stages of fruit formation. Among the insect species attacking ber *Zizyphus jujube,* caterpillars of the *Thiacidas postica, Euproctis fraterna* and *Porthmologa paraelina,* and weevils, *Amblyrrhinus poricollis* and *Myllocerus* sp. fed on leaves and inflorescence. Other pest species identified include: Ber beetle, *Adoretus pallens* and *A. nitidus,* it fed on leaves, cutting round holes in them. In severe infestation a few trees were observed completely defoliated. Such trees were unable to set fruit due to the cessation of photosynthesis. The first manifestation of attack by Ber mite *Larvacarus transitans,* was the appearance of scale-like minute galls on the twigs. These grown and harden with time, attaining a size of about 9 x 4.5 mm. Each gall housed a mite, which damaged by sucking sap from the host with its chelicerae, retarding growth and reducing the fruit crop.

The most common insect which frequently attacked and severely damaged *jujube* fruits was fruit flies species such as *Bactrocera zonata, Bactrocera dorsalis* and *Carpomyia* species which were the greatest enemies of the produce, but these can attack other fruits as well. Qureshi et al. (1991) reported *B. zonata* as a significant pest in India and Pakistan. Waterhouse (1993) identified *Bactrocera zonata* as one of the five most important pests of agriculture in South East Asia. A leaf-eating caterpillar, *Porthmologa paraclina,* attacked the foliage. All moths and beetles can be attracted by light and might drown then in kerosenized water.

Sharma et al. (2006) reported that ber leaf hopper *Qadria pakistanica* (Ahmed) (Hemiptera: Cicadellidae) was active from end of October to February under Ludhiana conditions of Punjab, India. The maximum infestation of 9.6–10.5 nymphs/leaf was recorded. In a further study, Sharma (2010) recorded 14 species of insects attacking ber in Punjab, India.

Balikai (2009) reported that though as many as 130 species of insect pests have been recorded in India, only few species have attained the pest status and cause substantial economic damage to ber. Karuppaiah et al. (2010) reported seven insect pests such as fruit fly (*Carpomyia vesuviana* Costa), stone weevil (*Aubeus himalayanus* Voss), ber butter fly (*Tarucus theophrastus* Fabricius), leaf webber/rollers (*Synclera univocolis* Walker), bark eating caterpillar (*Indarbela* sp.), termite (*Odentotermes* sp.) and grey weevil (*Myllocerus* sp.) attacking ber in Hisar,India . The damage of leaf feeders viz., ber butterfly, leaf webber and grey weevil damage was more during June to September and stone weevil damage was noticed from October to February. The fruit fly was recorded from November to February. In case of natural enemies, the fruit fly parasitoids *Fopius* sp. and other hymenopterans, braconids (*Apanteles* sp.), icheuunimonid wasp; neuropterans, green lace wing (*Chrysoperla* sp.) and spiders also have also been reported during this period. Beside to ber fruit fly, the stone weevil and ber butter fly seems to be serious pests of ber in this region. Sarwar (2006) reported several species of insects attacking ber tree which included certain caterpillars, weevils, beetles, and mites as given in Table 9.3.

A brief account of some of the major insect pests attacking ber is as given below:

9.8.1.1 Fruitfly

Several species of fruit flies infest ber fruits. *They include Carpomyia vesuviana, Dacus correctus* and *D. dorsalis* (Basha 1952; 1953; Saen 1986). Among all the species, *Carpomyia vesuviana* causes damage as much as 80 % of the crop under severe infestations (Cherian and Sunderam 1941). It is the most serious, widely distributed pest of ber infesting all varieties. It is distributed in india, Pakistan, southern Italy, north-west indies, middle east, southern Europe. The larvae are creamy white. The adults are brownish yellow with brown longitudinal strips on the thorax, being surrounded on the sides and back with black spots. Infestation starts with the initiation of fruit setting. The adult female lays eggs singly by inserting its ovipositor in the young developing fruit. The larvae start hatching after 2–3 days of egg laying and start feeding on pulp. Normally, only one larva is found in one fruit. Infested fruits become deformed and their growth becomes checked. A large number of such fruits drop off. The larval stage lasts 9–12 days. The pupal period lasts about 2 weeks after which the adult fly 5–8 mm long, 3 mm broad emerges. There may be 2 or 3 generations of the pest during the active period (Batra 1953). Collect and destroy infested fruits and rake the soil round the trees frequently, specially during summer to expose and kill the larvae. During the maturity of fruits, if necessary, spraying should be done with 0.5 % Malathion + 0.05 % sugar solution at weekly intervals. Spray 500 ml of 30 EC in 300 l of water per acre during February-March

Table 9.3 Insect pests attacking ber *(Zizyphus jujube)* tree (Source: Sarwar 2006)

Common name	Scientific name	Order	
Fruit flies	*Bactrocera zonata, B. Dorsalis, Carpomyia* sp	Diptera	Larvae infest fruits
Ber beetle	*Adoretus pallens, A. nitius*	Coleoptera	Feed on leaves
Gray-hairy caterpillar	*Thiacidas postica, Euproctis fraternal*	Lepidoptera	Feed on leaves
Leaf-eating caterpillar	*Porthmologa paraclina*	Lepidoptera	Feed on leaves
Weevils	*Amblyrrhinus poricollis, Myllocerus* sp.	Coleoptera	Adult eat leaves
Ber mite	*Larvacarus transitans*	Acarina	
Spittle bug	*Machaerota planitae*	Hemiptera	Numphs eat leaves
Ber mealy bug	*Drosichiella tamarindus* Green	Hemiptera	Infest foliage
Leaf webber	*Tonica ziziphi*	Lepidoptera	Larvae web leaves
Bark eating caterpillar	*Indarbela tetraonis* Moore	Lepidoptera	Larvae eat bark and bore branches
Fruit borer	*Meridarches scyrodes* Meyr	Lepidoptera	larvae bore fruits
Leaf butterfly	*Tarucus Theophrastus* F.	Lepidoptera	Larvae feed on leaves
Hairy caterpillar	*Thiacidas postica* W.	Lepidoptera	Larvae defoliate
Tussock moth	*Dasychira mendosa* Hb.	Lepidoptera	Larvae defoliate
Flower beetles	*Adoretus sp.*	Diptera	Adult eat leaves and larvae eat roots
Grey leaf weevil	*Myllocerus transmarinus* Hbst.	Coleoptera	Adult eat leaves

starting from the stage when 70–80 % fruits attain pea size and then repeating the spray at 1-month intervals (Pareek and Nath 1996). Use of resistant varieties can help to overcome infestation (Singh and Vashishtha 1984). Besides, certain species of braconidae such as *Bracon fletcheri, Opius carpomyiae* and *Omphalina* sp. (Hymenoptera: Braconidae) have been recorded to parasitise fruit fly

9.8.1.2 Ber Butter Fly (Tarucus theophrastus Fab.)

Larvae of ber bitter fly feed on sprouting tender shoots, leaves and flower buds. Feeding on chlorophyll content of leaves makes them appear whitish in colour. It damages the leaves up to 25–40%. The pest is active from June to September. Spray of quinalphos (0.05 %) during the sprouting and onset of flowering and repeat the spray at 15 days interval would give better control of this pest.

9.8.1.3 Fruit Borer

Damage by larvae of the fruit borer moth *Meridarchis scyrodes* has been observed in southern and western India (Sonawane and Dorge 1971; Pareek and Nath 1996). The reddish larvae bore into the fruit and feed on the pulp. The moths are dark brown. Collection and destruction of fallen fruits and digging the orchard soil under the tree canopy have also given good control. Pareek and Nath 1996 recommended first spray at pea stage with Monocrotophos (0.03 %), second spray after 15 days with Fenthion (0.05 %) and a third spray 15 days after the second spray with 0.01 % Carbaryl for the control of fruit borer. *Microbracon* sp. (Hymenoptera: Braconidae), and *Opius carpomyiae* (Hymenoptera: Braconidae) have been found to parasitise the borer.

9.8.1.4 Bark Eating Caterpillar

The caterpillars of *Indarbela quadrinotata, I. watsoni* and *I. Tetraonis* (Coleoptera: Cerambycidae) feed on bark of ber trees. Heavy infestation by this pest stunts the trees and adversely affects fruit yield. Frassy galleries caused by the pest need to be removed and cleaned. Application of the solution, made up by mixing one litre of kerosene and 100 g soap in 9 l of water, to the holes has been found to effectively control the bark eating caterpillar.

Hairy Caterpillars

A number of species of hairy caterpillars (*Dasychira mendosa, Euproctis fraterna* (Lepidoptera: Lamantriidae), *Thiacidas postica* feed on the young leaves and fruits (Bhatnagar and Lakra 1992). As many as six generations each of 44–84 days' duration have been observed in a year. A spray of 0.05 % Monocrotophos and 0.2 % Carbaryl (Killex carbaryl 50WP) was found most effective in controlling the pest (Pareek and Nath 1996). The hairy caterpillar is parasitised by *Apanteles taprobanae* (Hymenoptera: Braconidae), *Brachymeria* sp. (Hymenoptera: Chalcididae), *Charops obtusus* and *Goryphus* sp. (Hymenoptera: Ichneumonidae) (Mani et al. 2001).

9.8.1.5 Chafer Beetle (Ber Beetle or Leaf Chafer)

Chafer beetles (*Adoretus decanus, A. kanarensis, A. stoliezkae, A. pallens, A. versutus*) (Coleoptera: Scarabaeidae) devour ber leaves mainly during the night. Larvae hatch out in 1 week and feed on roots and vegetation. Adults emerge with the onset of rains. There is only one generation per year. Beetles can be controlled effectively by spraying 0.2 % Carbaryl 50WP and 0.05 % Monocrotophos (Pareek and Nath 1996). The light traps are quite effective in trapping the beetles. The beetles can be trapped and killed by dropping them into water containing kerosene. Raking around the trees is helpful in exposing the hibernating grubs and killing them.

9.8.1.6 Leaf Webber (Synclera Univocolis Walker)

The newly hatched caterpillar attacks unopened and partially opened leaves and leaf folding with silken threads. The larvae consumes green matter by scrapping, leaving behind the papery epidermis. The pest is active from August to September

Application of quinalphos 2 ml/liter or carbaryl 2 g/liter during new vegetative flush and repeat the spray after 3 weeks found effective.

9.8.1.7 Lac Insect

the lac insects, *Kerria lacca* and *K. sindica is a serious pest of ber* (Li and Hu 1994). It sucks the sap from the branches, and ultimately kills the tree. An infestation of 5000 nymphs/100 cm twigs caused a loss of 52.5–58.5 % fruit yield (Lakra and Kher 1990). Infected twigs should be cut off at the time of annual pruning and destroyed. After pruning, the trees should be sprayed with 0.1 % Dimethoate.

9.8.1.8 Mites

Mites have been reported to produce galls in floral buds of ber thereby hindering fruit production. The mite, *Eriophes cernuus is most serious gall forming mite distributed throughout india* Yamdagni and Gill 1968; Mukherjee et al. 1994) Treatment with 0.04 % dicofol have been reported give best results. The fungus, *Fusarium demicellulare* is found on the galls which helps to check the development and spread of galls during the initial stages (Singh and Singh 1978).

Bibliography

Abrol DP (1999) Pollination importance value of bees. Insect Environ 5:140–141

Abrol DP (2000) Honeybees visiting *Guava* (*Psidium guajava* L.) flowers. Environ Entomol 2(4):531–535

Abrol DP (2013) Annual progress report All India coordinated project on honeybees and pollinators. CCSHAU, Hisar, 50p

Ackerman WL (1961) Flowering, pollination, self-sterility and seed development of Chinese jujube. Proc Am Soc Hort Sci 77:265–269

Adgaba N, Awad AM, Al-Ghamdi AA, Alqarni AS, Radloff SE (2012) Nectar of *Ziziphus spina-christi* (l.) willd (Rhamnaceae): dynamics of secretion and potential for honey production. J Apic Sci 56(2):49–59

Agrawal A, Pati AK (2002) Increment in the size of an inanimate object in the case of bagworm moth Eumeta crameri westwood (Lepidoptera: Psychidae). Curr Sci 83:71–73

Agrawal A, Pati AK (2003) Larval case renovation – a unique behaviour in bagworm moth, *Eumeta crameri* Westwood. Curr Sci 85(12):1675

Ambrose JT Jr, Schultheis SB, Bambara Mangum W (1995) An evaluation of commercial bee attractants in the pollination of cucumbers and watermelons. Am Bee J 135:267–272

Ananthakrishnan TN (1971) Thrips (Thysanoptera) in agriculture, horticulture and forestry: diagnosis, bionomics and control. J Sci Ind Res 30:113–146

Anonymous (2011) IPM – schedule for mango. 44 pp. http://www.nhm.nic.in/Horticulture/IPM%20for%20Mango.pdf

Atwal AS, Dhaliwal GS (2008) Agricultural pests of south Asia and their management. Kalyani Publishers, New Delhi

Atwal AS, Dhaliwal GS (2010) Agricultural pests of south asia and their management. Kalyani Publishers, Ludhiana, 616 pp

Bạch T (2008) Mùa táo ở một vùng quê ven sông Dinh (Jujube production in a village near Dinh river) [online]. Available from http://agriviet.com/?comp=tcnewsandid=3941. Accessed 1 Nov 2008

Badiyala SD, Garg R (1990) Studies on pollination and fruit production by *Apis mellifera* L. in seven cultivars of *Litchi chinensis* Sonn. Indian Bee J 52:28–30

Bagle BG (1992) Incidence and control of fruitfly (*Carpomyia vesuviana* Costa) of ber (*Ziziphus mauritiana* Lamk.). Indian J Plant Protect 20(2):205–207

Bakhshi JC, Singh P (1974) The ber – a good choice for semi-arid and marginal soils. Indian Hort 19:27–30

Baldwin EG (1916) Perfect pollination of citrus groves. Gleanings Bee Cult 44:269–271

Balikai RA (2009) Insect pest status of ber (*Ziziphus mauritiana* Lamarck) in India and their management strategies. Acta Hort (ISHS) 840:461–474

Barbagallo S, Cravedi P, Pasqualini E, Patti I (1997) Aphids of the principal fruit-bearing crops. Bayer, Milan, 123 pp

Basha JMG (1952) Experiments on the control of the fruit borers of jujube (*Ziziphus* spp.) – *Carpomyia vesuviana* Costa and *Maridarchis scyrodes* Meyrick in South India. Indian J Entomol 14(3):229–238

Batra HN (1953) Biology and control of *Dacus diversus* Conquillet and *Carpomyia vesuviana* costa and important notes on other fruit flies in India. Indian J Agr Sci 23:87–112

Batten DJ (1986) Towards an understanding of reproductive failure in lychee (*Litchi chinensis* Sonn.). Acta Hort 175:79–83

Bhatia R, Sharma R, Agnitotri RP (2000) Incidence, varietal preference and control of fruit borer, conopomorpha cramerella (Lepidoptera: Gracillariidae) on litchi (litchi chinensis) in Himachal Pradesh. Indian J Agr Sci 70(5):301–304

Bhatnagar P, Lakra RK (1992) Biology and control of hairy caterpillar, *Euproctis fraterna* Moore (Lepidoptera: Lymontriidae) on jujube (*Ziziphus mauritiana* Lamk.). In *bio-ecology and control of insects pests*: proceedings of the national symposium on growth, development and control technology of insect pests, pp 150–155

Blackman RL, Eastop VF (2000) Aphids on the World's crops: an identification and information guide, 2nd edn. John Wiley & Sons, England, 466 pp

Boonithee A, Juntawong N, Pechhacker H, Huttinger E (1991) Floral visits to select crops by four Apis species and Trigona sp. In Thailand. In: van Heemert C, de Ruijter A (eds) Proceedings of the sixth international symposium on pollination. Tilburg, The Netherlands, pp 74–80

Burger DW (1985a) Fruit production of caged 'star ruby' grapefruit trees. J Rio Grande Val Hort Soc 38:15–18

Burger DW (1985b) Pollination eff ects on fruit production of 'star ruby' grapefruit (*Citrus paradisi* Macf.). Sci Hort (Amsterdam) 25:71–76

Butani DK (1977) Pests of litchi in India and their control. Fruits 32(4):269–273

Butcher FG (1956) Bees pollinate lychee blooms. Proc Florida Lychee Grow Assoc 3:59–60

Butcher FG (1957a) Pollinating insects on lychee blossoms. Fla State Hort Soc Proc 70:326–328

Butcher FG (1957b) Bees pollinate lychee blooms. Fla Lychee Grow Assoc 1956 Yearbook Proc 3:59–60

Butcher FG (1957c) Pollinating insects on lychee blossoms. Proc Florida State Hort Soc 70:326–328

Butcher FG (1958) Pollinating insects on lychee blossoms. Fla Lychee Grow Assoc 1957 Yearbook Proc 4:39–41

Cameron JW, Frost HB (1968) Genetics, breeding, and nucellar embryony. In: Reuther W (ed) The citrus industry, vol 2. University of California Press, Berkeley, pp 325–366

Cameron JW, Cole D Jr, Nauer EM (1960) Fruit size in relation to seed number in the Valencia orange and some other citrus varieties. Amer Soc Hort Sci Proc 76:170–180

Castro MS, Araujo VML (1998) Visita de abelhas às flores da goiabeira (*Psidium guajava* L.). In: Anais do XV Congresso Brasileiro de Fruticultura; 1998; Poços de Caldas; Brasil. Poços de Caldas; 1998

Chacoff NP, Aizen MA (2006) Edge effects on flower-visiting insects in grapefruit plantations bordering premontane subtropical forest. J Appl Ecol 43:18–27

Chacoff NP, Aizen MA (2007) Pollination requirements of pigmented grapefruit (*Citrus paradisi* Macf.) from Northwestern Argentina. Crop Sci 47(3):1143–1150

Chadha KL, Rajpoot MS (1969) Studies on floral biology, fruit set and its retention and quality of some lychee varieties. Indian J Hort 26:124–129

Chandrashekar GS, Sattigi HN (2009) Influence of bee attractants on bee pollination on seed quality and yields in radish. Karnataka J Agr Sci 22(4):777–788

Chaturvedi RB (1965) Preliminary studies in the sex distribution, pollination and fruit development in litchi (Litchi chinensis Sonn.). Allahabad Farmer 39:49–51

Chen Y (1993) Apiculture in China. Agricultural Publishing House, Beijing, 180 pp

Cherian MC, Sunderam CV (1941) Proceedings of the 20th Indian science congress, III: 191

Childers CC, McCoy CW, Nigg HN, Stansly PA, Rogers ME (2007) Florida citrus pest management guide: rust mites, spider mites and other phytophagous mites. ENY-603, UF, University of Florida, IFAS, Extension, Gainesville, FL 32611, USA

Chia CL, Hamilton RA, Evans DO (1997) Lychee. Univ. Hawaii C/T/A/H/R Hort. Commod. Fact Sheet #1. 4 pp

Choudhary JS, Prabhakar SC, Moanaro, Das B, Kumar S (2013) Litchi stink bug (Tessaratoma javanica) outbreak in Jharkhand, India, on litchi. Phytoparasitica 41:73–77

Chundawat BS, Srivastava HC (1978) Haryana Men Ber ki Kheti. Directorate of Extension, Haryana Agricultural University, Hisar, p 38 [in Hindi]

Coit JE (1915) Citrus fruits. The Macmillan Co., New York, 520 pp

Crane E (1990) Bees and beekeeping: science practice and world resources. Heinemann Newnes, Oxford, 614 pp

Crane E, Walker P (1984) The impact of pest management on bees and pollination. Tropical Development and Research Institute, London

Currie RW, Winston ML, Slessor KN, Mayer DF (1992) Effect of synthetic queen mandibular pheromone sprays on pollination of fruit crop by honey bees (Hymenoptera: Apidae). J Eco Ent 85(4):1293–1299

Das CS, Choudhury KR (1958) Floral biology of litchi (Litchi chinensis Sonn.). South Indian Hort 6:17–22

de Villiers EA (1990a) Die beheer van vrugtevlieë by lietsjies. Yearbook South African Litchi Grow Assoc 3:29–30

de Villiers EA (1990b) Coconut bugs on avocados. Farming in South Africa, Avocados H.5. 2 pp

de Villiers EA (1992a) Control of the litchi moth, Argyroploce peltastica Meyr. Yearbook South African Litchi Grow Assoc 4:8

de Villiers EA (1992b) Fruit fly. In: the cultivation of litchis. Agr Res Council South Africa Bull 425:56–58

Degani C, Stern RA, El-Batsri R, Gazit S (1995) Pollen parent effect on the selective abscission of Mauritius and Floridian lychee fruits. J Am Soc Hort Sci 120:523–526

DeGrandi-Hoffman G (1987) The honeybee pollination component of horticultural crop production systems. Hort Rev 9:237–249

Delaplane KS, Mayer DF (2000) Crop pollination by bees. CABI, Wallingford, p 344

Devi KR, Atluri JB, Reddi CS (1989) Pollination ecology of *Ziziphus mauritiana* (Rhamnaceae). Proc Indian Acad Sci (Plant Sci) 99(3):223–240

Dhaliwal JS (1975) Insect Pollination in Ber(Zizyphus mauritiana Lamk. Curr Sci 44(14):527

Dhaliwal HS, Srivastava S, Adlakha RL (1977) Insect pollination of litchi, *Litchi chinensis* Sonn., in the valley areas of the Indian Himalayas. Proceedings of the XXVIth international apicultural congress, Adelaide, 396

Dhankar NK (1999) Studies on insect pollination in Citrus spp. M.Sc. Thesis, CCS Haryana Agricultural University, Hisar, Haryana, 52p

du Toit AP (1990) Pollination research: a missing link in subtropical fruit production. Acta Hort 275:239–243
du Toit AP (1994) Pollination of avocados, mangoes and litchis. Inligtingsbulletin Instituut vir Tropiese en Subtropiese Gewasse 262:7–8
du Toit AP, Swart DJ (1995) Pollination and fruit set in Bengal litchi during the 1993 flowering season at Politsi: second report. Yearbook South African Litchi Grow Assoc 7:31–32
Dulta RC, Verma LR (1987) Role of insect pollinators on yield and quality of apple fruit. Indian J Hort 44:274–279
Duncan LW (1999) Nematode diseases of citrus. In: Timber LW, Duncan LW (eds) Citrus health management. APS Press, St. Paul, pp 136–148
Duncan LW (2005) Nematode parasites of citrus. In: Luc M, Sikora RA, Bridge J (eds) Plant parasitic nematodes in subtropical and tropical agriculture. CAB International, Wallingford, pp 437–466
Duncan LW, Cohn E (1990) Nematode parasites of citrus. In: Luc R, Sikora RA, Bridge J (eds) Plant parasitic nematodes in sub-tropical agriculture. Cab International, Wallingford, pp 321–346
Dutcher JD (2007) A review of resurgence and replacement causing pest outbreaks in IPM. In: Ciancio A, Mukerji KG (eds) General concepts in integrated pest and disease management. Springer, Dordrecht, pp 27–43
Eardley CD, Mansell MW (1996) The natural occurrence of insect pollinators in a litchi orchard. Yearbook South African Litchi Grow Assoc 8:27–29
Elmstorm GW, Maynard DN (1991) Attraction of honeybees to the watermelon with bee attractant. Proc Florida State Hort Soc 103:130–133
EPPO/CABI (1997) Bemisia tabaci. In: Quarantine pests for Europe, 2nd edn. CAB International, Wallingford, pp 121–127
Fahn A (1949) Studies in the ecology of nectar secretion. Palest J Bot Jerusalem Ser 4:207–224
Fahn A (1990) Plant anatomy, 4th edn. Pergamon Press, New York, 611 pp
Francke R, Jorge A, Mathieu JM (1969) Effects of insect pollinators on the production of valencia oranges: the honey bee and its effect on citrus production. Agronomia (Monterrey) 122: 7 pp. [In Spanish.]
Franco AMA, Hill JK, Kitschke C, Collingham YC, Roy DB, Fox R, Huntley B, Thomas CD (2006) Impacts of climate warming and habitat loss on extinctions at species' low-latitude range boundaries. Glob Chang Biol 12:1545–1553
Free JB (1993) Insect pollination of crops, 2nd edn. Academic, London
Galan Sauco V (1989) Litchi cultivation. Food and Agriculture Organization of the United Nations, Rome, 136 p
Galil J, Zeroni M (1967) On the pollination of *Zizyphus spina–christi* (L.) Willd. in Israel. Israel J Bot 16:71–77
Gill DS (1970) Studies on the floral biology and fruit development of ber (Z. mauritiana Lamk.) growing in the acid zone of Haryana. M.Sc. Thesis, HAU, Hisar
Glukhov MM (1955) Honey plants. 512 pp. Izd. 6, Perer. i Dop. Moskva, Gos. Izd-vo Selkhoz Lit-ry. [In Russian.]
Gómez-Torres ML, Nava DE, Parra JRP (2012) Life table of Tamarixia radiata (Hymenoptera: Eulophidae) on Diaphorina citri (Hemiptera: Psyllidae) at different temperatures. J Econ Entomol 105:338–343
Gorakh S, Vishal N, Pandey SD, Ray PK, Singh HS (2012) The litchi. Food and Agricultural Organization of the United Nations, New Delhi, 219 pp
Gould WP, Hennessey MK, Pena J, Castineiras A, Nguyen R, Crane J (1999) Nonhost status of lychees and longans to Caribbean fruit fly (Diptera: Tephritidae). J Econ Entomol 92:1212–1216
Grajales-Conesa J, Meléndez Ramírez V, Cruz-López L, Sánchez D (2013) Abejas nativas en huertos de naranja (*Citrus sinensis*) y de limón (*C. limon*) en Yucatán, México. Acta Zoológica Mexicana (n s) 29(2):437–440
Groff GW (1943) Some ecological factors involved in successful lychee culture. Proc Florida State Hort Soc 56:134–155

Grové T, de Beer MS, Dreyer S, Steyn WP (1999a) Monitoring fruit flies in litchi orchards. Yearbook South African Litchi Grow Assoc 10:34–36
Grové T, Steyn WP, de Beer MS (1999b) Exposure of different litchi cultivars to Mediterranean fruit fly, Ceratitis capitata (Weidemann) and the Natal fruit fly, Ceratitis rosa Karsch. Yearbook South African Litchi Grow Assoc 10:37–38
Grové T, Steyn WP, de Beer MS (1999c) Exposure of natal fruit fly Ceratitis rosa Karsch to litchi export conditions. Yearbook South African Litchi Grow Assoc 10:39–40
Gupta D, Bhatia R, Sharma NK, Chandel JS, Sharma R (1997) Incidence and management of red rust of litchi in lower hills of Himachal Pradesh. Pest Manag Hort Ecosyst 3:70–74
Gyi MM, Dikshit AK, Lal OP (2003a) Residue studies and bioefficacy of lambda-cyhalothrin and beta-cyfluthrin in ber (*Ziziphus mauritiana* Lamarck) fruits. Pestic Res J 15(1):26–27
Gyi MM, Lal OP, Dikshit AK, Sharma VP (2003b) Efficacy of insecticides for controlling ber fruit fly. Ann Plant Protect Sci 11(1):152–153
Hall DG, Childers CC, Eger JE et al (2007) Binomial sampling to estimate rust mite (Acari:Eriopphyidae) densities on orange fruit. J Econ Entomol 100:233–240
Hameed SF, Sharma DD, Agarwal ML (1992) Integrated pest management in litchi. Proceedings of national seminar on recent developments in litchi production held at R.A.U., Pusa, p 38
Hameed SF, Singh PP, Singh SP (2001) Pests. In: Chauhan KS (ed) Litchi: botany, production, utilization. Kalyani Publishers, Ludhiana, 228 pp
Hamilton RA, Seagrave-Smith H (1954) Growing guava for processing. University of Hawaii Press, Honolulu, 19 pp
Han S, Liu W, Chen Q (1999) Mass releasing anastatus japonicas to control tessaratoma papillosa in Hong Kong. Chinese J Biol Control 15(2):54–56
Hardin MR, Benrey B, Coll M, Lamp WO, Roderick GK, Barbosa P (1995) Arthropod pest resurgence: an overview of potential mechanisms. Crop Prot 14:3–18
Hassanein MH, Ibrahim MM (1959) Studies on the importance of insects, especially the honey bee in pollination of citrus in egypt. Agr Res Rev 37(3):390–409
Higo HA, Winston M, Slessor KN (1995) Mechanism by which honeybee (Hymenoptera: Apidae) queen pheromone sprays enhance pollination. Ann Ent Soc Am 88:366–373
Hoy MA, Nguyen R (2001) Classical biological control of Asian citrus psylla. Citrus Ind 81:48–50
Huang CC, Chang K, Sand, Chu YI (1994) Damage and population fluctuation of the litchi fruit borer, conopomorpha sinensis Bradley, in Chia-Nan district, Taiwan. Plant Protect Bull, Taipei 36(2):85–95
Hume HH (1957) Citrus fruits. Macmillan, New York
Inserra RN, Stanley JD, Bannon JHO (2005) Nematode quarantine and certification programmes implemented in Florida. Nematol Mediterr 33:113–123
Jothi BD, Tandon PL (1993) Insect pollinator activity on ber (Zizyphus mauritiana Lamk.). Pollination in tropics. Proceedings of the international symposium on pollination in tropics, 8 13 Aug 1993, Bangalore, p 210–211
Jyothi JVA (2004) Pollination of guava (*Psidium guajava* L.) by Honeybees (*Apis cerana indica* F.) at east Godavari District, Andhrapradesh. Environ Ecol 22(2):365–368
Kalmath S, Sattigi SN (2002) Use of Attractants in maximizing the quantitative parameters of onion (*Allium cepa* L.) seed. Indian Bee J 64(3–4):11–15
Karuppaiah V, More TA, Sivalingam PN, Khan H, Bagle BG, Ahlawat VP, Sehrawat SK, Partap PS (2010) Prevailing insect pests of ber (*Ziziphus mauritiana* Lamk) and their natural enemies in hot arid ecosystem. Haryana J Hort Sci 39(3/4):214–216
Khan KSAR (1929) Pollination and fruit formation in litchi *(Nephelium litchi,* Camb). Agric J India 24:183–187
Khurana AD, Gupta OP (1972) Bark eating caterpillars pose a serious threat to fruit trees. Indian Farmers Digest 5:51–52
Kihara H (1951) Triploid watermelons. Amer Soc Hort Sci Proc 58:217–230
King J, EM Exley, Vithange V (1989) Insect pollination for yield increases in lychee. In: D.J. Batten (ed). Proceedings of the 4th Austral. conference tree nut crops. Exotic Fruit Growers Assn., Lismore, Austral, p 142–145

Kitroo A, Abrol DP (1996) Studies on pollen carrying capacity and pollination efficiency of honey bees visiting litchi flowers. Indian Bee J 58:55–57

Klein A-M, Vaissière BE, Cane JH, Steffan-Dewenter I, Cunnigham SA, Kremen C, Tscharntke T (2007) Importance of pollinators in changing landscapes for world crops. Proc R Soc 274:303–313

Krezdorn AH (1970) Pollination requirements of citrus. In The Indispensable Pollinators, Ark. Agr Ext Serv Misc Pub 127:211–218

Krezdorn AH (1986a) Citrus fl owering, fruit set and development. Citrus Research and Education Center, Lake Alfred

Krezdorn AH (1986b) Pollination and related factors aff ecting fruit quality. Citrus Research and Education Center, Lake Alfred

Kumar S (1990) Studies on insect pollination in 'ber', Zizyphus mauritiana Lamk. M.Sc. Thesis. Haryana Agricultural University, Hisar

Kumar R, Lenin JK (1998) Studies on foraging behaviour of honey bees and pollination in orange, citrus reticulata Blanco. Indian Bee J 60(2):106–107

Kumar R, Chaudhary OP, Lenin JK (1996) Studies on pollination of litchi (Nephilium litchi Camb.) in Dehradunn India. Indian Bee J 58:9–12

Kumar H, Mittal V, Gupta A, Singh CP (2008) Population dynamics and seasonal occurrence of tessaratoma javanica Thunberg in litchi orchards. Ann Plant Protect Sci 16:70–73

Kumar R, Kumar V (2014) Impact of pollination by European honey bee, Apis mellifera L. on the yield and quality of litchi (Litchi chinensis Sonn.) fruits in India. Pest Manag Hortic Ecosystems 20(2):127–132

Kumar V, Kumar A, Nath V (2011) Emerging pests and diseases of litchi (Litchi chinensis Sonn.). Pest Manag Hort Ecosyst 17:11–13

Kumar A, Srivastava K, Patel RK, Nath V (2014a) Management of litchi fruit borer and litchi mite using bio-rational approaches under subtropics of bihar. Ecoscan VI:285–289

Kumar V, Venkataramireddy P, Anal AKD, Nath V (2014b) Outbreak of the looper, perixera illepidaria (Lepidoptera: Geometridae) on litchi, Litchi chinensis(Sapindales: Sapindaceae) – a new pest record from India. Florida Entomol 97(1):22–29

Lakra RK, Kher S (1990) Effect of incidence of lac insect, *Kerria lacca* (Kerr) on bearing and quality of jujube fruits in Haryana. Indian J Plant Protect 18(1):125–127

Lakshmi K, Mohana Rao G (1998) Plants for bees: guava. Bee World 79(3):135–137

Lall BS, Mallik SN (1976) Binomics and control of litchi leaf roller (Platypepla aprobola Meyer). Lepidoptera :Tortricidae. Proceedings of the national academy of sciences, India, p 22–23

Lall BS, Rahman MF (1975) Studies on the bionomics and control of the erinose mite, eriophyes litchi Keifer, Acarina: Eriophydae. Pesticides 9(11):49–54

Lall BS, Sharma DD (1976) Studies on the bionomics and control of cacao moth (Acrocercops cramerella Sn. Gracillariidae: Lepidoptera). Proceedings of the National Academy of Sciences, India, pp. 22–23.

Lee RF, Lehman PS, Navarro L (1999) Nursery practices and certification programs for bud wood and root stock. In: Timber LW, Duncan LW (eds) Citrus health management. APS Press, St. Paul, pp 35–46

Leksawasdi P, Kumchu C (1991) Mass rearing and releasing of the parasitoid Anastatus sp Kasetsart. J Nat Sci 25:47–53

Levin DA, Anderson WW (1970) Competition for pollinators between simultaneously flowering species. Am Nat 104(939):455–467

Li J, Hu X (1994) Rejuvenation and utilization of wild *Ziziphus mauritiana*. Forest Res 7(2):224–226

Liang WG, Huang MD (1994) Influence of citrus orchard ground cover plants on arthropod communities in China. Agr Ecosyst Environ 50:29–37

Liu DG, Tan BL (1999) Studies on the parasitoid community in a complex litchi orchard. Natural Enemies Insects 21:134–139

Liu MJ, Zhao ZH (2008) Germplasm resources and production of jujube in China. Proceedings of the international Jujube symposium, Baoding, China, 21–25 Sep 2008. (840). 25–32

Liu M, Zhao Z (2009) Germplasm resources and production of jujube in China. ISHS Acta Hort 840:25–32

Liu P, Liu MJ, Zhao ZH, Liu XY, Wang JR, Yan C (2009) Investigation on the characteristics of fruiting and seed development in chinese jujube (*Ziziphus jujuba* MILL.). Acta Hort (ISHS) 840:209–214 http://www.actahort.org/books/840/840_26.htm

Loper GM, Waller GD, Berdel RL (1976) Effect of flower age on sucrose content in nectar of citrus. HortScience 11:416–417

Lu FP, Zhao DX, Liu YP, Wang AP, Chen Q (2006) Toxicity of neem seed extract to tessaratoma papillosa (Drury) relative to its allozyme genotypes. Acta Entomologica Sinica 49:241–246

Lyrene PM (1983) Flowering and fruiting of chinese jujubes in Florida. HortScience 18:208–209

Malo SE, Campbell CW (1968) The guava. Fla Agr Ext Sen Fruit Crops Fact Sheet 4, 2 pp

Mani M, Gopalakrishnan C, Krishnamoorthy A (2001) Natural enemies of ber hairy caterpillar *Thiacidas postica* Walker (Lepidoptera: Noctuidae) in Karnataka. Entomon 26(3/4):313–315

Mann GS, Bindra OS (1976) Evaluation of different jujube (*Ziziphusmauritiana*) cultivars for resistance to the ber fruitfly. Punjab Hort J 16(1/2):64–67

Mann GS, Bindra OS (1977) Field incidence of daucus dorsalis on different cultivars of peaxgh(prunus persica) at Ludhiana. J Punjab Agri univ 14(4):507–508

Manzoorul-Haq, Rafie-Ul-Din M, Ghaffar A (1978) Effect of insect pollination on fruit bearing in kinnow mandarin citrus reticulata and physical and chemical properties of the fruit. J Apic Res 17(1):47–49

Martin JH (1987) An identifi cation guide to common whitefl y pest species of the world (Homoptera, Aleyrodidae). Trop Pest Manag 33:298–322

Martinez-Ferrer MT, Ripolles JL, Garcia-Mari F (2006) Enumerative and binomial sampling plans for citrus mealybug (Homoptera: Pseudococcidae) in citrus groves. J Econ Entomol 99:993–1001

Mathur AC, Tandon PL (1974) Litchi mite can be controlled. Indian Hort 19(1):11–12

Mayer DF, Britt RL, Lundon JD (1989) Evaluation of bee scent as a honeybee attractant. Am Bee J 129:41–42

McCoy CW, Selhime AG (1974) The fungal pathogen *Hirsutella thompsonii* and its potential use for control of the citrus rust mite in Florida. Proc Int Citrus Congr, Murcia 2:521–527

McConchie CA, Batten DJ (1991) Fruit set in lychee (Litchi chinensis Sonn.). Variations between flowers, panicles and trees. Aust J Agr Res 42:1163–1172

McGregor SE (1976) Citrus. p 149–157. In Insect pollination of cultivated crop plants. Agric Handb 496. U.S. Gov Print Office, Washington, DC

Mchedlishvili GI (1962) Pollination of citrus trees by bees. Pchelovodstvo 39(9):17. [In Russian.]

Menzel CM (1984) The pattern and control of reproductive development in lychee: a review. Sci Hort 22:333–345

Menzel CM (1991) Litchi. In: Verheij WM, Coronel RE (eds) Plant resources of South-East Asia, vol 2. Edible fruit and nuts. Pudoc, Wageningen, pp 191–195

Menzel CM, Simpson DR (1991) Effect of temperature and leaf water stress on panicle and flower development of litchi (Litchi chinensis Sonn.). J Hort Sci 66:335–344

Menzel CM, Simpson DR (1992a) Growth, flowering and yield of lychee cultivars. Sci Hort 49:243–254

Menzel CM, Simpson DR (1992b) Flowering and fruit set in lychee (Litchi chinensis Sonn.) in subtropical Queensland. Aust J Exp Agr 32:105–111

Meyer R (1995) Three new crops with proven market potential [online]. Available from http://www.newcrops.uq.edu.au/acotanc/papers/meyer.htm Accessed 23 May 2012

Mishra CB (1912) Litchi leaf curl. Indian Agr J 7:93

Mishra RK (1980) Gall formation and control of the erinose mite Eriophyes litchi. Proceedings of international symposium of TOBCIWPRS on intergated control in agriculture and forestry, Vienna, 436 pp

Mishra PK, Yazdani SS (1999) Population dynamics of forager honey bees on litchi, litchi chinensis Gaertn. Sonn blooms in north Bihar. Indian Bee J (1997) 593: 154–157

Mishra RM, Gupta P, Yadav GP (2004) Intensity and diversity of flower-visiting nsects in relation to plant density of *Zizyphus mauritiana* Lamk. Trop Ecol 45(2):263–270

Mitchell WC (1973) Insect and mite pests of guava. Hawaii Agr Exp Station Misc Publ 111:8–10

Mohanta M, Rahman A (1997) Studies on pollination of litchi (*Litchi chinensis* Soon.) by honey bees (*Apis dorsata* F.). Ind Bee J 59(2):74–75

Motial VS (1964) Fruit-set studies in sweet lime. Indian Acad Sci Proc Sect B 60(6):371–379

Mukherjee IN, Singh PK, Singh J (1994) Incidence and control of jujube gall mite (*Eriophyes cernus*) at Varanasi. Indian J Agr Sci 64(5):343–345

Muniappan R, Cruz J, Bamba J (2002) Invasive plants and their control in Micronesia. Micronesica 6(Supplement):47–49

Muniappan R, Shepard BM, Carner GR, Ooi PA-C (2012) Arthropod pests of horticultural crops in tropical asia. CABI, Wallingford, p 168

Mustard MJ, Liu S, Nelson RO (1953) Observations of floral biology and fruit setting in lychee varieties. Proc Florida State Hort Soc 66:212–220

Nadia TDL, Machado IC, Lopes AV (2007) Reproductive phenology and pollination system of *Ziziphus joazeiro* Mart. (Rhamnaceae): the role of *Apis mellifera* and autochthonous floral visitors as pollinators. Acta Bot Bras 21(4):835–845

Nakata S (1956) Lychee flowering and girdling. Hawaii Farm Sci 4(3):4–5

Nalawadi UG, Farooqi AA, Reddy MA, Gubbaiah N, Nalini AS (1973) Studies on the floral biology of Guava (Psidium guajava L) variety Lucknow-49 (Sardar). Mysore J Agric Sci 7(1):24–37

Naumann K, Winston ML, Slessor KN, Smirle MJ (1994) Synthetic honeybee (Hymenoptera: Apidae) queen mandibular gland pheromone applications affect pear and sweet cherry pollination. J Econ Ent 87:1595–1599

Nguyen CM (2008) Trong tao (cultivation of jujube). Agricultural Publishing House, Hanoi

Nguyen R, Hamon AB (1993) Citrus blackfly *Aleurocanthus woglumi* Ashby (Homoptera: Aleyrodidae). Entomology Circular 360. Florida Deptartment of Agriculture and Consumer Services, Division of Plant Industry. http://www.doacs.state.fl.us/pi/enpp/ento/entcirc/ent360.pdf

Nuru A, Awad AM, Al-Ghamdi AA, Alqarni AS, Radloff SE (2012) Nectar of *Ziziphus spina-christi* (L.) wild (Rhamnaceae): dynamics of secretion and potential for honey production. J Apic Res 56(2):49–59

Pandey RS, Yadava RPS (1970) Pollination of litchi (Litchi chinensis) by insects, with special reference to honeybees. J Api Res 9:103–105

Pareek OP, Nath V (1996) Ber. In: Coordinated fruit research in Indian arid zone – a two decades profile (1976–1995). National Research Centre for Arid Horticulture, Bikaner, p 9–30

Pareek S, Mukherjee S, Paliwal R (2007) Floral biology of ber – a review. Agr Rev 28(4):277–28

Partap U (2000) Foraging behaviour of *Apis cerana* on citrus (*Citrus sinensis* var. Red Junar) and its impact on fruit production. In: Matsuka M, Verma LR, Wongsiri S, Shrestha KK, PartapU (eds). Asian bees and beekeeping: progress of research and development, proceedings of the fourth AAA international conference 23–28 Mar 1998, Kathmandu New Delhi,: Oxford and IBH Publishing Co. Pvt. Ltd; pp 174–177

Partap U, Shukla AN, Verma LR (2000) Comparative foraging behaviour of *Apis cerana* and *Apis mellifera* in pollinating peach and plum flowers in Kathmandu Valley Nepal. In: Proceedings of the 4th Asian apicultural association conference, Kathmandu

Pateel MC, Sattigi HN (2007) Effect of different attractants on attracting the bees to cucumber (*Cucumis sativus* L.) crop. Karnataka J Agr Sci 20(4):761–763

Patel BH, Upadhyay VR, Murlidharan CM, Judal GS (1989) Comparative efficacy of different insecticides against ber fruitfly, *Carpomyia vesuviana* Costa. Indian J Plant Protect 17(1):39–45

Paull RE, Duarte O (2012) Tropical fruits. 2nd ed. (Crop production science in horticulture series; no. 20) CAB International CABI Nosworthy Way Wallingford Oxfordshire OX10 8DE UK. p 91–122

Pivovaro SZ (1974) Studies of the floral biology and the influence of growth regulators on fruit set and drop of Litchi chinensis Sonn. MSc thesis, Hebrew University of Jerusalem, Rehovot
Popli S, Kathpal TS, Lakra RK (1980) Note on dissipation of malathion residues from jujube fruits and leaves. Indian J Agr Sci 50(2):191–192
Prakash DS, Baskaran S, Mohandoss A (1993) Insect pollination of the guava, *Psidium guajava* Linn. Comp Physiol Ecol 18(2):60–64
Purseglove JW (1968) Tropical plants cotyledons 1, cotyledons 2. Longmans, London
Qureshi ZA, Hussain T, Siddiqui QH (1991) Relative preference of mango varieties by *Dacus zonatus* and *D. dorsalis*. Pak J Zool 23:85–87
Qureshi R, Waheed A, Arshad M, Umbreen T (2009) Medico-ethnobotanical inventory of Tehsil Chakwal, Pakistan. Pak J Bot 41(2):529–538
Ragumoorthi KN, Arumugam R (1992) Chemical control of ber fruitfly (*Carpomyia vesuviana* Costa). Indian J Plant Protect 20(1):32–36
Rajagopal D, Eswarappa G (2005) (Ed A J S Raju) Pollination potentiality of honeybees in increasing productivity of guava in Karnataka. In Changing trends in pollen spore research 2005 pp. Advances in Pollen Spore Research Vol. XXII: 131–141
Rama Devi K, Atluri JB, Reddi Subba C (1989) Pollination ecology of ecology of Ziziphus mauritiana Lamk. (Rhamnaceae). Proc Indian Acad Sci (Plan Sci) 99(3)):233–239
Randhawa GS, Bajwa BS, Bakshi JC (1961) Studies on the behaviour of seedlings of some rootstocks in nursery. Indian J Hort 18:71–80. Rao, 1997
Ray PS, Chhonkar VS (1981) Pollination and fruit development in Guava. South Indian Hortic 29(3):134–137
Robinson FA, Krezdorn AH (1962) Pollination of the orlando tangelo. Amer Bee J 102:132–133
Roubik DW (1995) Pollination of cultivated plants in the tropics. FAO Agric Serv
Saen T (1986) Host plants of the fruitflies in Thailand. King-Mongkut's Agrl J 4(1):1–15
Sanford M (1992) Pollination of citrus by honey bees. RF-AA092, Florida cooperative Extension Service. University of Florida, 6 p
Sanford MT(2003) Pollination of citrus by honey bees. University of Florida, IFAS Extension [Online]. Available: http://edis.ifas.ufl.edu/aa092. 25 June 2009
Sarwar M (2006) Incidence of insect pests on Ber (Zizyphus jujube) tree. Pakistan J Zool 38(4):261–263
Saxena DK, Rawat RR (1968) A study on the incidence, varietal performance and biology of the ber fruitfly in Madhya Pradesh. Madras Agr J 55:328–333
Schneider H (1968) The anatomy of citrus. In: Reuther W (ed) The citrus industry, vol 2. University of California Press, Berkeley, pp 1–23
Schreiner IH, Nafus DM (1992) Changes in a moth community mediated by biological control of the dominant species. Environ Entomol 21:664–668
Sehgal OP (1961) Studies on morphology and blossom biology in guava (Psidium guajava L.) M.Sc. Thesis. IARI, New Delhi
Sharma A (1992) Seasonal incidence of *Larvacarus transitans* on *Ziziphus* in Rajasthan. Ann Arid Zone 3(3):231–232
Sharma DR (2010) Current scenario of management of fruit pests in Punjab. Haryana J Hort Sci 39(1/2):110–120
Sharma DD, Kumar H (1986) How to control bark-eating caterpillars. Indian Hort 31(1):25
Sharma DD, Singh SP, Akhauri RK (1986) Relationship between the population of Aceria litchi Keifer on litchi and weather factors. Indian J Agr Sci 56(1):59–63
Sharma DR, Bal JS, Randhawa JS (2006) Studies on activity, varietal reaction and control of her leaf hopper, Horticultural Society of Haryana, Hisar, India. Haryana J Hort Sci 35(3/4):200–201
Sihag RC, Abrol DP (1986) Correlation and path coefficient analysis of environmental factors influencing flight activity of *Apis florea* F. J Apic Res 25:202–208
Singh MP (1971) Occurrence of platypepla aprobola Meyr. (Tortricidae: Lepidoptera) on litchi (Litchi chinensis) in North Bihar. Indian J Entomol 33(1):98. Special issue, Vol. VI: 285–289

Singh H (1974) Apoderus blandus faust (Coleoptera: Curculionidae)- a new pest of litchi (Litchi Chinensis Sonn). Indian J Entomol 36:239–240

Singh H (1978) Pest complex of litchi in Dehradun and Saharanpur district of Uttar Pradesh. Indian J Entomol 40(4):464

Singh MP (1984) Studies on the activity of some insect pollinators on jujube (Zizyphus mauritiana Lamk). Entomon 9(3):177–180

Singh R, Chopra SK (1998) Flower visitors of litchi (Litchi chinensis Sonn.) and their role in pollination and fruit production. Pest Manag Econ Zool 6:1–5

Singh J, Raghuraman M (2011) Emerging scenario of important mite pests in North India, Zoosymposiim 6: 170–179 In: G. J. de Moraes and H. Proctor [eds.]. Zoosymposium 6: 1–304, Acarology XIII, Proc Intl Congress

Singh LB, Singh UP (1954) The litchi. Superintendent, Printing and Stationary, Allahabad

Singh UP, Singh HB (1978) Occurrence of *Fusarium demicellulare* on living galls of *Ziziphus mauritiana* in India. Mycologia 70(5):1126–1129

Singh MP, Vashishtha BB (1984) Field screening of some ber cultivars for resistance to ber fruitfly, *Carpomyia vesuviana* Costa. Indian J Plant Protect 12(1):55–56

Singh KK, Chadha KL, Gupta MR (1973) Ber cultivation in Punjab. Punjab Agricultural University, Ludhiana

Singh JP, Jaiswal AK, Monobrullah MD, Patamajhi P (2009) Bioefficacy of insecticides against pentatomid bugs, tessaratoma javanicaThunb: a sporadic pest of kusum, schleichera oleosa. Indian J Entomol 71:259–273

Singh HS, Vishal N, Singh AA, Pandey SD (2012) Litchi: preventive practices and curative measures. Satish Serial Publishing House, Delhi, 479 pp

Smith D, Peña JE (2002) Tropical citrus pests. In: Peña JE, Sharp JL, Wysoki M (eds) Tropical fruit pests and pollinators: biology, economic importance, natural enemies, and control. CAB International, Wallingford, pp 57–101

Somerville D, White B (2005) 'Pollination of apples by honeybees'. *Agnote* DAI-132. NSW Department of Primary Industry

Sonawane BR, Dorge SK (1971) On the biology and control of fruit borer on ber. Poona Agr Coll Mag 61(1/2):1–4

Song JH, Kim SN, Riu KZ (2003) Spatial dispersion and sampling of adults of citrus red mite, Panonychus citri (McGregor) (Acari: Tetranychidae) in citrus orchard in autumn season. Korean J Appl Entomol 42:29–34 (in Korean)

Soost RK (1963) Citrus pollination. Calif Citrog 48:447–452

Soost RK (1969) The incompatibility gene system in citrus. Proc Int Citrus Symp 1:189–190

Soost RK (1987) Breeding citrus –genetics and nuclear embryony. In: Abbott AJ, Atkin RK (eds) Improving vegetatively propagated crops. Academic, London, pp 83–110

Soost RK, Cameron JW (1975) Citrus. In: Janick J, Moore JN (eds) Advances in fruit breeding. Purdue University Press, West Lafayette, pp 507–540

Souza DTM, Cauto RHN (2002) Pollination by insects in three orange [*Citrus sinensis* (L.) Osbeck] cultivars. *Cientifica*, 30(1–2):79–87. http://edis.ifas.ufl.edu/aa092. 25 June 2009

Stern RA, Gazit S (1996) Lychee pollination by the honeybee. J Amer Soc Hort Sci 120(1):152–157

Stern RA, Gazit S, El-Batsri R, Degani C (1993) Pollen parent effect on outcrossing rate, yield, and fruit characteristics of floridian and mauritius lychee. J Amer Soc Hort Sci 118(1):109–114

Tandon P, Arya HC (1979) Effect of growth regulators on carbohydrate metabolism of *Ziziphus jujuba* gall and normal stem tissue in culture. Biochemic and Physiologic der Pflanzen 174(9):772–779

Teaotia SS, Chauhan RS (1964) Flowering, pollination, fruit set and fruit drop studies in ber (Zizyphus mauritiana Lamk). Indian J Hort 21:40–50

Toledo, de AAR, Ruvolotakasusuki MCC, Baitala TV, Costa-Maia FM, Pereira HL, Halak AL, Chambó ED, Malerbo-Souza DT (2013) Pollination by honeybees (Apis mellifera L.) in orange (Citrus sinensis L. Osbeck). Scientia Agraria Paranaensis 12(4):236–246

Trehan KN (1956) Brief notes on crop pests and their control in the Punjab (India). JBombay Nat Hist Soc 54(3):581–626

Vacante V (2010) Citrus mites. CABI Publishing, Wallingford, p 378
Van den Berg MA, Greenland J (2000) *Tamarixia dryi*, a parasitoid of the citrus psylla, *Trioza erytreae*: a review. Afr Plant Protect 6(1):25–28
Van Horn CW, Todd FE (1954) Bees, bouquets and better tangerines. Prog Agr Ariz 6(1):11
Van Gundy SD (1982) Nematodes. In: Page AL, Miller RH, Keeney DR (eds) Methods of soil analysis. Part 2. American Society of Agronomy, Madison, pp 1124–1128
Vansell GH, Watkins WG, Bishop RK (1942) orange nectar and pollen in relation to bee activity. J Econ Ent 35:321–323
Vashistha BB, Pareek OP (1979) Flower morphology, fruit set and fruit drop in some ber (Ziziphus mauritiana Lamk.) cultivars. Ann Arid Zone 18:165–169
Vasquez J, Delgado C, Ferrero DM (2002) Les insectes nuisibles au goyavier (*Psidium guajava* L.) en Amazonie péruvienne. Fruits, Paris 57(5):323–334
Vayssieres JF (1997) Enquête sur les arthropods ravageurs et auxiliaires du litchi à la Réunion. Réport au Département des Productions Fruitières et Horticoles. CIRAD-FLHOR, Ile de la Rèunion, 40 pp
Verdejo-Lucas S, Sorribas FJ, Ornat C, Galeano M (2003) Evaluating *Pochonia chlamydosporia* in a double-cropping system of lettuce and tomato in plastic houses infested with *Meloidogyne javanica*. Plant Pathol 52:521–528
Verma AN, Khurana AD (1974) Further new host records of Indarbela species (Lepidoptera : Metarbelidae). Haryana Agr Univ J Res 4(3):253–254
Verma AN, Singh R (1974) Note on the screening of jujube (*Ziziphus mauritiana* Lamk.) cultivars for incidence of *Indarbela quadrinotata* Walker. Haryana J Hort Sci 3(1–2):62–64
Verma AN, Singh R (1975) Chemical control of bark eating caterpillar, *Indarbela quardinotata* (Walker) by painting the bark with insecticides. Haryana J Hort Sci 4(3/4):145–149
Verma AN, Singh R, Khurana AD (1972) Chemical control of hairy caterpillar. Indian J Agr Sci 42(10):928–931
Villiers ED, Grove T (2000) Guava pests. Neltropika Bull 7–10
Viraktamath S, Anagoudar (2002) Influence of bee attractants in enhancing pollination and yield parameters in *Cucumis sativus* L. Indian Bee J 64(1–2):23–27
Vu HC (2000) Trong cay An qua o Viet Nam (Cultivation of Fruit Trees in Vietnam). Agricultural Publishing House, Hanoi
Wafa AK, Ibrahim SH (1960) Effect of the honeybee as a polllnating agent on the yield of orange. Elfelaha (Jan–Feb), 18 pp. Cairo University, Egypt. [In Arabic.] AA-448/63
Waite GK (1992a) Pest management in lychees. Australian Rural Industries Research and Development Corporation, Canberra, 42 pp
Waite GK (1992b) Integrated pest management in lychee. In: Fullelove GD (ed) Proceedings of the third national lychee seminar. Australian Lychee Growers' Association, Bundaberg, pp 39–43
Waite GK, Hwang JS (2002) Pests of litchi and longan. In: Peña JE, Sharp JL, Wysoki M (eds) Tropical fruit pests and pollinators: biology, economic importance, natural enemies and control. CABI Publishing, Wallingford, p 331
Waite GK, McAlpine JD (1992) Honey bees as carriers of lychee erinose mite eriophyes litchii (Acari: Eriophyiidae). Exp Appl Acarol 15:299–302
Wang B, Qiu Y (1997) Growth and fruiting. In: Zhang ZW, Yuan PY, Wang BQ, Qui YP (eds) Litchi pictorial narration of cultivation. Pomology Research Institute Guangdong Academy of Agricultural Sciences, Guangzhou, pp 66–88
Wanichkul K, Noppapun U (2009) Floral biology and fruit development of *Ziziphus mauritiana* Lamk. 'Bomple'. pp 312–318. In Proceedings of the 47th Kasetsart university annual conference, Kasetsart, 17–20 Mar 2009. Kasetsart University, Bangkok, (Only the abstract seen)
Waterhouse DF (1993) The major arthropod pests and weeds of agriculture in Southeast Asia. ACIAR, Canberra
Webber HJ (1930) Influence of pollination on set of fruit in citrus. Calif Citrog 15(7):304, 322–323
Webber HJ (1946) Cultivated varieties of citrus. In: Webber HJ, Batchelor LD (eds) The citrus industry, vol 1. Univ. California Press, Berkeley, pp 569–584

Webster KW, Cooper P, Mound LA (2006) Studies on Kelly's citrus thrips, *Pezothrips kellyanus* (Bagnall) (Thysanoptera: Thripidae): sex attractants, host associations and country of origin. Aust J Entomol 45:67–74

Weiser J, Muma MH (1966) Entomophthora floridana n.s. (Phycomycetes: Entomophthoraceae), a parasite of the Texas citrus mite Tetranychus banksi. Fl Entomol 49:155–159

William H (2002) Notes on economic plants. Econ Bot 56(2):198–206

Winston ML, Slessor KN (1993) Application of queen honeybee mandibular pheromone for beekeeping and crop pollination. Bee world 74:111–128

Wongsiri S, Chen P (1995) The effects of agricultural development on honey bees in Thailand. Bee World 76:3–5

Wright N (1937) Pollination and the seediness of marsh grapefruit. Agr Soc Trinidad Tobago Proc 37:51–60

Yamdagni R, Gill AS (1968) Gall formation in floral buds of ber – a new threat to ber industry. Punjab Hort J 8:259

Yang CK, Luo QH (1999) A new genus and species of gall midge (Diptera: Cecidomyiidae) infesting litchi from China. Entomotaxonomia 21:129–132

Yokoyama KM, Wanitprapha K, Nakamoto ST, Chia CL (1991) Lychee. University Hawaii C/T/A/H/R Econ. Fact Sheet #12:4 pp

Zavrashvili RM (1964) Bees and the citrus crop. Pchelovodstvo 84(8):19. [In Russian.] AA-347/66

Zavrashvili RM (1967a) Importance of nectar in flowers of mandarin orange unshiu. Akad Nauk Gruz SSR Soobshch [Tiflis] 45(1):205–212, [In Georgian, Russian summary.] AA-789/70

Zavrashvili RM (1967) Influence of bees on the yield of citrus trees on the commercial plantations of Georgia. In 21st international apicultural congress proceedings, College Park Md Aug, pp 450–451. [In Russian, English summary]

Zhang ZW, Yuan PY, Wang BQ, Qui YP (1997) Litchi pictorial narration of cultivation. Pomology Research Institute, Guangdong Academy of Agricultural Sciences, Guangzhou, 189 pp

Chapter 10
Plantation Crops

Plantation crops are defined as a group of commercial crops perennial in nature, cultivated extensively in tropical and subtropical situations in a large and contiguous areas. They include coconut, areca nut, oil palm, cocoa, cashew nut, tea, coffee and rubber.

10.1 Pollination of Plantation Crops

10.1.1 Pineapple (Ananas comosus)

Ananas comosus var comosus (L.) Merr. (Bromeliaceae), flowers are borne on a terminal inflorescence which consists of 50–200 individual hermaphrodite flowers with tubular corolla. Flowers are normally self-sterile and fruit development is parthenocarpic.

10.1.1.1 Pollinators

Wind pollination is not known to occur in pine apple as the pollen is very sticky. It is reported to be pollinated by honeybees, native bees, ants, humming birds and beetles. Westerkamp and Gottsberger 2000 found that hummingbirds were the main pollinator of pine apple in America, whereas (Py et al. 1987) reported that cultivars are self-incompatible, forming fruits without seeds. Kudom and Kwapong (2010) reported that fourteen species of butterflies and one species of ants were the main insect floral visitors as well as four species of sunbirds (Tables 10.1 and 10.2). *Apis mellifera adansonii* Linnaeus, Xylocopa calens Lepeletier and Dactylurina staudingeri Gribodo were seen in the pineapple farm collecting nectar from flowers of weeds and other plants but not foraging on pineapple flowers. In addition, *A. mellifera, Musca domestica, M. sorbens* and *Chrysomya sp.* were seen collecting juice from over-ripped and rotten fruits of *A. comosus*.

D.P. Abrol, *Pollination Biology, Vol. 1*, DOI 10.1007/978-3-319-21085-8_10

Table 10.1 Insects visiting *A. comosus* flowers in Central Region of Ghana (Kudom and Kwapong 2010)

Order	Family	Scientific name
Lepidoptera	Nymphalidae	Precis sophia Suffert Charaxes tiridate Cramer Palla decius Cramer Danaus chrysippus Linnaeus
	Pieridae	Melanitis leda Linnaeus Mylothris chloris Fabricius Catopsilia florella Fabricius Belonois calypso Drury
	Papilionidae	Appias sylvia Fabricius Graphium adamastor Boisduval
	Acraeidae	Graphium pylades Fabricius Papilio demodocus Esper Acraea natalica Boisduval Acraea pharsalus Ward
Hymenoptera	Formicidae	*Lasius fuliginosus* Latreille

Table 10.2 Birds species found on A. comosus flowers in the Central Region of Ghana (Kudom and Kwapong 2010)

Order	Family	Scientific name
Apodiformes	Nectariniidae	Cinnyris sp
		Anabathmis sp Cyanomitra sp Chalcomitra sp

10.1.1.2 Pollination Requirements

Flowers that were exempted from their visit were still able to produce fruits that were as good as the fruits that developed from flowers that received a lot of floral visitors and there were no seeds from either set of fruits. Since nectar removal by floral visitors has no effect on fruit development intercropping pineapple with other crops can be mutually beneficial. Improved weed management as a result of intercropping would increase fruit yield in pineapple while the other crops benefit through the abundance and diversity of pollinators. Westerkamp and Gottsberger (2000) studies have established that pollination is not important in pineapple cultivation. In addition, mature fruits can be harvested early to prevent secondary infection and ultimately lower economic value of fruits as a result of *Apis mellifera* feeding behaviour on the fruit.

10.1.1.3 Pests of Pineapple

Pine apple does not suffer much on account of pest damage but insect as vectors of virus disease take a heavy toll. In India only mealy bugs are serious pests of pineapple and thrips, slug, caterpillar, black palm beetle and termites are of minor importance. Most common pests infesting pine apple are mealy bugs, scales and mites. Mealy bugs play an important role in transmission of pineapple closterovirus causing wilt in plants. Which spreads to other plants through wind or ants (Pegg 1993). Besides, several species of nematodes such as Root-knot nematode (*Meloidogyne javanica*) and root-lesion nematode (*Pratylenchus brachyurus*) and Reniform nematode (*Rotylenchulus reniformis*) are very common in all pineapple growing areas. Spiral nematode (*Helicotylenchus dihystera*) and ring nematode (*Criconemella ornata*) are also found in pine apple growing areas but they do very little damage (Stirling 1993). Soil-inhabiting arthropods, Symphyla or symphylids (*Hanseniella* spp.) also occur throughout pine apple growing areas in different countries (Waite 1993). They feed on plant root tips and hairs (Carter 1963; Py et al. 1987) resulting in many short branching roots and decrease in yields (Lacoeuilhe 1977; Kehe 1979). The white grubs of several beetle species such as *Scarabaeidae* (*Antitrogus mussoni*, *Anoplognathus porosus*, *Rhopaea magnicornis*, *Lepidiota squamulata L. grata*, *L. noxia*, and *L. gibbifrons*) are reported to feed on pineapple roots (Waite 1993) in Australia thereby enabling the secondary plant pathogens to enter the plant (Carter 1967). Fungal pathogens like *Pythium* and rootknot nematode may also infect the injured plant (Carter 1967).

Pineapple plants are also attacked by false spider mite, *Dolichotetranychus floridanus* (Rohrbach and Johnson 2003). which are more damaging under dry conditions (Waite 1993). The other pests attacking pine apples include grasshopper, cane weevil borer, African black beetle, pine apple mite and common army worm (Waite 1993). Among bird crows cause extensive damage in pine apple growing areas by feeding on fruit flesh (Table 10.3).

10.1.2 Coconut **Cocos nucifera** *L., Family Palmaceae*

The coconut flower is monoecious, having both staminate and pistillate florets on the same many-branched inflorescence. Flowering occurs on the plant throughout the year (Fig. 10.1).

10.1.2.1 Pollination Requirements

Sholdt and Mitchell (1967) reported that for good fruit set pollen must be transferred from staminate to the pistillate flowers. However, it does not matter whether the pollen transferred is from the same or different plant. Moore (2001) recommended placement of two honeybee colonies per ha. in conut orchards. Wrigley (1969) however stated that stated that self-fertilization in a dwarf coconut plants is a normal process.

Table 10.3 Common insect pests of pineapple

S.No	Common name	Scientific name	Order	Family	Distribution	Nature of damage	Management
1	Mealy bug	Pseudococcus bromiliae Bouche	Hemiptera	Pseudococcidae	India	Infest leaf base and roots	To avoid the incidence, plant healthy suckers. Before planting dip in fentrotion or phenthoate (0.05 %) emulsion for 15 min. Apply phorate granules @ 1.75 kg/ha, 3 months after planting. Remove grass and other monocot weeds which serve additional hosts for the mealy bug. Destroy ant colonies to check the spread of mealy bugs
2	Mealy bug	P. brevipes Cockerell	Hemiptera	Pseudococcidae	India	Infest leaf base and roots	-do-
3	Coconut black beetle	Orcytes rhinoceros (L.)	Coleoptera	Scarabacidae	India	Adults attack shoots	Do not allow the fruits to ripen in the field. Birds and rats damaged fruits should be plucked and dumped in the pits

4.	Termites	Odontotermes spp.	Isoptera		India	Termite infestation is noticed in plantain in sandy soils	Do not retain dried suckers in the field. Regular irrigation minimizes infestation. If required drench the field with chlorpyriphos (0.5 %)
5	The pine apple Thrips	*Thrips tabaci* Lindermann	Thysanoptera	Thripidae	Cosmopolitan	The damage is done both by nymphs and adults	Spray 250 ml malathion 50 EC in 250 l of water per acre as soon as the pest appears
6	The slug caterpillar	*Parasa lepida*	Lepidoptera	Cochilidiidae	India, Malaysia, Srilanka, south east asia	Young larvae feed gregariously by scrapping the under surface of the leaves and cause drying up of the foliage. As they grow, they get scattered and feed on entire leaves and cause defoliation of the host plant	Spray 250 ml methyl parathion 50 EC in 250 l of water per acre. Dust 10 % BHC

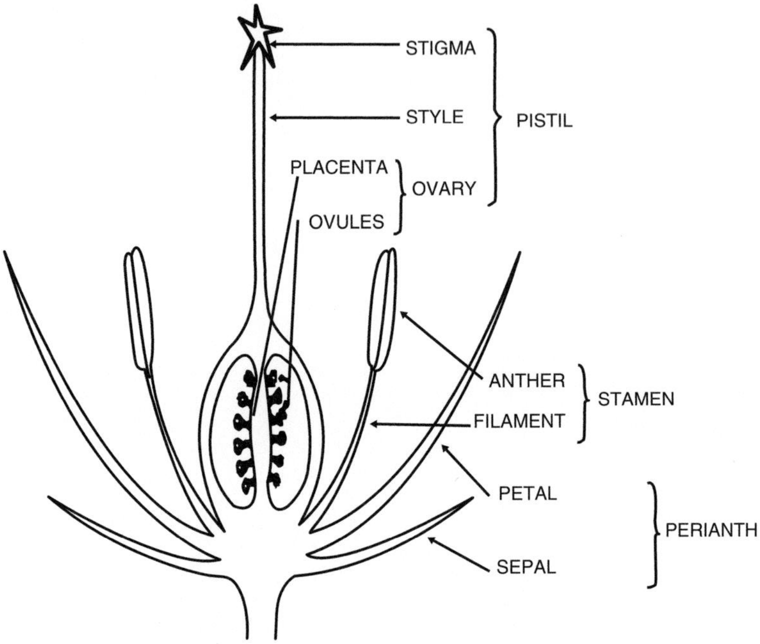

Fig. 10.1 Typical structure of coconut flower

10.1.2.2 Pollinators

Studies on the agents involved in transferring the pollen from the staminate to the pistillate flowers in coconut is still not fully elucidated, but the transfer is required irrespective of flowering habits of the plant. The flower cannot fertilize itself. Several agents such as wind, birds, mites, and insects, including ants, bees, earwigs, flies, and wasps have been mentioned as cross-pollinating agents of the coconut (Davis 1954), however, their effectiveness vary with the situations. Knuth (1904) and Knuth et al. (1904) remarks that the coconut is pollinated through the agency, of wind, but quotes Fr. Dahl who noticed the birds *Charmosyna subplacens Scl.* . Mull. and *Cinnyris frenata* Salvad as the frequent visitors' of coconut flowers in the Bismarck Archipelago. Hunger (1920) states that the coconut is pollinated by wind as well as by insects, and among these figure wasps, bees, flies, beetle and ants. Aldaba (1921) working in the Philippines found so little pollen carried by wind from one tree to another he attaches very little importance to Cross-pollination by this agent. The principal insects observed by him as probable pollinating agents are the house fly (*Musca domestica* Linn.) several species of *Lucila* (Diptera), V*espa luctosa* Sauss *Sarcophaga* sp, *Rhynchium atrum* Sauss., *Apis indica, Trigona biroi* (Hymenoptera) and *Prioionecerus caeruleipennis* Perty (Coleoptera).

Sampson (1923) from the peculiar structure of the flower and the honey glands infers that nature has intended that the coconut flowers should be fertilised by the aid if insects, Burkill (1919) has noted *Apis dorsata and A. indica* on coconuts in Singpore, but remarks that this genus is often found in the Malay Peninsula on "palms, overwhelmingly on male flowers, or on flowers in their male stage, obtaining food without giving what would seem to be an adequate return" and that only *Apis indica* has been seen behaving in that manner in Singapore. The insects that seem to do yeoman service in pollination or cross-pollination of coconuts in Singapore are some species of *Melipona* (the dammer bees), *Apis dorsata,* and some *Muscidae* principally *Musca* very near *nebulo* (the common, Oriental housefly), *Lucilia* sp., and *Pycnosomia* sp. These were seen visiting freely both the male and female flowers, though the flies seem to engage themselves more in sipping the honey from the female flowers than in feeding on the pollen or honey in the male flowers. *Apis indica* was a rare visitor to the male flowers, but this bee does not give an adequate return for the food it obtains from the flowers of palms (Burkill 1919). Wasps do not appear as useful as the bees in pollination as they visit the flowers mostly for the purpose of preying on the bees and other insects which are usually beneficial to the pollination of palms. *Vespa cincta was* frequently observed hunting insects and only on one occasion it was observed alighting on male flowers.

Many different species of insects are found on the coconut palm flowers (*Cocos nucifera* L.). A list of these provides a useful starting point for studies of pollination. These lists are available from only a few countries where the coconut is grown. Aldaba (1921) lists 8 species found in the Philippines; Liyanage (1957) lists 13 from Ceylon; Lever (1933, 1937) and Pagden and Lever (1935) list 25 from the British Solomon Islands, and Furtado (1924) lists 5 found in Singapore. Fifty-one different species of insects were collected from the flowers of the coconut palm. The insects found most often on both flowers were the honey bee *Apis mellifera* L.; black earwig Chelisoches morio (F.); wasps *Polistes exclamans* (Vier.), *P. olivaceus* (DeGeer) and P. macaensis (F.); ants Paratrechina (Prenolepis) longicornis (Latr.) and Pheidole megacephala (F.). The chloropid fly, Oscinella Formosa Becker, is reported from Hawaii for the first time (Lance 1966).

The large numbers of insects associated with coconut palm flowers may be due to the presence of nectar and pollen, or both. Since both kinds of flowers produce nectar, the presence of most of the insects on staminate flowers may be attributed to preference for pollen as food. Also, the staminate flowers are more abundant than the pistillate flowers.

The coconut palm (*Cocos nucifera* L.) is a cross pollinated perennial crop through the agency of wind and insects. Insects play a major role in cross pollination of coconut flowers. Among insects, three species of honey bees in varied proportions were recorded for effective cross pollination. It is a good source of pollen and nectar throughout the year. The secretion of nectar by the glands situated in the male and female flowers of coconut has been recorded by Huggins (1928a, b) and Narayana (1937). The groove present in the pollen indicated that the type adapted to be carried by the insect (Menon and Pandalai 1958). Among insects, honey bees, bumble bee, house fly, ants were recorded as pollinators in coconut palm. Petch (1913) stated that the main mode of pollination was through insects including ants. In Philippines, little pollen is carried by wind from one tree to another and pollination is chiefly effected by insects, the most important among them-being the house fly *(Musca domestica)*, several species of *Lasila* (Diptera) and *Apis indica* (Aldaba 1921). Subsequently Furtado (1924) listed other-possible agents in coconut pollination. The black ant (*Camponotus compressus* F.) was-the other important pollinator in coconut palm. Henry Louis and Chelladurai (1984) recorded the various species of insect pollinators and their frequency of visits in relation to matured male and female flowers in coconut. Sadakathulla and Ramachandran (1991) found that the conventional light trap in coconut garden trapped more honey bees. In a study conducted at Coconut Research Station, Veppankulam, South India during 1992, Sadakathulla (1993) reported three species of honey bees viz., Indian bee *Apis cerana indica* F. little bee, *Apis florae* F. and dammer bee, *Melipona iridipennis* D. visiting coconut flowers in different proportions in the inflorescence of 47 coconut genotypes comprising 28 varieties and 19 hybrids. The number of Indian bees visiting was 2.5 times more than the dammer bee. The number of honey bees visiting the hybrids was 36 % more than the varieties.

10.1.2.3 Pollination Recommendations and Practices

Studies have established that presence of bees could increase the production of coconut crop (Sholdt and Mitchell 1967; Morse and Laigo 1969). However, little information is available on the number of colonies required for optimum production of the crop.

10.1.2.4 Pests of Coconut

Insects and Other Pests

More than 750 species of insects associate with coconut, of which 165 are peculiar to this tree. Only a few are serious pests. *Oryctes* species, especially the rhinoceros beetle (*O. rhinoceros*), burrow into the petiole and penetrate the young immature leaves, sometimes causing death of the palm through secondary bacterial or fungal infection. The palm weevil *Rhynchophorus palmarum* serves as the vector for the red-ring disease nematode, while other *Rhynchophorus* species cause damage by laying eggs in the petiole and foliage. Scales insects (e.g. *Aspidiotus destructor*) attack the leaflets and cause loss of tree vigor. Leaf-eating and leaf-mining beetles (*Brontispa* and *Promecotheca*) and caterpillars (*Lepidoptera*) can cause severe damage to leaves if not controlled.

Total loss of fruit production in otherwise healthy palms can be caused by infestation of fruit-sucking bugs: *Promecotheca* in Africa and *Amblypelta* in the Solomon Islands. The coconut eriophyid mite, *Aceria guerreronis*, which seriously reduces fruit size, was first recognized in Mexico in the 1960s and has subsequently been reported elsewhere in Latin America, all over the Caribbean and West Africa, and is currently active in India and Sri Lanka. In localized areas, monkeys, squirrels and rats will chew on nursery seedlings, young palms and immature nuts; 5-month-old nuts are particularly preferred for their water content. Rats can cause heavy losses. Despite its close association with coconut in Melanesia and Polynesia, the robber crab (*Birgus latro*) is not a pest. It can climb the trunk of the coconut palm but does so to escape danger, rather than to feed off the nuts. The real relationship between this land-living crab and coconut is to enable the crab's short-lived aquatic larval stage to disperse long distances to other islands, by floating. In addition to various insects and other vertebrate pests, weed control is essential during plantation establishment. Failure to control weeds in the 2–4 m area around each palm can lead to shorter leaf length and fewer new leaves produced per year, fewer open bunches and fewer nuts set per tree. The larger diameter (4 m) circle is more beneficial for palms of 39–50 months old. The open canopy of coconut intercepts only about 40–50 % of the incident light. The little change occurs in canopy spread with age and limited root spread makes this an ideal tree for intercropping. Numerous crops are grown under coconut: fodder grass and other pastures for grazing, coffee, cacao, pineapple, banana, maize, mango, citrus, ginger, yams, sweet potatoes, beans, peanuts and medicinal, aromatic and spice crops such as black pepper which can increases the income of small growers.

In addition to the pathogens, there are number of other invertebrate (nematodes, arthropods) and vertebrates (rats, birds, etc.) pests which attack the crop in its various stages of development (Lever 1969; Krantz et al. 1978).

10.1.3 Cashew **Anacardium occidentale** *L., Family Anacardiaceae*

Cashew *is* native to northeast Brazil. The major producers are India, Brazil, Nigeria, Indonesia, Guinea-Bissau, Benin, Vietnam and Tanzania (FAO 2003).

10.1.3.1 Pollination

Both male and hermaphrodite flowers occur on the same inflorescence. The stigma is receptive as soon as the flower opens. The stigma is receptive for only 1 day (Madhava Rao and Vazir Hassan 1957). The flower produces an abundance of nectar, which is highly attractive to flies, bees, ants, and other insects (Morton 1961; Free 1970) (Fig. 10.2).

10.1.3.2 Pollination Requirements

The hermaphrodite flowers are self-fertile but bagged flowers set no fruit, and when hand pollinated a set of about five fruits per inflorescence was obtained (Northwood 1966). Many earlier studies have suggested that wind and many insect species, such

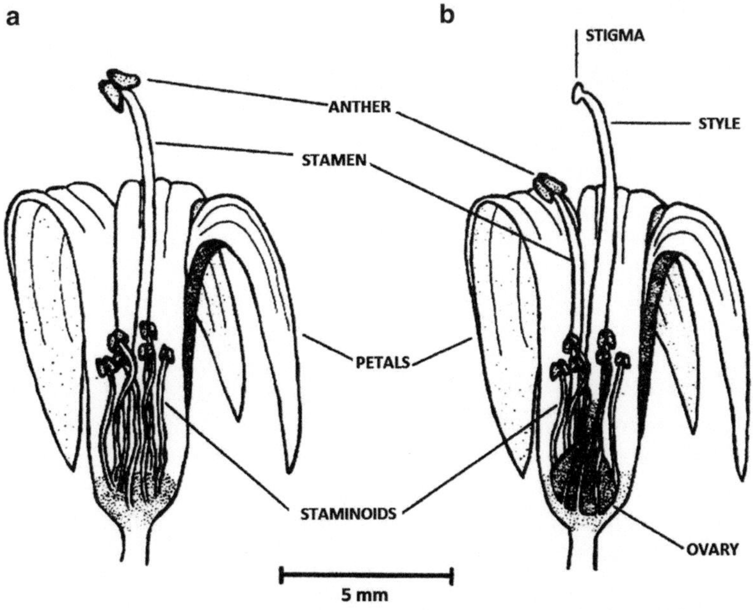

Fig. 10.2 Cross-sectional diagram of a male (**a**) and an hermaphrodite (**b**) cashew flower, highlighting the relative positions of the stamen and staminoids of each and of the stigma of the hermaphrodite flower

as ants, wasps and honey bees, are the pollinating agents of cashew (Bigger 1960; Damodaran et al. 1966; Free and Williams 1976; Reddi 1991; Freitas and Paxton 1996). However, their importance has not been critically determined. In commercial orchards, low yiels in cashew has been attributed to poor pollination.

10.1.3.3 Pollinators

A number of insect species have been recorded visiting cashew flowers such as ants, bees, butterflies and wasps (Freitas and Paxton 1996). The most frequent visitors of cashew flowers in natural habitats were *A. mellifera* and *C. tarsata* (Freitas and Paxton 1998). Both bee species showed foraging behaviour conducive to effective pollination; flower constancy, timing of visits in relation to hermaphrodite flower's anthesis, touching of anther and stigma in the same area of the body, systematic movement between young flowers, and great numbers of cashew pollen grains on their bodies.

10.1.3.4 Pollination Recommendations and Practices

The concentration of honeybee colonies has been reported to enhance fruit setting in cashew, however, there are no recommendations on the number of colonies required for pollination purposes.

10.1.3.5 Insect Pests of Cashew

Cashew is attacked by around 180 species of insect and non-insect pests in India resulting in substantial yield loss (Ayyar 1942; Sundararaju 1993b; Maruthadurai et al. 2012; Bhat and Rupa 2012). The most important pests which cause damage are the cashew stem borer, tea mosquito bug and bark borer (Bhat and Raviprasad 2007). The other pests include leaf and blossom webbers, shoot tip caterpillar, and apple and nut borers.

Cashew Stems and Root Borers *Plocaederus ferrugineus* L. (Coleoptera: Cerambycidae)

Cashew stem and root borers are serious pests in all major cashew growing areas of India. They cause damage up to the extent of 1–19 % every year (Abraham 1958; Sundararaju and Bakthavatsalam 1990; Mohapatra and Mohapatra 2004). *Plocaederus ferrugineus* L. (Coleoptera: cerambycidae) is the primary species, reported (Abraham 1958; Uthaiah et al. 1989; Rai 1984). Infestation up to 40 % is reported in different periods by this internal tissue borer and severely attacked trees die within a period of 2 years causing substantial tree loss (Maruthadurai, et al. 2012). *P. ferrugineus* was reported as the pest responsible for the sudden death of cashew trees in Nigeria and *Spondias mombin* (purple mombin) was identified as an alternative preferred host plant of this pest (Anikwe et al. 2007) with infestation rate of 36.3 % (Asogwa et al. 2009). The grubs are white elongate, cylindrical and apodus, and when full grown measure 7.5 cm. the adult is a medium sized dark brown beetles.

Biology

The mated females lay eggs under loose bark of tree trunk. The newly hatched grubs start feeding on soft tissue, bore into bark and make tunnels. The opening of tunnel is plugged with excreta and reddish mass of chewed fiber. When the grubs are full fed, they tunnel their way to the root region. Pupation takes place in calcareous shell formed by the grubs for this purpose. The life cycle occupies more than a month, thus, comprising several overlapping generations in a year.

Symptoms of Damage

The grubs feed inside the tree trunk or branches making tunnels and damaging the cambial tissue and stopping the flow of sap. This results in weakening of the plant and resulting in plant death, if the infestation continues. Fifteen years old plants are often seen infested with this pest. The infested trees tilt due to injury on anchoring roots (Maruthadurai et al. 2012)

Integrated Pest Management Techniques

Phytosanitation is an important component of CSRB infestation. To achieve this, the dead trees and trees which are beyond recovery (those having more than 50 % of bark circumference damage and / or showing yellowing of the canopy) should be uprooted and removed from the plantation before and after monsoon, since they serve as natural inoculum for multiplication and spread of the pest (Raviprasad et al. 2009; Sundararaju and Bakthavatsalam 1990).

Curative treatment of the infested trees in initial stages of infestation can save the tree. All the trees in the plantation are to be examined for symptoms of infestation by CSRB at the tree bases at monthly interval especially during harvest (January–May). In case external infestation symptoms are noticed, the CSRB grubs should be mechanically removed by careful chipping the bark. The injured portion as well as base of the tree and exposed root should be swabbed/drenched with chlorpyriphos 0.2 % solution (Raviprasad et al. 2009; Sundararaju and Bakthavatsalam 1994; Bhat and Rupa 2012). The newly planted grafts should be trained to have branching at a height of 0.75–1.00 m from ground level for facilitating better inspection and adopting pest management techniques effectively. In Nigeria, application of coal tar–kerosene mixture in ratio 1:2 to healthy trees within infested plots gave 100 % prophylactic efficacy at 12 months after treatment. Chemical control using Propoxur (a carbamate) and physical control by poking of holes created by the borer using a wire and killing of the insect are reported giving 98 % success in pest management (Anikwe et al. 2007). Bhat and Raviprasad (1996) reported mycosis of grubs of CSRB with entomopathogens viz., *Metarhizium anisopliae* and *Beauveria bassiana*. Mixing the spawn with organic matter like FYM, neem cake and cashew apple are practiced to enhance the spore load under the field condition. The eggs of CSRB were parasitized by *Avetiniella batocerae* (Bhat and Raviprasad 2007).

Tea Mosquito Bug

The tea mosquito bugs (TMB), *Helopeltis* spp. are the most serious foliage and fruit pests of cashew in India causing 20–60 % yield loss to the crop. Extensive review of literature containing information on species distribution, nature and extent of damage, biology, natural enemies, host plants and management of the pest are published by Devasahayam and Nair (1986), Sundararaju and Bakthavatsalam (1994), Sundararaju (1996) and Sundaraju and Sundarababu (1999). Out of the 41 species of *Helopeltis* recorded from tropics, only three species (*H. antonii,* Sign., *H. theivora* Wat. and *H. bradyi* Wat.) are confined to India. *H. antonii* occurs in southern and eastern India, whereas, *H. theivora* exists in west coast, Western Ghat and North East India. *H. bradyi* is found in West Coast and Western Ghat regions. Cashew and cocoa are damaged by all the three species. *H. antonii* also damages guava, ber and drumstick (Sundararaju and Sundarababu 1999). Damage of *H. antonii* was first reported on Cashew in India by Ayyar (1932). The record of *H. bradyi* Wat.

occurring on cocoa, guava and cashew was brought out by Stonedahl (1991) and Sundararaju (1996). Stonedahl (1991) documented literature on oriental species of *Helopeltis*.

Under west coast condition, three species of *Helopeltis* (*H. antonii* S., *H. theivora* W. and *H. bradyi* W.) infest cashew, whereas in other regions, only one species (*H. antonii*) exists on cashew. However, in all regions, *H. antonii* is the dominant species infesting cashew causing severe damage. *H. antonii* is distributed in most of cashew growing regions of Kerala, Karnataka, Goa, Maharashtra (Pillai et al. 1976); Tamil Nadu (Abraham 1958; Pillai et al. 1976); Goa (Sundararaju 1979) and Orissa (Jena et al. 1985).

Nature of Damage

On cashew, the nature of damage caused by *H. antonii* has been described by various workers (Abraham 1958; Pillai and Abraham 1975; Sathiamma 1978; Ambika and Abraham 1979; Pillai 1979; Devashyam and Nair 1986; Satapathy 1993; Sundraraju 1996). Both nymphs and adults suck sap from tender shoots, panicles and immature nuts and apples. The typical feeding damage symptom is the formation of necrotic lesion around the point of stylet insertion by the bug. The lesions on shoots and panicles coalesce and ultimately result in shoot blight or blossom blight. During outbreak situation, the entire flush dries up and the trees present a scorched appearance. This pest has got potential to cause cent percent loss in yield. However, on an average an yield loss of about 30 % is caused as a result of damage by this pest. Losses in nut yield of 25–50 % have been reported from Karnataka, Goa, Kerala and West Bengal (CCRS 1966; Desai, et al. 1977; Abraham and Nair 1981; Chatterjee 1989a).

Seasonal Incidence

The tea mosquito bug has a continuous cycle of generations throughout the year (Stonedahl 1991; Sundararaju 1996). The buildup of pest population commenced in October–November synchronising with the emergence of new flushes in cashew. The pest population reached its peak in January–February when trees are in full blossom. Young trees are more susceptible to its attack because of the availability of succulent growth than the older ones (Pillai et al. 1984; Rai 1984; Sundararaju 1984; Stonedahl 1991; Satapathy 1993; Sundararaju 1996).

Biology

The female bug is orange across shoulders and the male is almost black. Soon after the emergence, the adults mate and the female commences egg laying with 2 days. The female inserts the elongate sausage-shaped eggs in the tender shoots in groups of two or more and hairs from each egg protrude outside. A female may lay eggs upto 50. The incubation period is 5–7 days (Ambika and Abraham 1979; Satapathy

1993; Sundararaju 1996). The nymphs complete their development in about 2 weeks after passing through five moults. The life cycle is completed in about a month (25–32 days) and in a year there may be several generations (Stonedahl 1991; Sundararaju 1996). The winter is passed in hibernation as adult.

Varietal Resistance

All the recommended cashew varieties are susceptible to this pest. However, the mid season/late season flowering cashew varieties are able to escape from the severity of the pest to certain extent. Cashew accessions viz. VTH 153, Kunthur 24, Goa 11/6, VTH 153/1, 9/78 and 51 different cashew types in Karnataka, accession No. 665 in Kerala and BLA – 39–4 in West Bengal were reported as least susceptible to *H. antonii* (Ambika et al. 1979; Ghosh and Chatterjee 1993; NRCC 1988; Uthaiah et al. 1989). The cashew accession, Goa 11/6 with yield of 2 t/ha under unsprayed situations is reported to have only moderate level of pest incidence. This accession has been released as a variety "Bhaskara" from the Directorate of Cashew Research, Puttur (Sundararaju et al. 2006). A field screening methodology for cashew genotypes against TMB on a 0–4 scoring grid on the basis of number and nature of lesions was developed by Beevi and Mohapatro (2007) and Mohapatro (2008a)

Integrated Pest Management Techniques

Natural enemies are the egg parasitoids viz., *Erythmelus helopeltidis* Gahan. (Hymenoptera: Mymaridae) *Telenomus* sp. (Laricis group) (Scelionidae) *Chaetostricha* sp. (Trichogrammatidae) and *Gonatocerus* sp. nr. *bialbifuniculatus* Subba Rao reported on this pest from West Coast regions while, *Ufens* sp. is the egg parasitoid reported from East Coast of India. The build of this pest is naturally regulated through these egg parasitoids (Devashayam 1989; Sundararaju 1993a; Sundararaju 1996). They also report that parasitoid activities are naturally promoted under favourable weather condition (increased minimum temperature and relative humidity) during vulnerable period (November–February). The solitary nymphal endoparasitoids, *Leiophron* spp. (Hymenoptera: Braconidae) appear to be most promising in Indonesia and Africa (CIBC 1983), but the activities of hyper parasitoids limit their potentiality in biological control. *Crematogaster wroughtonii* Forel (Formicidae) has been recorded as a predator on nymphs of *Helopeltis* spp. Several species of spiders, *Hyllus* sp., *Oxyopes sehireta*, *Phidippus patch* and *Matidia* sp. five species of reduviid bugs (*Sycanus collaris* Fab), *Sphedanolestes signatus* Dist. and *Endochus inornatus* Stal., *Irantha armipes* Stal. and *Occamus typicus* Dist. have also been recorded as predators of the tea mosquito bug. Use of *Chrysoperla* sp. as predator of *H. theivora* has been reported by Somchoudhury et al. (1993). *Aspergillus flavus* and *A. tamarii* were confirmed to be pathogenic to *H. antonii* (Ambika and Abraham 1979; Devashayam and Nair 1986; Sundararaju 1984).

Chemical pesticides: Insecticides viz. carbaryl (0.1–0.15 %), quinalphos (0.05 %), dimethoate (0.05 %), phosalone (0.07 %), phosphomidan (0.03 %) were reported to be effective against *H. antonii* (Muthu and Bhaskaran 1980; Nair and

Abraham 1982, 1984; Samiayyan and Palaniswamy 1984; Singh and Pillai 1984; Sundararaju 1984; AICRPC 1989; Chatterjee 1989b; Gadase et al. 1991; 1993). Dust formulations of carbaryl (5 % and 10 %), phosalone (4 %), malathion (5 %) and quinalphos (1.5 %) were also found to be effective against *H. antonii* (Nair and Abraham 1983; Hiremath, et al. 1987; Bakthavatsalam, et al. 1993; Mohapatro 2008b)

Even though, many insecticides and plant products (botanicals) were effective, their ovicidal action is limited (Raviprasad et al. 2005). In the endemic areas it is recommended to spray three times with any of these insecticides during most vulnerable periods of crop coinciding with flushing, flowering and fruiting stage based on pest incidence. Spraying of recommended insecticides will be remunerative if the trees are giving economical yield (>2.0 kg/tree). Although, cashew is an insect pollinated crop, spraying of these insecticides during flowering season did not influence the fruit set (Pillai et al. 1984; Rai 1984; Sundararaju and Sundarababu 1999). Proper surveillance during flushing, flowering and fruiting period is essential for the management of TMB. Pest management treatment has to initiate within 5 days when 5–10 % of the flushes show pest incidence. Lambda cyhaolothrin (0.003 %) can be applied at flushing stage followed by L-cyhalothrin (0.003 %) or carbaryl (0.1 %) at flowering stage. The third spary of carbaryl (0.1 %) should be given at full fruit set stage to protect cashew from TMB (Bhat and Raviprasad 2007).

The Juvenile hormone analogues (farnesyl methyl ether @0.5 ug/nymph and graded doses of ZR 512 and MV 678) cause abnormalities on *H. antonii* when applied on fifth instar nymphs (Ambika and Abraham 1984). In western Africa also among the biological constraints, damage by the mirid bugs, Helopeltis sp., causes substantial yield losses in cashew depending on the variety, location and season (ICIPE 2013). Cashew farmers in Masasi, Tanzania have successfully used the African weaver ant to combat insect pests (Olotu et al. 2012; Masasi 2013). The other insect pests causing considerable damage on cashew are leaf and blossom webbers, shoot tip caterpillars which damage shoots and blossoms and shoot tips

Leaf and Blossom Webbers

The cashew shoots bearing fresh flushes and flowers get attacked by two species of leaf and shoot webbing caterpillars, *Lamida* (= *Macalla*) *moncusalis* Wlk. (Lepidoptera: Pyraustidae) and *Orthaga exvinaceae* Hamps. (Lepidoptera: Noctuidae). The symptoms of infestation are the presence of webs on terminal portions with clumped appearance and drying of webbed shoot/ inflorescences. Galleries of silken webs reinforced with plant scraps on the terminal portions of new shoots and blossom and castings indicate the presence of caterpillars (Rao et al. 2002).

Shoot Tip Caterpillars

The tiny yellowish to greenish brown larvae of the moth, *Hypotima (=Chelaria) haligramma* M. (Lepidoptera: Gelechidae) damage the shoot tips and inflorescences. The tender shoot tips are infested occasionally up to 25 to 35 mm leading to drying up of the shoot tips. This pest is regularly reported from the East coast tracts (Mohapatra et al. 1998).

Thrips

Both phyllophilous and anthophilous thrips species infesting *Anacardium occidentale* are highly polyphagous enabling consistency of thrips population. *Rhipiphorothrips cruentatus* (Hood), *Selenothrips rubrocinctus* (Giard) and *Retithrips syriacus* (Mayet) infesting leaves tend to show patterns of infestation in relation to water stress as well as the distribution of nutrients. Sizeable populations of inflorescence infesting *Scirtothrips dorsalis* (Hood), *Thrips hawaiiensis* (Morgan) and *Haplothrips ganglbaueri* (Schmutz) cause premature shedding of flowers (Ananathakrishnan 1985).

Mealy Bugs

Several species of mealy bugs attack the cashew plantations which include *Ferrisia virgata, Planococcus lilacinus, Planococcus citri and Phenococcus solenopsis.* Among all these species, the former species such as *Ferrisia virgata* is most serious causing extensive damage. The nymphs and adults suck sap from the tender plant parts results in withering of growing shoots, inflorescence and developing fruits. Continuous monitoring and early detection of infestation are essential to manage this pest. Fallen leaves under the tree canopy should be collected and burnt to avoid further spread of the pest. Spraying of dichlorvas 76 WSC 0.2 % (@ 2.5 ml / lit) or dimethoate 30 EC 0.05 % (@ 1.75 ml / lit) in combination with fish oil resin soap @ 20 g per litre of water reduces bug incidence effectively. Some of the predators like *Chrysoperla cornea, Menochilus sexmaculatus, Coccinella septumpunctata,* and *Scymnus coccivora* are very effective in cotrolling the pest (Maruthadurai et al. 2012).

10.2 Pollination Coffee and Cacao

10.2.1 Coffee Coffea *spp., Family Rubiaceae*

The coffee flowers are white in colour and fragrant occurring in clusters of 2 to 20 in the leaf axils. Pollen is shed immediately after the flower opens, and the stigma is immediately receptive. Both nectar and pollen are attractive to many kinds of insects. The pollen produced is in very small quantities and not sticky as a consequence pollen can be transported both by wind as well as insects.

10.2.1.1 Pollination Requirements

Wellman (1961) stated that *C. arabica* is self-fertile, yet at times some insect pollination occurs but it is not necessary. However, he stated that the other two species, *C. canephora* and *C. liberica*, are self-sterile and require action of wind or insects. Amaral (1952) showed that caged *C. arabica* plants produced 39 % less coffee than open plants. Later, Amaral (1960) found that fruit setting on the protected branches was 61.7 %, whereas on branches visited by bees it was 75.3 %, indicating a slightly beneficial effect. Lower (1911) claimed that honey bees are of great value in pollinating coffee in Puerto Rico during rainy seasons. T

10.2.1.2 Pollinators

In several studies, honeybees have been recognized as the most important pollinating insect visiting the coffee flowers (Raw and Free 1977; Roubik 2002b). In Costa Rica meliponine bees were reported as important pollinators (Ricketts 2004). Philpott et al. (2006) reported the role of ants in pollination of coffee (*Coffea arabica*) agroecosystems.

10.2.1.3 Pollination Recommendations and Practices

Coffee plantations are greatly benefitted by bee pollination. However, there are no recommendations on the number of colonies required for effective pollination.

10.2.1.4 Coffee and Cacao Pollination Systems

The genus Coffea (Rubiaceae) is native to tropical and subtropical Africa. Two coffee species are important crops in many tropical countries: Highland coffee, Coffea arabica L is grown throughout Latin America, Central and East Africa, India, and Southeast Asia. C. arabica is tetraploid and selfcompatible (Free 1993). Autonomous self pollination occurs when pollen is shed from stamens on to the stigma. The relative benefits of cross pollination have recently been examined (Klein et al. 2003c). Fruit weight is approximately 25 % greater when pollinators have access to flowers (Manrique and Thiemann 2002; Roubik 2002a; Ricketts et al. 2004). Because flowering is induced by heavy rain, coffee blooms synchronously for short periods annually. The primary pollinators are social bees (Roubik 2002a; Ricketts 2004; Veddeler et al. 2006; Bos et al. 2007) followed by less abundant solitary bees (Klein et al. 2003a).

Lowland coffee is grown in West and Central Africa, throughout Southeast Asia and in Brazil. It is self-incompatible. Cross pollination is mediated by insects and less efficiently by wind (Willmer and Stone 1989; Free 1993; Klein et al. 2003c). Like C. arabica, flowering is synchronous and induced by rain, but scattered flowering can occur over several months. The pollinator community is similar to that of highland coffee. During mass-flowering, social bees (honey and stingless bees) are the most abundant flower visitors, but when only single coffee plants are flowering, solitary bees are dominant and social bees are often absent (Willmer and Stone 1989).

10.2.2 Cacao, **Theobroma cacao** *L. (Sterculiaceae)*

Cacao, Theobroma cacao L. (Sterculiaceae) is native to Central and South America. It is grown as a crop throughout the humid tropics, with the most important cacao-exporting countries being Ivory Coast, Ghana, and Indonesia (Young 1994; International Cocoa Organization 2005). Cacao is generally self-incompatible, although a few self-compatible varieties exist (Falque et al. 1996). All varieties are strictly dependent on insect pollination, because the anthers with their sticky pollen are each enclosed by a folded petal. Flowers are cauliflorous and are present year-round, although peaks may occur in early wet seasons (Bos et al. 2007). The primary pollinators are biting midges of the family Ceratopogonidae, especially those of the genus Forcipomyia (Free 1993; Young 1994). Cacao pollinators depend on moist habitats and long dry seasons can decimate pollinator populations (Young 1982; Ruf 1995) (Fig. 10.3).

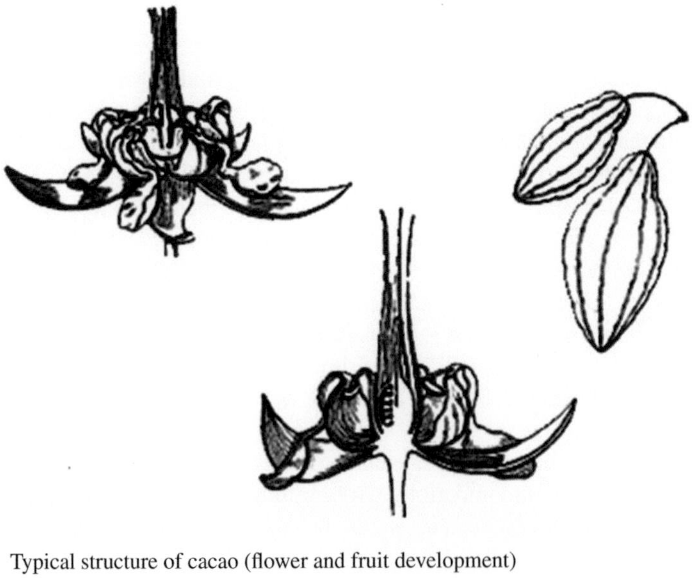

Fig. 10.3 Typical structure of cacao (flower and fruit development)

Pollination studies of local management and landscape effects on tropical plantation crops have examined C. arabica in Panama (Roubik 2002a), Indonesia (Klein et al. 2003a), Costa Rica (Ricketts 2004; Ricketts et al. 2004), and Brazil (DeMarco and Coehlo 2004) and C. canephora in Indonesia (Klein et al. 2003b). Complex agroforestry systems, in terms of plant diversity, were found to support higher bee diversity and bee visits to coffee flowers compared to monocultures or simple shaded systems (Klein et al. 2003a, b; DeMarco and Coehlo 2004; Veddeler et al. 2006). Similarly, such landscape effects on pollination have been examined for three tropical plantation crops in Australia: macadamia (Macadamia integrifolia; Heard and Exley 1994; Blanche et al. 2006) atemoya (Annona squamosa L. A. cherimola Mill.; Blanche and Cunningham 2005), and longan (Dimocarpus longan; Blanche et al. 2006), and for grapefruit (Citrus paradisi) in Argentina (Chacoff and Aizen 2006). These studies all found fewer insect visits to flowers in locations more isolated from non-crop vegetation, but the pattern of change depended upon the suite of insect visitors, which varies by crop and location. For longan (Blanche et al. 2006) and macadamia (Heard and Exley 1994) in Australia, C. arabica in Costa Rica (Ricketts 2004), and grapefruit in Argentina (Chacoff and Aizen 2006), native social stingless bees declined in abundance with increasing distance from native forest more rapidly than did Apis mellifera.

Although some solitary species declined with distance from forest in coffee (Klein et al. 2003c) and grapefruit (Chacoff and Aizen 2006), many solitary species responded more to local phenomena, in particular light intensity and plant diversity (Klein et al. 2003b). Solitary bees (e.g., Heriades and Paracella in the Megachilidae, Nomia in the Halictidae, and Ceratina in the Anthophoridae) often have small flight ranges and depend on local nesting sites (e.g., open soil, dead wood) and particular floral resources (Wcislo and Cane 1996; Kremen et al. 2002; Klein et al. 2003b). Therefore solitary bees are often especially affected by local agroforestry management practices like pruning of shade trees and weeding (Klein et al. 2003b). They further found that very small cavity-nesting bees (e.g., stingless bees were most abundant within 600 m of the forest margin) and large bees (e.g., giant honey bee Apis dorsata) decreased significantly in abundance with forest distance. Only the abundance of the very large bees (carpenter and large leaf-cutting bees) was not significantly related to forest distance. Very large bees even tended to be more abundant in sites far from forest showing that response to isolation from forest can be linked to body size. The pattern considering species richness in each size-class groups was similar, but the steepness of the relationship differed slightly.

The midges that pollinate cacao and the nitidulid beetles that pollinate atemoya breed in rotting fruits and similar substrates in the litter layer, and are expected to have smaller flight ranges than bees. For example, the most important cacao pollinators, ceratopognid flies can only fly up to 6 m, although their foraging range can be enlarged up to 35 m by wind (Decazy et al. 1981). Hence, in plantations where farmers remove leaf litter, pollinator populations are expected to depend on adjacent habitats with suitable breeding substrate. The dispersal range and behavior of the atemoya pollinators remains largely unknown, although it has been shown that their diversity and abundance benefit from proximity to rain forest (Blanche and

Cunningham 2005), possibly because farmers generally remove old fruits and other waste from plantations and thus, as for cacao, unwittingly diminish pollinator habitat within their plantations. These examples suggest that particularly small, generalist pollinating insects can be also influenced by the landscape matrix.

Most plantation crops are open to a range of pollinators, and therefore are pollinated by at least a few species even when relatively isolated from natural habitats in a highly agricultural environment. For plants that attract a range of pollinators, the community of visitors at any one point in time and space is the net effect of taxon-specific responses to habitat characteristics at different scales. However, tropical crop species with a narrow range of specialized pollinators, such as Brazil nut (Bertholletia excelsa) and oil palm (Elaeis guineensis), are likely to suffer more extreme pollinator shortage (Klein et al. 2007). Ricketts (2004) found that the pollination deficit in C. arabica was greatest in locations with low visitor diversity and that a decline between years in the dominant visitor to C. arabica was compensated by other flower-visiting bees, but only in locations near forest vegetation. There is a great, yet unexploited potential, to compare coffee and cacao pollination in agroforestry settings in their native and nonnative ranges to study biogeography of plant-pollinator interactions and assess generality of the relationship between different pollinator assemblages and their functional role.

10.2.2.1 Pests of Cocoa (*Theobroma cacao* L., Sterculiaceae)

Cocoa is South American in origin. More than 70 % of the world cocoa production comes from West Africa. Indonesia is the major producer of cocoa in Asia, followed by Papua New Guinea, Malaysia, India, the Philippines, and the Solomon Islands. Cocoa is grown both on a plantation scale and in smallholder agriculture in Asia. Cocoa pod borer and tea mosquito are the major pests of cocoa in Asia.

Common Name Cocoa pod borer

Scientific Name *Conopomorpha cramerella* (Snellen)

Order Lepidoptera

Family Gracillariidae

Distribution This pest occurs in almost all countries in South and Southeast Asia where hosts are available.

Life Cycle The flat, oval, 0.5 mm eggs are deposited by female moths onto the surface of pods. The larva hatches in about 7 days and immediately bore into the pod after tunneling through the floor of the egg. There are 4–6 instars which last 14–18 days when they develop in cocoa. Fully mature, 12 mm long larvae exit the pod and pupate on the pod, in green or dried leaves, or in other debris. The pupal stage lasts up to 8 days. Moths are nocturnal and rest during the day under leaves or branches.

Damage The cocoa pod borer is a key pest of cocoa and a major constraint to cocoa production in Southeast Asia. Damage is caused by the tunneling larvae that cause undersized, poor quality beans. In addition, the beans stick together and are hard to remove from the pulp.

Natural Enemies More than 24 different parasitoid species have been recorded from this pest from Malaysia, the Philippines, and Sri Lanka (Ooi 1988). Most were pupal parasitoids from the family Ichneumonidae; these include *Diaglytidea roepkei*, *Photoptera erythronota*, *Goryphus javanicus*, *Goryphus fasciatipennis*, *Goryphus* sp.

10.2.2.2 Pests of Other Crops

Paraphylax fasciatipennis, three other *Paraphylax* spp. and *Xanthopimpla* sp. Other parasitoids were from the families Trichogrammatidae, Braconidae, Ceraphronidae, Chalcidae, Encyrtidae, Eulophidae, Elasmidae, and Phoridae. Ooi (1988) has provided a list of the species of egg, larval, and pupal parasitoids. At least one parasitoid, *Trichogrammatoidea bactrae fumata*, has been studied for its possible use in a biological control program. There is a significant complex of generalist predators that feed on the eggs and pupae which include *Oecophylla smaragdina, Iridomyrmex anceps, Anoplolepis longipes, Crematogaster* sp., etc. Besides several species of fungi have been found to infect cocoa pod borer larvae and pupae *Beauveria bassiana being the most effective.*

Common Names Cocoa mirid or capsid, mosquito bug, tea mosquito

Scientific Names *Helopeltis theivora* (*Helopeltis theobrome*) Waterhouse and *Helopeltis schoutedeni* Reuter

Order Hemiptera

Family Miridae

Distribution South and Southeast Asia and Africa

Hosts Cocoa, mango, cashew, guava, tea, *Acalypha* spp., Japanese cherry, neem, and many others.

Life Cycle The 5–7 mm long adult mirids deposit eggs just below the surface of the pod or into tender stems at the rate of about 4–10 eggs per day and a female will lay about 200 eggs during her lifetime. Eggs hatch in about 7 days, and nymphs pass through 5 instars in 18 days. Populations peak during the rainy season (Sundararaju and SundaraBabu 1999; Hazarika et al. 2008).

Damage Nymphs and adults feed similarly by sucking sap from the plant. In guava, flower buds and fruits are attacked. Insect density and cocoa crop loss relationships were studied in Malaysia. Very young fruit (cherelles) were most seriously affected; one feeding by the mirid caused these fruit to abort. Even if young pods survive after feeding by the mirid, they were smaller and contain smaller seeds.

More mature pods are not as much affected by mirid feeding based on results from caging the insects onto the cocoa pods. Damage to pods at any stage may provide a route for invasion by plant pathogens. In addition, these mirids feed on the plants' soft growing tips and leaf petioles, which may result in dieback and canopy degeneration. However, the major damage is to young developing pods.

Natural Enemies *Beauveria bassiana* has been isolated from this insect in Malaysia and laboratory studies showed that it grew well on adult bugs. In plots where the black cocoa ant, *Dolichoderus thoracicus*, was removed by insecticides, damage by the cocoa mirids was significantly higher, which indicated the importance of these ants as predators. Other studies have shown that the weaver ant, *Oecophylla smaragdina*, can provide almost complete control of *H. theivora*. Some of the predators reported from India were *Chrysoperla cornea*, *Oxyopes* sp., *Plexippus* sp., and *Marpissa* sp.

Other major pests are:
Coffee borer, *Zeuzera coffeae*
Cocoa bark borer, *Squomura disciplaga*
White moth cicada, *Lawana conspersa*
Cluster caterpillar or army worm, *Spodoptera litura*

Other pests of regional importance: *Darna diducta* (Snellen), *Darna trima* (Moore), *Parasa philepida* (Cramer), *Setora nitens* Walker (Lep.: Limacodidae), *Conogetes* (*Dichocrocis*) *punctiferalis* (Guenee) (Lep.: Pyralidae), *Pagodiella hekmeyeri* Heylaerts (Lep.: Psychidae), *Rastrococcus spinosus* (Robinson) (Hem.: Pseudococcidae), and *Valanga nigricornis* (Burmeister) (Orthoptera: Acrididae).

10.2.2.3 Pests of Coffee (*Coffea arabica* L. and *Coffea canephora* Pierre, Rubiaceae)

Coffea arabica originated in the highlands of Ethiopia, southeastern Sudan, and northern Kenya. *Coffea canephora* (robusta) originated in the equatorial lowland forests of Africa. Arabica coffee grows better at 1,000–2,000 m and robusta thrives below 700 m in the tropics. Currently, coffee is planted on over 10 million hectares in 50 or more countries (Vega 2008a, b). Arabica coffee is superior in quality and accounts for 75 % of the world's production. Coffee is one of the most traded commodities in the world, and it is an important export crop of many developing countries. The major pest of coffee is the coffee berry borer. Coffee stem borer is important in some regions in Asia.

Coffee Berry Borer *Hypothenemus hampei* (Ferrari) (Coleoptera: Scolytidae)

Distribution Africa, Central and South America, India, Indonesia, Malaysia, the Philippines, Thailand, Laos, Vietnam, Cambodia, Sri Lanka, Tahiti, New Caledonia, Phonpei, Saipan, Kona (Hawaii)

Life Cycle The adult female cuts a hole in the tip of the berry as it begins to ripen and then tunnels into and lays eggs inside the berry. Duration of egg, larval, and pupal stages are 5–9, 10–26, and 4–9 days, respectively. The developmental period from egg-laying to adult is about 25–35 days (Waterhouse 1998).

Damage Considerable damage is caused when young berries are bored by the adults, as they fall from the plant. Larvae develop in seeds and these infested seeds either do not pass through the curing and grading process or, if they do, they will go to the lowest grade. Crop losses of 40–90 % have been reported in plantations where control measures are not adopted (Le Pelley 1968).

Natural Enemies Important parasitoids are *Prorops nasuta*, *Cephalonomia stephanoderis*, *Phymastichus coffea*, and *Heterospilus coffeicola. Prorops nasuta* was introduced to Java in 1923, and Sri Lanka in 1938, but it did not establish (Le Pelley 1968). It was also introduced to Latin America, the Caribbean, Oceania, and Asia with limited success (Vega et al. 2009). *Cephalonomia stephanoderis* was introduced from West Africa to more than 20 countries (Vega et al. 2009). *Phymastichus coff ea*, collected in Africa, was introduced to many countries in Latin America and to India.

Coffee Stem Borer, *Xylotrechus quadripes* Chevrolat (Coleoptera: Cerambycidae)

Distribution India, Nepal, Southeast Asia

Hosts Coffee, teak, *Olea dioica*

Life Cycle Eggs are elongate, oval, 1.25 mm in length and are laid in groups of one to ten in cracks and crevices in the bark. They are milk white when laid and later turn slightly yellow. The egg stage lasts 9–15 days. The grub on hatching is white, and later turns yellow. The grub bores into the bark for about 2 months and then into the wood. The larval stage lasts for about 9 months and the pupal stage for 3 weeks to a month. After emergence, adults remain in the pupal chamber for 3–7 days. Adults are active fliers.

Damage The tunneling of the grubs results in the death of young plants and weakening of older plants.

Natural Enemies Parasitoids reported include *Metapelma* sp. in India, *Epixorides caerulescens* in Sri Lanka, and *Mysepyris grandiceps*, *Sclerodermus domesticus*, *Doryctes bistriatus*, *Doryctes brevipetiolus*, *Doryctes strioliger, Doryctes picticeps*, *Doryctes tristriatus*, *Pristodoryctes striativentris*, *Promiscolus sesquistriatus*, *Eurytoma xylotrechi*, *Pristaulacus nigripes* var. *duporti*, and *Paraglypta tubigera* in Southeast Asia. Le Pelley (1968) recommends a study of the full distribution of the borer and of its parasitoids throughout its range with the objective of introducing parasitoids to some countries such as India where only one parasitoid has been reported.

Clearwing Moth, Bee Hawkmoth, *Cephanodes hylas* L. (Lepidoptera: Sphingidae)

Distribution Asia and Africa

Hosts Coffee, *Gardenia* sp.

Life Cycle The adult female lays single spherical eggs on young leaves and is capable of laying about 90 eggs. The egg, caterpillar, and pupal stages last for 3, 21, and 13 days, respectively. Pupation takes place in the soil or in the debris beneath the plant. The pupal stage lasts for about 13 days.

Damage This pest causes defoliation of coffee bushes when there is an outbreak.

Natural Enemies Eggs are parasitized by *Ooencyrtus malayensis* and the caterpillar by *Exorista sorbillans* in Malaysia. The reduviid, *Sycanus leucomesus*, has been noted to feed on the caterpillar (Le Pelley 1968).

Coffee Borer, *Zeuzera coffeae* Nietner (Lepidoptera: Cossidae)

Distribution Asia

Hosts Coffee, tea, loquat, cotton, orange, guava, teak, and others

Life Cycle Eggs are laid in masses and an adult is capable of laying over 1,000 eggs. The egg stage lasts 10–11 days. Young larvae first bore into green young shoots and as they grow migrate to thicker branches. The caterpillar stage lasts from 81 to 151 days (Le Pelley 1968).

Damage This caterpillar is an important branch and stem borer of coffee and cocoa. The boring in the stems results in desiccation of shoots and eventual dying. This insect was considered to be a serious pest of coffee in India in the early 1900s; however, it is no longer considered a serious pest.

Natural Enemies The parasitoids, *Amyosoma leuzerae* from Indonesia and India and *Bracon* sp., *Iphiaulax* sp., and *Glyptomorpha* sp. from Malaysia, have been recorded.

Other major pests are:

Fruit flies, *Bactrocera* spp., White moth cicada, *Lawana conspersa*, Pink hibiscus mealybug, *Maconellicoccus hirsutus*, Citrus mealybug, *Planococcus citri*, Rose beetles or chafer beetles, *Adoretus compressus* and *Adoretus sinicus*, Nett le caterpillar, *Parasa philepida* (*lepida*) (See Pests of Beans), Black citrus aphid, *Toxoptera aurantii*

10.3 Pollinators of Tea

10.3.1 Tea

Camellia sinensis (L.) O. *kuntze*, family Theaceae (Fig. 10.4)

The tea flowers occur either solitary or in clusters of two to four flowers. They are fragrant about 2.5 to 4.0 cm in length. They have five to seven white or pink-tinged petals and numerous stamens.

10.3.1.1 Pollination Requirements

Tea is self-sterile and depends upon pollinating insects (Purseglove 1968; Wight and Barua 1939; Wu 1967). Kutubidze (1958) reported that planned pollination resulted in more, larger, and heavier capsules. Bakhtadze (1932) reported that plants isolated from insect visits had an 85–95 % reduction in their seed set. Kutubidze (1964) recommended that steps be taken to obtain additional cross-pollination for increasing yield and quality of tea.

10.3.1.2 Pollinators

Wickramaratne and Vitarana (1985) reported that many different types of insects frequented the seed garden throughout the year. Diptera were the most numerous and were found to be efficient pollinators of tea flowers. Bees were also effective pollinators but were only present seasonally and in smaller numbers. Lack of pollinating insects is not likely to be a major factor limiting seed set of tea in Sri Lanka.

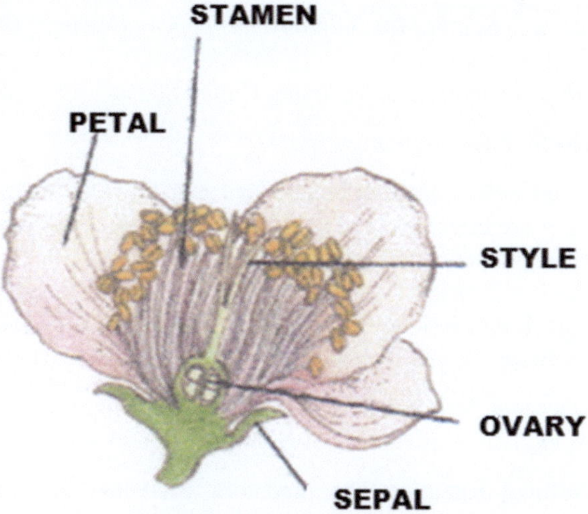

Fig. 10.4 Longitudinal section of tea flower (McGregor 1976)

10.3.1.3 Pollination Recommendations and Practices

No recommendations

10.3.1.4 Pests of Tea (*Camellia sinensis* L., Theaceae)

Tea is a native of South and Southeast Asia. China, India, and Sri Lanka are the major tea producers in Asia. Tea is grown in the hills up to 1,500 m in the tropics. The important pests of tea are tea mosquitoes and mites.

Scolytid Beetle, Shot Hole Borer, *Xyleborus fornicatus* Eichhoff (Coleoptera: Scolytidae)

Distribution South and Southeast Asia

Life Cycle This borer is a shiny dark brown to black, 2 mm long beetle and is known to attack several species of plants. It is considered to be a serious pest of tea in Sri Lanka. Mated females bore into the branches and inoculate the fungus, *Monacrosporium ambrosium*. When the bore hole reaches the center of the branch, the beetle forms an irregular breeding chamber and fills it with eggs. It will continue to expand the chamber and fill it with progeny which can be found at different stages of development.

Damage Construction of galleries causes direct damage to the branches and also allows accumulation of wood rot which debilitates the plant.

Natural Enemies No information is available.

Tea Red Spider Mite, *Oligonychus coffeae* (Nietner) (Acarina: Tetranychidae)

Distribution Asia, Australia, United States

Hosts Polyphagous

Life Cycle The adult female lays an average of 90 eggs at the rate of 4–6 per day. Eggs are bright red and 0.11 mm in diameter. The developmental stages include a sixlegged larva, a protonymph, and a deutonymph. The adult female is elliptical in shape, crimson in colour, and 0.3–0.4 mm in length. The male is smaller, 0.25–0.3138 mm in length, and the abdomen is tapered to a point at the posterior end. Overlapping generations are found on the host leaves (Cranham 1966).

Damage Affected leaves show bronze coloration on the upper side and close examination reveals webbing, red mites, and white cast skins.

Natural Enemies The predators, *Stethorus gilviforn*, *Micromus timidus*, and *Chrysoperla carnea* have been reported to feed on these mites (Das et al. 2010).

Purple Mite, *Calacarus carinatus* (Green) (Acarina: Eriophyidae)

Distribution India, Sri Lanka, Indonesia, Indochina

Life Cycle This mite is mostly found on older leaves and on both surfaces of leaves. The adult female mite is deep purple in color with five white waxy longitudinal ridges on the abdomen. The body is elongate and spindle shaped, and slightly broader at the anterior end. Adult females have two pairs of legs. Adult males are much shorter than females. Eggs are fl at and there are three developmental stages and the life cycle is completed in 6–13 days.

Damage Affected leaves show a purplish bronze discoloration and white cast skins.

Natural Enemies None reported.

Other major pests are: Tea mosquito, *Helopeltis theivora*, Red and black flat mite, *Brevipalpus phoenicis*, Yellow tea mite, broad mite, *Polyphagotarsonemus latus*, Rose beetles, chafer beetles, *Adoretus compressus*, *Adoretus sinicus*, White moth cicada, *Lawana conspersa*, Nettle caterpillar, *Parasa philepida* (*lepida*) Tea tortrix, coffee leaf roller, *Homona coff earia* Brown citrus aphid and Black citrus aphid, *Toxoptera citricidus* and *Toxoptera aurantii*

Other pests of minor importance are: Looper caterpillars, *Buzura suppressaria* Guen., *Hyposidra talaca* (Walker), and *Hyposidra infi xaria* Walker (Lep.: Geometridae), Red slug caterpillar, *Eterusia magnifi ca* Butl. (Lep.: Limacodidae), Flush worm, *Cydia leucostoma* Meyrick (Lepidoptera: Tortricidae), Leaf roller, *Caloptelia theivora* Walsingham (Lepidoptera: Gracillariidae), Leafh opper (*Empoasca flavescens* Fabr.) (Hem.: Cicadellidae), and thrips (*Mycterothrips setiventris* (Bagnall) and *Scirtothrips dorsalis* Hood) (Thysanoptera: Thripidae).

10.3.1.5 Management of Tea Insect Pests

Various components for the management of tea pests are discussed below:

Cultural Control

Pruning out heavily infested branches should be carried out to improve control by increasing air circulation and improving pesticide penetration into the plant. Shade crops should be used to establish the orchards as tea growing under shade trees tends to have less leafhoppers problem. Mulching also helps to increase humidity and therefore reduce leafhopper populations. The use of light traps could be employed to reduce the population of the devastating tea pests. The clustered leaf bunches on branches and affected twigs should be physically collected and destroyed. Omole (1983) recommended the digging out of the ground-crickets and handpicking of the bagworms, *Acanthopsyche spp* in tea bushes. Some aspects of termite cultural control in tea is by prevention of termite establishment in wounds and dead branches on mature shrubs so as to avoid spread of infestation to healthy stem. The attack of tea mosquito, *Helopeltis schoutedeni* occurs suddenly and scouting is important for its effective control. Tea should not be grown near other hosts for this pest.

Chemical Control

Adult scales are very resistant to insecticides; therefore the insecticide treatment should be timed to correspond to the first emergence of crawlers. Oil sprays are effective against crawlers and adult scales. Spraying with Imidacloprid at 0.25 ml/ltr (Rabindra 2012) and other selected insecticides within the acceptable standards are also recommended for the IPM strategy to control the pests.

Biological Control

The lady bird beetle *Chilomenes liniata* F. (Coccinelidae) has been observed to attack some insect pests of tea on the Mambilla plateau of Nigeria. The larval and adult stages of the predator feed on bugs and aphids on tea (Omole 1983).

Use of Resistance Varieties

Omole (1985a, 1986), identified some resistant tea clones at Kusuku (CRIN) clonal Gene-pool. Some of the resistant clones are clone 35, 68 and 143 which were resistant to scale insects ***Ceroplastes* spp**. The corresponding damage values for clone 35 were about 7.6 % in 1985 and 2.8 % in 1986. The remaining two clones 68 and 143, which appeared very resistant in 1985, as they were not attacked at all in that year were found to be susceptible during the 1986 season. This inconsistency in the susceptibility or tolerance of these tea clones in the same locality and within the short period of 2 years showed that tolerance or resistance of tea clones to insect pests can be apparent or temporary (Omole 1985b, 1986). It can break down or be modified by edaphic and environmental factors such as the nutrient status of host plants, prevailing weather conditions such as drought, relative humidity, insolation and wind speed. Finally, clonal tolerance may also be modified by essential cultural operation such as leaf plucking, pruning, fertilizer application and irrigation in tea cultivation and production.

Bibliography

Abraham EV (1958) Pests of cashew (*Anacardium occidentale*) in South India. Ind J of Agric Sci 28:531–544

Abraham CC, Nair GM (1981) Effective management of the tea mosquito bugs for breaking the yield barriers in cashew. Cashew Causerie 3(1):6–7

AICRPC (1989) All India co-ordinated research project on cashew. National Research Centre for Cashew. Puttur, Karnataka, (Mimeographed), pp. 88

Alam, MM (1976) A preliminary investigation into the damage caused by coconut mite in Grenada. Report Caribbean Agricultural Research and Development Institute, Barbados, Oct 1976.

Alam MM (1988) Coconut mite and its natural enemies in St. Vincent. Proc 24th Carib. Food Crops Soc 24:164–170

Aldaba VC (1921) The pollination of coconut. Philip Agr 10(5):195–208

Aldaba VC (1960) Influence of insects on pollination of caturra coffee. Rev De Agr (sao paulo) 35(2):139–147 [in Portuguese, english summary.]

Aldaba VC (1972) Insect pollination of coffea arabica l., and radius of action of apis mellifera linnaeus 1758, in the collection of pollen in coffee plantations in bloom. 82 pp., dept. Ent., "lutz de quiroz" superior school of agr., sao paulo univ., piracicaba s.p. Brazil. [in Portuguese, summary and general conclusions in english.]

Alexandra-Maria K, Ingolf S-D, Teja T (2003) Bee pollination and fruit set of Coffea arabica and C. canephora (Rubiaceae). Am J Bot 90(1):153–157

Amaral E (1952) [Essay on the influence of *Apis mellifera* L. On the pollination of the coffee plant (preliminary note).] Esc Super De agr "lutz de queiroz" (Sao Paulo, Brazil) Bul 9, 6 pp. [in Portuguese.]

Amaral E (1960) Ação dos insetos na polinização do *cafeeiro caturra*. Revista de Agric 35:139–147

Ambika B, Abraham CC (1979) Bioecology of *Helopletis antonii* Sign. (Miridae: Hemiptera) infesting cashew trees. Entomon 4:335–342

Ambika B, Abraham CC (1984) In: Rao EVVB, Khan HH (Eds) Cashew research and development, Indian Society for Plantation Crops, CPCRI, Kasaragod, pp 111–115

Ambika B, Abraham CC, Vidyadharan KK (1979) Relative susceptibility of cashew types of infestation by Helopeltis antonii Siq. (Hemiptera:Miridae). In the proceedings of 2nd annual symposium plantation crops, Kerala University, p 513–516

Ananthakrishnan TN (1985) Host relationship and damage potential of thrips infesting cashew. Acta Hort (ISHS) 108:131–134

Anikwe C, Okelana FA, Otuonye HA, Hammed LA, Aliyu OM (2007) The integrated management of an emerging insect pest of cashew: a case study of the cashew root and stem borer, Plocaederus ferrugineus in Ibadan, Nigeria. J Agric, Fore Soc Sci 5(1), ISSN: 1597–0906

Anonymous (1916) Bees and pollination. Planters' Chron, Bangalore 9(46):572

Anonymous (1985) Pests of economic importance to the Caribbean coconut industry. In: Proceedings of the coconut mite control workshop, St. Lucia, 19–21 Jun 1985

Araújo JP, Silva VV (eds) (1995) Cajucultura: modernas técnicas de produção. EMBRAPA/ CNPAT, Fortaleza

Asogwa EU, Anikwe JC, Ndubuaku TCN, Okelana FA, Hammed LA (2009) Host plant range and morphometrics descriptions of an emerging insect pest of cashew, *Plocaederus ferrugineus* L (Coleoptera: Cerambycidae) in Nigeria: a preliminary report. Int J Sustain Crop Prod 4(3):27–32

Ayyar,TVR (1932) Bulletin-cashew. Agricultural Department of Madras No. 27, pp 95

Ayyar TVR (1942) Insect enemies of cashew nut plant *(Anacardrum occidentale)* in South India. Mad Agric J 30:223–226

Bakhtadze K (1932) Pollination of tea, camelia sinensis, in Georgia. Subtropics 2(12):63–80

Bakthavatsalam N, Sundarababu D, Bhat PS (1993) The Cashew 7(3):12–13

Bartholomew DP, Paull RE, Rohrbach KG (2003) The pineapple: botany, production and uses. CABI Publishing, Wallingford

Beach JH (1984) The reproductive biology of the Peach or "Pejibayc" Palm (Dactris gasipaes) and a Wild Congener (D. porschiana) in the Atlantic lowlands of Costa Rica. Principes 28:107–119

Beevi SP, Mohapato GK (2007) A new field scoring methodology for cashew genotypes against tea mosquito bug, Helopeltis antonii Sign. J Plant Crop 35(3):139–145

Bhat PS, Raviprasad TN (1996) Pathogenicity of entomopathogenic fungi against cashew stem and root borer *Plocaederus ferruginues* L. (Coleoptera: Cerambycidae). J Plant Crop 24:265–271

Bhat PS, Raviprasad TN (2007) Management of tea mosquito bug, Helopeltis antonii Sign. With newer insecticide/products. National Seminar on Research Development and Marketing of Cashew 20–21st November 2007 p 56

Bhat MG, Rupa TR (2012) Current trends in cashew research in India and future focus pp19–27. In: Radhakrishnan B, Rajkumar R, Palani MN, Premkumar R, Ilango RV, Kumar MS, Chandrashekara KN, (eds) Siuvenir, PLACROSYM XX,UPASI Tea Research Foundation, Tamil Nadu, 37 p

Bigger M (1960) *Selenothrips rubrocinctus* Giard and the floral biology of cashew in Tanganyika. The East Afri Agricul J 25:229–234

Blanche KR, Cunningham SA (2005) Rainforest provides pollinating beetles for atemoya crops. J Econ Entomol 98:1193–1201

Blanche KR, Ludwig JA, Cunningham SA (2006) Proximity to rainforest enhances pollination and fruit set in orchards. J Appl Ecol 43:1182–1187

Bos M, Veddeler D, Bogdanski AK, Klein AM, Tscharntke T, Steffan-Dewenter I, Tylianakis J (2007) Caveats to quantifying ecosystem services: fruit abortion blurs benefits from crop pollination. Ecol Appl 17:1841–1849

Broadley RH (1993) Birds and mammals. In: Broadley RH, Wassman RC, Sinclair E (eds) Pineapple: pests and disorders. Queensland Department of Primary Industries, Brisbane, pp 30–32

Burkill IH (1919) Some notes on the pollination of flowers in the Botanic Gardens, Singapore, and in the other parts of the Malay Peninsula. Gardens' Bull Straits Settlement 2:165–170

Cabrera RI, Dominguez D (1987) *Hirsutella nodulosa* fungus, a new parasite for *E. guerreronis* coconut mite. Ciencia y Technica en la Agricultura. Citricus y otros frutales (Cuba) 10:41–51

Cardona IZ,. Potes SA (1971) Coconut scab or excoriation (*Cocos nucifera*) L. In Columbia. [*A. guerreronis*]. Acta Agron Palmira 21:133–139. (Burley and Styles, 1976)

Carter W (1963) Mealybug wilt of pineapple: a reappraisal. Ann N Y Acad Sci 105:741–746

Carter W (1967) Insects and related pests of pineapple in Hawaii. Pineapple Research Institute of Hawaii, Honolulu, p 105

Carvalho A, Krug CA (1950) Pollinating agents for coffee (*Coffea arabica*). Bragantia 9:11 [In Portuguese.]

Carvalho A, Ferwerda FP, Frahm-Leliveld JA, and others (1969) Coffee. In: Ferwerda FP, Wit F (eds) Outlines of perennial crop breeding in the tropics, H. Veenman and Zonen, N. V., Wageningen, pp 189–241

CCRS (1966) Central cashew research station, annual progress report 1965–66. Ullal, Karnataka

Chacoff NP, Aizen MA (2006) Edge effects on flowervisiting insects in grapefruit plantations bordering premontane subtropical forest. J Appl Ecol 43:18–27

Chatterjee ML (1989a) Insect pests on cashew in West Bengal and status of some important pests. Cashew 3(3):19–20

Chatterjee ML (1989b) A study on cashew flower thrips infestation and their control by different insecticides. The Cashew 3(4):12

Child R (1974) Coconuts, 2nd edn. Longman Group, London

CIBC (1983) Commonwealth Institute of Biological Control. Biocontrol News and Info 4:7–11

Cranham JE (1966) Tea pests and their control. Annu Rev Entomol 11:491–514

Damodaran VK, Abraham J, Alexander KM (1966) The morphology and biology of the cashew flower. II: anthesis, dehiscence, receptivity of stigma, pollination, fruit-set and fruit-development. Agric Res J Kerala 4(2):78–84

Damon A (2000) A review of the biology and control of the coffee berry borer, Hypothenemus hampei (Coleoptera: Scolytidae). Bull Entomol Res 90:453–465

Das S, Roy S, Mukhopadhyay A (2010) Diversity of arthropod natural enemies in the tea plantations of North Bengal with emphasis on their association with tea pests. Curr Sci 99:1457–1463

Davis JA (1954) Mysteries of cross-pollination. Indian cent. Coconut com. Ernakulam, Bul. 7:226–227

Dantas de Araujo CZ (1994) The biological interactions between ants and cashew-tree. In: Lenoir A, Arnold G, Lepage M, eds, Proceedings of the 12th congress of the international union for the study of social insects, Paris, Françe. Paris, Université Paris Nord, p 176

De Jong TJ, Waser NM, Klinkhamer PGL (1993) Geitonogamy: the neglected side of selfing. Trends Ecol Evol 8(9):321–325

Decazy B, Massoux F, Tschiendji C, and Misse C (1981) Ulilisation des radio-isotopes (32P) pour L'e'tude des agents pollinisateurs du cacaoyer dans les conditions ecologiques du sud-Cameroun. In: Proceedings of the sixth international Cocoa research conference, Caracas, Venezuela, pp 419–430

DeMarco P, Coehlo FM (2004) Services performed by the ecosystem: forest remnants influence agricultural cultures' pollination and production. Biodivers Conserv 13:245–1255

Desai MA, Kulkarni SV, Rodrigues A (1977) Report on the marketing survey on cashew in Goa. Government of Goa, Deman and Diu, pp 11–17

Devasahayam S (1989) *Erythmelus helopeltidis* Gahan (Hymenoptera: mymaridae) a new parasite of *Helopeltis antonii* Sign. On cashew. J Bombay Nat Hist Soc 86:113

Devasahayam S, Nair CPR (1986) The mosquito bug, Helopeltis antonii Sign. on cashew in India. J Plant Crop 14:1–10

Devreux M, Vallaeys G, Pochet P, Gilles A (1959) Research on the self-sterility of robusta coffee (coffea Canephora pierre). Pub Lnst Nat Agron Congo Belge: Ser Sci 78:44. 4262 (abstract) plant breed. 29(4):837

Dixie G, Sergeant A (1998) The future for the Ghanaian horticultural export industry, Accord Associates, Rushbrook Dorset DT 11-8NZ UK

Dwomoh EA, Afun JV, Ackonor JB, Agene VN (2009) Investigations on *Oecophylla longinoda* (Latreille) (Hymenoptera: Formicidae) as a biocontrol agent in the protection of cashew plantations. Pest Manag Sci 65(1):41–46

Dwomoh EA, Ackonorand JB, Afun JVK (2012) Survey of insect species associated with cashew (*Anacardium occidentale* Linn.) and their distribution in Ghana. J Agricul Res Fish 1(1):006–016

Entwistle PF (1972) Pests of cocoa, 1st edn. Longman, London

Falque M, Lesdalons C, Eskes AB (1996) Comparison of two cacao (Theobroma cacao L.) clones for the effect of pollination intensity on fruit set and seed content. Sex Plant Reprod 9:221–227

FAO (2003) Food and Agriculture Organization of the United Nations. *FAOSTAT statistics database.* Rome, UN Food and Agriculture Organization. http://apps.fao.org. Accessed Apr 2013

Ferwerda FP (1936) Pollination in coffee species grown in the Dutch East Indies. Zuchter 8:92–102, [In German.] See crane and green 1948 for abstract

Ferwerda FP (1948) Coffee breeding in java. Econ Bot 2(3):258–272

Ferwerda FP (1951) [fruit drop in robusta coffee and its relation to Pollenization and fertilization.] Vakblad voor biologen 31:123–130. [in Dutch.] Abstract in euphytica 1(3):232, 1952

Free JB (1970) Insect pollination of crops, 1st edn. Academic Press, London

Free JB (1993) Insect pollination of crops, 2nd edn. Academic Press, London

Free JB, Williams IH (1976) Insect pollination of *Anacardium occidentale* L., *Mangifera indica* L., *Blighia sapida* Koenig and *Persea americana* Mill. Trop Agricul 53(2):125–139

Freitas BM (1994) Beekeeping and cashew in North-eastern Brazil: the balance of honey and nut production. Bee World 75(4):160–168

Freitas BM (1995a) The pollination efficiency of foraging bees on apple (*Malus domestica* Borkh) and cashew (*Anacardium occidentale L.*). Ph.D. thesis, University of Wales, Wales

Freitas BM (1995b) Does *Borreria verticillata* compete with cashew (*Anacardium occidentale*) for pollination by honeybee? In: Proceedings of the XXXIV international congress of Apimondia; Lausanne, 15–19 Aug 1995, p 260–264

Freitas BM, Paxton RJ (1996) The role of wind and insects in cashew (*Anacardium occidentale*) pollination in NE Brazil. J Agric Sci 126:319–326

Freitas BM, Paxton RJ (1998) A comparison of two pollinators: the introduced honey bee *Apis mellifera* and an indigenous bee *Centris tarsata* on cashew *Anacardium occidentale* in its native range of NE Brazil. J Appl Ecol 35:109–121

Furtado CX (1924) A study of the coconut flower and its relation to fruit production. Gard Bul 3(7–8):261–273 [Singapore]

Gadase SK, Dumbre RB, Kharat SB (1991) Evaluation of some synthetic pyrethroids against tea mosquito, Helopeltis antonii Signoret on cashew. Cashew 5(2):14–15
Gadase SK, Dumbre RB, Kharat SB (1993) Evaluation of some insecticides for the control of tea mosquito on cashew. Cashew 7(2):13–14
Ghosh SN, Chatterjee ML (1993) Effect of nitrogen, phosphorus and potassium on incidence of tea mosquito and yield in cashew. Environ Ecol 11:337–339
Griffith R (1982a) Extract from "A report on a visit to St. Vincent, West Indies, from 26 March to 2 April 1982", pp 1–5
Griffith R (1982b) The development of a programme for the control of the coconut mite (*E. guerreronis*, Keifer) in the coconut groves of the island, Ministry of Agriculture, Lands and Research Division, Centeno [Trinidad and Tobago], 11 p
Griffith R (1984) The problem of the coconut mite, *E. guerreronis* Keifer, in the coconut groves of Trinidad and Tobago. Carib Food Crops Soc 20:128–23. Also published in 1995, J of the Agric Soc of Trin and Tob 85, pp 81–99
Haarer AE (1962) Modern coffee production. Leonard Hill, London, 495 pp
Hall RA (1981) Mite damage and possibilities of its control in Mexico. Unpublished report
Hall RA (1986) Experiments with the pathogenic fungus, *Hirsutella thompsonii*, against the coconut mite, *Eriophyes guerreronis*, in St. Lucia and other Windward Islands: findings, conclusions and recommendations. Project TCP/RLA/4510. Unpublished report
Hall RA, Hussey NW, Mariau D (1980) Resultatus d'unen enquete concernanties agents de lutte biologique contre l'acarien du cocotier *E. guerreronis*. Oleagineux (France) 35:395–400
Harder LD, Barrett SCH (1995) Mating cost of large floral displays in hermaphrodite flowers. Nature 373:512–515
Harler CR (1995) The culture and marketing of tea. Oxford University Press, London
Hatjina F (1998) Hive-entrance fittings as a simple and cost-effective way to increase cross-pollination by honey bees. Bee World 79:71–80
Hazarika LK, Bhuyan M, Hazarika BN (2008) Insect pests of tea and their management. Annu Rev Entomol 54:267–284
Hazarika LK, Bhuyan M, Hazarika BN (2009) Insect pests of tea and their management. Annu Rev Entomol 54:267–284. doi:10.1146/annurev.ento. 53. 103106.093359
Heard TA, Exley EM (1994) Diversity, abundance and distribution of insect visitors to macadamia flowers. Environ Entomol 23:91–100
Hearer AE (1958) Modern coffee production. Leonard Hills, New York
Hedstrom I (1986) 'Pollen carders of *Cocos nucifera* L. (palmae) in Costa Rica and Ecuador (Neotropical region). Rev Bioi Trop 34:297–301
Hedstrom I (1987) Fruit flies (Diptera, Tephritidae) infesting common guaya (*Psidium guaava* L.) (Myrtaceae) in Ecuador. Rev Biol Trop 35:373–374
Henderson A (1985) Pollination of *Socratea exorrhiza* and *Iriartea ventricosa*. Principes 29:64–71
Henry Louis I, Chelladurai M (1984) Nature and frequency of insects pollinating the coconut palm (*Cocos nucifera* L.). Indian Cocon J 15(8):1–6
Hernandez Roque F (1978) Combate quimico del eriofido del cocotero *A. (Eriophyes) guerreronis* (K) en la costa de Guerrero. Agric Tecn Mex (Mex) 23:23–38
Hiremath IG, Khan MM, Kumar DP (1987) The Cashew 1(2):8–9
Howard FW, Abreu-Rodriguez E (1991) Tightness of the perianth of coconuts in relation to infestation by coconut mites. Fl Entomol 74:358–361
Howard FW,Abreu Rodriguez E, Denmark HA (1990) Geographical and seasonal distribution of the coconut mite, *A. guerreronis* (Acari: Eriophyidae), Puerto Rico and Florida, USA. J Agric Univ PR, 74:237–251
Huggins HD (1928a) Pollination and crop production. Br Guiana Agric J 1:90–95
Huggins HD (1928b) Pollination and crop production. Br Guiana Agric J 1:164–169
Hunger FWT (1920) *Cocos nucifera*. Scheltema and Holkema's Boekhandel, Amsterdam, 518 pp
Hussey NW (1975) Report on mission to Grenada and Jamaica under the Caribbean Technical Assistance arrangements to investigate mite damage on coconuts, 29 October–15 November 1975, Cited by Hall 1981

Hutchinson JL (2012) Impact of honeybees on coffee pollination in Jamaica, West Indies. M.Sc. thesis, Humboldt State University Jamaica, West Indies

ICIPE (2013) Integrated management of major insect pests and diseases of cashew in East and Western Africa, Fact sheet, The International Centre of Insect Physiology and Ecology (ICIPE) http://www.giz.de/Themen/en/dokumente/giz2011-1-en-factsheet-icipe-integrated-management-insect-pests.pdf. Accessed April 2013

International Cocoa Organization (2005) Annual report for 2003/2004. International Cocoa Organization (ICCO), London

Jay SC (1986) Spatial management of honeybees on crops. Ann Rev Entomol 31:49–65

Jena BC, Patnaik NC, Satapathy CR (1985) Insect fauna of cashew nut in Orissa. Sci Cult 51:385–386

Johnson LK, Hubbel SP (1974) Aggression and competition among sting-Iess bees: field studies. Ecology 55:120–127

Julia JF, Mariau D (1979) New research on the coconut mite *E. guerreronis* (K), in the Ivory Coast. Oléagineux 34:181–189, France

Kehe M (1979) Les symphyles en culture d' ananas en Cote-d'Ivoire. In: Congres sur Lutte contre les Insectes en Milieu Tropical. Chambre de Commerce et d'Industrie de Marseille, Marseille, pp 441–445

Kevan PG (1975) Pollination and environmental conservation. Environ Conser 2:293–298

Khoo KC, Ibrahim Y, Maelzer DA, Lim TK. 1982. Entomofauna of cashew in West Malaysia. In: Proceedings of the international conference in plant protection in Tropics; 1982; Kuala Lumpur; Malaysia. Kuala Lumpur; pp 289–294

Klein AM, Steffan-Dewenter I, Tscharntke T (2003a) Fruit set of highland coffee increases with the diversity of pollinating bees. Proc Royal Soc London, B 270:955–961

Klein AM, Steffan-Dewenter I, Tscharntke T (2003b) Pollination of Coffea canephora in relation to local and regional agroforestry management. J Appl Ecol 40:837–845

Klein AM, Steffan-Dewenter I, Tscharntke T (2003c) Bee pollination and fruit set of Coffea arabica and C. canephora (Rubiaceae). Am J Bot 90:153–157

Klein AM, Vaissiere BE, Cane JH, Steffan-Dewenter I, Cunningham SA, Kremen C, Tscharntke T (2007) Importance of pollinators in changing landscapes for world crops. Proceedings of the Royal society of London. Bull Am Meteorol Soc 274:303–313

Knuth P (1904) Handbuch der Blütenbiologie III, 1. Engelmann, Leipzig

Knuth P, Appel O, Loew E (1904/1905) Handbuch der Blu¨tenbiologie. III.1/III.2. Engelmann, Leipzig, Germany

Krantz J, Schmutterer K, Koch W (eds) (1978) Disease, pests, and weeds in tropical crops. Wiley, Chichester, 666 p

Kremen C, Williams NM, Thorp R (2002) Crop pollination form native bees at risk from agricultural intensification. Proc Natl Acad Sci U S A 99:16812–16816

Krug CA (1935) Hybridization of coffee, a preliminary study of flowering habits and methods of crossing. J Hered 26:325–330

Krug CA, Mendes JET, Carvalho A, Mendes AJT (1950) [A new type of coffee.] Bragantia 10(1):11–25. [in portuguese, english summary.]

Kudom AA, Kwapong PK (2010) Floral visitors of *ananas comosus* in ghana: a preliminary assessment. J Pollin Ecol 2(5):27–32

Kutubidze VV (1958) [Intravarietal supplementary pollination of the tea plant with a pollen mixture.] Agrobiologiya 4:53–56. [In Russian.] Abstract In Plant Breeding Abs 29(4): no. 4254, p 835, 1959

Kutubidze VV (1964) [Methods of the production of tea hybrid seeds, the increase of its yield and the improvement of its quality.] Makharadze. vses. nauch.-issled. lnst. chaya. [subtropical crops.] 4:39–42. [In Russian, English summary.]

Lacoeuilhe JJ (1977) *Symphyles sur ananas en* Martinique. Document IRFA, non publiee

Lance SL (1966) Insects associated with the flowers of the coconut palm, Cocos nucifera L., in Hawaii XIX, No. 2, Sept 1966, 293

Le P, Le Pelley RH (1968) Pests of coffee. Longmans, Green, London, 590 pp

Lever RJAW (1933) Status of economic entomology in the British Solomon Islands. Bull Entomol Res 24:253–256
Lever RJAW (1937) Economic insects and biological control in the British Solomon Islands. Bull Entomol Res 28:325–331
Lever RJAW (1969) Pests of the coconut palm. FAO Agricultural Studies, No. 77, Rome 190 p
Liyanage DV (1957) Insects on coconut inflorescences. Ceylon Coconut Q 8(1–2):39
Lower WV (1911) Beekeeping in Puerto Rico. Puerto Rico Agr Expt Sta Cir 13, 31 pp
Madhava Rao VN, Vazir Hassan M (1957) Preliminary studies on the floral biology of cashew (Anacardium occidentale Linn.). Indian J Agr Sci 27:277–288
Manrique AJ, Thimann RE (2002) Coffee (Coffea arabica) pollination with Africanized honeybees in Venezuela. Interciencia 27:414–416
Mariau D (1969) *Aceria guerreronis*: Récent ravageur de la cocoteraie Dahoméenne. Oléagineux 24:209–272
Mariau D (1977) *A. (Eriophyes) guerreronis*: an important pest of African and American coconut groves. Oléagineux 32:101–109
Mariau D (1986) Behaviour of *E. guerreronis* Keifer with respect to different varieties of coconut. Oléagineux (France) 41:499–505
Mariau D, Julia JF (1970) Acariasis caused by *A. guerreronis* (Keifer), pest of the coconut palm. Oléagineux 25:459–464
Mariau D, Tchibozo HM (1973) Trials of chemical control of *A. guerreronis* (Keifer). (Coconuts). Oléagineux 28:133–135
Martyn EB (1930) Annual report of Botany and Mycology Division (1929). Agric J Brit Guy 3:226–233. Cited by Hall 1981
Maruthadurai R, Desai AR, Prabhu HRC, Singh NP (2012) Insect pests of cashew and their management. Technical Bulletin No. 28, ICAR Research Complex for Goa, Old Goa
Masasi (2013) Masasi farmers use weaver ants to combat insect pests in cashew http://masasifarmers.webs.com. Accessed April 2013
Mayne WW (1934) Annual report of the coffee scientific officer 1933–34 Mysore Coffee Expt. Sta Bul 12
Mcdonald JH (1930) Coffee growing: with special reference to East Africa. East Africa, London, 205 pp
McGregor SE (1976) Insect Pollination of Cultivated Crop Plants. USDA Agriculture handbook 496:93–98
Medina-Gaud S, Abreu E (1986) The coconut palm eriophyid mite, *E. guerreronis* (Keifer) in Puerto Rico (Acarina: Eriophyidae). J Agric Univ PR 70:223–224. Mariau (1969)
Mendes AJTL (1949) Cytological observations in coffee. Bragantia 9(1–4):25–34 [In Portuguese, English summary.]
Mendes AJTL (1961) Speed of pollen tube penetration in coffee arabica. Bragantia 20(1):495–502 [In Portuguese, English summary.]
Menon KPV, Pandalai KM (1958) The coconut palm – a monograph. Indian Cent Coconut Com, Ernakulam, 384 pp
Mohapatra LN, Mohapatra RN (2004) Distribution, intensity and damage of cashew stem and root borer (*Plocaederus ferrugineus* L.) in Orissa. Indian J Entomol 66(1):4–7
Mohapatra LN, Behara AK, Satapthy CR (1998) Influence of the environmental factors on the cashew nut shoot tip caterpillar, *Hypotima haligramma* Meyr. Cashew Bull 35:17–18
Mohapatro GK (2008a) A refined damage rate scoring method for the tea mosquito bug infestation in cashew. Indian J Entomol 70(3):273–277
Mohapatro GK (2008b) Evaluation of insecticides for control of tea mosquito bug, *Helopeltis antonii* and other insect pests of cashew. Indian J Entomol 70(3):217–222
Monaco LC, Carvalho A (1969) Coffee genetics and breeding in Brazil. Shell Pub Health Agr News 12(2):74–77
Montealegre MR (1946) The fertilization of coffee flowers. Revista del Iust De Defensa del Cafe de Costa Rica 15:337–340 [In spanish.]
Moore D (1986) Bract arrangement in the coconut fruit in relation to attack by the Coconut Mite *E. guerreronis* Keifer. Trop Agric 63:285–288

Moore D (2001) Insects of palm flowers and fruits. In: Howard FW, Moore D, Giblin-Davis RM, Abad RG (eds) Insects on palms. CAB International, Wallingford, pp 233–266
Moore D, Alexander L (1987) Aspects of migration and colonization of the coconut palm by the coconut mite, *E. guerreronis* (Keifer) (Acari: Eriophyidae). Bull Ent Res 77:641–650
Moore D, Alexander L (1990) Resistance of coconuts in St Lucia to attack by the coconut mite E guerreronis Keifer. Trop Agric 67:33–36
Moore D, Alexander L, Hall RA (1989) The coconut mite, *E. guerreronis* Keifer in St. Lucia: yield losses and attempt to control it with acaricide, polybutene and *Hirsutella* fungus. Trop. Pest Manag 35:83–89
Moore D, Ridout MS, Alexander L (1991) Nutrition of coconuts in St Lucia and relationship with attack by coconut mite A guerreronis Keifer. Trop Agric 68:41–44
Morse RA, Laigo EM (1969) The potential and problems of beekeeping in the Philippines. Bee world 50(1):9–14
Morton JF (1961) The cashew's brighter future. Economy Botany 15:57–78
Muthu M, Baskaran P (1980) Performance of quinalphos against the insect pests of cashew (*Anacardium occidentale* L.). Proceedings of the Second Annual Symposium on Plantation Crops, pp 474–483. [Plantation Crops Symposium II, PLACROSYM II, held on June 26–29, 1979, in Ootacamund, India]
Mutter NES, Bigger M (1961) Cashew. Tanganyika Min. Agr Bul 11, 5 pp
Myers N, Mittermeier RA, Mittermeier CG, da Fonseca GAB, Kent J (2000) Biodiversity hotspots for conservation priorities. Nature 403:853–858
Nair GM, Abraham CC (1982) Agric Res J Kerala 20(2):41–48
Nair GM, Abraham CC (1983) Agric Res J Kerala 21(1):21–26
Nair GM, Abraham CC (1984) Agric Res J Kerala 22:118–123
Narayana GV (1937) On the nectar secretion of coconut flowers, *Cocos nucifera* Linn. Proc Indian Acad Sci 6(4):224
Nogueira-neto P, Carvalho A, Filho HA (1959) The effect of the exclusion of pollinating insects on the yield of bourbon coffee. Bragantia 18:441–468 [In Portuguese, English summary.]
Northwood PJ (1966) Some observations on flowering and fruit setting in the cashew *Anacardium occidentale* L. Trop Agr 431:35–42
NRCC (1988) (National Research Centre for Cashew) Annual Report for 1987, Puttur, Karnataka, pp 41
Ochse JJ, Soule MJ, Dijkman MJ, Wehlburg C (1961) Tropical and subtropical agriculture, vol 2. The Macmillan Co, New York
Olotu MI, du Plessis H, Seguni ZS, Maniania NK (2012) Efficacy of the African weaver ant Oecophylla longinoda (Hymenoptera: Formicidae) in the control of Helopeltis spp. (Hemiptera: Miridae) and Pseudotheraptus wayi (Hemiptera: Coreidae) in cashew crop in Tanzania. J Pest Manag Sci 69(8):911–918
Olschewski R, Tscharntke T, Benı'tez PC, Schwarze S, Klein AM (2006). Economic evaluation of pollination services and pest management comparing coffee landscapes in Ecuador and Indonesia. Ecol Soc 11(1):7. http://www.ecologyandsociety.org/vol11/iss1/art7/
Olvera-Fonseca S (1986) El acaro cuasante de la "ro ~ na del cocotero" en Veracruz, Mexico. Folia Entomol Mex No 67:45–51
Omole MM (1983) Insects and diseases associated with tea on the Mambilla plateau. Ann Report CRIN Ibadan Nigeria, pp 51–53
Omole MM (1985a) Pests of arabica coffee on the Mambilla Plateau. Ann Report CRIN, Ibadan, pp 43–46
Omole MM (1985b) Protection of tea nurseries and plantation against pests and diseases. Ann Report CRIN, Ibadan, pp 88–91
Omole MM (1986) Protection of tea nurseries and plantations against pests and Diseases. Ann Report CRIN, Ibadan, pp 53–66
Ooi PAC (1988) Laboratory studies *of Tetrastichus sokolowskii*. Entomophaga 33:145–152
Ortega A, Rodriguez JV, Caribay CV (1965) Investigaciones preliminaries sobre el eriofido del fruto del cocotero, *Aceria guerreronis* Keifer, en la costa Grande de Guerrero. Revista Agric Tecn Mex 11:222–226

Otterbein HD (1988) Studies on the biology, importance and control of *E. guerreronis* on *Cocos nucifera* in Costa Rica. Kiel (Germany, FR), 153 p

Pagden HT, Lever RJAW (1935) Insects of the coconut palm and the present position of the coconut problem in the British Solomon Islands Protectorate. British Solomon Islands Protectorate Agr Gaz 3(1):2–22

Pegg KG (1993) Diseases. In: Broadley RH, Wassman RC, Sinclair E (eds) Pineapple: pests and disorders. Queensland Department of Primary Industries, Brisbane, pp 21–29

Petch T (1913) The flowers of coconut palm. Trop Agric 41:449–455

Petty G (1978.) *H.12 Pineapple pests*: *White Grubs in pineapple*. Pineapple Series H: Diseases and Pests, Government Printer, Pretoria, Republic of South Africa, 3 pp

Philpott SM, Uno S, Maldonado J (2006) The importance of ants and high-shade management to coffee pollination and yield in Chiapas. Mexico Biod Cons 15:487–501

Pillai GB (1979) Pests. In: Nair MK, Rao EVVB, Nambiar KKN, Nambiar MC (eds) Cashew (Anacardium occidentale). Central Plantation Crops Research Institute, Kasaragod, pp 55–72

Pillai GB, Abraham VA (1975) Indian Cashew J 10(1):5–7

Pillai GB, Dubey DP, Singh V (1976) Pests of cashew and their control in India – A review of current status. J Plant Crops 4:37–50

Pillai GB, Singh V, Dubey OP, Abraham VA (1984) Seasonal abundance of tea mosquito bug, *Helopltis antonii* on cashew in relation to meteorological factors. In: Roa EVVB, Khan HH (eds) Cashew research and development. Indian Society for Plantation Crops, CPCRI, Kasaraghod, pp 103–110

Preiss JA, Mimler M, Klein A-M, Schwarze S, Tscharntke T, Steffan-Dewenter I (2007) Linking deforestation scenarios to pollination services and economic returns in coffee agroforestry systems. Ecol Appl 17:407–417. doi:10.1890/ 05–1795

Purseglove JW (1968) Tropical crops, Dicotyledons. Longmans, London, 719 p

Py C, Lacoeville JJ, Teisson C (1987) The pineapple, cultivation and uses. Maisonneuve et Larose, Paris

Rabindra RJ (2012) Sustainable pest management in tea: prospects and challenges. Two and a Bud 59:1–10

Rai PS (1984) Handbook of cashew pests. Research co Publications, Delhi

Rao AR, Naidu VG, Prasad PR (2002) Biology of cashew shoot and blossom Webber (Lamida moncusalis Walker) studies on the. Indian J Plant Protec 30(2):167–171

Raviprasad TN, Sundararaju D, Bhat PS (2005) Efficacy of botanicals against *Helopeltis antonii* Sign. infesting cashew. The Cashew XIX(4):9–14

Raviprasad TN, Bhat PS, Sundararaju D (2009) Integrated pest management approaches to minimize incidence of cashew stem and root borers (*Plocaederus* spp.). J Plant Crops 37(3):185–189

Raw A, Free JB (1977) Pollination of coffee (Coffea arabica) by honeybees. Trop Agric 54:365–370

Red Ring Research (1983) Workshop on agricultural research policies and management in the Caribbean United Nations Economic Commission for Latin American sub-regional office for the Caribbean, 26–30 September 1983. Unpublished

Reddi EUB (1987) Under-pollination: a major constraint of cashewnut production. Proc Indian Natl Sci Academy B 53(3):249–252

Reddi EUB (1991) Pollinating agent of cashew-wind or insects? Indian Cashew J 20(4):13–18

Ricketts TH (2004) Tropical forest fragments enhance pollinator activity in nearby coffee crops. Conserv Biol 18:1262–1271

Ricketts TH, Daily GC, Ehrlich PR, Michener CD (2004) Economic value of tropical forest to coffee production. Proc Natl Acad Sci U S A 101:12579–12582

Rohrbach KG, Johnson M (2003) The pineapple: botany, production and uses. Bartholomew DP et al. (eds). CABI Publishing, Wallingfor

Roubik DW (2002a) Feral African bees augment neotropical coffee yield. In: Kevan P, Imperatriz Fonseca VL (eds) Pollinating bees – The conservation link between agriculture and nature. Ministry of Environment, Brasilia, pp 255–266

Roubik DW (2002b) The value of bees to the coffee harvest. Nature 417:708

Rudin J (1942) Bees as pollination agents in fruit trees. Rev Inst Defensa cafe de costa rica 12(96):490–491 [In Spanish.]

Ruf F (1995) Booms et crises du cacao. CIRARD-SAR, Montpellier

Sadakathulla S (1993) Evaluation of honey bee visits on coconut genotypes. Cocos 9:47–50

Sadakathulla S, Ramachandran TK (1991) Light traps kill honey bees. Bee Keeping and Development, UK, Sep. 1991 (20):8

Sakimura K (1966) A brief enumeration of pineapple insects in Hawaii. In: Proc. VI-5 Div. Meeting (Agriculture-Plant Protection), The XI Pacific Science Congress, Tokyo, Japan, Sept. 1, 1966. pp 1–7

Samiayyan K, Palaniswamy KP (1984) Cashew Causerie 6(4):9–11

Sampson HC (1923) The coconut palm. 262 pp. J. Bale, sons, and Danielson, Ltd., London Sarangamath PA, Shantappa PB, Kulkarni, DS, 1976. Varietal differences in nut and copra characters. In: Nayar NM (ed) 1982. Coconut research and development – Proc Internat Symp on Coconut Res and Dev Kerala, 27–31 Dec 1976. Wiley Eastern Limited, pp 181–185

Satapathy CR (1993) Bioecology of major insect pests of cashew (*Anacardium occidentale* Linn.) and evaluation of certain pest management practices. Ph.D. thesis, U.A.S., Bangalore, p 224

Sathiamma B (1978) Occurrence of insect pests on cashew. Cashew Bull 15(4):9–10

Schliesske J (1988) On the gall mite fauna (Acari: Eriophyoidae) of *Cocos nucifera* L. in Costa Rica. Nachrichtenblatt des Deutschen Pflanzenschutzdienstes (Germany, F.R.) 40:124–127

Sein F, Jr (1923) [Bees in coffee plantations.] Puerto Rico Insular Sta Cir 79, 6 pp. [In Spanish]

Sein F Jr (1959) Do bees help coffee? Hacienda 55:36–50 [Ln Spanish.]

Sholdt LL, Mitchell WA (1967) The pollination of *Cocos nucifera* L. in Hawaii. Trop Agricul 44:133–142

Simura T, Oosone K (1956) Studies on the fertilization of the tea plant. Ikushugaku Zasshi-jap. J Breeding 6:11–14 [In Japanese, English abstract.]

Singh V, Pillai GB (1984) Field evaluation of the efficacy of four insecticides in the control of tea mosquito. In: Bhaskara Rao EVV, Hameed Khan H (eds) *Proceedings of the international cashew symposium*, Cochin, p 301

Somchoudhury AK, Samanta A, Dhar P (1993) In: Proceedings international symposium on tea science and human health pp. 330–338. (Teatech, 1993). Tea Research Association, India. Annals of Plant Protection Sciences 17(1):65–68

Stirling GR (1993) Nematodes. In: Broadley RH, Wassman RC, Sinclair E (eds) Pineapple: pests and disorders. Queensland Department of Primary Industries, Brisbane, pp 21–29

Stonedahl GM (1991) The oriental species of Helopeltis (Heteroptera : Miridae): a review of economic literature and guide to identification. Bull Entomol Res 81:465–490

Sundararaju D (1979) A note on major pest problem of cashew, coconut and arecanut and their control in Goa. In: Proceedings of PLACROSYM II, Ootacamund, 26–29 June 1979, pp 474–483, 513–516

Sundararaju D (1984) Studies on cashew pests and their natural enemies in Goa. J Plant Crops 12:38–46

Sundararaju D (1993a) Studies on the parasitoids of the mosquito bug, *Helopeltis antonii* Sign. (Hymenoptera: mymaridae) on cashew with special reference to *Telenomous* sp. (Hymenoptera: Scelionidae). J Bio Control 7:68

Sundararaju D (1993b) Compilation of recently recorded and some new pests of cashew in India. The Cashew 7:15–19

Sundararaju D (1996) Studies on *Helopletis* spp. with reference to *Helopeltis antonii* Sign.in Tamil Nadu. Ph.D. thesis, T.N. A.U., Coimbatore

Sundararaju D, Bakthavatsalam N (1990) Cashew pest management for coastal Karnataka. The Cashew 4:3–6

Sundararaju D, Bakthavatsalam N (1994) Pests of cashew. In: Chadha KL, Rethinam P (eds) Advances in horticulture, Vol.10. Mathotra Publishing House, New Delhi, pp 759–785

Sundararaju D, Sundarababu PC (1999) *Helopeltis* spp. (Heteroptera: Miridae) and their management in plantation and horticultural crops of India. J Plant Crops 27(3):155–174

Sundararaju D, Yadukumar N, Bhat PS, Raviprasd TN, Venkattakumar R, Dixit S (2006) Yield performance of "Bhaskara" cashew variety in coastal Karnataka. J Plant Crops 34(3):216–219

Thankamma Pillai PK, Pillai GB (1975) Note on shedding of immature fruits in cashew. Indian J Agr Sci 45(5):233–234
Uthaiah BC, Rai PS, Khan MM, Hiremath IG, Kumar DP (1989) Pre bearing performance of some cashew type in coastal Karnataka. The Cahew 3:9–11
Van Mele P (2008) Biological control with the weaver ant Oecophylla Longinoda in Africa: a review of research and development efforts to link farmers to organic markets. Outlooks Pest Manag 19(4):180–183
Veddeler D, Klein AM, Tscharntke T (2006) Contrasting responses of bee communities to coffee flowering at different spatial scales. Oikos 112:594–601
Veddeler D, Olschewski R, Tscharntke T, Klein AM (2008) The contribution of non-managed social bees to coffee production: new economic insights based on farm-scale yield data. Agroforest Syst 73:109–114
Vega FE (2008a) The rise of coffee. Am Sci 96:138–145
Vega FE (2008b) Insect pathology and fungal endophytes. J Invertebr Pathol 98:277–279
Vega FE, Infante F, Castillo A, Jaramillo J (2009) The coffee berry borer, *Hypothenemus hampei* (Ferrari) (Coleoptera: Curculionidae): a short review, with recent findings and future research directions. Terr Arthr Rev 2:129–147
Velez M, Bustillo AE, Posada FJ (2000) Predacion sobre Hypothenemus hampei, (Ferrari) de las hormigas Solenopsis spp., Pheidole spp., y Dorymyrmex spp. durante el secado del cafe'. In: 500 Ve' lez M, Bustillo AE, Posada FJ (eds), Resu' menes XXVII Congreso. Sociedad Colombiana de Entomologı'a, Medellı'n, Colombia, pp 17
Velez M, Bustillo AE, Posada FJ (2001) Hormigas de la zona central cafetera y perspectivas de su uso en el control de Hypothenemus hampei (Ferrari) (Coleoptera: Scolytidae). In: Ve' lez M., Bustillo AE, Posada FJ (eds), Resu' menes XXVIII Congreso. Sociedad Colombiana de Entomologı'a, Pereira, Colombia, pp 51
Waite GR (1993) Pests. In: Broadley RH, Wassman RC, Sinclair E (eds) Pineapple: pests and disorders. Queensland Department of Primary Industries, Brisbane, pp 21–29
Waterhouse DF (1998) Biological control of insect pests: Southeast Asian prospects, ACIAR monograph series, no. 51. Australian Center for International Agricultural Research, Canberra, 548 pp
Wcislo WT, Cane JH (1996) Floral resource utilization by solitary bees (Hymenoptera: Apoidea) and exploitation of their stored foods by natural enemies. Annu Rev Entomol 41:257–286
Wellman FL (1961) Coffee; botany cultivation and utilization. Leonard Hill, London, 488 pp
Westerkamp C, Gottsberger G (2000) Diversity pays in crop pollination. Crop Sci 40:1209–1222
Wickramaratne MRT, Vitarana SI (1985) Insect pollination of tea (camellia sinensis l.) in Sri Lanka. Trop Agr 62(3):243–247
Wight W, Barua PK (1939) The tea plant industry, some general principles. Indian Tea Assoc Tocklai Expt Sta Memo 7, 13 pp
Willc A, Michencr CD (1973) The nest architecture of stinglcss becs with special reference of those of Costa Rica (Hymenoptcra; Apidae). Rev Biol Trop 21(1):1–278
Willmer PGG, Stone N (1989) Incidence of entomophilous pollination of lowland coffee (*Coffea canephora*); the role of leaf cutter bees in Papua New Guinea. Entomol Exp Appl 50:113–124
Wrigley G (1969) Tropical agriculture. Frederick A. Praeger, NY and Washington, 376 pp.
Wu CT (1967) Studies on the percentages of fruit-setting of self- and cross-pollination and its relation to some economic characters of the f1 hybrids in tea plant. J Agr Assoc China 59:24–39
Young AM (1982) Effects of shade cover and availability of midge breeding sites on pollinating midge populations and fruit set in two cocoa farms. J Appl Ecol 19:47–63
Young AM (1994) The chocolate tree. Smithsonian Institution Press, Washington, DC
Zimmermann A (1928) [Pollination of coffee trees.] His kaffee, ed. 2, pp 31–34. Deutscher auslandverlag, Hamburg. [In German.]

Chapter 11
Research Needs for Better Fruit Productivity

11.1 Introduction

Pollination is an essential ecosystem service and a prerequisite for fertilization and fruit/seed set. The relationship between plants and pollinator is a mutualistic one in which pollinators get food in the form of nectar and pollen and in turn pollinate the plants. Sprengel (1793) published his findings on the interactions between flowers and insects and stated that transfer of pollen from stamens to the stigma was necessary for seeds and fruit production. Since then much has been known and it is established that more than 300,000 species of animals depend upon floral resources (Buchmann and Nabhan 1996).

It is universally recognized that the seed and fruit production of crops, including tree crops for can be enhanced by extending pollination biology. Indians knew the role of pollination and events leading to fertilization of plants from prehistoric times. Madhuvana or gardens for bees were very common during the period of Ramayana (Belavadi 1993). The higher fruit set and better quality can be achieved in honeybee pollinated flowers compared to those in open pollinated. The sustainable development of agriculture through higher seed set/fruit set can be achieved by encouraging beekeeping by honeybee management, promotion of bee flora and integration of beekeeping programmes with social/agroforestry. The research and development needs for indigenous honeybee *Apis cerana* improvement and its utilization for effective pollinator in mountainous areas involve development in rearing technology of bees, encouraging migratory beekeeping and including the courses for capacity development in school curricula.

11.2 Constraints in Promoting Honeybee for Pollination

Horticultural crops attract not only a large number of insect pollinators especially the honey bees for nectar and pollen but also other insects which feed on flowers, leaves, fruits thereby, causing serious economic losses (Shrestha and Shrestha

D.P. Abrol, *Pollination Biology, Vol. 1*, DOI 10.1007/978-3-319-21085-8_11

2000). This requires the application of insecticides to combat the pest which poses serious problem for the foraging activity of honeybees and developing brood (Thapa 2003). The protection of pollinators, including honeybees from pesticides for pest control on the one hand and the role of honeybees (*Apis* spp.) for crop pollination on the other have become essential components of modern agriculture. Unfortunately, these two practices are not always compatible, as honeybees are susceptible to many of commonly used pesticides used for the control of insect pests (Abrol 1996). The major constraint confronting pollinator-plant interaction is the indiscriminate and excessive use of pesticides for controlling insect-pests (Zhong et al. 2004). The loss of honeybees directly effect beekeeping through loss of honey production and indirectly the crop production due to inadequate pollination. Reduction of population of these beneficial insects due to insecticides, therefore, incurs significant environmental, ecological and economic costs (Khan and Dethe 2004).

Generally, fumigative action of insecticides used under field conditions is of much shorter durations then the effect of contact and stomach poison. A prolong repellent effect will deprive the flowers of pollination benefits of insect visits, while a short repellency will deter the insect pollinators from visiting the treated bloom for a brief period and thereafter, allow them to resume their foraging activity (with minimum residual hazards) without compromising with the yield potential of the crop. Honeybees because of their long hours of working on the crop flowers for pollen and nectar collection are at more risk to poisoning. Furthermore, modern agricultural practices have resulted in rapid decline of wild bee pollinators by destruction of their nesting sites. Even the toxicity of pesticides is not same to honeybees and wild bees as some insecticides are more toxic to one species of wild bee than the other (Baker and Gyawali 1994; Sharma 1994).

One of the most important factors for ensuring sufficient supply of bee colonies is to ensure supply of food through continuous bee forage. Beekeeping in any specific area cannot develop without an understanding of availability of bee floral plants. For migratory beekeeping, special floral calendars for the different foraging zones along the migration route are should be available. Preparation of a floral calendar for any specific area requires complete observation of the seasonal changes in the vegetation patterns and/or agroecosystems of the area, the foraging behaviour of the bees, and the manner in which the honeybee colonies interact with their floral environment. The preparation of floral calendars requires several years of regular observations and refinement of the information obtained. In order to assess the beekeeping potential of an area it is necessary to know about the composition of the vegetation in the locality, bee forage value of individual species and the usual flowering periods of important bee forage plants. Sequence of flowering of different plant species in the area has to be observed to find out the seasons when a large amount of nectar is available, which helps in storage of surplus honey that can be harvested by the beekeeper.

Apis cerana as compared to the European honeybee *Apis mellifera* forages on a wide variety of resources. Even the resources that provide small quantities of rewards are worked upon resulting in out crossing of a large spectrum of crops/ flowers. The Asiatic hive bee *Apis cerana* does not forage over long distances staying within a kilometer or so of the hive. The information on floral sources provides sound Maintenance practices for growth to its floral preferences and foraging range are discussed.

11.3 Insect Flower Visitors and Crop Pollinators

A number of agents are necessary to effect pollination. These includes a biotic (wind, water, gravity etc.) and biotic (insects, bats, birds, snails and slugs etc.). But among all of them, the insects are the most important with bees (*Apis* spp.) contributing more than 80 % of the total pollination carried out by insects. Honeybees have attributes that makes them valuable for crop pollination. These can be marshalled in adequate numbers at desired places. Their body parts are modified to effect pollination. Their wide host range enables them to pollinate much type of crops. They work for longer durations and are also less affected by adverse climatic conditions as compared to other insect pollinators. Their foraging behaviour patterns are highly favourable to qualify them as most efficient pollinators of crops. In addition to pollination services, colonies can be managed to produce honey, bee wax, propolis, royal jelly etc. which are all saleable products. Out of four *Apis* species available in India, only two indigenous honeybee, *Apis cerana* Fabr. and the exotic *Apis mellifera* L are domesticated whereas the other two *Apis florea* L. and *Apis dorsata* Fabr. are wild and nest in nature. Techniques now have been developed for the transport and placement later bees at desired locations at desired time. The forage preference, pollinating efficiency and density of bee species vary depending on the crop. The number of foragers on particular crop at any moment of time is determined by the colony size, weather factors (temperature, wind velocity, relative humidity, solar radiation etc.) and floral attractiveness (colour, odour) and nectar and pollen concentration of flower.

11.4 Decline in Native Pollinators

Native pollinators have witnessed an unprecedented decline in recent years. Factors responsible for this decline include habitat loss, increasing monoculture and negative impacts of pesticides and herbicides. The introduction of the European honey bee, *Apis mellifera* had also a negative effect on native *Apis cerana* because of competition for nectar resources and spread of diseases and harmful mites. In isolated mountain areas of Afghanistan, Pakistan, Nepal and India, subsistence farmers are totally dependent on their own resources for their survival. Due to environmental degradation as well as poor pollination, the quantity and quality of many life-saving mountain crops is declining significantly, making survival increasingly difficult and forcing people to migrate to the plains.

11.5 Decline in Fruit and Seed Production

The negative impact of declining pollinator intensity is visible in Himachal Pradesh of India, Jammu and Kashmir as well as in mountain areas of Afghanistan and China. Despite increasing agronomic inputs, fruit production has declined in recent years. In fact, the negative effects of these agronomic inputs on pollinators is one of the major causes of pollination failure and hence the observed declines in productivity. For example, apple cultivation in Himachal Pradesh in India, though it initially gave significant economic gains, has resulted in a loss of agricultural biodiversity and a decline in natural insect pollinators. In this area, farmers are now compelled to rent colonies of honeybees for pollinating their apple orchards. In China, farmers have resorted to hand pollination of their apples and pears, as there are not enough natural insect pollinators to ensure a proper fruit setting. Awareness about the use and function of honeybees is lacking, and the beekeepers in this area hesitate to let their bees into this fruit-producing valley because of the serious overuse of pesticides in apple orchards. In Pakistan, disappointed farmers are cutting down their apple trees due to low yields caused by insufficient pollination.

11.6 Pollination of Agricultural Crops

The plants requiring pollination or benefitted from pollination are presented in Table 11.1 (Abrol 1997). Insect and other organisms play major role in boosting agricultural production by significantly increasing the yields of crops, vegetables, fruits and seeds through visiting flowers and helping in pollination (Abrol 1993).

The self-pollinated crop species occupy less than 15 % and the remaining are cross-pollinated crops that need help of pollinating agents, wind, water or insects for fertilization. Some crops also exhibit often cross-pollinated nature. The genetic architecture of such crops is intermediate between self- and cross-pollinated species (Pokhrel 2006; Partap 2001).

Table 11.1 Classification of crop species based on their natural mode of pollination aspects

Self-pollinated but benefited by insect visits	Cross-pollinated plants and highly benefited by insect visits
Fruit Trees: apricot, citrus, peach etc.	*Fruit trees*: apple, avocado, banana, cherry, date palm, fig, coconut, papaya, plum, loquat, strawberry, almond, niger, mango, pear, blackberry, raspberry, chestnut, hazelnut etc.

11.7 Research Needs for Better Fruit Productivity

Honey bees has long been recognized as the most important and dependable pollinator of wide variety of crops. However, with recent decline in number of managed bee colonies, there is a need to diversify pollinators to accomplish pollination. Diversification of pollinators will help to improve biodiversity conservation, agricultural sustainability and crop security. The main problem with the latter group is that their pollination activity is irregular and unreliable. Even among Hymenoptera, all are not equally good pollinators and it is only Apoidea which are highly diverse structurally and behaviourally as well as taxonomically (Kevan and Baker 1983). Apoid bees which form a diverse group of Hymenoptera are well adapted for pollination. The solitary bees are also valuable pollinators but not dependable because their population fluctuates greatly from year to year. Bumble bees are the most efficient pollinators for pollen transfer because of their big size and can work at extremely low temperatures from dawn to dark when other pollinating insects are not active (Batra 1994).

On temperate fruit crops, the domesticated honeybees have been found to constitute high proportion of insect pollinators and there is no doubt that these are the only pollinators which can be relied upon for effective pollination and its population can be managed as per requirements of the crops easily. However, potential dangers have been found whenever one species has been used and relied upon for pollination. Outbreak of Thai sac brood viral disease of *A. cerana* during early eighties of the last century in this country resulted sudden collapse of managed honeybee population when more than 95 % of *A. cerana* colonies perished. Under such compelling circumstances need for diversification and conservation of pollinators is realized. Moreover, in cold desert areas like Spiti valley of Himachal Pradesh where honeybees do not occur in nature, the pollination is being accomplished in highly cross pollinated crop of apple by other native pollinators such as *Nomia* sp., *Halictus* sp., *Ceratina hieroglyphia, Xylocopa* sp., *Bombus* sp., Colletidae, *Prosopis* sp. (Hymenoptera); *Eristalis tenax, Scaeva pyrastis, Eupeodes frequens* and *Syrphus* sp. (Diptera).

11.8 Conservation of Pollinators

Deep and intensive cultivation has destroyed nesting habitat of wild bee pollinators. This has resulted in disappearance of native pollinators in many areas where intensive agriculture was adopted. Indiscriminate application of insecticides has resulted in devastating setbacks to non-target useful fauna mainly pollinators and biocontrol agents. For increasing the number of native pollinators development of habitat management programs has been suggested by many workers based on sound ecological principles. Further, there is a need for diversification bee forage plants due to excessive monoculture which can be achieved through plantation of multipurpose flora under social and agroforestry programmes. Artificial nesting sites have also been suggested for conservation and propagation of potential pollinators like bumble bees and alkali bees. The domesticated indigenous hive bee, *A. cerana* also needs conservation by encouraging traditional as well as beekeeping in modern hives. Considering the increasing global need of insect pollination and decline in the pollinator community, non-*Apis* bees along with honey bees hold immense importance. In order to formulate the conservation policy of any species, one needs considerable amount of available information regarding its habit and habitat. Keeping this fact in mind, there is an urgent need to generate a baseline source of overall information regarding non-*Apis* bees providing ecosystem services as effective pollinators of various plant species. Documentation of diversity and occurrence of non-*Apis* bees across different landscapes may help understand the insect pollinator services in various ecosystems across the area. Management protocols are needed to increase the use of wild pollinator species as alternatives to honey bees, either through domestication of alternative pollinator species and/or habitat management to encourage the presence of wild pollinators.

11.9 Current Status and Future Pollination Needs

There is an urgent need for development and adoption of policies to ensure sufficient population of pollinators for different crops in different geographical areas. Techniques for rearing of solitary bees and bumblebees need to be improved, promote taxonomic research and encourage expansion of beekeeping industry. Conservation of pollinators requires reorientation land use policies to encourage habitats for wild and managed bees. Natural vegetation needs to be encouraged to minimize disturbances which destroy nest sites. Extensive research on ecology and taxonomy of bees and factors affecting their population build up such as food and nest site availability, parasites, diseases, predators should be carried out. Studies on foraging range of pollinators, their temporal and spatial distribution and responses to quality and quantity of food resources must be known.

Considering the importance of bees on crop pollination, more coordinated research is needed. In India the total cultivated areas is 189 million hectares and at least 1/3 area is under entomophilous crops which require pollination. At a very modest rate of 3 colonies/hectare, 567 million colonies of honeybees are needed as against the merely the one million colonies at present. It is very necessary to survey different agro-climatic zones to determine availability of resources so that can be supplemented with forage.

The lean pollen gathering period can be corrected by planting suitable pollen and nectar yielding flora of that period. Pesticides applications wherever possible can be modified as soil application instead of spray, can avoid drift to nearby hives. The selection and breeding of honeybees characteristics best suited for pollination of different crops seems promising to enhance pollinating effectiveness.

Development and use of insects other than hive bees takes a long time and need major research and extension efforts before such insects can be reared and managed for pollination of crops. The main constraints to promoting managed pollination by using honeybees and other pollinators are lack of awareness and understanding among farmers, extension workers, planners and policy-makers about the importance of pollinators and pollination, lack of integrating pollination in agricultural development packages, scarcity of managed colonies of honeybees, and lack of knowledge about conservation, rearing and use of pollinators and their pollination behaviour.

The responsibility of scientists is to transmit the knowledge of pollinator utilization to the wider public, business and policy audiences in a simple and understandable form and develop novel practical solutions to specific problems. Policy makers on the other hand need to facilitate the investigators by providing clear priorities and resources to allow the research community to build appropriate knowledge base. It is therefore, within the capacity of scientists, policy makers and stake holders to understand mitigate the threats to pollination. The important researchable issues include:

1. Pollinators biodiversity, characterization and taxonomy, multiplication, conservation and domiciliation of pollinators.
2. Prepare protocol for multiplication of pollen vectors
3. Prepare bee maps in the country with respect to the distribution, availability of different bee species in different crops in time and space.
4. A data base on pollinators, their abundance, behaviour, floral rewards and pollinator plant interaction needs to be generated.
5. The role of honeybees in crop pollination needs to be given due importance.
6. Non-*Apis* bees, pollinators such as stingless bees, carpenter bees, bumblebees, megachilids and many other bees need to be promoted for crop pollination.
7. In addition to cross pollination, the role of insects in self pollinated crops need to be worked out.
8. Role of other insects in increasing crop yields qualitatively and quantitatively.
9. Use of pollinators for garden and green house crops
10. Bee lure, attractants, repellents to be used to increase the attractiveness to the crop or repel the bees from insecticidal treated crop.
11. Safety of pollinators from pesticides, diseases, parasites and predators and from adverse climatic conditions.
12. Fast and safe multiplication /domiciliation of pollinators both for green house and open fields.
13. Need to extend technology through demonstration, farmers participatory trials and HRD
14. Surveillance of bee diseases, pests and predators in various eco-geographical zones. Enforce strict quarantine, isolation, certification of disease free status measures through legislation. Create a network of laboratory facilities for the identification, testing and control of bee diseases and pests.

15. Capacity building and awareness training programmes should be organized in terms of management plans, training manuals, honey festivals, seminars, conference for farmers, fruit growers, extension workers, NGO.s and policy makers to promote migratory beekeeping as full-time occupation among rural people and beekeepers.
16. In order to induce beekeepers to migrate colonies during prolonged dearths, or for different flows or for pollination, a subsidy should be provided to cover the expenditure of migration.
17. Management protocols are needed to increase the use of wild pollinator species as alternatives to honey bees, either through domestication of alternative pollinator species and/or habitat management to encourage the presence of wild pollinators.
18. The economic and ecological importance of pollinators and the issue of their declines around the world have not been recognized in most mainstream research and development efforts. Apparently most people, including farmers and policy makers, are generally unaware of the services pollinators provide to natural and agro ecosystems.
19. To effectively address this issue it is necessary to bring pollination concerns into the policy, research and development mainstream through promoting their integration into agricultural research policies, extension and outreach activities. There is a dearth of non-technical literature for promoting awareness among planners and policy makers.
20. Public outreach is key to pollinator protection, conservation, and restoration. Develop and communication outreach capacity in the form of multilingual manuals on pollinator conservation and restoration for farmers. Disseminate information on pollination in agricultural environments through data bases, websites, and networks.
21. Establish a national advisory group on pollinator conservation and to create an information network on pollinator conservation and a directory of pollinator experts. Development of guidelines for policy makers and for farmers is the need of the hour.
22. There is a need to assess the state of scientific and indigenous knowledge on pollinator conservation and to develop and update global and national lists of threatened pollinator species.
23. Educate the public on the importance of pollinators
24. Raise awareness of the pollination crisis
25. Train the next generation of researchers and taxonomists; and
26. Support national plans for the conservation of bees and increase the awareness of governments, industry and the public.
27. Use of electronic media-All India Radio Station for broadcasting talks on the role of honeybees in promoting pollination.

The research areas for encouraging pollinators and enhancing pollination include:

Orchard Layout and Cross-pollination Orchard lay out is most important for effective pollination. In orchards where cross-pollination is required, polliniser varieties should be adjacent to each other and must flower at the same time. When planning the orchard layout, if the spacing of trees across rows is greater than along rows, the bees will tend to work more along rows than across the rows. If the polliniser trees are planted in separate rows, bees may be inclined to forage along the row only. The transfer of pollen from trees in one row to an adjacent row is minimal and fruit set may be disappointing.

Optimizing Pollination of Fruit Crops with Bees Investment in pollinators is essential for reaching the potential of many fruit crops. To reach this potential, growers need to know how to optimize pollination of their particular crop(s) and varieties being grown. The recommended densities of honeybee hives for optimal yields as discussed elsewhere with each fruit crop should be used.

Strategies for Effective Pollination In some of the fruit crops, bees start working on flowers as soon as they open. This helps early fertilization of ovules before they lose vigour. The honey bee colonies should be moved 2–3 days before the flowering begins. In case of fruit crops with small flowers, a different strategy needs to be adopted. For instance in case of blueberry, colonies should be moved in the crop after 5 % bloom and before 25 % of full bloom.

Breeding to Increase Pollination If cross pollination is necessary select and breed varieties which are highly attractive to bees. There is need to increasing support for pollination research to encourage expansion of beekeeping. This will enhance increased food, fibre, fruit and seed production.

Management of Honeybee Colonies (Moving Colonies) The colonies should be moved to a target crop when bloom is 5–10 %. Placing the hive with in 0.5 km radius increases the crop pollination. Colonies placed near crops collect more pollen and nectar, spend less time collecting load of pollen and nectar, the number of flights increases for both types with proximity to the floral source. In general, 3–5 colonies/hectare should be placed equidistant from each other within the crop. Weeds or other non target crops should be eliminated or mowed when in flower, to avoid competition for foraging bees. Use of pollen dispensers and disposal pollination units can enhance pollination efficiency.

11.10 Extension Strategies for Pollination

The changing agricultural/horticultural scenario in developing countries, with large areas being changed to monoculture at the expense of massive deforestation is adversely affecting the pollinator population. An awareness and scientific research component is therefore necessary in the present day conditions explain the role of different clienteles and extension agencies for promoting pollination through the application of apiculture.

There are three basic approaches in extension strategy (1) consultancy, (2) promotional, and (3) participatory which are not mutually exclusive but complimentary to each other.

11.11 Consultancy Approach

This is the first step in the extension strategy and involves bringing about awareness and providing advice.

11.12 Raising Public Awareness

Farmers orchardists are quick to see the destruction caused to their crops by pests and diseases or by inadequate fertilization. The losses resulting due to inadequate pollination usually go unnoticed. Extension agencies therefore need to make the farmers aware about the benefits of pollination which can be done through.

11.13 Demonstration

The benefits of bee pollination to fruit crops can be demonstrated to farmers by placing colonies in their orchards at the time of flowering of crops and then comparing the quality and quantity of yield in orchards receiving pollination with those deprived of pollinators.

Impact Generation Large scale movement and placement of honeybee colonies at sensitive locations to create maximum impact on farmers/orchardists as to the benefits of bee pollination.

Providing Advice Honeybees are an important bio-input for quantity and quality of seed set and fruit production. The orchardist and farmers need to be advised about the number of colonies placed, time of placement, plan of placement, direction of placement for pollination including safety from pesticides.

11.14 Promotional Approach

In this step, the extension agency makes its own decision on what to do and how to do it on the basis of area specific views and needs of the farmers/orchardists. This involves:

Development need based technologies: of New technologies developed by research scientists are improved and indigenous technologies developed for easy acceptance among the rural people. Some such technologies developed for honeybee pollination include the three modernized wall hive, straw hive, mud hive and orchard hive.

11.15 Participatory Approach

This is a partnership between n the clients and the extension agency and together they work out the modalities for filling the knowledge gaps through:

Training and Education One of the most important constraints in beekeeping development has been the absence of trained and experienced personnel. Most of the agriculture training programmes in existence today were developed when modern beekeeping was introduced into the country. Changes in the art and practice of beekeeping in recent years are not reflected in these training programmes and even some beekeeping courses are quite out of date. Agriculture development programmes should give priority to training of beekeepers, supervisors and other extension officials for generating competent human resource for imparting the skill and practice of honey bee keeping as a vocation and its use in enhancing pollination percentage of fruits, vege and high value cash crops thereof. Training of extension workers should get top priority, because they form the link between the research laboratories and the beekeepers.

Both short term location specific and long term on-campus training programmes are required to be organized on management of honeybee colonies for pollination. The location specific trainings are required to be focused on floral morphology, insect visitation for pollination, knowledge of bee flora and integration of apiculture supporting trees in social forestry plantations. On-campus trainings should be of 2 week duration encompassing the components of pollination biology to be dovetailed with curricula on commercial apiculture. Strategies for marketing of honey and value addition of apiary honeys should also be taught. The production of unifloral honeys from medicinal plant species, honeys from forest plant species that do not have any agricultural chemicals, etc. possess export potentials. The development of apiculture as a commercial entrepreneurship will automatically favour the pollination percentage.

11.16 Feed Back and Monitoring

Problems in the implementation of any scheme are resolved on the basis of information given by the participants. The extension agency regularly monitors the adoption and success of the pollination strategy involving honeybees imparted to the clients through number of beehives/improvised hives developed in villages and yield data of crop. Facility for providing honeybees on rent for those who do not wish taking up apiculture as a vocation will be another key strategy for promoting pollination.

11.17 Relevance of Technologies to Developing Countries

In the developing countries much vaunted green revolution that encouraged the food sufficiency at any cost has wrought considerable environmental degradation. Increasing dependence on intensive practices, pesticides and other chemicals as well as deforestation had an adverse effect on nesting and mating sites of natural pollinators. Extensive use of farm machinery has driven them out necessitating the need for their conservation for optimum crop yields. Some countries have successfully increased their crop yield through management and augmentation of pollinators other than honeybees. A classical example is pollination of oil palm in Malaysia by introduction of weevil *Elaeidobius kamerunicus* from west Africa to Malaysia in 1981. There it spread immediately resulting in millions of dollors of profit to the industry. Today Malaysia is the global leader in production (58 %) and exports (78 %) of oil palm. In India, not much attention has been made to pollinator management. Countries like Mexico, Argentina and China are far ahead in having number of honeybee colonies and raising greater output of apiary products, enabling them to be established exporters of such products for decades whereas India does not feature on export map.

Bibliography

Abrol DP (1993) Insect pollination and crop production in Jammu and Kashmir. Curr Sci 65(3):265–269

Abrol DP (1996) Bee keeping in integrated rural development. Ind Farm 46(2):17–22

Abrol DP (1997) Bees and beekeeping in India. Kalyani Publishers, Ludhiana, p 450

Abrol DP (2009) Bees and beekeeping in India, 2nd edn. Kalyani Publishers, Ludhiana, 719 pp

Abrol DP, Andotra RS (2003) Relative toxicity of some insecticides to *A. mellifera* L. J Asia Paci Entomol 6(2):235–237

Alam MZ, Quadir MA (1986) Role of honeybee in fruit and seed setting of bottle-gourd, Lagenaria siceraria (Mol) Standl. Punjab Veg Grower 21:32–34

Anderson DL (2008) Surveillance of parasites and diseases of honeybees in Papua New Guinea and Indonesia. CSIRO/DAFF Report September 2008. 41pp. published by ACIAR GPO Box 1571 Canberra ACT 2601 Australia

Anderson DL, Morgan MJ (2007) Genetic and morphological variation of bee-parasitic Tropilaelaps mites (Acari: Laelapidae): New and re-defined species. Exp Appl Acarol 43:1–24

Anderson DL, Trueman JWH (2000) *Varroa jacobsoni* (Acari: Varroidae) is more than one species. Exp Appl Acarol 24:165–189

Anderson D, Annand N et al (2010) Control of Asian honeybees in the Solomon Islands, Australian Government. Australian Centre for International Agricultural Research, Canberra

Baker SL, Gyawali BK (1994) Promoting proper pesticide use: obstacles and opportunities for an integrated pest management programme in Nepal. HMG/MOA/Winrock Intl, Kathman-du, Nepal, 125 pp

Banaszak J (1995) Changes in fauna of wild bees in Europe. Pedagogical University, Bydgoszcz

Batra SWT (1981) Biological control in agroecosystems. Science 215:134–139

Batra SWT (1985) Bees and pollination in our changing environment. Apidologie 26:361–370

Batra S (1994) Diversity with pollen bees. Am Bee J 134:591–594
Belavadi VV (1993) Knowledge of pollination in ancient India. Curr Sci 65(3):193–194
Benedek P (2003a) Bee pollination of fruit trees: recent advances and research perspective-I. J Apic Sci 47(1):87–94
Benedek P (2003b) Bee pollination of fruit trees: recent advances and research perspective-II. J Apic Sci 47(1):95–100
Brown BJ, Mitchell RJ, Graham SA (2002) Competition for pollination between an invasive species (purple loosestrife) and a native congener. Ecology 83:2328–2336
Buchmann S, Nabhan G (1996) The forgotten pollinators. Island Press, Washington, DC
Butz-Huryn VM (1997) Ecological impacts of introduced honey bees. Q Rev Biol 72:275–296
Chittka L, Schürkens S (2001) Successful invasion of a floral market. Nature 411:653
Connor IJ (1969) Honey bee pollination requirements of hybrid cucumbers *Cucumis sativus* L. M. Sc. thesis, Mich State Univ, 150 pp
Corbet SA, Williams IH, Osborne JL (1991) Bees and the pollination of crops and wild flowers in the European Community. Bee World 72:47–59
Crane E (1991) *Apis* species of tropical Asia as pollinators and some rearing methods for them. Acta Hort 288:29–48
Cuningham SA, Fitzgibben F, Heard TA (2002) The future of pollinators for Australian agriculture. Agriculture 53(8):893–900
Day MC (1991) Towards the conservation of Aculeate Hymenoptera in Europe. Convention on the conservation of European wildlife and natural habitats. Council of Europe Press, Strasbourg, Nature and Environment Series 51
Dicks LV et al. (2012) Identifying key knowledge needs for evidence-based conservation of wild insect pollinators: a collaborative cross-sectoral exercise. Insect conservation and diversity, 2012. Online Early DOI: 10.1111/j.1752-4598.2012.00221.x
Dulta PC, Verma LR (1987) Role of insect pollinators on yield and quality of apple fruit. Indian J Hort 44:274–279
Dyer FC, Seeley TD (1991) Dance dialects and foraging range in three Asian honey bee species. Behav Ecol Sociobiol 28:227–233
Evans JD, Pettis JS, Hood WM, Shimanuki H (2003) Tracking an invasive honey bee pest: mitochondrial DNA variation in North American small hive beetles. Apidologie 34:103–109
Faegi K, Van der Pijl L (1979) The principles of pollination ecology, 3rd edn. Pergamon Press, Oxford
Falk S (1991) A review of scarce and threatened bees, wasp and ants of great Britain. Research and survey in nature conservation no. 35, Nature Conservancy Council, UK
Free JB (1993) Insect pollination of crops, 2nd edn. Academic Press, Harcourt Brace Jovanovich Publication, London, 684 p
Gupta JK, Goyal NP, Sharma JP, Gautam DR (1993) The effect of placement of varying numbers of *Apis mellifera* colonies on the fruit set in apple orchards having different proportions of pollinisers. In: Veeresh GK, Uma Shankar R, Ganeshaiah KN (eds) Proceedings of the international symposium on pollination in the tropics. International Union for Studies on Social Insects, India
Gupta JK, Rana BS, Sharma HK, (2000). Pollination of kiwifruit in Himachal Pradesh. In: Matsuka M, Verma LR, Wongsiri S, Shrestha KK, Uma Partap (eds) Asian bees and beekeeping: progress of research and development. Proceedings of the fourth international conference, 23–28 March 1998, Kathmandu, Oxford and IBH Publishing Co Pvt. Ltd., New Delhi, 274 pp
Hill JK, Thomas CD, Fox R, Telfer MG, Willis SG, Asher J, Huntley B (2002) Responses of butterflies to twentieth century climate warming: implications for future changes. Proc R Soc London (B) 269:2163–2171
Inoue T, Adri S, Salmah S (1990) Nest site selection and reproductive ecology of the Asian bees, *Apis cerana indica* in central Sumatra. In: Sakagami SF, Ohgushi R, Roubik DW (eds) Natural history of social wasps and bees in equatorial Sumatra, Sapporo, Hokkaido University Press. *Interactions*. Cambridge, Massachusetts, Harvard University Press, pp 219–232

Jennersten O (1988) Pollination of *Dianthus deltoides* (Caryophillaceae): effects of habitat fragmentation on visitation and seed set. Conserv Biol 2:359–366
Jodha NS, Shrestha S (1993) Sustainable and more productive mountain agriculture: problems and prospects. In: Mountain environment and development – Part 3 (Thematic Papers), Kathmandu, ICIMOD, pp 1–65
Kafle GP (1992) Salient features of beekeeping in Nepal. In: Verma LR (ed) Honeybees in the mountain agriculture. Oxford and IBH, New Delhi, pp 155–162
Kearns CA, Inouye DW, Waser NM (1998) Endangered mutualisms: the conservation of plant-pollinator interactions. Ann Rev Ecolo Sys 29:83–112
Kenmore P, Krell R (1998) Global perspectives on pollination in agriculture and agroecosystem management. In: International workshop on the conservation and sustainable use of pollinators in agriculture, with emphasis on Bees, Sao Paulo, 7–9 October 1998
Kevan PG (1975) Forest application of the insecticide fenitrothion and its effect on wild bee pollinators (Hymenoptera: Apoidea) of lowbush blueberries (Vaccinium SPP.) in Southern New Brunswick, Canada. Biol Conserv 7(4):301–309
Kevan PG, Baker HG (1983) Insects as flower visitors and pollinators. Annu Rev Entomol 28:407–453
Khan RB, Dethe MD (2004) Median lethal time of new pesticides to foragers of honeybee. Pestology 28(1):28–29
Kölreuter JG (1761) D. Joseph Gottlieb Kölreuter's Vorläufige Nachricht von einigen das Geschlecht der Pflanzen betreffenden Versuchen und Beobachtungen. Gleditschischen Handlung, Leipzig
Kreuss A, Tscharntke T (2002) Grazing intensity and the diversity of grasshoppers, butterflies, and trap nesting bees and wasps. Conserv Biol 16:1570–1580
Kuang BY, Kuang HO (2002) The biology of the honeybee. Kunming Science and Technology, Kunming (in Chinese)
Levin MD (1989) Honeybee pollination of egg plant (*Solanum melongena* L.). The Thirty-first Intl Agric Cong, August 1987, Warsaw, Poland. Apimondia Pub House, Bucharest, pp 344–348
Mayer DF (2003) Bee kills from pesticides and the expert witness. Am Bee J 143(7):563
Morse RA, Calderone NW (2000) The value of honey bees as pollinators of US crops in 2000. Electronic Document Downloaded from the Internet Page http://bee.airoot.com/beeculture/pollination2000
Neupane KR (2001) Foraging preference of honeybee species to selected horticultural crops. M. Sc. thesis, Submitted to IAAS, Rampur, Chitwan, pp 95
Nieuwhof M (1969) Cole crops. Leonard Hill, London, 353 pp
O'Toole C (1993) Diversity of native bees and agroecosystems. In: LaSalle J, Gauld I (eds) Hymenoptera and biodiversity. Commonwealth Agricultural Bureau International, London, pp 69–106
O'Toole C (1994) Who cares for solitary bees? In: Metheson A (ed) Forage for bees in an agricultural landscape. International Bee Research Association, Cardiff, pp 47–56
Oldroyd BP, Wongsiri S (2006) Asian honey bees: biology, conservation and human interactions. Harvard University Press, Cambridge, 340 pp
Oldroyd BP, Rinderer TE, Wongsiri S (1992) Pollen resource partitioning by *Apis* dorsata, A. *cerana*, A. andreniformis and *A. florea* in Thailand. J Apic Res 31:3–7
Ollerton J, Winfree R, Tarrant S (2011) How many flowering plants are pollinated by animals? Oikos 120:321–326
Osborne JL, Williams IH, Corbet SA (1991) Bees, pollination and habitat change in the European Community. Bee World 72:99–116
Otis GW, Kralj J (2001) Parasitic brood mites not present in America. In: Webster TC, Delaplane KS (eds) Mites of the honeybee. Dadant, Hamilton, pp 251–252
Partap U (1998a) Successful pollination of apples in Himachal Pradesh. Beekee Dev 48:6–7
Partap T (1998b) Agricultural sustainability challenges in upland areas of semi arid and humid Asia. In: Proceedings of the study meeting on sloping land agriculture and natural resources management, Tokyo, Asian Productivity Organisation, pp 39–84

Partap T (1999) Sustainable land management in marginal mountain areas of the Himalayan region. Moun Res Dev 19:251–260

Partap U (2000a) Foraging behaviour of *Apis cerana* on citrus (*Citrus sinensis* var. Red Junar) and its impact on fruit production. In: Matsuka M, Verma LR, Wongsiri S, Shrestha KK, Uma Partap (eds) Asian bees and beekeeping: progress of research and development. Proceedings of the fourth AAA international conference, Kathmandu, Oxford and IBH Publishing Co. Pvt. Ltd, New Delhi, 23–28 March 1998, 274 pp

Partap U, (2000b) Pollination of strawberry by the Asian hive bee, *Apis cerana* F. In: Matsuka M, Verma LR, Wongsiri S, Shrestha KK, Uma Partap (eds) Asian bees and beekeeping in Asia: progress of research and development. Proceedings of the fourth AAA international conference, Kathmandu. Oxford and IBH Publishing Co. Pvt. Ltd. New Delhi, 23–28 March 1998, 274pp

Partap U (2001) Warning signals from the Apple Valleys. Video Film, 31 minutes and 7 seconds, VHS Format. ICIMOD, Kathmandu

Partap U, Partap T (1997). Managed crop pollination: the missing dimension of mountain agricultural productivity. Mountain Farming Systems' Discussion Paper Series No. MFS 97/1, Kathmandu, ICIMOD

Partap U, Partap T (2000) Pollination of apples in China. Beekee Dev 54:6–7

Partap U, Partap T (2002) Warning signals from Apple Valleys of the HKH region: productivity concerns and pollination problems. ICIMOD, Kathmandu, pp 106

Partap U, Verma LR (1992) Floral biology and foraging behaviour of Apis cerana on lettuce crop and its impact on seed production. Prog Hort 24:42–47

Partap U, Verma LR (1994) Pollination of radish by *Apis cerana*. J Apic Res 33:237–241

Partap U, Shukla AN, Verma LR (2000a) Pollination of peach and plum by *Apis cerana*. In: Matsuka M, Verma LR, Wongsiri S, Shrestha KK, Partap U (eds) Asian bees and beekeeping: progress of research and development. Oxford and IBH, New Delhi, pp 171–173

Partap U, Shukla AN, Verma LR (2000b) Comparative attractiveness of broad-leaf mustard, cauliflower and radish to *Apis cerana* in Kathmandu Valley of Nepal. In: Proceedings of the 4th Asian apicultural association international conference, Kathmandu, pp 203–205

Partap U, Shukla AN, Verma LR, (2000c) Impact of *Apis cerana* pollination on fruit quality and yield in peach and plum in the Kathmandu Valley of Nepal. In: Matsuka M, Verma LR, Wongsiri S, Shrestha KK, Uma Partap (eds) Asian bees and beekeeping: progress of research and development. Proceedings of the fourth AAA international conference, Kathmandu, Oxford and IBH Publishing Co. Pvt. Ltd., New Delhi, 23–28 March 1998, 274 pp

Pokhrel S (2006) Status and management of domesticated and wild honeybees (*Apis* spp.) in Chitwan, Nepal. Ph.D. dissertation, TU, Institute of Agriculture and Animal Sciences, Rampur, Chitwan, Nepal, 243 pp

Potts SG, Vulliamy B, Dafni A, Ne'eman G, O'Toole C, Roberts S, Willmer PG (2003) Response of plant-pollinator communities following fire: changes in diversity, abundance and reward structure. Oikos 101:103–112

Punchihewa RWK (1994) Beekeeping for honey production in Sri Lanka: management of Asiatic hive honeybee Apis cerana in its natural tropical monsoonal environment. Sri Lanka Department of Agriculture and Canadian International Development Agency, Sri Lanka

Rasmont P (1988) Monographie écologique et zoogéographique des bourdons de France et de Belgique (Hymenoptera, Apidae, Bombinae). Ph.D. thesis, Faculté des Sciences Agronomique de l'Etat, Gembloux

Robinson WE, Morse RA (1989) The value of honeybees as pollinators of US crops. Am Bee J 129(1):477–487

Ruttner F (1988) The genus Apis, biogeography and taxonomy of honeybees. Springer, Berlin, pp 3–11

Schmid-Hempel P, Durrer S (1991) Parasites, floral resources and reproduction in natural populations of bumblebees. Oikos 62:342–350

Schultheis JR et al. (1998) Commercial production of pickling and slicing cucumbers in North Carolina. Pub NoAG-552.

Seeley TD, Seeley RH, Aratanakul P (1982) Colony defence strategies of the honeybees in Thailand. Eco Mono 52:43–63
Sharma KC (1994) Current experiences and practices in pesticide use in the Bagmati zone. ADPI series #9, ICIMOD, Kathmandu, Nepal
Sharma D Relative toxicity of some insecticides to adult foragers and brood of *Apis mellifera* L. Ph.D. thesis. SKUAST-Jammu, 225 pp
Shrestha JB, Shrestha KK (2000) Beekeeping in Nepal: problems and potentials. In: M. Matsuka, L.R. Verma, S. Wongsiri, K. K. Shrestha, and U. Partap (eds.) Asian bees and beekeeping: progress of research and development. Fourth Asian apicultural association international conference Kathmandu, Nepal, Oxford and IBH Publishing Co. Pvt. Ltd., New Delhi, pp 262–265
Singh MP, Singh KI, Devi CS, (2000) Role of *Apis cerana* pollination on yield and quality of rapeseed and sunflower crops. In: Matsuka M, Verma LR, Wongsiri S, Shrestha KK, Uma Partap (eds) Asian bees and beekeeping in Asia: progress of research and development. Proceedings of the fourth AAA international conference, Kathmandu, Oxford and IBH Publishing Co. Pvt. Ltd., New Delhi, 23–28 March 1998, 274pp
Sorthia BK, Chari MS (1985) Toxicity of some insecticides to honeybees, Apis florea F and Apis mellifera L. J Entomol Res 92(2):195–197
Sprengel C (1793) Das entdeckte Geheimnis der Natur im Bau und in der Befruchtung der Srivastava, V. K. 1969. Studies on the floral biology of *Abelmoschus esculentus* (L.) Moench. Indian J Hort 26:165–170
Steffan-Dewenter I, Tscharntke T (1999) Effects of habitat isolation on pollinator communities and seed set. Oecologia 121:432–440
Steffan-Dewenter I, Tscharntke T (2000) Resource overlap and possible competition between honey bees and wild bees in central Europe. Oecologia 122:288–296
Thapa RB (2003) Pesticide pollution and integrated pest management. In: Neupane FP (ed) Integrated pest management in Nepal proceedings of the national seminar, Himalayan Resource Institute, New Baneshwor, Kathmandu, Nepal, 25–26 September 2002, pp 175–194
Thomson D (2004) Competitive interactions between the invasive European honey bee and native bumble bees. Ecology 85:458–470
Travis MJ (2003) Climate change and habitat destruction: a deadly anthropogenic cocktail. Proc R Soc London (B) 270:467–473
United Nations Food and Agriculture Organization (FAO) (1961) Agricultural and horticultural seeds. FAO Agro-studies 55, 531 pp
Verma LR, Partap U (1993) The Asian Hive Bee, Apis cerana, as a pollinator in vegetable seed production. ICIMOD, Kathmandu, 52pp
Verma LR, Partap U (1994) Foraging behaviour of *Apis cerana* on cabbage and cauliflower and its impact on seed production. J Api Res 33:231–236
Watanabe ME (1994) Pollination worries rise as honey bees decline. Science 265:1170
Westrich P (1989) Die Wildbienen Baden-Württembergs. Ulmer, Stuttgart
Williams PH (1986) Bumble bees and their decline in Britain. Central Association of Beekeepers, Essex
Williams IH (1996) Aspects of bee diversity and crop pollination in the European Union. In: Matheson A, Buchmann,SL, O'Toole C, Westrich P, Williams IH (eds) The conservation of bees, Linnaean Society Symposium Series 18. Academic Press, London, pp 210–226
Williams IH, Martin AP, White RP (1987) The effect of insect pollination on plant development and seed production in winter oilseed rape (*Brassica napus* L.). J Agricul Sci Cambridge 109:135–139
Williams IH, Corbet SA, Osborne JL (1991) Beekeeping, wild bees and pollination in the European Community. Bee World 72:170–180
Zhong HE, Mark L, Sheve P, Cate B (2004) Minimizing the import if the mosquito adulterine noted or honeybees, *Apis mellifera* (Hymenoptera: Apidae): Aerial Ultra volume application using a high-pressure nozzle system. J Econ Entomol 97(1):1–7

MIX
Papier aus verantwortungsvollen Quellen
Paper from responsible sources
FSC® C105338

If you have any concerns about our products,
you can contact us on
ProductSafety@springernature.com

In case Publisher is established outside the EU,
the EU authorized representative is:
Springer Nature Customer Service Center GmbH
Europaplatz 3, 69115 Heidelberg, Germany

Printed by Libri Plureos GmbH
in Hamburg, Germany